Elasticity and Fluid Dynamics

Elasticity and Fluid Dynamics

Volume 3 of *Modern Classical Physics*

KIP S. THORNE *and* **ROGER D. BLANDFORD**

PRINCETON UNIVERSITY PRESS

Princeton and Oxford

Requests for permission to reproduce material from this work
should be sent to permissions@press.princeton.edu

Published by Princeton University Press
41 William Street, Princeton, New Jersey 08540
6 Oxford Street, Woodstock, Oxfordshire OX20 1TR

press.princeton.edu

All Rights Reserved
ISBN (pbk.) 978-0-691-20734-6
ISBN (e-book) 978-0-691-21557-0

British Library Cataloging-in-Publication Data is available

Editorial: Ingrid Gnerlich and Arthur Werneck
Production Editorial: Mark Bellis
Text and Cover Design: Wanda España
Production: Jacqueline Poirier
Publicity: Matthew Taylor and Amy Stewart
Copyeditor: Cyd Westmoreland

This book has been composed in MinionPro, Whitney, and Ratio Modern by Windfall Software, Carlisle, Massachusetts, using ZzTEX

Printed on acid-free paper.

Printed in China

10 9 8 7 6 5 4 3 2 1

A NOTE TO READERS

This book is the third in a series of volumes that together comprise a unified work titled *Modern Classical Physics*. Each volume is designed to be read independently of the others and can be used as a textbook in an advanced undergraduate- to graduate-level course on the subject of the title or for self-study. However, as the five volumes are highly complementary to one another, we hope that reading one volume may inspire the reader to investigate others in the series—or the full, unified work—and thereby explore the rich scope of modern classical physics.

To Carolee and Liz

CONTENTS

T2 Track Two; see page xvii

17   Compressible and Supersonic Flow   875

18   Convection   917

BOXES

Long, long ago, when we (the authors) were students, some basic understanding of continuous media—fluids and elastic solids—was expected of us, and there were physics courses in which we could learn it. That is no longer true, but we believe it should be, and this book is designed to facilitate such courses. *Why?*

Fluids and elastic solids are central to everyday life; this has always been so. And they are also central to modern physics, science more broadly, and technology. This centrality has intensified in recent years—so much so that a basic understanding of the behavior of elastic solids and fluids should be part of the repertoire of every physicist and engineer and of most other natural scientists.

We humans encounter elastic solids almost everywhere—for example, in the earth's crust, mountains, skyscrapers, bridges, and Venus flytrap flowers; and in their occasional sudden transitions to a new equilibrium: earthquakes, avalanches, the buckling and collapse of skyscrapers on 9/11, the swinging collapse of bridges, and the snap-shut of the Venus flytrap to capture and eat an insect.

Elastic solids play crucial roles in modern scientific research and technology—for example, in the development and applications of new materials such as carbon nanotubes; in high-precision physics experiments with torsion pendula (e.g., to test the inverse square law of gravity and thereby search for evidence of macroscopic higher dimensions); in micro cantilevers for nanotechnology and for tests of the quantum theory of measurement; and in experiments and technology with strands of DNA, which, despite their molecular-scale nature, can behave like an elastic body. Many physics experiments have been set back because physicists lacked a basic understanding of elasticity and failed to communicate well with engineers who had it.

Like solids, fluids are central to everyday life. We encounter them, for example, in tornados; ocean currents; breaking ocean waves; the flight of airplanes and the air turbulence that buffets airplanes; the aerodynamically controllable, erratic flight of baseballs, golf balls, and cricket balls; the design and thrust of rocket engines; the production, propagation, and reception of music from an orchestra; the formation of soap bubbles and their oscillation and bursting; boiling water on a hot stove; and breaking ocean waves.

Fluids play central roles in astrophysics, geophysics, meteorology, oceanography, biophysics, medicine, chemistry, and engineering; and in technologies for aeronautics, rocketry, space flight, wind turbines, oil pipelines, and heating and cooling systems, among others.

Fluids exhibit fascinating phenomena and behaviors—many of them nonlinear—that add great richness to their roles in nature and in science and technology; for example, vorticity and its persistence and transport; viscosity-induced boundary layers near solid surfaces, the forces they exert (e.g., lift and drag on airplane wings), and their control by the surface's shape and roughness; frictionless (inviscid) flows when viscosity is low, and the huge influence that friction-induced boundary layers can exert on inviscid flows (e.g., poloidal circulation in a stirred teacup and the resulting rapid mixing of cream into the flow); shock waves and their formation; turbulence, its universal Kolmogorov spectrum due to the cascade of energy from large scales to small, and several universal routes toward the onset of turbulence, and those routes' surprising connection to chaotic mathematical maps.

Among the most pressing issues facing humanity today are the existential threats associated with energy and climate. Here, all aspects of elasticity and fluid dynamics come into play, both in understanding our human predicament and in assessing and implementing complementary remedial strategies. Designing better wind turbines and safer nuclear reactors and understanding the ramifications of geo-engineering provide three out of countless examples. Many physics students of today will, assuredly, be employed in these endeavors and will have much to contribute.

In this book, we study most of the phenomena described above, and many more, as applications of the fundamental principles of elasticity and fluid dynamics. However, our central focus is not these examples. Rather it is the fundamental principles themselves.

WHY ELASTICITY AND FLUID DYNAMICS TOGETHER, IN ONE BOOK?

Because there is great commonality between them. Both deal with *continuum physics*, so the concepts of the two fields, and their mathematical tools and modes of reasoning, are very similar and interrelated. However, each is readily understood without the other. Readers who wish to focus on just fluid dynamics (the second part of the book) can easily understand that part without first reading the elasticity part and vice versa.

QUANTUM PHYSICS IN THIS BOOK

This book deals primarily with *classical* continuum physics. Nevertheless, we make frequent reference to quantum mechanical concepts and phenomena, and we sometimes use quantum concepts and techniques in the classical domain. This is because classical physics arises from quantum physics as an approximation, and sometimes the imprints left on classical physics by its quantum roots are so strong that classical phenomena are most powerfully discussed and analyzed in quantum language.

OUR GEOMETRIC APPROACH TO PHYSICS

We (the authors) embrace a coordinate-independent, geometric approach to physics in our roles as teachers, mentors, and research scientists. The essence of this approach is that all the laws of physics must be expressible in coordinate-independent, geometric language (in terms of coordinate-independent scalars, vectors, tensors, and associated differential operators). This geometric language has great power. It usually elucidates the physics of a situation more clearly than coordinate-based language, and it often circumvents lengthy, coordinate-based calculations. For those students and other readers who are not familiar with this geometric approach, we have provided a detailed introduction to it in a long appendix (that is Chap. 1 of MCP).

It is by no means necessary for readers to study the appendix before plunging into the body of this book. Many or most readers who ignore it will get along just fine without it, at least most of the time. When readers do run into difficulty, they can easily dip into the appendix for clarification. However, familiarity with the appendix will give readers a deeper understanding of and appreciation for our approach to a number of important ideas and results in continuum physics.

RELATIVITY

In this book, we have chosen to focus almost entirely on nonrelativistic (Newtonian) continuum physics. However, from time to time, we step into the relativistic domain of near-light speeds and relativistically large stresses. For example, Chap. 13 contains a section on relativistic fluid dynamics without viscosity or heat conduction, and elsewhere, we point readers to references and textbooks where other aspects of relativistic continuum physics can be found.

GUIDANCE FOR READERS

The amount and variety of material covered in this book may seem overwhelming. If so, keep in mind that

- *the primary goals of this book* are to teach the fundamental principles of elasticity and fluid dynamics, to illustrate those principles in action, and through our illustrations to give the reader some physical and intuitive understanding of elastic solids and fluids.

We do not seek to teach a mastery of the many illustrative applications contained in the book, but rather to convey the spirit of how to apply the basic concepts of continuum physics in a wide range of venues. To help students and readers who feel overwhelmed, we have labeled as "Track Two" sections that can be skipped on a first reading, or skipped entirely—but are sufficiently interesting that many readers may choose to browse or study them. Track-Two sections are labeled by the symbol T2 .

We have aimed this book at advanced undergraduates and first- and second-year graduate students, of whom we expect only (1) a typical physics or engineering student's facility with applied mathematics, and (2) a typical undergraduate-level understanding of classical mechanics, electromagnetism, elementary thermodynamics, and (occasionally) elementary quantum mechanics. We also target working scientists and engineers who want to improve their understanding of elasticity or fluid dynamics.

This book contains eight chapters, two on elasticity and six on fluid dynamics. These can be covered in a one-semester course on continuum physics, or in a one-quarter course, though one quarter may require omitting Track-Two material. The six fluid chapters are readily suitable for a one-quarter or one-semester course in fluid dynamics. We expect the book also to be used as supplementary reading in many other courses (e.g., on astrophysics, geophysics, aerodynamics, and the design of high-precision experiments in fundamental physics).

This book is the third of five volumes that together constitute a single treatise, *Modern Classical Physics* (or "MCP," as we shall call it). The full treatise was published in 2017 as an embarrassingly thick single book. (The electronic edition is a good deal lighter.) For readers' convenience, we have placed, at the end of this volume, the Table of Contents, Preface, and Acknowledgments of MCP. The five separate textbooks of this decomposition are

- Volume 1: *Statistical Physics*,
- Volume 2: *Optics*,
- Volume 3: *Elasticity and Fluid Dynamics*,
- Volume 4: *Plasma Physics*, and
- Volume 5: *Relativity and Cosmology*.

These individual volumes are much more suitable for human transport and for use in individual courses than their one-volume parent treatise, MCP.

The present volume is enriched by extensive cross-references to the other four volumes—cross-references that elucidate the rich interconnections of various areas of physics.

In this and the other four volumes, we have retained the chapter numbers from MCP and, for the body of each volume, MCP's pagination. In fact, the body of this volume and its appendix are identical to the corresponding MCP chapters, aside from corrections of errata (which are tabulated at the MCP website http://press.princeton .edu/titles/MCP.html) and a small amount of updating that has not changed pagination. For readers' cross-referencing convenience, a list of the chapters in each of the five volumes appears immediately after this Preface.

EXERCISES

Exercises are a major component of this volume, as well as of the other four volumes of MCP. The exercises are classified into five types:

1. *Practice.* Exercises that provide practice at mathematical manipulations (e.g., of tensors).

2. *Derivation.* Exercises that fill in details of arguments skipped over in the text.

3. *Example*. Exercises that lead the reader step by step through the details of some important extension or application of the material in the text.

4. *Problem*. Exercises with few, if any, hints, in which the task of figuring out how to set up the calculation and get started on it often is as difficult as doing the calculation itself.

5. *Challenge*. Especially difficult exercises whose solution may require reading other books or articles as a foundation for getting started.

We urge readers to try working many of the exercises—especially the examples, which should be regarded as continuations of the text and which contain many of the most illuminating applications. Exercises that we regard as especially important are designated by **.

UNITS
Throughout this volume, we use SI units, as is customary today in fluid dynamics.

BRIEF OUTLINE OF THIS BOOK
In Chap. 11, on *elastostatics*, we introduce the basic concepts of elasticity theory: stress and strain in solids, and for elastic solids, their linear stress-strain relationship, mediated by elastic moduli. From these we deduce the Navier-Cauchy equation of elastostatic equilibrium for solids that are not dynamical, and we explore a wide range of applications, including (among others): a torsion pendulum and a microcantilever for high-precision physics experiments, the engineering of cantilever bridges, the possibility of a tether stretching from Earth's surface up to a geosynchronous satellite, the Venus flytrap, and experiments to measure the properties of DNA molecules. Most important are applications that illustrate concepts or techniques that occur widely elsewhere in physics, science, and technology: (1) the *buckling* of a longitudinally compressed beam (or of the 9/11 skyscrapers), which illustrates both mathematical catastrophe theory and the bifurcation of equilibria when instabilities set in; and (2) the reduction of the elastostatic equations from 3 dimensions to 2 when dealing with thin plates, and from 3 dimensions to 1 when dealing with beams, rods, and wires.

In Chap. 12, on elastodynamics, we allow a stressed, elastic solid to oscillate dynamically, and we explore the resulting many types of elastodynamic waves; for example, (1) seismic waves in Earth (compressional waves, shear waves, and surface waves) and their coupling to one another at interfaces (e.g., between Earth's crust and mantle); (2) sound waves used in ultrasonic technology for imaging inside solids, such as the human body; (3) various waves and modes of oscillation on rods, beams, and strings, and how they change as applied stresses are changed (most importantly, as increasing compression pushes a beam through a bifurcation of equilibria, and buckling ensues). We also explore the quantum domain for oscillatory modes of an elastic solid, and the nature and properties of the oscillations' quanta—*phonons*. And we elucidate, for elastodynamics, analogs of computational tools widely used in electrodynamics: separation of variables, Green's functions, and the geometric optics approximation.

We begin our study of fluids (gases like air, and liquids like water) in Chap. 13, on the foundations of fluid dynamics. We start with a discussion of the physical nature of fluids and then develop the concepts and equations for a fluid in *hydrostatic equilibrium*, in the presence of gravity and possibly steady rotation, so gravitational forces, centrifugal forces, and external pressures are balanced by the fluid's internal pressure; and the fluid does not become dynamical. In applications, we discover how complex and interesting hydrostatic configurations can be (e.g., the structures of Earth's atmosphere, stars, planets, and especially binary star systems). As the foundation for transitioning from statics to dynamics, we develop with great care the geometric laws for conservation of mass, momentum, and angular momentum when a fluid is allowed to become increasingly complex: by adding to its pressure—step by step—external gravity, viscous stresses, diffusive heat flow, and self-gravity. The resulting richness is far too much to handle in applications, all at once. Fortunately, almost every practical application is dominated by only one or two (or at most, three) of these phenomena; so in the remainder of this book, we specialize to situations where most of them can be neglected. Our first—and very important—specialization is to a fluid whose density is idealized as constant (an excellent approximation, as we shall show, when the flow is highly subsonic) and that has pressure and shear viscosity but no heat diffusion and usually no gravity. In this case, as we deduce, the fluid dynamics is controlled by mass conservation and the *Navier-Stokes equation* (the balancing of viscous stresses, pressure, and fluid acceleration).

In Chap. 14, on vorticity, we explore this Navier-Stokes equation in action for various subsonic flows and discover the remarkable properties of vorticity (the curl of the fluid's velocity field) and its great value for physical intuition and computational analysis. By analogy with magnetic field lines, we introduce *vortex lines* that are divergence free and so never end, except on boundaries of the fluid. We find that vortex lines are typically frozen into the fluid and conserved in time, but they can diffuse through the fluid due to viscosity and can be created at boundaries (e.g., by a spoon stirring tea in a teacup) and by density gradients that don't line up with pressure gradients. Among our important applications are (1) tornados, whirlpools, vortex tubes streaming off the tips of airplane wings, and viscosity-induced boundary layers near solid surfaces; (2) the profound influence of boundary layers on a wing's lift and drag and on bulk flows (e.g., in a bathtub with drain open and in a stirred teacup); (3) how winds pile up mounds of water called "gyres" in the ocean, and how the weight of those gyres drives deep ocean currents (such as the Gulf stream that warms the east coast of North America); (4) instabilities in shear flows and their generation of billow clouds and turbulence in the stratosphere; and (5) the enormous atmospheric viscous drag on small dust particles trying to fall to Earth after giant volcanic eruptions or nuclear explosions, and the viscously suspended particles' long-lasting influence on the atmosphere's opacity and resulting global warming.

In Chap. 15, we explore turbulence—still in the domain of subsonic (constant-density) flows with no heat diffusion, and so still using the Navier-Stokes equation. Turbulence is the most complex, irreproducible, and ill understood of all fluid dynamical phenomena. We describe the observed and measured properties of turbulence and then develop a semi-quantitative model for weak turbulence. We use this model to explain and explore, among other things: (1) the Kolmogorov spectrum for turbulent energy as a function of frequency and wavelength (a spectrum of considerable importance, e.g., in geophysics, astrophysics, and observational astronomy); (2) jets of one fluid (e.g., smoke from a smokestack) entering a quiescent other fluid (still air); (3) wakes behind moving objects (e.g., boats and airplanes); and (4) turbulent boundary layers (e.g., on an airplane wing) and the influence of the turbulence on lift and drag.

In recent years, experiments have revolutionized our understanding of the onset of turbulence as the parameters of a flow slowly change. They suggest, as we discuss, that there are only a few discrete routes toward turbulence. Those same routes also show up in mathematical maps that possess strange attractors and chaotic behavior. We present and discuss several of those maps and describe their great similarity to the onset of turbulence in fluid dynamic experiments.

In Chap. 16, on waves, we explore a wide variety of wave phenomena that occur in subsonic (constant-density), heat-diffusion-free fluids (the domain of the Navier-Stokes equation), usually with negligible influence from viscosity and turbulence, except for slowly damping the waves. Our examples are chosen to illustrate general properties of waves and of their interactions with a fluid flow. The wave phenomena we study include, among others, (1) waves on the surface of a pond and the influence of the water depth on their dispersion (different wave speeds at different frequencies); (2) ocean waves and their growth and breaking as they near the shore; (3) waves induced on a flowing river by submerged rocks; (4) waves influenced by surface tension (which we otherwise ignore), including capillary waves around small insects that ride on the surface of water or around a toy boat; and (5) Rossby waves in a rotating fluid, where centrifugal forces play a restoring-force role.

We pay special attention to solitary waves (solitons), in which nonlinear growth is counterbalanced by dispersion in a stable manner that enables the wave to maintain its shape without change as it travels long distances; and to sound waves in a fluid (with tiny oscillatory changes of density as well as pressure). We use sound waves to explore wave generation and especially the forces that a departing wave exerts back on its source—*radiation reaction;* and also to explore the phenomenon of *runaway solutions* that often plague radiation-reaction equations (e.g., in electrodynamics and also here): how runaway solutions arise, and why they are unphysical manifestations of errors in the mathematics.

In Chap. 17, on compressible and supersonic flows, we abandon viscosity (i.e., assume it is negligible) in order to explore, without viscosity's complications, flows whose speeds are near sonic, transonic, and/or supersonic, and the unavoidable density variations that they produce. We develop a remarkable mathematical tool, called *Riemann invariants*, for analyzing such flows in one space dimension, and another, called *self-similarity*, to gain deep insights into them. We explore in detail (1) transitions from subsonic flow to supersonic (e.g., at the throat that divides a rocket engine from its out-flaring nozzle); (2) the shock fronts (near discontinuities) that inevitably occur in transitions from supersonic to subsonic and that very quickly convert the energy of bulk fluid flow into fast, random molecular motions (i.e., heat); and (3) the nonlinear steepening of very strong sound waves in a fluid to form shocks. Among our applications are breaking ocean waves; sonic booms—manifestations of shock fronts created by supersonic airplanes and by space shuttles reentering Earth's atmosphere; fireballs ignited by atomic bomb detonations; astrophysical supernova remnants (hot, expanding bubbles in interstellar gas, created and driven by supernova explosions); the accretion of interstellar gas onto a black hole, which transitions from subsonic to supersonic and then near-light speed; and the bow shock front that forms when the supersonic solar wind impinges on a planet.

Finally, in Chap. 18, on convection, we allow for significant heat diffusion, but in highly subsonic flows—for example, heat diffusion in the cooling of nuclear reactors by a coolant liquid flowing through pipes. Heat diffusion can alter the temperature distribution of a fluid and thereby, via thermal expansion, alter the fluid's density distribution, which in turn can influence the effects of buoyancy, which can drive or suppress convection. The convection, like diffusion, transports heat; so heat diffusion and convection can compete and cooperate in complex ways that we explore in this chapter, in a variety of applications, such as (1) in stars, where heat generated at the star's center by thermonuclear burning must be transported to the star's surface and there is radiated away; and (2) at the onset of convection and then boiling in a pan of liquid heated from below (e.g., on a kitchen stove); this is a venue for insightful experiments about routes to turbulence (see Chap. 15). In Chap. 18, we also explore the influence on fluids of changes in buoyancy driven by the diffusion of salt or other heavy chemical constituents; for example, salt fingers near the surface of the ocean on a calm day, which are driven and retarded by the combined diffusive effects of salt and heat, and frictional effects of viscosity.

There is one more force, not treated in this book, that can strongly influence fluid dynamics: a magnetic field coupled to electric current in a fluid with substantial electric conductivity. The study of this is called *magnetohydrodynamics*, and it is the subject of MCP Chap. 19. We have included that Chap. 19 in Volume 4 (*Plasma Physics*) of our MCP series, instead of in this Volume 3, because plasmas are by far the most important venue today for magnetohydrodynamics.

Volume 4: Plasma Physics

Volume 5: Relativity and Cosmology

PART IV

ELASTICITY

Although ancient civilizations built magnificent pyramids, palaces, and cathedrals and presumably developed insights into how to avoid their collapse, mathematical models for this (the theory of elasticity) were not developed until the seventeenth century and later.

The seventeenth-century focus was on a beam (e.g., a vertical building support) under compression or tension. Galileo initiated this study in 1632, followed most notably by Robert Hooke[1] in 1660. Bent beams became the focus with the work of Edme Mariotte in 1680. Bending and compression came together with Leonhard Euler's 1744 theory of the buckling of a compressed beam and his derivation of the complex shapes of very thin wires, whose ends are pushed toward each other (*elastica*). These ideas were extended to 2-dimensional thin plates by Marie-Sophie Germain and Joseph-Louis Lagrange in 1811–1816 in a study that was brought to full fruition by Augustus Edward Hugh Love in 1888. The full theory of 3-dimensional, stressed, elastic objects was developed by Claude-Louis Navier and by Augustin-Louis Cauchy in 1821–1822. A number of great mathematicians and natural philosophers then developed techniques for solving the Navier-Cauchy equations, particularly for phenomena relevant to railroads and construction. In 1956, with the advent of modern digital computers, M. J. Turner, R. W. Clough, H. C. Martin, and L. J. Topp pioneered finite-element methods for numerically modeling stressed bodies. Finite-element numerical simulations are now a standard tool for designing mechanical structures and devices, and, more generally, for solving difficult elasticity problems.

These historical highlights cannot begin to do justice to the history of elasticity research. For much more detail see, for example, the (out-of-date) long introduction in Love (1927); and for far more detail than most anyone wants, see the (even more out-of-date) two-volume work by Todhunter and Pearson (1886).

---

1. One of whose many occupations was architect.

Despite its centuries-old foundations, elasticity remains of great importance today, and its modern applications include some truly interesting phenomena. Among those applications, most of which we shall touch on in this book, are

1. the design of bridges, skyscrapers, automobiles, and other structures and mechanical devices and the study of their structural failure;

2. the development and applications of new materials, such as carbon nano-tubes, which are so light and strong that one could aspire to use them to build a tether connecting a geostationary satellite to Earth's surface;

3. the design of high-precision physics experiments with torsion pendula and microcantilevers, including brane-worlds-motivated searches for gravitational evidence of macroscopic higher dimensions of space;

4. the creation of nano-scale cantilever probes in atomic-force microscopes;

5. the study of biophysical systems, such as DNA molecules, cell walls, and the Venus fly trap plant; and

6. the study of plate tectonics, quakes, seismic waves, and seismic tomography in Earth and other planets. Much of the modern, geophysical description of mountain building involves buckling and fracture within the lithosphere through viscoelastic processes that combine the principles of elasticity with those of fluid dynamics discussed in Part V.

Indeed, elastic solids are so central to everyday life and to modern science that a basic understanding of their behavior should be part of the repertoire of every physicist. That is the goal of Part IV of the book.

We devote just two chapters to elasticity. The first (Chap. 11) will focus on *elastostatics:* the properties of materials and solid objects that are in static equilibrium, with all forces and torques balancing out. The second (Chap. 12) will focus on *elastodynamics*: the dynamical behavior of materials and solid objects that are perturbed away from equilibrium.

11

Elastostatics

Ut tensio, sic vis

ROBERT HOOKE (1678)

11.1 Overview

11.1

From the viewpoint of continuum mechanics, a *solid* is a substance that recovers its original shape after the application and removal of any small stress. Note the requirement that this be true for *any* small stress. Many fluids (e.g., water) satisfy our definition as long as the applied stress is isotropic, but they will deform permanently under a shear stress. Other materials (e.g., Earth's crust) are only solid for limited times but undergo plastic flow when a small stress is applied for a long time.

We focus our attention in this chapter on solids whose deformation (quantified by a *tensorial strain*) is linearly proportional to the applied, small, *tensorial stress*. This linear, 3-dimensional stress-strain relationship, which we develop and explore in this chapter, generalizes Hooke's famous 1-dimensional law, which states that if an elastic wire or rod is stretched by an applied force F (Fig. 11.1a), its fractional change of length (its strain) is proportional to the force, $\Delta\ell/\ell \propto F$.

Hooke's law

Hooke's law turns out to be one component of a 3-dimensional stress-strain relation, but to understand it deeply in that language, we must first define and understand the *strain tensor* and the *stress tensor*. Our approach to these tensors follows the geometric, frame-independent philosophy introduced in Chap. 1. Some readers may wish to review that philosophy and mathematics by rereading or browsing Chap. 1.

We begin our development of elasticity theory in Sec. 11.2 by introducing, in a frame-independent way, the vectorial displacement field $\boldsymbol{\xi}(\mathbf{x})$ inside a stressed body (Fig. 11.1b), and its gradient $\boldsymbol{\nabla}\boldsymbol{\xi}$, whose symmetric part is the strain tensor $\mathbf{S}$. We then express the strain tensor as the sum of an expansion Θ that represents volume changes and a shear $\boldsymbol{\Sigma}$ that represents shape changes.

In Sec. 11.3.1, we introduce the stress tensor, and in Sec. 11.3.2, we discuss the realms in which there is a linear relationship between stress and strain, and ways in which linearity can fail. In Sec. 11.3.3, assuming linearity, we discuss how the material resists volume change by developing an opposing isotropic stress, with a stress/strain ratio that is equal to the *bulk modulus K*. We discuss how the material also

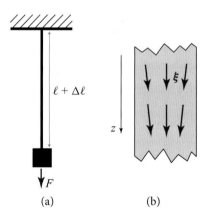

FIGURE 11.1 (a) Hooke's 1-dimensional law for a rod stretched by a force F: $\Delta\ell/\ell \propto F$. (b) The 3-dimensional displacement vector $\boldsymbol{\xi}(\mathbf{x})$ inside the stretched rod.

resists a shear-type strain by developing an opposing shear stress with a stress/strain ratio equal to twice the shear modulus 2μ. In Sec. 11.3.4, we evaluate the energy density stored in elastostatic strains; in Sec. 11.3.5, we explore the influence of thermal expansion on the stress-strain relationship; and in Sec. 11.3.6, we discuss the atomic-force origin of the elastostatic stresses and use atomic considerations to estimate the magnitudes of the bulk and shear moduli. Then in Sec. 11.3.7, we compute the elastic force density inside a linear material as the divergence of the sum of its elastic stresses, and we formulate the law of elastostatic stress balance (the Navier-Cauchy equation) as the vanishing sum of the material's internal elastic force density and any other force densities that may act (usually a gravitational force density due to the weight of the elastic material). We discuss the analogy between this elastostatic stress-balance equation and Maxwell's electrostatic and magnetostatic equations. We describe how mathematical techniques common in electrostatics can also be applied to solve the Navier-Cauchy equation, subject to boundary conditions that describe external forces.

In Sec. 11.4, as a simple example, we use our 3-dimensional formulas to deduce Hooke's law for the 1-dimensional longitudinal stress and strain in a stretched wire.

When the elastic body that one studies is very thin in two dimensions compared to the third (e.g., a wire or rod), we can reduce the 3-dimensional elastostatic equations to a set of coupled 1-dimensional equations by taking moments of the elastostatic equations. We illustrate this technique in Sec. 11.5, where we treat the bending of beams and other examples.

Elasticity theory, as developed in this chapter, is an example of a common (some would complain far too common) approach to physics problems, namely, to linearize them. Linearization may be acceptable when the distortions are small. However, when deformed by sufficiently strong forces, elastic media may become unstable to small displacements, which can then grow to large amplitude, causing rupture. We study an example of this in Sec. 11.6: the buckling of a beam when subjected to a sufficiently large longitudinal stress. Buckling is associated with *bifurcation of equilibria,* a phenomenon that is common to many physical systems, not just elastostatic ones. We illustrate bifurcation in Sec. 11.6 using our beam under a compressive load, and we explore its connection to catastrophe theory.

In Sec. 11.7, we discuss dimensional reduction by the method of moments for bodies that are thin in only 1 dimension, not two, such as plates and thin mirrors. In such bodies, the 3-dimensional elastostatic equations are reduced to 2 dimensions. We illustrate our 2-dimensional formalism by the stress polishing of telescope mirrors.

Because elasticity theory entails computing gradients of vectors and tensors, and practical calculations are often best performed in cylindrical or spherical coordinate systems, we present a mathematical digression in Track-Two Sec. 11.8—an introduction to how one can perform practical calculations of gradients of vectors and tensors in the orthonormal bases associated with curvilinear coordinate systems, using the concept of a connection coefficient.

As illustrative examples of both connection coefficients and elastostatic force balance, in Track-Two Sec. 11.9 and various exercises, we give practical examples of solutions of the elastostatic force-balance equation in cylindrical coordinates using two common techniques of elastostatics and electrostatics: separation of variables (text of Sec. 11.9.2) and Green's functions (Ex. 11.27).

11.2 Displacement and Strain

We begin our study of elastostatics by introducing the elastic displacement vector, its gradient, and the irreducible tensorial parts of its gradient. We then identify the strain as the symmetric part of the displacement's gradient.

11.2.1 Displacement Vector and Its Gradient

Elasticity provides a major application of the tensorial techniques we developed in Chap. 1. Label the position of a *point* (a tiny bit of solid) in an unstressed body, relative to some convenient origin in the body, by its position vector $\mathbf{x}$. Let a force be applied, so the body deforms and the point moves from $\mathbf{x}$ to $\mathbf{x} + \boldsymbol{\xi}(\mathbf{x})$; we call $\boldsymbol{\xi}$

displacement vector

the point's *displacement vector* (Fig. 11.1b). If $\boldsymbol{\xi}$ were constant (i.e., if its components in a Cartesian coordinate system were independent of location in the body), then the body would simply be translated and would undergo no deformation. To produce a deformation, we must make the displacement $\boldsymbol{\xi}$ change from one location to another. The most simple, coordinate-independent way to quantify those changes is by the gradient of $\boldsymbol{\xi}$, $\nabla\boldsymbol{\xi}$. This gradient is a second-rank tensor field, which we denote by $\mathbf{W}$:

$$\boxed{\mathbf{W} \equiv \nabla\boldsymbol{\xi}.} \tag{11.1a}$$

This tensor is a geometric object, defined independently of any coordinate system in the manner described in Sec. 1.7. In slot-naming index notation (Sec. 1.5), it is denoted

$$W_{ij} = \xi_{i;j}, \tag{11.1b}$$

where the index j after the semicolon is the name of the gradient slot.

In a Cartesian coordinate system the components of the gradient are always just partial derivatives [Eq. (1.15c)], and therefore the Cartesian components of $\mathbf{W}$ are

$$W_{ij} = \frac{\partial \xi_i}{\partial x_j} = \xi_{i,j}. \tag{11.1c}$$

(Recall that indices following a comma represent partial derivatives.) In Sec. 11.8, we learn how to compute the components of the gradient in cylindrical and spherical coordinates.

In any small neighborhood of any point $\mathbf{x}_o$ in a deformed body, we can reconstruct the displacement vector $\boldsymbol{\xi}$ from its gradient $\mathbf{W}$ up to an additive constant. Specifically, in Cartesian coordinates, by virtue of a Taylor-series expansion, $\boldsymbol{\xi}$ is given by

$$\xi_i(\mathbf{x}) = \xi_i(\mathbf{x}_o) + (x_j - x_{oj})(\partial \xi_i / \partial x_j) + \dots$$
$$= \xi_i(\mathbf{x}_o) + (x_j - x_{oj})W_{ij} + \dots. \tag{11.2}$$

If we place the origin of Cartesian coordinates at $\mathbf{x}_o$ and let the origin move with the point there as the body deforms [so $\boldsymbol{\xi}(\mathbf{x}_o) = 0$], then Eq. (11.2) becomes

$$\xi_i = W_{ij}x_j \quad \text{when } |\mathbf{x}| \text{ is sufficiently small.} \tag{11.3}$$

We have derived this as a relationship between components of $\boldsymbol{\xi}$, $\mathbf{x}$, and $\mathbf{W}$ in a Cartesian coordinate system. However, the indices can also be thought of as the names of slots (Sec. 1.5) and correspondingly, Eq. (11.3) can be regarded as a geometric, coordinate-independent relationship among the vectors and tensor $\boldsymbol{\xi}$, $\mathbf{x}$, and $\mathbf{W}$.

In Ex. 11.2, we use Eq. (11.3) to gain insight into the displacements associated with various parts of the gradient $\mathbf{W}$.

11.2.2 Expansion, Rotation, Shear, and Strain

In Box 11.2, we introduce the concept of the irreducible tensorial parts of a tensor, and we state that in physics, when one encounters an unfamiliar tensor, it is often useful to identify the tensor's irreducible parts. The gradient of the displacement vector, $\mathbf{W} = \nabla \boldsymbol{\xi}$, is an important example. It is a second-rank tensor. Therefore, as discussed in Box 11.2, its irreducible tensorial parts are its trace $\Theta \equiv \mathrm{Tr}(\mathbf{W}) = W_{ii} = \nabla \cdot \boldsymbol{\xi}$, which is called the deformed body's *expansion* (for reasons we shall explore below); its symmetric, trace-free part $\boldsymbol{\Sigma}$, which is called the body's *shear*; and its antisymmetric part $\mathbf{R}$, which is called the body's *rotation*:

irreducible tensorial parts of a tensor

expansion, shear, and rotation

$$\Theta = W_{ii} = \nabla \cdot \boldsymbol{\xi}, \tag{11.4a}$$

$$\Sigma_{ij} = \frac{1}{2}(W_{ij} + W_{ji}) - \frac{1}{3}\Theta g_{ij} = \frac{1}{2}(\xi_{i;j} + \xi_{j;i}) - \frac{1}{3}\xi_{k;k}\, g_{ij}, \tag{11.4b}$$

$$R_{ij} = \frac{1}{2}(W_{ij} - W_{ji}) = \frac{1}{2}(\xi_{i;j} - \xi_{j;i}). \tag{11.4c}$$

Here g_{ij} is the metric, which has components $g_{ij} = \delta_{ij}$ (Kronecker delta) in Cartesian coordinates, and repeated indices [the ii in Eq. (11.4a)] are to be summed [Eq. (1.9b) and subsequent discussion].

We can reconstruct $\mathbf{W} = \nabla \boldsymbol{\xi}$ from these irreducible tensorial parts in the following manner [Eq. (4) of Box 11.2, rewritten in abstract notation]:

$$\nabla \boldsymbol{\xi} = \mathbf{W} = \frac{1}{3}\Theta \mathbf{g} + \boldsymbol{\Sigma} + \mathbf{R}. \tag{11.5}$$

Let us explore the physical effects of the three separate parts of $\mathbf{W}$ in turn. To understand expansion, consider a small 3-dimensional piece $\mathcal{V}$ of a deformed body (a *volume element*). When the deformation $\mathbf{x} \to \mathbf{x} + \boldsymbol{\xi}$ occurs, a much smaller element of area[1] $d\boldsymbol{\Sigma}$ on the surface $\partial \mathcal{V}$ of $\mathcal{V}$ gets displaced through the vectorial distance $\boldsymbol{\xi}$ and in the process sweeps out a volume $\boldsymbol{\xi} \cdot d\boldsymbol{\Sigma}$. Therefore, the change in the volume element's volume, produced by $\boldsymbol{\xi}$, is

$$\delta V = \int_{\partial \mathcal{V}} d\boldsymbol{\Sigma} \cdot \boldsymbol{\xi} = \int_{\mathcal{V}} dV \nabla \cdot \boldsymbol{\xi} = \nabla \cdot \boldsymbol{\xi} \int_{\mathcal{V}} dV = (\nabla \cdot \boldsymbol{\xi})\, V. \tag{11.6}$$

---

1. Note that we use $\boldsymbol{\Sigma}$ for a vectorial area and $\boldsymbol{\Sigma}$ for the shear tensor. There should be no confusion.

In quantum mechanics, an important role is played by the *rotation group*: the set of all rotation matrices, viewed as a mathematical entity called a group (e.g., Mathews and Walker, 1970, Chap. 16). Each tensor in 3-dimensional Euclidean space, when rotated, is said to generate a specific *representation* of the rotation group. Tensors that are "big" (in a sense to be discussed later in this box) can be broken down into a sum of several tensors that are "as small as possible." These smallest tensors are said to generate *irreducible representations* of the rotation group. All this mumbo-jumbo is really simple, when one thinks about tensors as geometric, frame-independent objects.

As an example, consider an arbitrary second-rank tensor W_{ij} in 3-dimensional, Euclidean space. In the text W_{ij} is the gradient of the displacement vector. From this tensor we can construct the following "smaller" tensors by linear operations that involve only W_{ij} and the metric g_{ij}. (As these smaller tensors are enumerated, the reader should think of the notation used as the basis-independent, frame-independent, slot-naming index notation of Sec. 1.5.1.) The smaller tensors are the contraction (i.e., trace) of W_{ij},

$$\Theta \equiv W_{ij} g_{ij} = W_{ii}; \tag{1}$$

the antisymmetric part of W_{ij},

$$R_{ij} \equiv \frac{1}{2}(W_{ij} - W_{ji}); \tag{2}$$

and the symmetric, trace-free part of W_{ij},

$$\Sigma_{ij} \equiv \frac{1}{2}(W_{ij} + W_{ji}) - \frac{1}{3}g_{ij} W_{kk}. \tag{3}$$

It is straightforward to verify that the original tensor W_{ij} can be reconstructed from these three smaller tensors plus the metric g_{ij} as follows:

$$W_{ij} = \frac{1}{3}\Theta g_{ij} + R_{ij} + \Sigma_{ij}. \tag{4}$$

One way to see the sense in which Θ, R_{ij}, and Σ_{ij} are smaller than W_{ij} is by counting the number of independent real numbers required to specify their components in an arbitrary basis. (Think of the index notation as components on a chosen basis.) The original tensor W_{ij} has three × three = nine components (W_{11}, W_{12}, W_{13}, W_{21}, … W_{33}), all of which are independent. By contrast, the scalar Θ has just one. The antisymmetric

(continued)

BOX 11.2. (continued)

tensor R_{ij} has just three independent components, R_{12}, R_{23}, and R_{31}. Finally, the nine components of Σ_{ij} are not independent; symmetry requires that $\Sigma_{ij} \equiv \Sigma_{ji}$, which reduces the number of independent components from nine to six; being trace-free, $\Sigma_{ii} = 0$, reduces it further from six to five. Therefore, (five independent components in Σ_{ij}) + (three independent components in R_{ij}) + (one independent component in Θ) $= 9 =$ (number of independent components in W_{ij}).

The number of independent components (one for Θ, three for R_{ij}, and five for Σ_{ij}) is a geometric, basis-independent concept: It is the same, regardless of the basis used to count the components; and for each of the smaller tensors that make up W_{ij}, it is easily deduced without introducing a basis at all (think here in slot-naming index notation): The scalar Θ is clearly specified by just one real number. The antisymmetric tensor R_{ij} contains precisely the same amount of information as the vector

$$\phi_i \equiv -\frac{1}{2}\epsilon_{ijk}R_{jk}, \tag{5}$$

as can be seen from the fact that Eq. (5) can be inverted to give

$$R_{ij} = -\epsilon_{ijk}\phi_k; \tag{6}$$

and the vector ϕ_i can be characterized by its direction in space (two numbers) plus its length (a third). The symmetric, trace-free tensor Σ_{ij} can be characterized geometrically by the ellipsoid $(g_{ij} + \varepsilon\Sigma_{ij})\zeta_i\zeta_j = 1$, where ε is an arbitrary number $\ll 1$, and ζ_i is a vector whose tail sits at the center of the ellipsoid and whose head moves around on the ellipsoid's surface. Because Σ_{ij} is trace-free, this ellipsoid has unit volume. Therefore, it is specified fully by the direction of its longest principal axis (two numbers) plus the direction of a second principal axis (a third number) plus the ratio of the length of the second axis to the first (a fourth number) plus the ratio of the length of the third axis to the first (a fifth number).

Each of the tensors Θ, R_{ij} (or equivalently, ϕ_i), and Σ_{ij} is irreducible in the sense that one cannot construct any smaller tensors from it, by any linear operation that involves only it, the metric, and the Levi-Civita tensor. Irreducible tensors in 3-dimensional Euclidean space always have an odd number of components. It is conventional to denote this number by $2l + 1$, where the integer l is called the "order of the irreducible representation of the

(continued)

rotation group" that the tensor generates. For Θ, R_{ij} (or equivalently, ϕ_i), and Σ_{jk}, l is 0, 1, and 2, respectively. These three tensors can be mapped into the spherical harmonics of order $l = 0$, 1, 2; and their $2l + 1$ components correspond to the $2l + 1$ values of the quantum number $m = -l, -l + 1 \ldots, l - 1, l$. (For details see, e.g., Thorne, 1980, Sec. II.C.)

In physics, when one encounters a new tensor, it is often useful to identify the tensor's irreducible parts. They almost always play important, independent roles in the physical situation one is studying. We meet one example in this chapter, another when we study fluid dynamics (Chap. 13), and a third in general relativity (Box 25.2).

Here we have invoked Gauss's theorem in the second equality, and in the third we have used the smallness of $\mathcal{V}$ to infer that $\boldsymbol{\nabla} \cdot \boldsymbol{\xi}$ is essentially constant throughout $\mathcal{V}$ and so can be pulled out of the integral. Therefore, the fractional change in volume is equal to the trace of the stress tensor (i.e., the expansion):

expansion as fractional volume change

$$\boxed{\frac{\delta V}{V} = \boldsymbol{\nabla} \cdot \boldsymbol{\xi} = \Theta.} \qquad (11.7)$$

See Fig. 11.2 for a simple example.

shearing displacements

The shear tensor $\boldsymbol{\Sigma}$ produces the shearing displacements illustrated in Figs. 11.2 and 11.3. As the tensor has zero trace, there is no volume change when a body undergoes a pure shear deformation. The shear tensor has five independent components (Box 11.2). However, by rotating our Cartesian coordinates appropriately, we can transform away all the off-diagonal elements, leaving only the three diagonal elements Σ_{xx}, Σ_{yy}, and Σ_{zz}, which must sum to zero. This is known as a *principal-axis transformation*. Each element produces a stretch ($\Sigma_{..} > 0$) or squeeze ($\Sigma_{..} < 0$) along its

shear's stretch and squeeze along principal axes

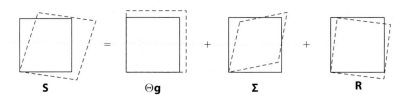

S = Θg + Σ + R

FIGURE 11.2 A simple example of the decomposition of a 2-dimensional distortion **S** of a square body into an expansion Θ, a shear $\boldsymbol{\Sigma}$, and a rotation **R**.

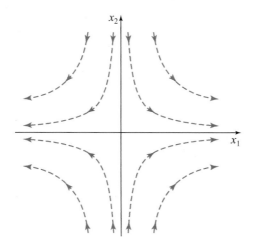

FIGURE 11.3 Shear in 2 dimensions. The displacement of points in a solid undergoing pure shear is the vector field $\boldsymbol{\xi}(\mathbf{x})$ given by Eq. (11.3) with W_{ji} replaced by Σ_{ji}: $\xi_j = \Sigma_{ji}x_i = \Sigma_{j1}x_1 + \Sigma_{j2}x_2$. The integral curves of this vector field are plotted in this figure. The figure is drawn using principal axes, which are Cartesian, so $\Sigma_{12} = \Sigma_{21} = 0$ and $\Sigma_{11} = -\Sigma_{22}$, which means that $\xi_1 = \Sigma_{11}x_1$ and $\xi_2 = -\Sigma_{11}x_2$; or, equivalently, $\xi_x = \Sigma_{xx}x$ and $\xi_y = -\Sigma_{xx}y$. The integral curves of this simple vector field are the hyperbolas shown. Note that the displacement increases linearly with distance from the origin. The shear shown in Fig. 11.2 is the same as this, but with the axes rotated counterclockwise by 45°.

axis,[2] and their vanishing sum (the vanishing trace of $\boldsymbol{\Sigma}$) means that there is no net volume change. The components of the shear tensor in any Cartesian coordinate system can be written down immediately from Eq. (11.4b) by substituting the Kronecker delta δ_{ij} for the components of the metric tensor g_{ij} and treating all derivatives as partial derivatives:

$$\Sigma_{xx} = \frac{2}{3}\frac{\partial \xi_x}{\partial x} - \frac{1}{3}\left(\frac{\partial \xi_y}{\partial y} + \frac{\partial \xi_z}{\partial z}\right), \quad \Sigma_{xy} = \frac{1}{2}\left(\frac{\partial \xi_x}{\partial y} + \frac{\partial \xi_y}{\partial x}\right), \quad (11.8)$$

and similarly for the other components. The analogous equations in spherical and cylindrical coordinates are given in Sec. 11.8.

The third term **R** in Eq. (11.5) describes a pure rotation, which does not deform the solid. To verify this, write $\boldsymbol{\xi} = \boldsymbol{\phi} \times \mathbf{x}$, where $\boldsymbol{\phi}$ is a small rotation of magnitude ϕ about an axis parallel to the direction of $\boldsymbol{\phi}$. Using Cartesian coordinates in 3-dimensional Euclidean space, we can demonstrate by direct calculation that the symmetric part of $\mathbf{W} = \nabla \boldsymbol{\xi}$ vanishes (i.e., $\Theta = \boldsymbol{\Sigma} = 0$) and that

rotation vector

$$R_{ij} = -\epsilon_{ijk}\phi_k, \quad \phi_i = -\frac{1}{2}\epsilon_{ijk}R_{jk}. \quad (11.9a)$$

---

2. More explicitly, $\Sigma_{xx} > 0$ produces a stretch along the x-axis, $\Sigma_{yy} < 0$ produces a squeeze along the y-axis, etc.

Therefore, the elements of the tensor **R** in a Cartesian coordinate system just involve the vectorial rotation angle $\boldsymbol{\phi}$. Note that expression (11.9a) for $\boldsymbol{\phi}$ and expression (11.4c) for R_{ij} imply that $\boldsymbol{\phi}$ is half the curl of the displacement vector:

$$\boldsymbol{\phi} = \frac{1}{2} \nabla \times \boldsymbol{\xi}. \tag{11.9b}$$

A simple example of rotation is shown in the last picture in Fig. 11.2.

Elastic materials resist expansion Θ and shear $\boldsymbol{\Sigma}$, but they don't mind at all having their orientation in space changed (i.e., they do not resist rotations **R**). Correspondingly, in elasticity theory a central focus is on expansion and shear. For this reason the symmetric part of the gradient of $\boldsymbol{\xi}$,

strain tensor
$$S_{ij} \equiv \frac{1}{2}(\xi_{i;j} + \xi_{j;i}) = \Sigma_{ij} + \frac{1}{3}\Theta g_{ij}, \tag{11.10}$$

which includes the expansion and shear but omits the rotation, is given a special name—the *strain*—and is paid great attention.

Let us consider some examples of strains that arise in physical systems.

1. Understanding how materials deform under various *loads* (externally applied forces) is central to mechanical, civil, and structural engineering. As we learn in Sec. 11.3.2, all Hookean materials (materials with strain proportional to stress when the stress is small) crack or break when the load is so great that any component of their strain exceeds ~ 0.1, and almost all crack or break at strains ~ 0.001. For this reason, in our treatment of elasticity theory (this chapter and the next), we focus on strains that are small compared to unity.

2. Continental drift can be measured on the surface of Earth using very long baseline interferometry, a technique in which two or more radio telescopes are used to detect interferometric fringes using radio waves from an astronomical point source. (A similar technique uses the Global Positioning System to achieve comparable accuracy.) By observing the fringes, it is possible to detect changes in the spacing between the telescopes as small as a fraction of a wavelength (~ 1 cm). As the telescopes are typically 1,000 km apart, this means that dimensionless strains $\sim 10^{-8}$ can be measured. The continents drift apart on a timescale $\lesssim 10^8$ yr, so it takes roughly a year for these changes to grow large enough to be measured. Such techniques are also useful for monitoring earthquake faults.

3. The smallest time-varying strains that have been measured so far involve laser interferometer gravitational-wave detectors, such as LIGO. In each arm of a LIGO interferometer, two mirrors hang freely, separated by 4 km. In 2015 their separations were monitored (at frequencies of ~ 100 Hz) to $\sim 4 \times 10^{-19}$ m, four ten-thousandths the radius of a nucleon. The associated strain is 1×10^{-22}. Although this strain is not associated with an elastic solid, it does indicate the high accuracy of optical measurement techniques.

Exercise 11.1 *Derivation and Practice: Reconstruction of a Tensor from Its Irreducible Tensorial Parts*

Using Eqs. (1), (2), and (3) of Box 11.2, show that $\frac{1}{3}\Theta g_{ij} + \Sigma_{ij} + R_{ij}$ is equal to W_{ij}.

Exercise 11.2 *Example: Displacement Vectors Associated with Expansion, Rotation, and Shear*

(a) Consider a $\mathbf{W} = \nabla\boldsymbol{\xi}$ that is pure expansion: $W_{ij} = \frac{1}{3}\Theta g_{ij}$. Using Eq. (11.3) show that, in the vicinity of a chosen point, the displacement vector is $\xi_i = \frac{1}{3}\Theta x_i$. Draw this displacement vector field.

(b) Similarly, draw $\boldsymbol{\xi}(\mathbf{x})$ for a $\mathbf{W}$ that is pure rotation. [Hint: Express $\boldsymbol{\xi}$ in terms of the vectorial angle $\boldsymbol{\phi}$ with the aid of Eq. (11.9b).]

(c) Draw $\boldsymbol{\xi}(\mathbf{x})$ for a $\mathbf{W}$ that is pure shear. To simplify the drawing, assume that the shear is confined to the x-y plane, and make your drawing for a shear whose only nonzero components are $\Sigma_{xx} = -\Sigma_{yy}$. Compare your drawing with Fig. 11.3.

11.3 Stress, Elastic Moduli, and Elastostatic Equilibrium

11.3.1 Stress Tensor

The forces acting in an elastic solid are measured by a second-rank tensor, the *stress tensor* introduced in Sec. 1.9. Let us recall the definition of this stress tensor.

Consider two small, contiguous regions in a solid. If we take a small element of area $d\Sigma$ in the contact surface with its positive sense[3] (same as the direction of $d\Sigma$ viewed as a vector) pointing from the first region toward the second, then the first region exerts a force $d\mathbf{F}$ (not necessarily normal to the surface) on the second through this area. The force the second region exerts on the first (through the area $-d\Sigma$) will, by Newton's third law, be equal and opposite to that force. The force and the area of contact are both vectors, and there is a linear relationship between them. (If we double the area, we double the force.) The two vectors therefore will be related by a second-rank tensor, the stress tensor $\mathbf{T}$:

$$\boxed{d\mathbf{F} = \mathbf{T} \cdot d\Sigma = \mathbf{T}(\ldots, d\Sigma); \quad dF_i = T_{ij}d\Sigma_j.}$$ (11.11) **stress tensor**

Thus the tensor $\mathbf{T}$ is the net (vectorial) force per unit (vectorial) area that a body exerts on its surroundings. Be aware that many books on elasticity (e.g., Landau and Lifshitz, 1986) define the stress tensor with the opposite sign to that in Eq. (11.11). Also be careful not to confuse the shear tensor Σ_{jk} with the vectorial infinitesimal surface area $d\Sigma_j$.

---

3. For a discussion of area elements including their positive sense, see Sec. 1.8.

We often need to compute the total elastic force acting on some finite volume $\mathcal{V}$. To aid in this, we make an important assumption, discussed in Sec. 11.3.6: the stress is determined by local conditions and can be computed from the local arrangement of atoms. If this assumption is valid, then (as we shall see in Sec. 11.3.6), we can compute the total force acting on the volume element by integrating the stress over its surface $\partial\mathcal{V}$:

$$\mathbf{F} = -\int_{\partial\mathcal{V}} \mathbf{T} \cdot d\mathbf{\Sigma} = -\int_{\mathcal{V}} \mathbf{\nabla} \cdot \mathbf{T} dV, \tag{11.12}$$

where we have invoked Gauss's theorem, and the minus sign is included because by convention, for a closed surface $\partial\mathcal{V}$, $d\mathbf{\Sigma}$ points out of $\mathcal{V}$ instead of into it.

Equation (11.12) must be true for arbitrary volumes, so we can identify the *elastic force density* $\mathbf{f}$ acting on an elastic solid as

elastic force density

$$\boxed{\mathbf{f} = -\mathbf{\nabla} \cdot \mathbf{T}.} \tag{11.13}$$

In elastostatic equilibrium, this force density must balance all other volume forces acting on the material, most commonly the gravitational force density, so

force balance equation

$$\boxed{\mathbf{f} + \rho\mathbf{g} = 0,} \tag{11.14}$$

where $\mathbf{g}$ is the gravitational acceleration. (Again, there should be no confusion between the vector $\mathbf{g}$ and the metric tensor $\mathbf{g}$.) There are other possible external forces, some of which we shall encounter later in the context of fluids (e.g., an electromagnetic force density). These can be added to Eq. (11.14).

Just as for the strain, the stress tensor $\mathbf{T}$ can be decomposed into its irreducible tensorial parts, a pure trace (the *pressure P*) plus a symmetric trace-free part (the *shear stress*):

pressure and shear stress

$$\mathbf{T} = P\mathbf{g} + \mathbf{T}^{\text{shear}}; \quad P = \frac{1}{3}\,\text{Tr}(\mathbf{T}) = \frac{1}{3}T_{ii}. \tag{11.15}$$

There is no antisymmetric part, because the stress tensor is symmetric, as we saw in Sec. 1.9. Fluids at rest exert isotropic stresses: $\mathbf{T} = P\mathbf{g}$. They cannot exert shear stress when at rest, though when moving and shearing, they can exert a viscous shear stress, as we discuss extensively in Part V (initially in Sec. 13.7.2).

In SI units, stress is measured in units of Pascals, denoted Pa:

Pascal

$$1\,\text{Pa} = 1\,\text{N m}^{-2} = 1\frac{\text{kg m s}^{-2}}{\text{m}^2}, \tag{11.16}$$

or sometimes in GPa $= 10^9$ Pa. In cgs units, stress is measured in dyne cm^{-2}. Note that $1\,\text{Pa} = 10\,\text{dyne cm}^{-2}$.

Now let us consider some examples of stresses.

1. Atmospheric pressure is equal to the weight of the air in a column of unit area extending above the surface of Earth, and thus is roughly $P \sim \rho g H \sim$

10^5 Pa, where $\rho \simeq 1\,\text{kg m}^{-3}$ is the density of air, $g \simeq 10\,\text{m s}^{-2}$ is the acceleration of gravity at Earth's surface, and $H \simeq 10\,\text{km}$ is the atmospheric scale height [$H \equiv (d \ln P/dz)^{-1}$, with z the vertical distance]. Thus 1 atmosphere is $\sim 10^5$ Pa (or, more precisely, 1.01325×10^5 Pa). The stress tensor is isotropic.

2. Suppose we hammer a nail into a block of wood. The hammer might weigh $m \sim 0.3\,\text{kg}$ and be brought to rest from a speed of $v \sim 10\,\text{m s}^{-1}$ in a distance of, say, $d \sim 3\,\text{mm}$. Then the average force exerted on the wood by the nail, as it is driven, is $F \sim mv^2/d \sim 10^4$ N. If this is applied over an effective area $A \sim 1\,\text{mm}^2$, then the magnitude of the typical stress in the wood is $\sim F/A \sim 10^{10}$ Pa $\sim 10^5$ atmosphere. There is a large shear component to the stress tensor, which is responsible for separating the fibers in the wood as the nail is hammered.

3. Neutron stars are as massive as the Sun, $M \sim 2 \times 10^{30}\,\text{kg}$, but have far smaller radii, $R \sim 10\,\text{km}$. Their surface gravities are therefore $g \sim GM/R^2 \sim 10^{12}\,\text{m s}^{-2}$, 10 billion times that encountered on Earth. They have solid crusts of density $\rho \sim 10^{16}\,\text{kg m}^{-3}$ that are about 1 km thick. In the crusts, the main contribution to the pressure is from the degeneracy of relativistic electrons (see Sec. 3.5.3). The magnitude of the stress at the base of a neutron-star crust is $P \sim \rho g H \sim 10^{31}$ Pa! The crusts are solid, because the free electrons are neutralized by a lattice of ions. However, a crust's shear modulus is only a few percent of its bulk modulus.

4. As we discuss in Sec. 28.7.1, a popular cosmological theory called *inflation* postulates that the universe underwent a period of rapid, exponential expansion during its earliest epochs. This expansion was driven by the stress associated with a *false vacuum*. The action of this stress on the universe can be described quite adequately using a classical stress tensor. If the interaction energy is $E \sim 10^{15}$ GeV, the supposed scale of grand unification, and the associated lengthscale is the Compton wavelength associated with that energy, $l \sim \hbar c/E$, then the magnitude of the stress is $\sim E/l^3 \sim 10^{97}(E/10^{15}\,\text{GeV})^4$ Pa.

5. Elementary particles interact through forces. Although it makes no sense to describe this interaction using classical elasticity, it is reasonable to make order-of-magnitude estimates of the associated stress. One promising model of these interactions involves *strings* with mass per unit length $\mu = g_s^2 c^2/(8\pi G) \sim 1\,\text{Megaton/fermi}$ (where Megaton is not the TNT equivalent!), and cross section of order the Planck length squared, $L_P^2 = \hbar G/c^3 \sim 10^{-70}\,\text{m}^2$, and tension (negative pressure) $T_{zz} \sim \mu c^2/L_P^2 \sim 10^{110}$ Pa. Here $\hbar$, G, and c are Planck's reduced constant, Newton's gravitation constant, and the speed of light, and $g_s^2 \sim 0.025$ is the string coupling constant.

6. The highest possible stress is presumably associated with spacetime singularities, for example at the birth of the universe or inside a black hole. Here the characteristic energy is the Planck energy $E_P = (\hbar c^5/G)^{1/2} \sim 10^{19}$ GeV, the lengthscale is the Planck length $L_P = (\hbar G/c^3)^{1/2} \sim 10^{-35}$ m, and the associated ultimate stress is $\sim E_P/L_P^3 \sim 10^{114}$ Pa.

11.3.2 Realm of Validity for Hooke's Law

In elasticity theory, motivated by Hooke's Law (Fig. 11.1), we assume a linear relationship between a material's stress and strain tensors. Before doing so, however, we discuss the realm in which this linearity is true and some ways in which it can fail.

For this purpose, consider again the stretching of a rod by an applied force (Fig. 11.1a, shown again in Fig. 11.4a). For a sufficiently small stress $T_{zz} = F/A$ (with A the cross sectional area of the rod), the strain $S_{zz} = \Delta\ell/\ell$ follows Hooke's law (straight red line in Fig. 11.4b). However, at some point, called the *proportionality limit* (first big dot in Fig. 11.4b), the strain begins to depart from Hooke's law. Despite this deviation, if the stress is removed, the rod returns to its original length. At a somewhat larger stress, called the *elastic limit*, that ceases to be true; the rod is permanently stretched. At a still larger stress, called the *yield limit* or *yield point*, little or no increase in stress causes a large increase in strain, usually because the material begins to flow plasticly. At an even larger stress, the *rupture point*, the rod cracks or breaks. For a ductile substance like polycrystalline copper, the proportionality limit and elastic limit both occur at about the same rather low strain $\Delta\ell/\ell \sim 10^{-4}$, but yield and rupture do not occur until $\Delta\ell/\ell \sim 10^{-3}$. For a more resilient material like cemented tungsten carbide, strains can be proportional and elastic up to $\sim 3 \times 10^{-3}$. Rubber is non-Hookean (stress is not proportional to strain) at essentially all strains; its proportionality limit is exceedingly small, but it returns to its original shape from essentially all nonrupturing deformations, which can be as large as $\Delta\ell/\ell \sim 8$ (the yield and rupture points).[4] Especially significant is that in almost all solids except rubber, the proportionality, elastic, and yield limits are all small compared to unity.

proportionality limit

elastic limit

yield point

rupture point

11.3.3 Elastic Moduli and Elastostatic Stress Tensor

In realms where Hooke's law is valid, there is a corresponding linear relationship between the material's stress tensor and its strain tensor. The most general linear equation relating two second-rank tensors involves a fourth-rank tensor known as the *elastic modulus tensor* **Y**. In slot-naming index notation,

elastic modulus tensor

$$T_{ij} = -Y_{ijkl}S_{kl}. \tag{11.17}$$

---

4. Rubber is made of long, polymeric molecules, and its elasticity arises from uncoiling of the molecules when a force is applied, which is a different mechanism than is found in crystalline materials (Xing, Goldbart, and Radzihovsky, 2007).

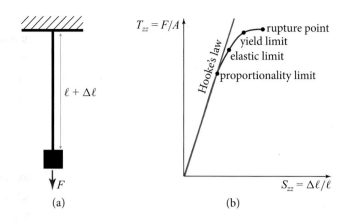

FIGURE 11.4 The stress-strain relation for a rod, showing special points at which the behavior of the rod's material changes.

Now, a general fourth-rank tensor in 3 dimensions has $3^4 = 81$ independent components. Elasticity can get complicated! However, the situation need not be so dire. There are several symmetries that we can exploit. Let us look first at the general case. As the stress and strain tensors are both symmetric, $\mathbf{Y}$ is symmetric in its first pair of slots, and we are free to choose it symmetric in its second pair: $Y_{ijkl} = Y_{jikl} = Y_{ijlk}$. There are therefore 6 independent components Y_{ijkl} for variable i, j and fixed k, l, and vice versa. In addition, as we will show, $\mathbf{Y}$ is symmetric under an interchange of its first and second pairs of slots: $Y_{ijkl} = Y_{klij}$. There are therefore $(6 \times 7)/2 = 21$ independent components in $\mathbf{Y}$. This is an improvement over 81. Many substances, notably crystals, exhibit additional symmetries, which can reduce the number of independent components considerably.

The simplest, and in fact most common, case arises when the medium is *isotropic.* In other words, there are no preferred directions in the material. This occurs when the solid is polycrystalline or amorphous and completely disordered on a scale large compared with the atomic spacing, but small compared with the solid's inhomogeneity scale.

If a medium is isotropic, then its elastic properties must be describable by scalars that relate the irreducible parts P and $\mathbf{T}^{\text{shear}}$ of the stress tensor $\mathbf{T}$ to those, Θ and $\mathbf{\Sigma}$, of the strain tensor $\mathbf{S}$. The only mathematically possible, linear, coordinate-independent relationship between $\{P, \mathbf{T}^{\text{shear}}\}$ and $\{\Theta, \mathbf{\Sigma}\}$ involving solely scalars is $P = -K\Theta$, $\mathbf{T}^{\text{shear}} = -2\mu\mathbf{\Sigma}$, corresponding to a total stress tensor

$$\mathbf{T} = -K\Theta\mathbf{g} - 2\mu\mathbf{\Sigma}. \qquad (11.18)$$

bulk modulus, shear modulus, and stress tensor for isotropic elastic medium

Here K is called the *bulk modulus* and μ the *shear modulus,* and the factor 2 is included for purely historical reasons. The first minus sign (with $K > 0$) ensures that the isotropic part of the stress, $-K\Theta\mathbf{g}$, resists volume changes; the second minus sign (with $\mu > 0$) ensures that the symmetric, trace-free part, $-2\mu\mathbf{\Sigma}$, resists shape changes (resists shearing).

Hooke's law (Figs. 11.1 and 11.4) can be expressed in this same stress-proportional-to-strain form. The stress, when the rod is stretched, is the force F that does the stretching divided by the rod's cross sectional area A, the strain is the rod's fractional change of length $\Delta\ell/\ell$, and so Hooke's law takes the form

$$F/A = -E\Delta\ell/\ell,$$

(11.19)

with E an elastic coefficient called *Young's modulus*. In Sec. 11.4, we show that E is a combination of the bulk and shear moduli: $E = 9\mu K/(3K + \mu)$.

In many treatments and applications of elasticity, the shear tensor $\mathbf{\Sigma}$ is paid little attention. The focus instead is on the strain S_{ij} and its trace $S_{kk} = \Theta$, and the elastic stress tensor (11.18) is written as $\mathbf{T} = -\lambda\Theta\mathbf{g} - 2\mu\mathbf{S}$, where $\lambda \equiv K - \frac{2}{3}\mu$. In these

Lamé coefficients

treatments μ and λ are called the *first and second Lamé coefficients* and are used in place of μ and K. We shall not adopt this viewpoint.

11.3.4 Energy of Deformation

Take a wire of length ℓ and cross sectional area A, and stretch it (e.g., via the "Hooke's-law experiment" of Figs. 11.1 and 11.4) by an amount ζ' that grows gradually from 0 to $\Delta\ell$. When the stretch is ζ', the force that does the stretching is [by Eq. (11.19)] $F' = EA(\zeta'/\ell) = (EV/\ell^2)\zeta'$; here $V = A\ell$ is the wire's volume, and E is its Young's modulus. As the wire is gradually lengthened, the stretching force F' does work

$$W = \int_0^{\Delta\ell} F'd\zeta' = \int_0^{\Delta\ell} (EV/\ell^2)\zeta'd\zeta'$$
$$= \frac{1}{2}EV(\Delta\ell/\ell)^2.$$

This tells us that the stored elastic energy per unit volume is

$$U = \frac{1}{2}E(\Delta\ell/\ell)^2.$$

(11.20)

To generalize this formula to a strained, isotropic, 3-dimensional medium, consider an arbitrary but small region $\mathcal{V}$ inside a body that has already been strained by a displacement vector field ξ_i and is thus already experiencing an elastic stress $T_{ij} = -K\Theta\delta_{ij} - 2\mu\Sigma_{ij}$ [Eq. (11.18)]. Imagine building up this displacement gradually from zero at the same rate everywhere in and around $\mathcal{V}$, so at some moment during the buildup the displacement field is $\xi_i' = \xi_i\epsilon$ (with the parameter ϵ gradually growing from 0 to 1). At that moment, the stress tensor (by virtue of the linearity of the stress-strain relation) is $T_{ij}' = T_{ij}\epsilon$. On the boundary $\partial\mathcal{V}$ of the region $\mathcal{V}$, this stress exerts a force $\Delta F_i' = -T_{ij}'\Delta\Sigma_j$ across any surface element $\Delta\Sigma_j$, from the exterior of ∂V to its interior. As the displacement grows, this surface force does the following amount of work on $\mathcal{V}$:

$$\Delta W_{\text{surf}} = \int \Delta F_i'd\xi_i' = \int (-T_{ij}'\Delta\Sigma_j)d\xi_i' = -\int_0^1 T_{ij}\epsilon\Delta\Sigma_j\xi_i'd\epsilon = -\frac{1}{2}T_{ij}\Delta\Sigma_j\xi_i.$$

(11.21)

The total amount of work done can be computed by adding up the contributions from all the surface elements of $\partial \mathcal{V}$:

$$W_{\text{surf}} = -\frac{1}{2} \int_{\partial \mathcal{V}} T_{ij} \xi_i d\Sigma_j = -\frac{1}{2} \int_{\mathcal{V}} (T_{ij}\xi_i)_{;j} dV = -\frac{1}{2}(T_{ij}\xi_i)_{;j} V. \quad (11.22)$$

In the second step we have used Gauss's theorem, and in the third step we have used the smallness of the region $\mathcal{V}$ to infer that the integrand is very nearly constant and the integral is the integrand times the total volume V of $\mathcal{V}$.

Does W_{surf} equal the elastic energy stored in $\mathcal{V}$? The answer is "no," because we must also take account of the work done in the interior of $\mathcal{V}$ by gravity or any other nonelastic force that may be acting. Although it is not easy in practice to turn gravity off and then on, we must do so in the following thought experiment. In the volume's final deformed state, the divergence of its elastic stress tensor is equal to the gravitational force density, $\boldsymbol{\nabla} \cdot \mathbf{T} = \rho \mathbf{g}$ [Eqs. (11.13) and (11.14)]; and in the initial, undeformed and unstressed state, $\boldsymbol{\nabla} \cdot \mathbf{T}$ must be zero, whence so must $\mathbf{g}$. Therefore, we must imagine growing the gravitational force proportional to ϵ just like we grow the displacement, strain, and stress. During this growth, with $\mathbf{g}' = \epsilon \mathbf{g}$, the gravitational force $\rho \mathbf{g}' V$ does the following amount of work on our tiny region $\mathcal{V}$:

$$W_{\text{grav}} = \int \rho V \mathbf{g}' \cdot d\boldsymbol{\xi} = \int_0^1 \rho V \mathbf{g} \epsilon \cdot \boldsymbol{\xi} d\epsilon = \frac{1}{2} \rho V \mathbf{g} \cdot \boldsymbol{\xi} = \frac{1}{2}(\boldsymbol{\nabla} \cdot \mathbf{T}) \cdot \boldsymbol{\xi} V = \frac{1}{2} T_{ij;j} \xi_i V.$$

$$(11.23)$$

The total work done to deform $\mathcal{V}$ is the sum of the work done by the elastic force (11.22) on its surface and the gravitational force (11.23) in its interior, $W_{\text{surf}} + W_{\text{grav}} = -\frac{1}{2}(\xi_i T_{ij})_{;j} V + \frac{1}{2} T_{ij;j} \xi_i V = -\frac{1}{2} T_{ij} \xi_{i;j} V$. This work gets stored in $\mathcal{V}$ as elastic energy, so the energy density is $U = -\frac{1}{2} T_{ij} \xi_{i;j}$. Inserting (for an isotropic material) $T_{ij} = -K\Theta g_{ij} - 2\mu \Sigma_{ij}$ and $\xi_{i;j} = \frac{1}{3}\Theta g_{ij} + \Sigma_{ij} + R_{ij}$ in this equation for U and performing some simple algebra that relies on the symmetry properties of the expansion, shear, and rotation (Ex. 11.3), we obtain

$$\boxed{U = \frac{1}{2} K\Theta^2 + \mu \Sigma_{ij} \Sigma_{ij}.} \quad (11.24)$$

elastic energy density at fixed temperature

Note that this elastic energy density is always positive if the elastic moduli are positive, as they must be for matter to be stable against small perturbations, and note that it is independent of the rotation R_{ij}, as it should be on physical grounds.

For the more general, anisotropic case, expression (11.24) becomes [by virtue of the stress-strain relation $T_{ij} = -Y_{ijkl}\xi_{k;l}$, Eq. (11.17)]

$$U = \frac{1}{2}\xi_{i;j} Y_{ijkl} \xi_{k;l}. \quad (11.25)$$

The volume integral of the elastic energy density given by Eq. (11.24) or (11.25) can be used as an action from which to compute the stress, by varying the displacement (Ex. 11.4). Since only the part of $\mathbf{Y}$ that is symmetric under interchange of the first and second pairs of slots contributes to U, only that part can affect the action-principle-derived stress. Therefore, it must be that $Y_{ijkl} = Y_{klij}$. This is the symmetry we asserted earlier.

Exercise 11.3 *Derivation and Practice: Elastic Energy*
Beginning with $U = -\frac{1}{2}T_{ij}\xi_{i;j}$ [text following Eq. (11.23)], derive Eq. (11.24) for the elastic energy density inside a body.

Exercise 11.4 *Derivation and Practice: Action Principle for Elastic Stress*
For an anisotropic, elastic medium with elastic energy density $U = \frac{1}{2}\xi_{i;j}Y_{ijkl}\xi_{k;l}$, integrate this energy density over a 3-dimensional region $\mathcal{V}$ (not necessarily small) to get the total elastic energy E. Now consider a small variation $\delta\xi_i$ in the displacement field. Evaluate the resulting change δE in the elastic energy without using the relation $T_{ij} = -Y_{ijkl}\xi_{k;l}$. Convert to a surface integral over $\partial\mathcal{V}$, and thence infer the stress-strain relation $T_{ij} = -Y_{ijkl}\xi_{k;l}$.

11.3.5

11.3.5 Thermoelasticity

In our discussion of deformation energy, we tacitly assumed that the temperature of the elastic medium was held fixed during the deformation (i.e., we ignored the possibility of any thermal expansion). Correspondingly, the energy density U that we computed is actually the physical free energy per unit volume $\mathcal{F}$, at some chosen temperature T_0 of a heat bath. If we increase the bath's and material's temperature from T_0 to $T = T_0 + \delta T$, then the material wants to expand by $\Theta = \delta V / V = 3\alpha\delta T$ (i.e., it will have vanishing expansional elastic energy if Θ has this value). Here α is its coefficient of linear thermal expansion. (The factor 3 is because there are three directions into which it can expand: x, y, and z.) Correspondingly, the physical-free-energy density at temperature $T = T_0 + \delta T$ is

coefficient of linear thermal expansion

elastic physical free energy

$$\mathcal{F} = \mathcal{F}_0(T) + \frac{1}{2}K(\Theta - 3\alpha\delta T)^2 + \mu\Sigma_{ij}\Sigma_{ij}. \tag{11.26}$$

The stress tensor in this heated and strained state can be computed from $T_{ij} = -\partial\mathcal{F}/\partial S_{ij}$ [a formula most easily inferred from Eq. (11.25) with U reinterpreted as $\mathcal{F}$ and $\xi_{i;j}$ replaced by its symmetrization, S_{ij}]. Reexpressing Eq. (11.26) in terms of S_{ij} and computing the derivative, we obtain (not surprisingly!)

$$T_{ij} = -\frac{\partial\mathcal{F}}{\partial S_{ij}} = -K(\Theta - 3\alpha\delta T)\delta_{ij} - 2\mu\Sigma_{ij}. \tag{11.27}$$

What happens if we allow our material to expand *adiabatically* rather than at fixed temperature? Adiabatic expansion means expansion at fixed entropy S. Consider a small sample of material that contains mass M and has volume $V = M/\rho$. Its entropy is $S = -[\partial(\mathcal{F}V)/\partial T]_V$ [cf. Eq. (5.33)], which, using Eq. (11.26), becomes

$$S = S_0(T) + 3\alpha K \Theta V.$$ (11.28)

Here we have neglected the term $-9\alpha^2 K \delta T$, which can be shown to be negligible compared to the temperature dependence of the elasticity-independent term $S_0(T)$. If our sample expands adiabatically by an amount $\Delta V = V \Delta \Theta$, then its temperature must go down by the amount $\Delta T < 0$ that keeps S fixed (i.e., that makes $\Delta S_0 = -3\alpha K V \Delta \Theta$). Noting that $T \Delta S_0$ is the change of the sample's thermal energy, which is $\rho c_V \Delta T$ (c_V is the specific heat per unit mass), we see that the temperature change is

$$\frac{\Delta T}{T} = \frac{-3\alpha K \Delta \Theta}{\rho c_V} \quad \text{for adiabatic expansion.}$$ (11.29)

temperature change in adiabatic expansion

This temperature change, accompanying an adiabatic expansion, alters slightly the elastic stress [Eq. (11.27)] and thence the bulk modulus K (i.e., it gives rise to an adiabatic bulk modulus that differs slightly from the isothermal bulk modulus K introduced in previous sections). However, the differences are so small that they are generally ignored. For further details, see Landau and Lifshitz (1986, Sec. 6).

11.3.6 Molecular Origin of Elastic Stress; Estimate of Moduli

11.3.6

It is important to understand the microscopic origin of the elastic stress. Consider an ionic solid in which singly ionized ions (e.g., positively charged sodium and negatively charged chlorine) attract their nearest (opposite-species) neighbors through their mutual Coulomb attraction and repel their *next* nearest (same-species) neighbors, and so on. Overall, there is a net electrostatic attraction on each ion, which is balanced by the short-range repulsion of its bound electrons against its neighbors' bound electrons. Now consider a thin slice of material of thickness intermediate between the inter-atomic spacing and the solid's inhomogeneity scale (Fig. 11.5).

Although the electrostatic force between individual pairs of ions is long range, the material is electrically neutral on the scale of several ions; as a result, when averaged

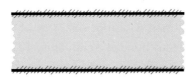

FIGURE 11.5 A thin slice of an ionic solid (between the dark lines) that interacts electromagnetically with ions outside it. The electrostatic force on the slice is dominated by interactions between ions lying in the two thin shaded areas, a few atomic layers thick, one on each side of the slice. The force is effectively a surface force rather than a volume force. In elastostatic equilibrium, the forces on the two sides are equal and opposite, if the slice is sufficiently thin.

TABLE 11.1: Density ρ; bulk, shear, and Young's moduli K, μ, and E, respectively; Poisson's ratio ν; and yield strain S_Y under tension, for various materials

Substance	ρ (kg m^{-3})	K (GPa)	μ (GPa)	E (GPa)	ν	S_Y	c_L (km s^{-1})	c_T (km s^{-1})
Carbon nanotube	1,300			~1,000		0.05		
Steel	7,800	170	81	210	0.29	0.003	5.9	3.2
Copper	8,960	130	45	120	0.34	0.0006	4.6	2.2
Rock	3,000	70	40	100	0.25	0.001	6.0	3.5
Glass	2,500	47	28	70	0.25	0.0005	5.8	3.3
Rubber	1,200	10	0.0007	0.002	0.50	~8	1.0	0.03
DNA molecule				0.3		~0.1		

Notes: The final two columns are the longitudinal and transverse sound speeds c_L, c_T, defined in Chap. 12. The DNA molecule is discussed in Ex. 11.12.

over many ions, the net electric force is short range (Fig. 11.5). We can therefore treat the net force acting on the thin slice as a surface force, governed by local conditions in the material. This is essential if we are to be able to write down a localized linear stress-strain relation $T_{ij} = -Y_{ijkl}S_{kl}$ or $T_{ij} = -K\Theta\delta_{ij} - 2\mu\Sigma_{ij}$. This need not have been the case; there are other circumstances where the net electrostatic force is long range, not short. One example occurs in certain types of crystal (e.g., tourmaline), which develop internal, long-range *piezoelectric* fields when strained.

Our treatment so far has implicitly assumed that matter is continuous on all scales and that derivatives are mathematically well defined. Of course, this is not the case. In fact, we not only need to acknowledge the existence of atoms, we must also use them to compute the elastic moduli.

magnitudes of elastic moduli

We can estimate the elastic moduli in ionic or metallic materials by observing that, if a crystal lattice were to be given a dimensionless strain of order unity, then the elastic stress would be of order the electrostatic force between adjacent ions divided by the area associated with each ion. If the lattice spacing is $a \sim 2$ Å $= 0.2$ nm and the ions are singly charged, then K and $\mu \sim e^2/4\pi\epsilon_0 a^4 \sim 100$ GPa. This is about a million atmospheres. Covalently bonded compounds are less tightly bound and have somewhat smaller elastic moduli; exotic carbon nanotubes have larger moduli. See Table 11.1.

On the basis of this argument, it might be thought that crystals can be subjected to strains of order unity before they attain their elastic limits. However, as discussed in Sec. 11.3.2, most materials are only elastic for strains $\lesssim 10^{-3}$. The reason for this difference is that crystals are generally imperfect and are laced with *dislocations*. Relatively small stresses suffice for the dislocations to move through the solid and for the crystal thereby to undergo permanent deformation (Fig. 11.6).

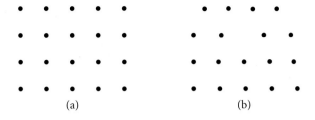

(a)　　　　　　　　　　(b)

FIGURE 11.6 The ions in one layer of a crystal. In subsequent layers, going into each picture, the ion distribution is the same. (a) This perfect crystal, in which the atoms are organized in a perfectly repeating lattice, can develop very large shear strains without yielding. (b) Real materials contain dislocations that greatly reduce their rigidity. The simplest type of dislocation, shown here, is the *edge dislocation* (with the central vertical atomic layer having a terminating edge that extends into the picture). The dislocation will move transversely, and the crystal thereby will undergo inelastic deformation when the strain is typically greater than $\sim 10^{-3}$, which is $\sim 1\%$ of the yield shear strain for a perfect crystal.

Exercise 11.5 *Problem: Order-of-Magnitude Estimates*

(a) What is the maximum size of a nonspherical asteroid? [Hint: If the asteroid is too large, its gravity will deform it into a spherical shape.]

(b) What length of steel wire can hang vertically without breaking? What length of carbon nanotube? What are the prospects for creating a tether that hangs to Earth's surface from a geostationary satellite?

(c) Can a helium balloon lift the tank used to transport its helium gas? (Purcell, 1983).

Exercise 11.6 *Problem: Jumping Heights*

Explain why all animals, from fleas to humans to elephants, can jump to roughly the same height. The field of science that deals with topics like this is called *allometry* (Ex. 11.18).

11.3.7 Elastostatic Equilibrium: Navier-Cauchy Equation

11.3.7

It is commonly the case that the elastic moduli K and μ are constant (i.e., independent of location in the medium), even though the medium is stressed in an inhomogeneous way. (This is because the strains are small and thus perturb the material properties by only small amounts.) If so, then from the elastic stress tensor $\mathbf{T} = -K\Theta\mathbf{g} - 2\mu\mathbf{\Sigma}$ and expressions (11.4a) and (11.4b) for the expansion and shear in terms of the displacement vector, we can deduce the following expression for the elastic force density $\mathbf{f}$ [Eq. (11.13)] inside the body:

elastic force density

$$\mathbf{f} = -\nabla \cdot \mathbf{T} = K\nabla\Theta + 2\mu\nabla \cdot \mathbf{\Sigma} = \left(K + \frac{1}{3}\mu\right)\nabla(\nabla \cdot \boldsymbol{\xi}) + \mu\nabla^2\boldsymbol{\xi}; \quad (11.30)$$

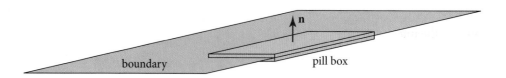

FIGURE 11.7 Pill box used to derive boundary conditions in electrostatics and elastostatics.

see Ex. 11.7. Here $\mathbf{\nabla} \cdot \mathbf{\Sigma}$ in index notation is $\Sigma_{ij;j} = \Sigma_{ji;j}$. Extra terms must be added if we are dealing with anisotropic materials. However, in this book Eq. (11.30) will be sufficient for our needs.

If no other countervailing forces act in the interior of the material (e.g., if there is no gravitational force), and if, as in this chapter, the material is in a static, equilibrium state rather than vibrating dynamically, then this force density will have to vanish throughout the material's interior. This vanishing of $\mathbf{f} \equiv -\mathbf{\nabla} \cdot \mathbf{T}$ is just a fancy version of Newton's law for static situations, $\mathbf{F} = m\mathbf{a} = 0$. If the material has density ρ and is pulled on by a gravitational acceleration $\mathbf{g}$, then the sum of the elastostatic force per unit volume and gravitational force per unit volume must vanish, $\mathbf{f} + \rho\mathbf{g} = 0$:

Navier-Cauchy equation for elastostatic equilibrium

$$\mathbf{f} + \rho\mathbf{g} = \left(K + \frac{1}{3}\mu \right) \mathbf{\nabla}(\mathbf{\nabla} \cdot \mathbf{\xi}) + \mu\nabla^2\mathbf{\xi} + \rho\mathbf{g} = 0. \qquad (11.31)$$

This is often called the *Navier-Cauchy equation*.[5]

When external forces are applied to the surface of an elastic body (e.g., when one pushes on the face of a cylinder) and gravity acts on the interior, the distribution of the strain $\mathbf{\xi}(\mathbf{x})$ inside the body can be computed by solving the Navier-Cauchy equation (11.31) subject to boundary conditions provided by the applied forces.

In electrostatics, one can derive boundary conditions by integrating Maxwell's equations over the interior of a thin box (a "pill box") with parallel faces that snuggle up to the boundary (Fig. 11.7). For example, by integrating $\mathbf{\nabla} \cdot \mathbf{E} = \rho_e/\epsilon_o$ over the interior of the pill box and then applying Gauss's law to convert the left-hand side to a surface integral, we obtain the junction condition that the discontinuity in the normal component of the electric field is equal $1/\epsilon_o$ times the surface charge density. Similarly, in elastostatics one can derive boundary conditions by integrating the elastostatic equation $\mathbf{\nabla} \cdot \mathbf{T} = 0$ over the pill box of Fig. 11.7 and then applying Gauss's law:

$$0 = \int_V \mathbf{\nabla} \cdot \mathbf{T} \, dV = \int_{\partial V} \mathbf{T} \cdot d\mathbf{\Sigma} = \int_{\partial V} \mathbf{T} \cdot \mathbf{n} \, dA = [(\mathbf{T} \cdot \mathbf{n})_{\text{upper face}} - (\mathbf{T} \cdot \mathbf{n})_{\text{lower face}}] A.$$

$$(11.32)$$

Here in the next-to-last expression we have used $d\mathbf{\Sigma} = \mathbf{n} \, dA$, where dA is the scalar area element, and $\mathbf{n}$ is the unit normal to the pill-box face. In the last term we have

---

5. It was first written down by Claude-Louis Navier (in 1821) and in a more general form by Augustin-Louis Cauchy (in 1822).

assumed the pill box has a small face, so $\mathbf{T} \cdot \mathbf{n}$ can be treated as constant and be pulled outside the integral. The result is the boundary condition that

$$\mathbf{T} \cdot \mathbf{n} \quad \text{must be continuous across any boundary;} \qquad (11.33)$$

boundary conditions for Navier-Cauchy equation

in index notation, $T_{ij}n_j$ is continuous.

Physically, this is nothing but the law of force balance across the boundary: the force per unit area acting from the lower side to the upper side must be equal and opposite to that acting from upper to lower. As an example, if the upper face is bounded by vacuum, then the solid's stress tensor must satisfy $T_{ij}n_j = 0$ at the surface. If a normal pressure P is applied by some external agent at the upper face, then the solid must respond with a normal force equal to P: $n_i T_{ij} n_j = P$. If a vectorial force per unit area $\mathcal{F}_i$ is applied at the upper face by some external agent, then it must be balanced: $T_{ij}n_j = -\mathcal{F}_i$.

Solving the Navier-Cauchy equation (11.32) for the displacement field $\boldsymbol{\xi}(\mathbf{x})$, subject to specified boundary conditions, is a problem in elastostatics analogous to solving Maxwell's equations for an electric field subject to boundary conditions in electrostatics, or for a magnetic field subject to boundary conditions in magnetostatics. The types of solution techniques used in electrostatics and magnetostatics can also be used here. See Box 11.3.

EXERCISES

Exercise 11.7 *Derivation and Practice: Elastic Force Density*
From Eq. (11.18) derive expression (11.30) for the elastostatic force density inside an elastic body.

Exercise 11.8 **Practice: Biharmonic Equation*
A homogeneous, isotropic, elastic solid is in equilibrium under (uniform) gravity and applied surface stresses. Use Eq. (11.30) to show that the displacement inside it, $\boldsymbol{\xi}(\mathbf{x})$, is biharmonic, i.e., it satisfies the differential equation

$$\boxed{\nabla^2 \nabla^2 \boldsymbol{\xi} = 0.} \qquad (11.34a)$$

Show also that the expansion Θ satisfies the Laplace equation

$$\boxed{\nabla^2 \Theta = 0.} \qquad (11.34b)$$

11.4 Young's Modulus and Poisson's Ratio for an Isotropic Material: A Simple Elastostatics Problem

11.4

As a simple example of an elastostatics problem, we explore the connection between our 3-dimensional theory of stress and strain and the 1-dimensional Hooke's law (Fig. 11.1).

BOX 11.3. METHODS OF SOLVING THE NAVIER-CAUCHY EQUATION

Many techniques have been devised to solve the Navier-Cauchy equation (11.31), or other equations equivalent to it, subject to appropriate boundary conditions. Among them are:

- Separation of variables. See Sec. 11.9.2.
- Green's functions. See Ex. 11.27 and Johnson (1985).
- Variational principles. See Marsden and Hughes (1986, Chap. 5) and Slaughter (2002, Chap. 10).
- Saint-Venant's principle. One changes the boundary conditions to something simpler, for which the Navier-Cauchy equation can be solved analytically, and then one uses linearity of the Navier-Cauchy equation to compute an approximate, additive correction that accounts for the difference in boundary conditions.[1]
- Dimensional reduction. This method reduces the theory to 2 dimensions in the case of thin plates (Sec. 11.7), and to 1 dimension for rods and for translation-invariant plates (Sec. 11.5).
- Complex variable methods. These are particularly useful in solving the 2-dimensional equations (Boresi and Chong, 1999, Appendix 5B).
- Numerical simulations on computers. These are usually carried out by the method of finite elements, in which one approximates stressed objects by a finite set of elementary, interconnected physical elements, such as rods; thin, triangular plates; and tetrahedra (Ugural and Fenster, 2012, Chap. 7).
- Replace Navier-Cauchy by equivalent equations. For example, and widely used in the engineering literature, write force balance $T_{ij;j} = 0$ in terms of the strain tensor S_{ij}, supplement this with an equation that guarantees S_{ij} can be written as the symmetrized gradient of a vector field (the displacement vector), and develop techniques to solve these coupled equations plus boundary conditions for S_{ij} [Ugural and Fenster (2012, Sec. 2.4); also large parts of Boresi and Chong (1999) and Slaughter (2002)].
- Mathematica or other computer software. These software packages can be used to perform complicated analytical analyses. One can then explore their predictions numerically (Constantinescu and Korsunsky, 2007).

1. In 1855 Barré de Saint-Venant had the insight to realize that, under suitable conditions, the correction will be significant only locally (near the altered boundary) and not globally. (See Boresi and Chong, 1999, pp. 288ff; Ugural and Fenster, 2012, Sec. 2.16, and references therein.)

Consider a thin rod of square cross section hanging along the $\mathbf{e}_z$ direction of a Cartesian coordinate system (Fig. 11.1). Subject the rod to a stretching force applied normally and uniformly at its ends. (It could just as easily be a rod under compression.) Its sides are free to expand or contract transversely, since no force acts on them: $dF_i = T_{ij} d\Sigma_j = 0$. As the rod is slender, vanishing of dF_i at its x and y sides implies to high accuracy that the stress components T_{ix} and T_{iy} will vanish throughout the interior; otherwise there would be a very large force density $T_{ij;j}$ inside the rod. Using $T_{ij} = -K\Theta g_{ij} - 2\mu\Sigma_{ij}$, we then obtain

$$T_{xx} = -K\Theta - 2\mu\Sigma_{xx} = 0, \tag{11.35a}$$

$$T_{yy} = -K\Theta - 2\mu\Sigma_{yy} = 0, \tag{11.35b}$$

$$T_{yz} = -2\mu\Sigma_{yz} = 0, \tag{11.35c}$$

$$T_{xz} = -2\mu\Sigma_{xz} = 0, \tag{11.35d}$$

$$T_{xy} = -2\mu\Sigma_{xy} = 0, \tag{11.35e}$$

$$T_{zz} = -K\Theta - 2\mu\Sigma_{zz}. \tag{11.35f}$$

From the first two of these equations and $\Sigma_{xx} + \Sigma_{yy} + \Sigma_{zz} = 0$, we obtain a relationship between the expansion and the nonzero components of the shear,

$$K\Theta = \mu\Sigma_{zz} = -2\mu\Sigma_{xx} = -2\mu\Sigma_{yy}; \tag{11.36}$$

and from this and Eq. (11.35f), we obtain $T_{zz} = -3K\Theta$. The decomposition of S_{ij} into its irreducible tensorial parts tells us that $S_{zz} = \xi_{z;z} = \Sigma_{zz} + \frac{1}{3}\Theta$, which becomes, on using Eq. (11.36), $\xi_{z;z} = [(3K + \mu)/(3\mu)]\Theta$. Combining with $T_{zz} = -3K\Theta$, we obtain Hooke's law and an expression for Young's modulus E in terms of the bulk and shear moduli:

$$\frac{-T_{zz}}{\xi_{z;z}} = \frac{9\mu K}{3K + \mu} = E. \tag{11.37}$$

Hooke's law and Young's modulus

It is conventional to introduce *Poisson's ratio* ν, which is minus the ratio of the lateral strain to the longitudinal strain during a deformation of this type, where the transverse motion is unconstrained. It can be expressed as a ratio of elastic moduli as follows:

$$\nu \equiv -\frac{\xi_{x;x}}{\xi_{z;z}} = -\frac{\xi_{y;y}}{\xi_{z;z}} = -\frac{\Sigma_{xx} + \frac{1}{3}\Theta}{\Sigma_{zz} + \frac{1}{3}\Theta} = \frac{3K - 2\mu}{2(3K + \mu)}, \tag{11.38}$$

Poisson's ratio

where we have used Eq. (11.36). We tabulate these relations and their inverses for future use:

$$\boxed{E = \frac{9\mu K}{3K + \mu}, \quad \nu = \frac{3K - 2\mu}{2(3K + \mu)}; \quad K = \frac{E}{3(1 - 2\nu)}, \quad \mu = \frac{E}{2(1 + \nu)}.}$$

$$\tag{11.39}$$

We have already remarked that mechanical stability of a solid requires that K, $\mu >$ 0. Using Eq. (11.39), we observe that this imposes a restriction on Poisson's ratio, namely that $-1 < \nu < 1/2$. For metals, Poisson's ratio is typically about 1/3, and the shear modulus is roughly half the bulk modulus. For a substance that is easily sheared but not easily compressed, like rubber (or neutron star crusts; Sec. 11.3.6), the bulk modulus is relatively high and $\nu \simeq 1/2$ (cf. Table 11.1). For some exotic materials, Poisson's ratio can be negative (cf. Yeganeh-Haeri, Weidner, and Parise, 1992).

Although we derived them for a square strut under extension, our expressions for Young's modulus and Poisson's ratio are quite general. To see this, observe that the derivation would be unaffected if we combined many parallel, square fibers together. All that is necessary is that the transverse motion be free, so that the only applied force is uniform and normal to a pair of parallel faces.

11.5

11.5 Reducing the Elastostatic Equations to 1 Dimension for a Bent Beam: Cantilever Bridge, Foucault Pendulum, DNA Molecule, Elastica

When dealing with bodies that are much thinner in 2 dimensions than the third (e.g., rods, wires, and beams), one can use the *method of moments* to reduce the 3-dimensional elastostatic equations to ordinary differential equations in 1 dimension (a process called *dimensional reduction*). We have already met an almost trivial example of this in our discussion of Hooke's law and Young's modulus (Sec. 11.4 and Fig. 11.1). In this section, we discuss a more complicated example, the bending of a beam through a small displacement angle. In Ex. 11.13, we shall analyze a more complicated example: the bending of a long, elastic wire into a complicated shape called an *elastica*.

dimensional reduction

Our beam-bending example is motivated by a common method of bridge construction, which uses cantilevers. (A famous historical example is the old bridge over the Firth of Forth in Scotland that was completed in 1890 with a main span of half a kilometer.)

The principle is to attach two independent beams to the two shores as cantilevers, and allow them to meet in the middle. (In practice the beams are usually supported at the shores on piers and strengthened along their lengths with trusses.) Similar cantilevers, with lengths of order a micron or less, are used in atomic force microscopes and other nanotechnology applications, including quantum-information experiments.

Let us make a simple model of a cantilever (Fig. 11.8). Consider a beam clamped rigidly at one end, with length ℓ, horizontal width w, and vertical thickness h. Introduce local Cartesian coordinates with $\mathbf{e}_x$ pointing along the beam and $\mathbf{e}_z$ pointing vertically upward. Imagine the beam extending horizontally in the absence of gravity. Now let it sag under its own weight, so that each element is displaced through a small distance $\boldsymbol{\xi}(\mathbf{x})$. The upper part of the beam is stretched, while the lower part is compressed, so there must be a *neutral surface* where the horizontal strain $\xi_{x,x}$ vanishes.

neutral surface

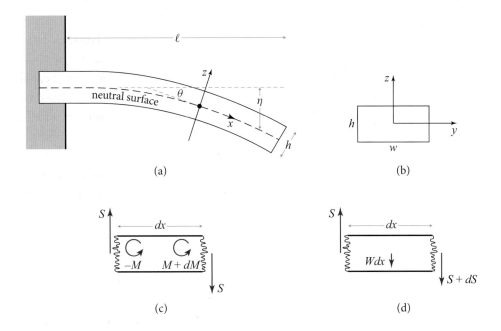

(a)

(b)

(c)

(d)

FIGURE 11.8 Bending of a cantilever. (a) A beam is held rigidly at one end and extends horizontally with the other end free. We introduce an orthonormal coordinate system (x, y, z) with $\mathbf{e}_x$ extending along the beam. We only consider small departures from equilibrium. The bottom of the beam will be compressed, the upper portion extended. There is therefore a neutral surface $z = 0$ on which the strain $\xi_{x,x}$ vanishes. (b) The beam has a rectangular cross section with horizontal width w and vertical thickness h; its length is ℓ. (c) The bending torque M must be balanced by the torque exerted by the vertical shear force S. (d) The shear force S must vary along the beam so as to support the beam's weight per unit length, W.

This neutral surface must itself be curved downward. Let its downward displacement from the horizontal plane that it occupied before sagging be $\eta(x)$ (> 0), let a plane tangent to the neutral surface make an angle $\theta(x)$ (also > 0) with the horizontal, and adjust the x and z coordinates so x runs along the slightly curved neutral plane and z is orthogonal to it (Fig. 11.8). The longitudinal strain is then given to first order in small quantities by

$$\xi_{x,x} = \frac{z}{\mathcal{R}} = z\frac{d\theta}{dx} \simeq z\frac{d^2\eta}{dx^2}, \tag{11.40a}$$

longitudinal strain

where $\mathcal{R} = dx/d\theta > 0$ is the radius of curvature of the beam's bend, and we have chosen $z = 0$ at the neutral surface. The 1-dimensional displacement $\eta(x)$ will be the focus for dimensional reduction of the elastostatic equations.

As in our discussion of Hooke's law for a stretched rod (Sec. 11.4), we can regard the beam as composed of a bundle of long, parallel fibers, stretched or squeezed along their length and free to contract transversely. The longitudinal stress is therefore

$$T_{xx} = -E\xi_{x,x} = -Ez\frac{d^2\eta}{dx^2}. \tag{11.40b}$$

We can now compute the horizontal force density, which must vanish in elasto-static equilibrium:[6]

$$f_x = -T_{xx,x} - T_{xz,z} = Ez\frac{d^3\eta}{dx^3} - T_{xz,z} = 0. \tag{11.40c}$$

method of moments

This is a partial differential equation. We convert it into a 1-dimensional ordinary differential equation by the *method of moments*: We multiply it by z and integrate over z (i.e., we compute its "first moment"). Integrating the second term, $\int zT_{xz,z}dz$, by parts and using the boundary condition $T_{xz} = 0$ on the upper and lower surfaces of the beam, we obtain

$$\frac{Eh^3}{12}\frac{d^3\eta}{dx^3} = -\int_{-h/2}^{h/2} T_{xz}\,dz. \tag{11.40d}$$

Using $T_{xz} = T_{zx}$, notice that the integral, when multiplied by the beam's width w in the y direction, is the vertical shear force $S(x)$ in the beam:

shear force, S

$$\boxed{S = \int T_{zx}dydz = w\int_{-h/2}^{h/2} T_{zx}dz = -D\frac{d^3\eta}{dx^3}.} \tag{11.41a}$$

Here

flexural rigidity or bending modulus of an elastic beam, D

$$\boxed{D \equiv E\int z^2 dydz \equiv EA\,r_g^2 = Ewh^3/12} \tag{11.41b}$$

is called the beam's *flexural rigidity*, or its *bending modulus*. Notice that, quite generally, D is the beam's Young's modulus E times the second moment of the beam's cross sectional area A along the direction of bend. Engineers call that second moment $A\,r_g^2$ and call r_g the radius of gyration. For our rectangular beam, this D is $Ewh^3/12$.

As an aside, we can gain some insight into Eq. (11.41a) by examining the torques that act on a segment of the beam with length dx. As shown in Fig. 11.8c, the shear forces on the two ends of the segment exert a clockwise torque $2S(dx/2) = Sdx$. This is balanced by a counterclockwise torque due to the stretching of the upper half of the segment and compression of the lower half (i.e., due to the bending of the beam). This *bending torque* is

bending torque, M

$$\boxed{M \equiv \int T_{xx}zdydz = -D\frac{d^2\eta}{dx^2}} \tag{11.41c}$$

6. Because the coordinates are slightly curvilinear rather than precisely Cartesian, our Cartesian-based analysis makes small errors. Track-Two readers who have studied Sec. 11.8 can evaluate those errors using connection-coefficient terms that were omitted from this equation: $-\Gamma_{xjk}T_{jk} - \Gamma_{jkj}T_{xk}$. Each Γ has magnitude $1/\mathcal{R}$, so these terms are of order $T_{jk}/\mathcal{R}$, whereas the terms kept in Eq. (11.40c) are of order T_{xx}/ℓ and T_{xz}/h. Since the thickness h and length ℓ of the beam are small compared to the beam's radius of curvature $\mathcal{R}$, the connection-coefficient terms are negligible.

on the right end of the segment and minus this on the left, so torque balance says $(dM/dx)dx = S dx$:

$$S = dM/dx; \tag{11.42}$$

see Fig. 11.8c. This is precisely Eq. (11.41a).

Equation (11.41a) [or equivalently, Eq. (11.42)] embodies half of the elastostatic equations. It is the x component of force balance $f_x = 0$, converted to an ordinary differential equation by evaluating its lowest nonvanishing moment: its first moment, $\int z f_x dy dz = 0$ [Eq. (11.40d)]. The other half is the z component of stress balance, which we can write as

$$T_{zx,x} + T_{zz,z} + \rho g = 0 \tag{11.43}$$

(vertical elastic force balanced by gravitational pull on the beam). We can convert this to a 1-dimensional ordinary differential equation by taking its lowest nonvanishing moment, its zeroth moment (i.e., by integrating over y and z). The result is

$$\frac{dS}{dx} = -W, \tag{11.44}$$

where $W = g\rho wh$ is the beam's weight per unit length (Fig. 11.8d).

weight per unit length, W

Combining our two dimensionally reduced components of force balance, Eqs. (11.41a) and (11.44), we obtain a fourth-order differential equation for our 1-dimensional displacement $\eta(x)$:

$$\frac{d^4\eta}{dx^4} = \frac{W}{D}. \tag{11.45}$$

elastostatic force balance equation for bent beam

(Fourth-order differential equations are characteristic of elasticity.)

Equation (11.45) can be solved subject to four appropriate boundary conditions. However, before we solve it, notice that *for a beam of a fixed length ℓ, the deflection η is inversely proportional to the flexural rigidity.* Let us give a simple example of this scaling. Floors in U.S. homes are conventionally supported by wooden joists of 2" by 6" lumber with the 6" side vertical. Suppose an inept carpenter installed the joists with the 6" side horizontal. The flexural rigidity of the joist would be reduced by a factor 9, and the center of the floor would be expected to sag 9 times as much as if the joists had been properly installed—a potentially catastrophic error.

Also, before solving Eq. (11.45), let us examine the approximations that we have made. First, we have assumed that the sag is small compared with the length of the beam, when making the small-angle approximation in Eq. (11.40a); we have also assumed the beam's radius of curvature is large compared to its length, when treating our slightly curved coordinates as Cartesian.[7] These assumptions will usually be valid, but are not so for the elastica studied in Ex. 11.13. Second, by using the method of moments rather than solving for the complete local stress tensor field, we have ignored the effects of some components of the stress tensor. In particular, when evaluating the bending torque [Eq. (11.41c)] we have ignored the effect of the T_{zx} component of the stress tensor. This is $O(h/\ell)T_{xx}$, and so our equations can only be accurate for fairly slender beams. Third, the extension above the neutral surface and the compression below the neutral surface lead to changes in the cross sectional shape of the beam. The fractional error here is of order the longitudinal shear, which is small for real materials.

The solution to Eq. (11.45) is a fourth-order polynomial with four unknown constants, to be set by boundary conditions. In this problem, the beam is held horizontal at the fixed end, so that $\eta(0) = \eta'(0) = 0$, where the prime denotes d/dx. At the free end, T_{zx} and T_{xx} must vanish, so the shear force S must vanish, whence $\eta'''(\ell) = 0$ [Eq. (11.41a)]; the bending torque M [Eq. (11.41c)] must also vanish, whence [by Eq. (11.42)] $\int S dx \propto \eta''(\ell) = 0$. By imposing these four boundary conditions $\eta(0) = \eta'(0) = \eta''(\ell) = \eta'''(\ell) = 0$ on the solution of Eq. (11.45), we obtain for the beam shape

displacement of a clamped cantilever

$$\eta(x) = \frac{W}{D}\left(\frac{1}{4}\ell^2 x^2 - \frac{1}{6}\ell x^3 + \frac{1}{24}x^4\right). \tag{11.46a}$$

Therefore, the end of the beam sags by

$$\eta(\ell) = \frac{W\ell^4}{8D}. \tag{11.46b}$$

Problems in which the beam rests on supports rather than being clamped can be solved in a similar manner. The boundary conditions will be altered, but the differential equation (11.45) will be unchanged.

Now suppose that we have a cantilever bridge of constant vertical thickness h and total span $2\ell \sim 100$ m made of material with density $\rho \sim 8 \times 10^3$ kg m^{-3} (e.g., steel) and Young's modulus $E \sim 100$ GPa. Suppose further that we want the center of the bridge to sag by no more than $\eta \sim 1$ m. According to Eq. (11.46b), the thickness of the beam must satisfy

$$h \gtrsim \left(\frac{3\rho g \ell^4}{2E\eta}\right)^{1/2} \sim 2.8 \text{ m}. \tag{11.47}$$

This estimate makes no allowance for all the extra strengthening and support present in real structures (e.g., via trusses and cables), and so it is an overestimate.

---

7. In more technical language, when neglecting the connection-coefficient terms discussed in the previous footnote.

Exercise 11.9 *Derivation: Sag in a Cantilever*

(a) Verify Eqs. (11.46) for the sag in a horizontal beam clamped at one end and allowed to hang freely at the other end.

(b) Now consider a similar beam with constant cross section and loaded with weights, so that the total weight per unit length is $W(x)$. What is the sag of the free end, expressed as an integral over $W(x)$, weighted by an appropriate Green's function?

Exercise 11.10 *Example: Microcantilever*

A microcantilever, fabricated from a single crystal of silicon, is being used to test the inverse square law of gravity on micron scales (Weld et al., 2008). It is clamped horizontally at one end, and its horizontal length is $\ell = 300\ \mu$m, its horizontal width is $w = 12\ \mu$m, and its vertical height is $h = 1\ \mu$m. (The density and Young's modulus for silicon are $\rho = 2,000\ \text{kg m}^{-3}$ and $E = 100\ \text{GPa}$, respectively.) The cantilever is loaded at its free end with a $m = 10\ \mu$g gold mass.

(a) Show that the static deflection of the end of the cantilever is $\eta(\ell) = mg\ell^3/(3D) = 9\ \mu$m, where $g = 10\ \text{m s}^{-2}$ is the acceleration due to gravity. Explain why it is permissible to ignore the weight of the cantilever.

(b) Next suppose the mass is displaced slightly vertically and then released. Show that the natural frequency of oscillation of the cantilever is $f = 1/(2\pi)\sqrt{g/\eta(\ell)} \simeq 170$ Hz.

(c) A second, similar gold mass is placed $100\ \mu$m away from the first. Estimate roughly the Newtonian gravitational attraction between these two masses, and compare with the attraction of Earth. Suggest a method that exploits the natural oscillation of the cantilever to measure the tiny gravitational attraction between the two gold masses.

The motivation for developing this technique was to seek departures from Newton's inverse-square law of gravitation on ~micron scales, which had been predicted if our universe is a membrane ("brane") in a higher-dimensional space ("bulk") with at least one macroscopic extra dimension. No such departures have been found as of 2016.

Exercise 11.11 *Example: Foucault Pendulum*

In any high-precision Foucault pendulum, it is important that the pendular restoring force be isotropic, since anisotropy will make the swinging period different in different planes and thereby cause precession of the plane of swing.

(a) Consider a pendulum of mass m and length ℓ suspended (as shown in Fig. 11.9) by a rectangular wire with thickness h in the plane of the bend (X-Z plane) and thickness w orthogonal to that plane (Y direction). Explain why the force that the wire exerts on the mass is approximately $-\mathbf{F} = -(mg\cos\theta_o + m\ell\dot{\theta}_o^2)\mathbf{e}_x$, where g is the acceleration of gravity, θ_o is defined in the figure, and $\dot{\theta}_o$ is the time derivative of θ_o due to the swinging of the pendulum. In the second term we have

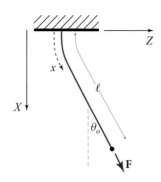

FIGURE 11.9 Foucault pendulum.

assumed that the wire is long compared to its region of bend. Express the second term in terms of the amplitude of swing $\theta_o^{\max}$, and show that for small amplitudes $\theta_o^{\max} \ll 1$, $\mathbf{F} \simeq -mg\mathbf{e}_x$. Use this approximation in the subsequent parts.

(b) Assuming that all along the wire, its angle $\theta(x)$ to the vertical is small, $\theta \ll 1$, show that

$$\theta(x) = \theta_o\big(1 - e^{-x/\lambda}\big), \tag{11.48a}$$

where λ (not to be confused with the second Lamé coefficient) is

$$\lambda = \frac{h}{(12\epsilon)^{1/2}}, \tag{11.48b}$$

$\epsilon = \xi_{x,x}$ is the longitudinal strain in the wire, and h is the wire's thickness in the plane of its bend. [Hint: The solution to Ex. 11.9 might be helpful.] Note that the bending of the wire is concentrated near the support, so this is where dissipation will be most important and where most of the suspension's thermal noise will arise (cf. Sec. 6.8 for discussion of thermal noise).

(c) Hence show that the shape of the wire is given in terms of Cartesian coordinates by

$$Z = [X - \lambda(1 - e^{-X/\lambda})]\,\theta_o \tag{11.48c}$$

to leading order in λ, and that the pendulum period is

$$P = 2\pi \left(\frac{\ell - \lambda}{g}\right)^{1/2}. \tag{11.48d}$$

(d) Finally, show that the pendulum periods when swinging along $\mathbf{e}_x$ and $\mathbf{e}_y$ differ by

$$\frac{\delta P}{P} = \left(\frac{h - w}{\ell}\right)\left(\frac{1}{48\epsilon}\right)^{1/2}. \tag{11.48e}$$

From Eq. (11.48e) one can determine how accurately the two thicknesses h and w must be equal to achieve a desired degree of isotropy in the period. A similar analysis can be carried out for the more realistic case of a slightly elliptical wire.

Exercise 11.12 *Example: DNA Molecule—Bending, Stretching, Young's Modulus, and Yield Point*

A DNA molecule consists of two long strands wound around each other as a helix, forming a cylinder with radius $a \simeq 1$ nm. In this exercise, we explore three ways of measuring the molecule's Young's modulus E. For background and further details, see Marko and Cocco (2003) and Nelson (2008, Chap. 9).

(a) Show that if a segment of DNA with length ℓ is bent into a segment of a circle with radius R, its elastic energy is $E_{\rm el} = D\ell/(2R^2)$, where $D = (\pi/4)a^4 E$ is the molecule's flexural rigidity.

(b) Biophysicists define the DNA's *persistence length* ℓ_p as that length which, when bent through an angle of 90°, has elastic energy $E_{\rm el} = k_B T$, where k_B is Boltzmann's constant and T is the temperature of the molecule's environment. Show that $\ell_p \simeq D/(k_B T)$. Explain why, in a thermalized environment, segments much shorter than ℓ_p will be more or less straight, and segments with length $\sim \ell_p$ will be randomly bent through angles of order 90°.

(c) Explain why a DNA molecule with total length L will usually be found in a random coil with diameter $d \simeq \ell_p \sqrt{L/\ell_p} = \sqrt{L\ell_p}$. Observations at room temperature with $L \simeq 17 \, \mu$m reveal that $d \simeq 1 \, \mu$m. From this show that the persistence length is $\ell_p \simeq 50$ nm at room temperature, and thence evaluate the molecule's flexural rigidity and from it, show that the molecule's Young's modulus is $E \simeq 0.3$ GPa; cf. Table 11.1.

(d) When the ends of a DNA molecule are attached to glass beads and the beads are pulled apart with a gradually increasing force F, the molecule begins to uncoil, just like rubber. To understand this semiquantitatively, think of the molecule as like a chain made of N links, each with length ℓ_p, whose interfaces can bend freely. If the force acts along the z direction, explain why the probability that any chosen link will make an angle θ to the z axis is $dP/d\cos\theta \propto \exp[+F\ell_p \cos\theta/(k_B T)]$. [Hint: This is analogous to the probability $dP/dV \propto \exp[-PV/(k_B T)]$ for the volume V of a system in contact with a bath that has pressure P and temperature T [Eq. (5.49)]; see also the discussion preceding Eq. (11.56).] Infer that when the force is F, the molecule's length along the force's direction is $\bar{L} \simeq L(\coth\alpha - 1/\alpha)$, where $\alpha = F\ell_p/(k_B T)$ and $L = N\ell_p$ is the length of the uncoiled molecule. Infer, further, that for $\alpha \ll 1$ (i.e., $F \ll k_B T/\ell_p \sim 0.1$ pN), our model predicts $\bar{L} \simeq \alpha L/3$, i.e. a linear force-length relation $F = (3k_B T/\ell_p)\bar{L}/L$, with a strongly temperature dependent spring constant, $3k_B T/\ell_p \propto T^2$. The measured value of this spring constant, at room temperature, is about 0.13 pN (Fig. 9.5 of Nelson, 2008). From this infer a value 0.5 GPa for the molecule's Young's modulus. This agrees surprisingly well with the 0.3 GPa deduced in part (c), given the crudeness of the jointed chain model.

(e) Show that when $F \gg k_B T/\ell_p \sim 0.1$ pN, our crude model predicts (correctly) that the molecule is stretched to its full length $L = N\ell_p$. At this point, its true elasticity should take over and allow genuine stretching. That true elasticity turns out to dominate only for forces $\gtrsim 10$ pN. [For details of what happens between 0.1 and 10 pN, see, e.g., Nelson (2008), Secs. 9.1–9.4.] For a force between ~10 and ~80 pN, the molecule is measured to obey Hooke's law, with a Young's modulus $E \simeq 0.3$ GPa that agrees with the value inferred in part (c) from its random-coil diameter. When the applied force reaches $\simeq 65$ pN, the molecule's double helix suddenly stretches greatly with small increases of force, changing its structure, so this is the molecule's yield point. Show that the strain at this yield point is $\Delta \ell / \ell \sim 0.1$; cf. Table 11.1.

Exercise 11.13 **Example: Elastica*

Consider a slender wire of rectangular cross section with horizontal thickness h and vertical thickness w that is resting on a horizontal surface, so gravity is unimportant. Let the wire be bent in the horizontal plane as a result of equal and opposite forces F that act at its ends; Fig. 11.10. The various shapes the wire can assume are called *elastica*; they were first computed by Euler in 1744 and are discussed in Love (1927, pp. 401–404). The differential equation that governs the wire's shape is similar to that for the cantilever [Eq. (11.45)], with the simplification that the wire's weight does not enter the problem and the complication that the wire is long enough to deform through large angles.

elastica

It is convenient (as in the cantilever problem, Fig. 11.8) to introduce curvilinear coordinates with coordinate x measuring distance along the neutral surface, z measuring distance orthogonal to x in the plane of the bend (horizontal plane), and y measured perpendicular to the bending plane (vertically). The unit vectors along the x, y, and z directions are $\mathbf{e}_x$, $\mathbf{e}_y$, $\mathbf{e}_z$ (Fig. 11.10). Let $\theta(x)$ be the angle between $\mathbf{e}_x$ and the applied force $\mathbf{F}$; $\theta(x)$ is determined, of course, by force and torque balance.

(a) Show that force balance along the x and z directions implies

$$F \cos \theta = \int T_{xx} dy dz, \qquad F \sin \theta = \int T_{zx} dy dz \equiv S. \qquad (11.49a)$$

(b) Show that torque balance for a short segment of wire implies

$$S = \frac{dM}{dx}, \qquad (11.49b)$$

where $M(x) \equiv \int z T_{xx} dy dz$ is the bending torque.

(c) Show that the stress-strain relation in the wire implies

$$M = -D \frac{d\theta}{dx}, \qquad (11.49c)$$

where $D = E w h^3 / 12$ is the flexural rigidity [Eq. (11.41b)].

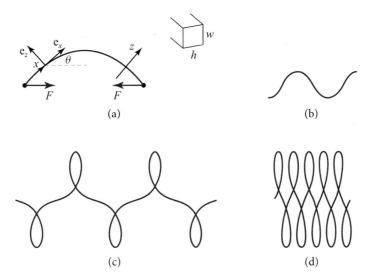

FIGURE 11.10 Elastica. (a) A bent wire is in elastostatic equilibrium under the action of equal and opposite forces applied at its two ends. x measures distance along the neutral surface; z measures distance orthogonal to the wire in the plane of the bend. (b)–(d) Examples of the resulting shapes.

(d) From the relations in parts (a)–(c), derive the following differential equation for the shape of the wire:

$$\boxed{\frac{d^2\theta}{dx^2} = -\frac{F\sin\theta}{D}.}$$
(11.49d)

This is the same equation as describes the motion of a simple pendulum!

(e) For Track-Two readers who have studied Sec. 11.8: Go back through your analysis and identify any place that connection coefficients would enter into a more careful computation, and explain why the connection-coefficient terms are negligible.

(f) Find one nontrivial solution of the elastica equation (11.49d) either analytically using elliptic integrals or numerically. (The general solution can be expressed in terms of elliptic integrals.)

(g) Solve analytically or numerically for the shape adopted by the wire corresponding to your solution in part (f), in terms of precisely Cartesian coordinates (X, Z) in the bending (horizontal) plane. Hint: Express the curvature of the wire, $1/\mathcal{R} = d\theta/dx$, as

$$\frac{d\theta}{dx} = -\frac{d^2 X}{dZ^2}\left[1 + \left(\frac{dX}{dZ}\right)^2\right]^{-3/2}.$$
(11.49e)

(h) Obtain a uniform piece of wire and adjust the force **F** to compare your answer with experiment.

11.6 Buckling and Bifurcation of Equilibria

So far, we have considered stable elastostatic equilibria and have implicitly assumed that the only reason for failure of a material is exceeding the yield limit. However, anyone who has built a house of cards knows that mechanical equilibria can be unstable, with startling consequences. In this section, we explore a specific, important example of a mechanical instability: *buckling*—the theory of which was developed long ago, in 1744 by Leonard Euler.

A tragic example of buckling was the collapse of the World Trade Center's Twin Towers on September 11, 2001. We discuss it near the end of this section, after first developing the theory in the context of a much simpler and cleaner example.

11.6.1 Elementary Theory of Buckling and Bifurcation

Take a new playing card and squeeze it between your finger and thumb (Fig. 11.11). When you squeeze gently, the card remains flat. But when you gradually increase the compressive force F past a critical value F_{crit}, the card suddenly buckles (i.e., bends), and the curvature of the bend then increases rather rapidly with increasing applied force.

To understand quantitatively the sudden onset of buckling, we derive an eigen-equation for the transverse displacement η as a function of distance x from one end of the card. (Although the card is effectively 2-dimensional, it has translation symmetry along its transverse dimension, so we can use the 1-dimensional equations of Sec. 11.5.) We suppose that the ends are free to pivot but not move transversely, so

$$\eta(0) = \eta(\ell) = 0. \tag{11.50}$$

For small displacements, the bending torque of our dimensionally reduced 1-dimensional theory is [Eq. (11.41c)]

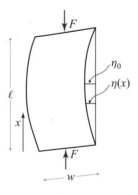

FIGURE 11.11 A playing card of length ℓ, width w, and thickness h is subjected to a compressive force F applied at both ends. The ends of the card are free to pivot.

$$M(x) = -D\frac{d^2\eta}{dx^2}, \tag{11.51}$$

where $D = wh^3E/12$ is the flexural rigidity [Eq. (11.41b)]. As the card is very light (negligible gravity), the total torque around location x, acting on a section of the card from x to one end, is the bending torque $M(x)$ acting at x plus the torque $-F\eta(x)$ associated with the applied force. This sum must vanish:

$$D\frac{d^2\eta}{dx^2} + F\eta = 0. \tag{11.52}$$

The eigensolutions of Eq. (11.52) satisfying boundary conditions (11.50) are

$$\eta = \eta_0 \sin kx, \tag{11.53a}$$

with eigenvalues

$$k = \left(\frac{F}{D}\right)^{1/2} = \frac{n\pi}{\ell} \quad \text{for nonnegative integers } n. \tag{11.53b}$$

Therefore, there is a critical force (first derived by Leonhard Euler in 1744), given by

$$F_{\text{crit}} = \frac{\pi^2 D}{\ell^2} = \frac{\pi^2 wh^3 E}{12\ell^2}. \tag{11.54}$$

critical force for buckling

When $F < F_{\text{crit}}$, there is no solution except $\eta = 0$ (an unbent card). When $F = F_{\text{crit}}$, the unbent card is still a solution, and there suddenly is the additional, arched solution (11.53) with $n = 1$, depicted in Fig. 11.11.

The linear approximation we have used cannot tell us the height η_0 of the arch as a function of F for $F \geq F_{\text{crit}}$; it reports, incorrectly, that for $F = F_{\text{crit}}$ all arch heights are allowed, and that for $F > F_{\text{crit}}$ there is no solution with $n = 1$. However, when nonlinearities are taken into account (Ex. 11.14), the $n = 1$ solution continues to exist for $F > F_{\text{crit}}$, and the arch height η_0 is related to F by

$$F = F_{\text{crit}}\left\{1 + \frac{1}{2}\left(\frac{\pi\eta_0}{2\ell}\right)^2 + O\left[\left(\frac{\pi\eta_0}{2\ell}\right)^4\right]\right\}. \tag{11.55}$$

The sudden appearance of the arched equilibrium state as F is increased through F_{crit} is called a *bifurcation of equilibria*. This bifurcation also shows up in the elasto-dynamics of the playing card, as we deduce in Sec. 12.3.5. When $F < F_{\text{crit}}$, small perturbations of the card's unbent shape oscillate stably. When $F = F_{\text{crit}}$, the unbent card is neutrally stable, and its zero-frequency motion leads the card from its unbent equilibrium state to its $n = 1$, arched equilibrium. When $F > F_{\text{crit}}$, the straight card is an unstable equilibrium: its $n = 1$ perturbations grow in time, driving the card toward the $n = 1$ arched equilibrium state.

A nice way of looking at this bifurcation is in terms of free energy. Consider candidate equilibrium states labeled by the height η_0 of their arch. For each value of η_0, give the card (for concreteness) the $n = 1$ sine-wave shape $\eta = \eta_0 \sin(\pi x/\ell)$. Compute the total elastic energy $E(\eta_0)$ associated with the card's bending, and subtract off the

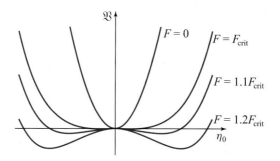

FIGURE 11.12 Representation of bifurcation by a potential energy function $\mathfrak{V}(\eta_0)$. When the applied force is small ($F \leq F_{\text{crit}}$), there is only one stable equilibrium. As the applied force F is increased, the bottom of the potential well flattens, and eventually (for $F > F_{\text{crit}}$) the number of equilibria increases from one to three, of which only two are stable.

work $F\delta X$ done on the card by the applied force F when the card arches from $\eta_0 = 0$ to height η_0. [Here $\delta X(\eta_0)$ is the arch-induced decrease in straight-line separation between the card's ends.] The resulting quantity, $\mathfrak{V}(\eta_0) = E - F\delta X$, is the card's *free energy*—analogous to the physical free energy $F = E - TS$ for a system in contact with a heat bath (Secs. 5.4.1 and 11.3.5), the enthalpic free energy when in contact with a pressure bath (Ex. 5.5h), and the Gibbs (chemical) free energy $G = E - TS + PV$ when in contact with a heat and pressure bath (Sec. 5.5). It is the relevant energy for analyzing the card's equilibrium and dynamics when the force F is continually being applied at the two ends. In Ex. 11.15 we deduce that this free energy is

free energy of a bent card or beam

$$\mathfrak{V} = \left(\frac{\pi \eta_0}{2\ell}\right)^2 \ell \left[(F_{\text{crit}} - F) + \frac{1}{4} F_{\text{crit}} \left(\frac{\pi \eta_0}{2\ell}\right)^2 \right] + O\left[F_{\text{crit}} \ell \left(\frac{\pi \eta_0}{2\ell}\right)^6 \right], \quad (11.56)$$

which we depict in Fig. 11.12.

At small values of the compressive force $F < F_{\text{crit}}$, the free energy has only one minimum $\eta_0 = 0$ corresponding to a single stable equilibrium, the straight card. However, as the force is increased through F_{crit}, the potential minimum flattens out and then becomes a maximum flanked by two new minima (e.g., the curve $F = 1.2F_{\text{crit}}$). The maximum for $F > F_{\text{crit}}$ is the unstable, zero-displacement (straight-card) equilibrium, and the two minima are the stable, finite-amplitude equilibria with positive and negative η_0 given by Eq. (11.55).

This procedure of representing a continuous system with an infinite number of degrees of freedom by just one or a few coordinates and finding the equilibrium by minimizing a free energy is quite common and powerful.

Thus far, we have discussed only two of the card's equilibrium shapes (11.53): the straight shape $n = 0$ and the single-arch shape $n = 1$. If the card were con-

higher order equilibria

strained, by gentle, lateral stabilizing forces, to remain straight beyond $F = F_{\text{crit}}$, then at $F = n^2 F_{\text{crit}}$ for each $n = 2, 3, 4, \ldots$, the nth-order perturbative mode, with

$\eta = \eta_0 \sin(n\pi x/\ell)$, would become unstable, and a new, stable equilibrium with this shape would bifurcate from the straight equilibrium. You can easily explore this for $n = 2$ using a new playing card.

These higher-order modes are rarely of practical importance. In the case of a beam with no lateral constraints, as F increases above F_{crit}, it will buckle into its single-arched shape. For beam dimensions commonly used in construction, a fairly modest further increase of F will bend it enough that its yield point and then rupture point are reached. To experience this yourself, take a thin meter stick, compress its ends between your two hands, and see what happens.

11.6.2 Collapse of the World Trade Center Buildings

Now we return to the example with which we began this section. On September 11, 2001, al-Qaeda operatives hijacked two Boeing 767 passenger airplanes and crashed them into the 110-story Twin Towers of the World Trade Center in New York City, triggering the towers' collapse a few hours later, with horrendous loss of life.

The weight of a tall building such as the towers is supported by vertical steel beams, called "columns." The longer the column is, the lower the weight it can support without buckling, since $F_{crit} = \pi^2 D/\ell^2 = \pi^2 EA(r_g/\ell)^2$, with A the beam's cross sectional area, r_g its radius of gyration along its bending direction, and ℓ its length [Eqs. (11.54) and (11.41b)].[8] The column lengths are typically chosen such that the critical stress for buckling, $F_{crit}/A = E(\pi r_g/\ell)^2$, is roughly the same as the yield stress, $F_{yield} \simeq 0.003E$ (cf. Table 11.1), which means that the columns' *slenderness ratio* is $\ell/r_g \sim 50$. The columns are physically far longer than $50r_g$, but they are anchored to each other laterally every $\sim 50r_g$ by beams and girders in the floors, so their effective length for buckling is $\ell \sim 50r_g$. The columns' radii of gyration r_g are generally made large, without using more steel than needed to support the overhead weight, by making the columns hollow, or giving them H-shaped cross sections. In the Twin Towers, the thinnest beams had $r_g \sim 13$ cm, and they were anchored in every floor, with floor separations $\ell \simeq 3.8$ m, so their slenderness ratio was actually $\ell/r_g \simeq 30$.

According to a detailed investigation (NIST, 2005, especially Secs. 6.14.2 and 6.14.3), the crashing airplanes ignited fires in and near floors 93–99 of the North Tower and 78–83 of the South Tower, where the airplanes hit. The fires were most intense in the floors and around uninsulated central steel columns. The heated central columns lost their rigidity and began to sag, and trusses then transferred some of the weight above to the outer columns. In parallel, the heated floor structures began

description of failure modes

---

8. As noted in the discussion, after Eq. (11.41b), Ar_g^2 is really the second moment of the column's cross sectional area, along its direction of bend. If the column is supported at its ends against movement in both transverse directions, then the relevant second moment is the transverse tensor $\int x^i x^j dx dy$, and the direction of buckling (if it occurs) will be the eigendirection of this tensor that has the smallest eigenvalue (the column's narrowest direction).

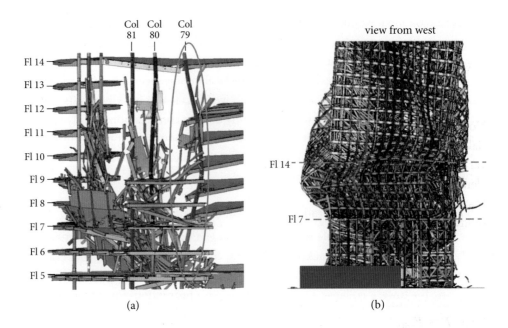

FIGURE 11.13 (a) The buckling of column 79 in building WTC7 at the World Trade Center, based on a finite-element simulation informed by all available observational data. (b) The subsequent buckling of the building's core. From NIST (2008).

to sag, pulling inward on the buildings' exterior steel columns, which bowed inward and then buckled, initiating the buildings' collapse. [This is a somewhat oversimplified description of a complex situation; for full complexities, see the report, NIST (2005).]

This column buckling was somewhat different from the buckling of a playing card because of the inward pull of the sagging floors. Much more like our playing-card buckle was the fate of an adjacent, 47-story building called WTC7. When the towers collapsed, they injected burning debris onto and into WTC7. About 7 hours later, fire-induced thermal expansion triggered a cascade of failures in floors 13–16, which left column 79 with little stabilizing lateral support, so its effective length ℓ was increased far beyond $50r_g$. It then quickly buckled (Fig. 11.13a) in much the same manner as our playing card, followed by column 80, then 81, and subsequently columns 77, 78, and 76 (NIST, 2008, especially Sec. 2.4). Within seconds, the building's entire core was buckling (Fig. 11.13b).

11.6.3   ### 11.6.3 Buckling with Lateral Force; Connection to Catastrophe Theory

Returning to the taller Twin Towers, we can crudely augment the inward pull of the sagging floors into our free-energy description of buckling, by adding a term $-F_{\mathrm{lat}}\eta_0$, which represents the energy inserted into a bent column by a lateral force F_{lat} when its center has been displaced laterally through the distance η_0. Then the free energy

(11.56), made dimensionless and with its terms rearranged, takes the form

$$\varphi \equiv \frac{\mathfrak{V}}{F_{\text{crit}}\ell} = \frac{1}{4}\left(\frac{\pi\eta_0}{2\ell}\right)^4 - \frac{1}{2}\left(\frac{2(F - F_{\text{crit}})}{F_{\text{crit}}}\right)\left(\frac{\pi\eta_0}{2\ell}\right)^2 - \left(\frac{2F_{\text{lat}}}{\pi F_{\text{crit}}}\right)\left(\frac{\pi\eta_0}{2\ell}\right).$$

(11.57)

Notice that this equation has the canonical form $\varphi = \frac{1}{4}a^4 - \frac{1}{2}za^2 - xa$ for the potential that governs a cusp catastrophe, whose state variable is $a = \pi\eta_0/(2\ell)$ and control variables are $z = 2(F - F_{\text{crit}})/F_{\text{crit}}$ and $x = (2/\pi)F_{\text{lat}}/F_{\text{crit}}$; see Eq. (7.72).[9] From the elementary mathematics of this catastrophe, as worked out in Sec. 7.5.1, we learn that although the lateral force F_{lat} will make the column bend, it will not induce a bifurcation of equilibria until the control-space cusp $x = \pm 2(z/3)^{3/2}$ is reached:

interpretation in terms of catastrophe theory

$$\frac{F_{\text{lat}}}{F_{\text{crit}}} = \pm\pi\left(\frac{2(F - F_{\text{crit}})}{3F_{\text{crit}}}\right)^{3/2}.$$

(11.58)

Notice that the lateral force F_{lat} actually delays the bifurcation to a higher vertical force, $F > F_{\text{crit}}$. However, this is not significant for the physical buckling, since the column in this case is bent from the outset, and as F_{lat} increases, it stops carrying its share of the building's weight and moves smoothly toward its yield point and rupture; Ex. 11.16.

11.6.4 Other Bifurcations: Venus Fly Trap, Whirling Shaft, Triaxial Stars, and Onset of Turbulence

11.6.4

This bifurcation of equilibria, associated with the buckling of a column, is just one of many bifurcations that occur in physical systems. Another is a buckling type bifurcation that occurs in the 2-dimensional leaves of the Venus fly trap plant; the plant uses the associated instability to snap together a pair of leaves in a small fraction of a second, thereby capturing insects for it to devour; see Fortere et al. (2005). Yet another is the onset of a lateral bend in a shaft (rod) that spins around its longitudinal axis (see Love, 1927, Sec. 286). This is called *whirling*; it is an issue in drive shafts for automobiles and propellers, and a variant of it occurs in spinning DNA molecules during replication—see Wolgemuth, Powers, and Goldstein (2000). One more example is the development of triaxiality in self-gravitating fluid masses (i.e., stars) when their rotational kinetic energy reaches a critical value, about 1/4 of their gravitational energy; see Chandrasekhar (1962). Bifurcations also play a major role in the onset of turbulence in fluids and in the route to chaos in other dynamical systems; we study turbulence and chaos in Sec. 15.6.

whirling shaft

9. The lateral force F_{lat} makes the bifurcation *structurally stable*, in the language of catastrophe theory (discussed near the end of Sec. 7.5) and thereby makes it describable by one of the generic catastrophes. Without F_{lat}, the bifurcation is not structurally stable.

For further details on the mathematics of bifurcations with emphasis on elasto-statics and elastodynamics, see, for example, Marsden and Hughes (1986, Chap. 7). For details on buckling from an engineering viewpoint, see Ugural and Fenster (2012, Chap. 11).

Exercise 11.14 *Derivation and Example: Bend as a Function of Applied Force*
Derive Eq. (11.55) relating the angle $\theta_o = (d\eta/dx)_{x=0} = k\eta_o = \pi\eta_o/\ell$ to the applied force F when the card has an $n = 1$, arched shape. [Hint: Consider the card as comprising many thin parallel wires and use the elastica differential equation $d^2\theta/dx^2 = -(F/D)\sin\theta$ [Eq. (11.49d)] for the angle between the card and the applied force at distance x from the card's end. The $\sin\theta$ becomes θ in the linear approximation used in the text; the nonlinearities embodied in the sine give rise to the desired relation. The following steps along the way toward a solution are mathematically the same as used when computing the period of a pendulum as a function of its amplitude of swing.]

(a) Derive the first integral of the elastica equation

$$(d\theta/dx)^2 = 2(F/D)(\cos\theta - \cos\theta_o), \tag{11.59}$$

where θ_o is an integration constant. Show that the boundary condition of no bending torque (no inflection of the card's shape) at the card ends implies $\theta = \theta_o$ at $x = 0$ and $x = \ell$; whence $\theta = 0$ at the card's center, $x = \ell/2$.

(b) Integrate the differential equation (11.59) to obtain

$$\frac{\ell}{2} = \sqrt{\frac{D}{2F}} \int_0^{\theta_o} \frac{d\theta}{\sqrt{\cos\theta - \cos\theta_o}}. \tag{11.60}$$

(c) Perform the change of variable $\sin(\theta/2) = \sin(\theta_o/2)\sin\phi$ and thereby bring Eq. (11.60) into the form

$$\ell = 2\sqrt{\frac{D}{F}} \int_0^{\pi/2} \frac{d\phi}{\sqrt{1 - \sin^2(\theta_o/2)\sin^2\phi}} = 2\sqrt{\frac{D}{F}}\, K[\sin^2(\theta_o/2)]. \tag{11.61}$$

Here $K(y)$ is the complete elliptic integral of the first type, with the parameterization used by Mathematica (which differs from that of many books).

(d) Expand Eq. (11.61) in powers of $\theta_o/2$ to obtain

$$F = F_{\text{crit}}\frac{4}{\pi^2}K^2[\sin^2(\theta_o/2)] = F_{\text{crit}}\left[1 + \frac{1}{2}\left(\frac{\theta_o/2}{2}\right)^2 + \dots\right], \tag{11.62}$$

from which deduce our desired result, Eq. (11.55).

Exercise 11.15 *Problem: Free Energy of a Bent, Compressed Beam*
Derive Eq. (11.56) for the free energy $\mathfrak{V}$ of a beam that is compressed with a force F and has a critical compression $F_{\text{crit}} = \pi^2 D/\ell^2$, where D is its flexural rigidity. [Hint: It

must be that $\partial V / \partial \eta_0 = 0$ gives Eq. (11.55) for the beam's equilibrium bend amplitude η_0 as a function of $F - F_{\text{crit}}$. Use this to reduce the number of terms in $\mathfrak{V}(\eta_0)$ in Eq. (11.56) that you need to derive.]

Exercise 11.16 *Problem: Bent Beam with Lateral Force*
Explore numerically the free energy (11.57) of a bent beam with a compressive force F and lateral force F_{lat}. Examine how the extrema (equilibrium states) evolve as F and F_{lat} change, and deduce the physical consequences.

Exercise 11.17 **Problem: Applications of Buckling—Mountains and Pipes*
Buckling plays a role in many natural and human-caused phenomena. Explore the following examples.

(a) *Mountain building.* When two continental plates are in (very slow) collision, the compressional force near their interface drives their crustal rock to buckle upward, producing mountains. Estimate how high such mountains can be on Earth and on Mars, and compare your estimates with their actual heights. Read about such mountain building in books or on the web.

(b) *Thermal expansion of pipes.* When a segment of pipe is heated (e.g., by the rising sun in the morning), it will expand. If its ends are held fixed, this can easily produce a longitudinal stress large enough to buckle the pipe. How would you deal with this in an oil pipeline on Earth's surface? In a long vacuum tube? Compare your answers with standard engineering solutions, which you can find in books or on the web.

Exercise 11.18 *Example: Allometry*
Allometry is the study of biological scaling laws that relate various features of an animal to its size or mass. One example concerns the ratio of the width to the length of leg bones. Explain why the width to the length of a thigh bone in a quadruped might scale as the square root of the stress that it has to support. Compare elephants with cats in this regard. (The density of bone is roughly 1.5 times that of water, and its Young's modulus is $\sim$20 GPa.)

11.7 Reducing the Elastostatic Equations to 2 Dimensions for a Deformed Thin Plate: Stress Polishing a Telescope Mirror

The world's largest optical telescopes (as of 2016) are the Gran Telescopio Canarias in the Canary Islands and the two Keck telescopes on Mauna Kea in Hawaii, which are all about 10 m in diameter. It is very difficult to support traditional, monolithic mirrors so that the mirror surfaces maintain their shape (their "figure") as the telescope slews, because they are so heavy, so for Keck a new method of fabrication was sought. The solution devised by Jerry Nelson and his colleagues was to construct the telescope out

11.7 Stress Polishing a Telescope Mirror   **609**

of 36 separate hexagons, each 0.9 m on a side. However, this posed a second problem: how to grind each hexagon's reflecting surface to the required hyperboloidal shape. For this, a novel technique called *stressed mirror polishing* was developed. This technique relies on the fact that it is relatively easy to grind a surface to a spherical shape, but technically highly challenging to create a nonaxisymmetric shape. So during the grinding, stresses are applied around the boundary of the mirror to deform it, and a spherical surface is produced. The stresses are then removed, and the mirror springs into the desired nonspherical shape. Computing the necessary stresses is a problem in classical elasticity theory and, in fact, is a good example of a large number of applications where the elastic body can be approximated as a thin plate and its shape analyzed using elasticity equations that are reduced from 3 dimensions to 2 by the method of moments.

stressed mirror polishing

For stress polishing of mirrors, the applied stresses are so large that we can ignore gravitational forces (at least in our simplified treatment). We suppose that the hexagonal mirror has a uniform thickness h and idealize it as a circle of radius R, and we introduce Cartesian coordinates with (x, y) in the horizontal plane (the plane of the mirror before deformation and polishing begin), and z vertical. The mirror is deformed as a result of a net vertical force per unit area (pressure) $P(x, y)$. This pressure is applied at the lower surface when upward (positive) and the upper surface when downward (negative). In addition, there are shear forces and bending torques applied around the rim of the mirror.

As in our analysis of a cantilever in Sec. 11.5, we assume the existence of a neutral surface in the deformed mirror, where the horizontal strain vanishes, $T_{ab} = 0$. (Here and below we use letters from the early part of the Latin alphabet for horizontal components $x = x^1$ and $y = x^2$.) We denote the vertical displacement of the neutral surface by $\eta(x, y)$. By applying the method of moments to the 3-dimensional stress-balance equation $T_{jk,k} = 0$ in a manner similar to our cantilever analysis, we obtain the following 2-dimensional equation for the mirror's shape (Ex. 11.19):

$$\boxed{\nabla^2(\nabla^2\eta) = P(x, y)/D.}$$

(11.63a)

elastostatic force balance for a bent plate on which a pressure P acts: shape equation

Here ∇^2 is the horizontal Laplacian: $\nabla^2\eta \equiv \eta_{,aa} = \eta_{,xx} + \eta_{,yy}$. Equation (11.63a) is the 2-dimensional analog of the equation $d^4\eta/dx^4 = W(x)/D$ for the shape of a cantilever on which a downward force per unit length $W(x)$ acts [Eq. (11.45)]. The 2-dimensional flexural rigidity that appears in Eq. (11.63a) is

$$\boxed{D = \frac{Eh^3}{12(1 - \nu^2)},}$$

(11.63b)

2-dimensional flexural rigidity

where E is the mirror's Young's modulus, h is its thickness, and ν is its Poisson's ratio. The operator $\nabla^2\nabla^2$ acting on η in the shape equation (11.63a) is called the *biharmonic operator;* it also appears in 3-dimensional form in the biharmonic equation (11.34a)

biharmonic operator

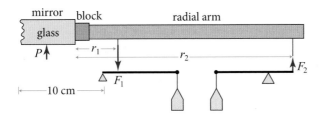

FIGURE 11.14 Schematic showing the mirror rim, a radial arm attached to it via a block, and a lever assembly used to apply shear forces and bending torques to the rim during stress polishing. (F_1 need not equal F_2, as there is a pressure P applied to the back surface of the mirror and forces applied at 23 other points around its rim.) The shear force on the mirror rim is $S = F_2 - F_1$, and the bending torque is $M = r_2 F_2 - r_1 F_1$.

for the displacement inside a homogeneous, isotropic body to which surface stresses are applied.

The shape equation (11.63a) must be solved subject to boundary conditions around the mirror's rim: the applied shear forces and bending torques.

The individual Keck mirror segments were constructed out of a ceramic material with Young's modulus $E = 89$ GPa and Poisson's ratio $\nu = 0.24$ (similar to glass; cf. Table 11.1). A mechanical jig was constructed to apply the shear forces and bending torques at 24 uniformly spaced points around the rim of the mirror (Fig. 11.14). The maximum stress was applied for the six outermost mirrors and was 2.4×10^6 N m^{-2}, 12% of the material's breaking tensile strength (2×10^7 N m^{-2}).

This stress polishing worked beautifully, and the Keck telescopes have become highly successful tools for astronomical research.

EXERCISES

Exercise 11.19 **Derivation and Example: Dimensionally Reduced Shape Equation for a Stressed Plate*
Use the method of moments (Sec. 11.5) to derive the 2-dimensional shape equation (11.63a) for the stress-induced deformation of a thin plate, and expression (11.63b) for the 2-dimensional flexural rigidity. Here is a step-by-step guide, in case you want or need it.

(a) Show on geometrical grounds that the in-plane strain is related to the vertical displacement by [cf. Eq. (11.40a)]

$$\xi_{a,b} = -z\eta_{,ab}. \tag{11.64a}$$

(b) Derive an expression for the horizontal components of the stress, T_{ab}, in terms of double derivatives of the displacement function $\eta(x, y)$ [analog of $T_{xx} = -Ezd^2\eta/dx^2$, Eq. (11.40b), for a stressed rod]. This can be done (i) by arguing on physical grounds that the vertical component of stress, T_{zz}, is much smaller than the horizontal components and therefore can be approximated as zero [an approximation to be checked in part (f)]; (ii) by expressing $T_{zz} = 0$ in terms of the strain and thence displacement and using Eqs. (11.39) to obtain

$$\Theta = -\left(\frac{1-2\nu}{1-\nu}\right) z\nabla^2\eta, \tag{11.64b}$$

where ∇^2 is the horizontal Laplacian; and (iii) by then writing T_{ab} in terms of Θ and $\xi_{a,b}$ and combining with Eqs. (11.64a) and (11.64b) to get the desired equation:

$$T_{ab} = Ez\left[\frac{\nu}{(1-\nu^2)}\nabla^2\eta\,\delta_{ab} + \frac{\eta_{,ab}}{(1+\nu)}\right]. \tag{11.64c}$$

(c) With the aid of Eq. (11.64c), write the horizontal force density in the form

$$f_a = -T_{ab,b} - T_{az,z} = -\frac{Ez}{1-\nu^2}\nabla^2\eta_{,a} - T_{az,z} = 0. \tag{11.64d}$$

Then, as in the cantilever analysis [Eq. (11.40d)], reduce the dimensionality of this force equation by the method of moments. The zeroth moment (integral over z) vanishes. Why? Therefore, the lowest nonvanishing moment is the first (multiply f_a by z and integrate). Show that this gives

$$S_a \equiv \int T_{za}dz = D\nabla^2\eta_{,a}, \tag{11.64e}$$

where D is the 2-dimensional flexural rigidity (11.63b). The quantity S_a is the vertical shear force per unit length acting perpendicular to a line in the mirror whose normal is in the direction a; it is the 2-dimensional analog of a stressed rod's shear force S [Eq. (11.41a)].

(d) For physical insight into Eq. (11.64e), define the bending torque per unit length (bending torque density) as

$$M_{ab} \equiv \int z T_{ab}dz, \tag{11.64f}$$

and show with the aid of Eq. (11.64c) that (11.64e) is the *law of torque balance* $S_a = M_{ab,b}$—the 2-dimensional analog of a stressed rod's $S = dM/dx$ [Eq. (11.42)].

(e) Compute the total vertical shear force acting on a small area of the plate as the line integral of S_a around its boundary, and by applying Gauss's theorem, deduce that the vertical shear force per unit area is $S_{a,a}$. Argue that this must be balanced by the net pressure P applied to the face of the plate, and thereby deduce the *law of vertical force balance*:

$$S_{a,a} = P. \tag{11.64g}$$

By combining this equation with the law of torque balance (11.64e), obtain the plate's bending equation $\nabla^2(\nabla^2\eta) = P/D$ [Eq. (11.63a)—the final result we were seeking].

(f) Use this bending equation to verify the approximation made in part (b) that T_{zz} is small compared to the horizontal stresses. Specifically, show that $T_{zz} \simeq P$ is $O(h/R)^2 T_{ab}$, where h is the plate thickness, and R is the plate radius.

Exercise 11.20 *Example: Paraboloidal Mirror*

Show how to construct a paraboloidal mirror of radius R and focal length f by stress polishing.

(a) Adopt a strategy of polishing the stressed mirror into a segment of a sphere with radius of curvature equal to that of the desired paraboloid at its center, $r = 0$. By comparing the shape of the desired paraboloid to that of the sphere, show that the required vertical displacement of the stressed mirror during polishing is

$$\eta(r) = \frac{r^4}{64 f^3},\tag{11.64h}$$

where r is the radial coordinate, and we only retain terms of leading order.

(b) Hence use Eq. (11.63a) to show that a uniform force per unit area

$$P = \frac{D}{f^3},\tag{11.64i}$$

where D is the flexural rigidity, must be applied to the bottom of the mirror. (Ignore the weight of the mirror.)

(c) Based on the results of part (b), show that if there are N equally spaced levers attached at the rim, the vertical force applied at each of them must be

$$F_z = \frac{\pi D R^2}{N f^3}.\tag{11.64j}$$

(d) Show that the radial displacement inside the mirror is

$$\xi_r = -\frac{r^3 z}{16 f^3},\tag{11.64k}$$

where z is the vertical distance from the neutral surface, halfway through the mirror.

(e) Hence show that the maximum stress in the mirror is

$$T_{\max} = \frac{(3+v)R^2 h E}{32(1-v^2)f^3},\tag{11.64l}$$

where h is the mirror thickness.

(f) Calculate the bending torque M that must be applied at each lever (Fig. 11.14). Comment on the limitations of this technique for making a thick, "fast" (i.e., $2R/f$ large) mirror.

11.8 Cylindrical and Spherical Coordinates: Connection Coefficients and Components of the Gradient of the Displacement Vector T2

Thus far in our discussion of elasticity, we have restricted ourselves to Cartesian coordinates. However, many problems in elasticity are most efficiently solved using cylindrical or spherical coordinates, so in this section, we develop some mathematical tools for those coordinate systems. In doing so, we follow the vectorial conventions of standard texts on electrodynamics and quantum mechanics (e.g., Jackson, 1999; Cohen-Tannoudji, Diu, and Laloë, 1977). We introduce an *orthonormal* set of basis vectors associated with each of our curvilinear coordinate systems; the coordinate lines are orthogonal to one another, and the basis vectors have unit lengths and point along the coordinate lines. In our study of continuum mechanics (Part IV, Elasticity; Part V, Fluid Dynamics; and Part VI, Plasma Physics), we follow this practice. When studying General Relativity (Part VII), we introduce and use basis vectors that are *not* orthonormal.

orthonormal basis vectors of cylindrical or spherical coordinates

cylindrical coordinates

Our notation for cylindrical coordinates is (ϖ, ϕ, z); ϖ (pronounced "pomega") is distance from the z-axis, and ϕ is the angle around the z-axis:

$$\varpi = \sqrt{x^2 + y^2}, \qquad \phi = \arctan(y/x). \tag{11.65a}$$

The unit basis vectors that point along the coordinate axes are denoted $\mathbf{e}_\varpi$, $\mathbf{e}_\phi$, and $\mathbf{e}_z$, and are related to the Cartesian basis vectors by

$$\mathbf{e}_\varpi = (x/\varpi)\mathbf{e}_x + (y/\varpi)\mathbf{e}_y, \quad \mathbf{e}_\phi = -(y/\varpi)\mathbf{e}_x + (x/\varpi)\mathbf{e}_y,$$

$$\mathbf{e}_z = \text{Cartesian } \mathbf{e}_z. \tag{11.65b}$$

spherical coordinates

Our notation for spherical coordinates is (r, θ, ϕ), with (as should be very familiar)

$$r = \sqrt{x^2 + y^2 + z^2}, \qquad \theta = \arccos(z/r), \qquad \phi = \arctan(y/x). \tag{11.66a}$$

The unit basis vectors associated with these coordinates are

$$\mathbf{e}_r = \frac{x}{r}\mathbf{e}_x + \frac{y}{r}\mathbf{e}_y + \frac{z}{r}\mathbf{e}_z, \quad \mathbf{e}_\theta = \frac{z}{r}\mathbf{e}_\varpi - \frac{\varpi}{r}\mathbf{e}_z, \quad \mathbf{e}_\phi = -\frac{y}{\varpi}\mathbf{e}_x + \frac{x}{\varpi}\mathbf{e}_y. \tag{11.66b}$$

Because our bases are orthonormal, the components of the metric of 3-dimensional space retain the Kronecker-delta values

$$\boxed{g_{jk} \equiv \mathbf{e}_j \cdot \mathbf{e}_k = \delta_{jk},} \tag{11.67}$$

which permits us to keep all vector and tensor indices down, by contrast with spacetime, where we must distinguish between up and down; cf. Sec. 2.5.[10]

---

10. Occasionally—e.g., in the useful equation $\epsilon_{ijm}\epsilon_{klm} = \delta^{ij}_{kl} \equiv \delta^i_k\delta^j_l - \delta^i_l\delta^j_k$ [Eq. (1.23)]—it is convenient to put some indices up. In our orthonormal basis, any component with an index up is equal to that same component with an index down: e.g., $\delta^i_k \equiv \delta_{ik}$.

In Jackson (1999), Cohen-Tannoudji, Diu, and Laloë (1977), and other standard texts, formulas are written down for the gradient and Laplacian of a scalar field, and the divergence and curl of a vector field, in cylindrical and spherical coordinates; one uses these formulas over and over again. In elasticity theory, we deal largely with second-rank tensors and will need formulas for their various derivatives in cylindrical and spherical coordinates. In this book we introduce a mathematical tool, *connection coefficients* Γ_{ijk}, by which those formulas can be derived when needed.

connection coefficients

The connection coefficients quantify the turning of the orthonormal basis vectors as one moves from point to point in Euclidean 3-space: they tell us how the basis vectors at one point in space are *connected to* (related to) those at another point. More specifically, we define Γ_{ijk} by the two equivalent relations

$$\nabla_k \mathbf{e}_j = \Gamma_{ijk} \mathbf{e}_i; \qquad \Gamma_{ijk} = \mathbf{e}_i \cdot (\nabla_k \mathbf{e}_j). \tag{11.68}$$

Here $\nabla_k \equiv \nabla_{\mathbf{e}_k}$ is the directional derivative along the orthonormal basis vector $\mathbf{e}_k$; cf. Eq. (1.15a). Notice that (as is true quite generally; cf. Sec. 1.7) the differentiation index comes *last* on Γ; and notice that the middle index of Γ names the basis vector that is differentiated. Because our basis is orthonormal, it must be that $\nabla_k(\mathbf{e}_i \cdot \mathbf{e}_j) = 0$. Expanding this expression out using the standard rule for differentiating products, we obtain $\mathbf{e}_j \cdot (\nabla_k \mathbf{e}_i) + \mathbf{e}_i \cdot (\nabla_k \mathbf{e}_j) = 0$. Then invoking the definition (11.68) of the connection coefficients, we see that Γ_{ijk} is antisymmetric on its first two indices:

$$\Gamma_{ijk} = -\Gamma_{jik}. \tag{11.69}$$

In Part VII, when we use nonorthonormal bases, this antisymmetry will break down.

It is straightforward to compute the connection coefficients for cylindrical and spherical coordinates from (i) the definition (11.68); (ii) expressions (11.65b) and (11.66b) for the cylindrical and spherical basis vectors in terms of the Cartesian basis vectors; and (iii) the fact that *in Cartesian coordinates the connection coefficients vanish* ($\mathbf{e}_x$, $\mathbf{e}_y$, and $\mathbf{e}_z$ do not rotate as one moves through Euclidean 3-space). One can also deduce the cylindrical and spherical connection coefficients by drawing pictures of the basis vectors and observing how they change from point to point. As an example, for cylindrical coordinates we see from Fig. 11.15 that $\nabla_\phi \mathbf{e}_\varpi = \mathbf{e}_\phi / \varpi$. A similar pictorial calculation (which the reader is encouraged to do) reveals that $\nabla_\phi \mathbf{e}_\phi = -\mathbf{e}_\varpi / \varpi$. All other derivatives vanish. By comparing with Eq. (11.68), we see that *the only nonzero connection coefficients in cylindrical coordinates are*

connection coefficients for orthonormal bases of cylindrical and spherical coordinates

$$\Gamma_{\varpi\phi\phi} = -\frac{1}{\varpi}, \quad \Gamma_{\phi\varpi\phi} = \frac{1}{\varpi}, \tag{11.70}$$

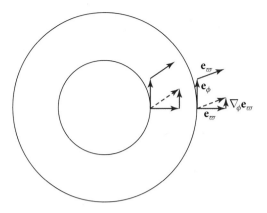

FIGURE 11.15 Pictorial evaluation of $\Gamma_{\phi\varpi\phi}$. In the rightmost assemblage of vectors we compute $\nabla_\phi \mathbf{e}_\varpi$ as follows. We draw the vector to be differentiated, $\mathbf{e}_\varpi$, at the tail of $\mathbf{e}_\phi$ (the vector along which we differentiate) and also at its head. We then subtract $\mathbf{e}_\varpi$ at the head from that at the tail; this difference is $\nabla_\phi \mathbf{e}_\varpi$. It obviously points in the $\mathbf{e}_\phi$ direction. When we perform the same calculation at a radius ϖ that is smaller by a factor 2 (left assemblage of vectors), we obtain a result, $\nabla_\phi \mathbf{e}_\varpi$, that is twice as large. Therefore, the length of this vector must scale as $1/\varpi$. By looking quantitatively at the length at some chosen radius ϖ, one can see that the multiplicative coefficient is unity: $\nabla_\phi \mathbf{e}_\varpi = \mathbf{e}_\phi/\varpi$. Comparing with Eq. (11.68), we deduce that $\Gamma_{\phi\varpi\phi} = 1/\varpi$.

which have the required antisymmetry [Eq. (11.69)]. Likewise, for spherical coordinates (Ex. 11.22), we have

$$\Gamma_{\theta r\theta} = \Gamma_{\phi r\phi} = -\Gamma_{r\theta\theta} = -\Gamma_{r\phi\phi} = \frac{1}{r}, \quad \Gamma_{\phi\theta\phi} = -\Gamma_{\theta\phi\phi} = \frac{\cot\theta}{r}. \tag{11.71}$$

These connection coefficients are the keys to differentiating vectors and tensors. Consider the gradient of the displacement, $\mathbf{W} = \nabla\boldsymbol{\xi}$. Applying the product rule for differentiation, we obtain

$$\nabla_k(\xi_j \mathbf{e}_j) = (\nabla_k \xi_j)\mathbf{e}_j + \xi_j(\nabla_k \mathbf{e}_j) = \xi_{j,k}\mathbf{e}_j + \xi_j \Gamma_{ljk}\mathbf{e}_l. \tag{11.72}$$

directional derivative along basis vector

Here the comma denotes the directional derivative, along a basis vector, of the components treated as scalar fields. For example, in cylindrical coordinates we have

$$\xi_{i,\varpi} = \frac{\partial \xi_i}{\partial \varpi}, \quad \xi_{i,\phi} = \frac{1}{\varpi}\frac{\partial \xi_i}{\partial \phi}, \quad \xi_{i,z} = \frac{\partial \xi_i}{\partial z}; \tag{11.73}$$

and in spherical coordinates we have

$$\xi_{i,r} = \frac{\partial \xi_i}{\partial r}, \quad \xi_{i,\theta} = \frac{1}{r}\frac{\partial \xi_i}{\partial \theta}, \quad \xi_{i,\phi} = \frac{1}{r\sin\theta}\frac{\partial \xi_i}{\partial \phi}. \tag{11.74}$$

Taking the ith component of Eq. (11.72) we obtain

$$\boxed{W_{ik} = \xi_{i;k} = \xi_{i,k} + \Gamma_{ijk}\xi_j.}$$

(11.75)

Here $\xi_{i;k}$ are the nine components of the gradient of the vector field $\boldsymbol{\xi}(\mathbf{x})$.

We can use Eq. (11.75) to evaluate the expansion $\Theta = \mathrm{Tr}\,\mathbf{W} = \boldsymbol{\nabla}\cdot\boldsymbol{\xi}$. Using Eqs. (11.70) and (11.71), we obtain

$$\Theta = \boldsymbol{\nabla}\cdot\boldsymbol{\xi} = \frac{\partial\xi_\varpi}{\partial\varpi} + \frac{1}{\varpi}\frac{\partial\xi_\phi}{\partial\phi} + \frac{\partial\xi_z}{\partial z} + \frac{\xi_\varpi}{\varpi}$$

$$= \frac{1}{\varpi}\frac{\partial}{\partial\varpi}\left(\varpi\xi_\varpi\right) + \frac{1}{\varpi}\frac{\partial\xi_\phi}{\partial\phi} + \frac{\partial\xi_z}{\partial z}$$

(11.76)

in cylindrical coordinates and

$$\Theta = \boldsymbol{\nabla}\cdot\boldsymbol{\xi} = \frac{\partial\xi_r}{\partial r} + \frac{1}{r}\frac{\partial\xi_\theta}{\partial\theta} + \frac{1}{r\sin\theta}\frac{\partial\xi_\phi}{\partial\phi} + \frac{2\xi_r}{r} + \frac{\cot\theta\,\xi_\theta}{r}$$

$$= \frac{1}{r^2}\frac{\partial}{\partial r}(r^2\xi_r) + \frac{1}{r\sin\theta}\frac{\partial}{\partial\theta}(\sin\theta\,\xi_\theta) + \frac{1}{r\sin\theta}\frac{\partial\xi_\phi}{\partial\phi}$$

(11.77)

in spherical coordinates, in agreement with formulas in standard textbooks, such as the flyleaf of Jackson (1999).

The components of the rotation are most easily deduced using $R_{ij} = -\epsilon_{ijk}\phi_k$ with $\boldsymbol{\phi} = \frac{1}{2}\boldsymbol{\nabla}\times\boldsymbol{\xi}$, and the standard expressions for the curl in cylindrical and spherical coordinates (Jackson, 1999). Since the rotation does not enter into elasticity theory in a significant way, we refrain from writing down the results. The components of the shear are computed in Box 11.4.

By a computation analogous to Eq. (11.72), we can construct an expression for the gradient of a tensor of any rank. For a second-rank tensor $\mathbf{T} = T_{ij}\mathbf{e}_i \otimes \mathbf{e}_j$ we obtain (Ex. 11.21)

$$\boxed{T_{ij;k} = T_{ij,k} + \Gamma_{ilk}T_{lj} + \Gamma_{jlk}T_{il}.}$$

(11.78)

components of gradient of a tensor

Equation (11.78) for the components of the gradient can be understood as follows. In cylindrical or spherical coordinates, the components T_{ij} can change from point to point as a result of two things: a change of the tensor $\mathbf{T}$ or the turning of the basis vectors. The two connection coefficient terms in Eq. (11.78) remove the effects of the basis turning, leaving in $T_{ij;k}$ only the influence of the change of $\mathbf{T}$ itself. There are two correction terms corresponding to the two slots (indices) of $\mathbf{T}$; the effects of basis turning on each slot get corrected one after another. If $\mathbf{T}$ had had n slots, then there would have been n correction terms, each with the form of the two in Eq. (11.78).

These expressions for derivatives of tensors are not required for dealing with the vector fields of introductory electromagnetic theory or quantum theory, but they are essential for manipulating the tensor fields encountered in elasticity. As we shall see in Sec. 24.3, with one further generalization, we can go on to differentiate tensors in

BOX 11.4. SHEAR TENSOR IN SPHERICAL AND CYLINDRICAL COORDINATES T2

Using our rules (11.75) for forming the gradient of a vector, we can derive a general expression for the shear tensor:

$$\Sigma_{ij} = \frac{1}{2}(\xi_{i;j} + \xi_{j;i}) - \frac{1}{3}\delta_{ij}\xi_{k;k}$$

$$= \frac{1}{2}(\xi_{i,j} + \xi_{j,i} + \Gamma_{ilj}\xi_l + \Gamma_{jli}\xi_l) - \frac{1}{3}\delta_{ij}(\xi_{k,k} + \Gamma_{klk}\xi_l). \quad (1)$$

Evaluating this in cylindrical coordinates using the connection coefficients (11.70), we obtain

$$\Sigma_{\varpi\varpi} = \frac{2}{3}\frac{\partial\xi_\varpi}{\partial\varpi} - \frac{1}{3}\frac{\xi_\varpi}{\varpi} - \frac{1}{3\varpi}\frac{\partial\xi_\phi}{\partial\phi} - \frac{1}{3}\frac{\partial\xi_z}{\partial z},$$

$$\Sigma_{\phi\phi} = \frac{2}{3\varpi}\frac{\partial\xi_\phi}{\partial\phi} + \frac{2}{3}\frac{\xi_\varpi}{\varpi} - \frac{1}{3}\frac{\partial\xi_\varpi}{\partial\varpi} - \frac{1}{3}\frac{\partial\xi_z}{\partial z},$$

$$\Sigma_{zz} = \frac{2}{3}\frac{\partial\xi_z}{\partial z} - \frac{1}{3}\frac{\partial\xi_\varpi}{\partial\varpi} - \frac{1}{3}\frac{\xi_\varpi}{\varpi} - \frac{1}{3\varpi}\frac{\partial\xi_\phi}{\partial\phi},$$

$$\Sigma_{\phi z} = \Sigma_{z\phi} = \frac{1}{2\varpi}\frac{\partial\xi_z}{\partial\phi} + \frac{1}{2}\frac{\partial\xi_\phi}{\partial z},$$

$$\Sigma_{z\varpi} = \Sigma_{\varpi z} = \frac{1}{2}\frac{\partial\xi_\varpi}{\partial z} + \frac{1}{2}\frac{\partial\xi_z}{\partial\varpi},$$

$$\Sigma_{\varpi\phi} = \Sigma_{\phi\varpi} = \frac{1}{2}\frac{\partial\xi_\phi}{\partial\varpi} - \frac{\xi_\phi}{2\varpi} + \frac{1}{2\varpi}\frac{\partial\xi_\varpi}{\partial\phi}. \quad (2)$$

Likewise, in spherical coordinates using the connection coefficients (11.71), we obtain

$$\Sigma_{rr} = \frac{2}{3}\frac{\partial\xi_r}{\partial r} - \frac{2}{3r}\xi_r - \frac{\cot\theta}{3r}\xi_\theta - \frac{1}{3r}\frac{\partial\xi_\theta}{\partial\theta} - \frac{1}{3r\sin\theta}\frac{\partial\xi\phi}{\partial\phi},$$

$$\Sigma_{\theta\theta} = \frac{2}{3r}\frac{\partial\xi_\theta}{\partial\theta} + \frac{\xi_r}{3r} - \frac{1}{3}\frac{\partial\xi_r}{\partial r} - \frac{\cot\theta\xi_\theta}{3r} - \frac{1}{3r\sin\theta}\frac{\partial\xi_\phi}{\partial\phi},$$

$$\Sigma_{\phi\phi} = \frac{2}{3r\sin\theta}\frac{\partial\xi_\phi}{\partial\phi} + \frac{2\cot\theta\xi_\theta}{3r} + \frac{\xi_r}{3r} - \frac{1}{3}\frac{\partial\xi_r}{\partial r} - \frac{1}{3r}\frac{\partial\xi_\theta}{\partial\theta},$$

$$\Sigma_{\theta\phi} = \Sigma_{\phi\theta} = \frac{1}{2r}\frac{\partial\xi_\phi}{\partial\theta} - \frac{\cot\theta\xi_\phi}{2r} + \frac{1}{2r\sin\theta}\frac{\partial\xi_\theta}{\partial\phi},$$

$$\Sigma_{\phi r} = \Sigma_{r\phi} = \frac{1}{2r\sin\theta}\frac{\partial\xi_r}{\partial\phi} + \frac{1}{2}\frac{\partial\xi_\phi}{\partial r} - \frac{\xi_\phi}{2r},$$

$$\Sigma_{r\theta} = \Sigma_{\theta r} = \frac{1}{2}\frac{\partial\xi_\theta}{\partial r} - \frac{\xi_\theta}{2r} + \frac{1}{2r}\frac{\partial\xi_r}{\partial\theta}. \quad (3)$$

any basis (orthonormal or nonorthonormal) in a curved spacetime, as is needed to perform calculations in general relativity.

Although the algebra of evaluating the components of derivatives such as in Eq. (11.78) in explicit form (e.g., in terms of $\{r, \theta, \phi\}$) can be long and tedious when done by hand, in the modern era of symbolic manipulation using computers (e.g., Mathematica, Matlab, or Maple), the algebra can be done quickly and accurately to obtain expressions such as Eqs. (3) of Box 11.4.

Exercise 11.21 *Derivation and Practice: Gradient of a Second-Rank Tensor* T2
By a computation analogous to Eq. (11.72), derive Eq. (11.78) for the components of the gradient of a second-rank tensor in any orthonormal basis.

Exercise 11.22 *Derivation and Practice: Connection in Spherical Coordinates* T2
(a) By drawing pictures analogous to Fig. 11.15, show that

$$\nabla_\phi \mathbf{e}_r = \frac{1}{r}\mathbf{e}_\phi, \qquad \nabla_\theta \mathbf{e}_r = \frac{1}{r}\mathbf{e}_\theta, \qquad \nabla_\phi \mathbf{e}_\theta = \frac{\cot\theta}{r}\mathbf{e}_\phi. \qquad (11.79)$$

(b) From these relations and antisymmetry on the first two indices [Eq. (11.69)], deduce the connection coefficients (11.71).

Exercise 11.23 *Derivation and Practice: Expansion in Cylindrical and Spherical Coordinates* T2
Derive Eqs. (11.76) and (11.77) for the divergence of the vector field $\boldsymbol{\xi}$ in cylindrical and spherical coordinates using the connection coefficients (11.70) and (11.71).

11.9 Solving the 3-Dimensional Navier-Cauchy Equation in Cylindrical Coordinates T2

11.9

11.9.1 Simple Methods: Pipe Fracture and Torsion Pendulum T2

11.9.1

As an example of an elastostatic problem with cylindrical symmetry, consider a cylindrical pipe that carries a high-pressure fluid (e.g., water, oil, natural gas); Fig. 11.16. How thick must the pipe's wall be to ensure that it will not burst due to the fluid's pressure? We sketch the solution, leaving the details to Ex. 11.24.

We suppose, for simplicity, that the pipe's length is held fixed by its support system: it does not lengthen or shorten when the fluid pressure is changed. Then by symmetry, the displacement field in the pipe wall is purely radial and depends only on radius: its only nonzero component is $\xi_\varpi(\varpi)$. The radial dependence is governed by radial force balance,

$$f_\varpi = K\Theta_{;\varpi} + 2\mu\Sigma_{\varpi j;j} = 0 \qquad (11.80)$$

[Eq. (11.30)].

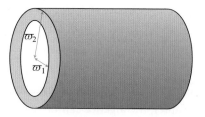

FIGURE 11.16 A pipe whose wall has inner and outer radii ϖ_1 and ϖ_2.

Because ξ_ϖ is independent of ϕ and z, the expansion [Eq. (11.76)] is given by

$$\Theta = \frac{d\xi_\varpi}{d\varpi} + \frac{\xi_\varpi}{\varpi}. \tag{11.81}$$

The second term in the radial force balance equation (11.80) is proportional to $\Sigma_{\varpi j;j}$ which—using Eq. (11.78) and noting that the only nonzero connection coefficients are $\Gamma_{\varpi\phi\phi} = -\Gamma_{\phi\varpi\phi} = -1/\varpi$ [Eq. (11.70)] and that symmetry requires the shear tensor to be diagonal—becomes

$$\Sigma_{\varpi j;j} = \Sigma_{\varpi\varpi,\varpi} + \Gamma_{\varpi\phi\phi}\Sigma_{\phi\phi} + \Gamma_{\phi\varpi\phi}\Sigma_{\varpi\varpi}. \tag{11.82}$$

Inserting the components of the shear tensor from Eqs. (2) of Box 11.4 and the values of the connection coefficients and comparing the result with expression (11.81) for the expansion, we obtain the remarkable result that $\Sigma_{\varpi j;j} = \frac{2}{3}\partial\Theta/\partial\varpi$. Inserting this into the radial force balance equation (11.80), we obtain

$$f_\varpi = \left(K + \frac{4\mu}{3}\right)\frac{d\Theta}{d\varpi} = 0. \tag{11.83}$$

Thus, inside the pipe wall, the expansion is independent of radius ϖ, and correspondingly, the radial displacement must have the form [cf. Eq. (11.81)]

$$\xi_\varpi = A\varpi + \frac{B}{\varpi} \tag{11.84}$$

for some constants A and B, whence $\Theta = 2A$ and $\Sigma_{\varpi\varpi} = \frac{1}{3}A - B/\varpi^2$. The values of A and B are fixed by the boundary conditions at the inner and outer faces of the pipe wall: $T_{\varpi\varpi} = P$ at $\varpi = \varpi_1$ (inner wall) and $T_{\varpi\varpi} = 0$ at $\varpi = \varpi_2$ (outer wall). Here P is the pressure of the fluid that the pipe carries, and we have neglected the atmosphere's pressure on the outer face by comparison. Evaluating $T_{\varpi\varpi} = -K\Theta - 2\mu\Sigma_{\varpi\varpi}$ and then imposing these boundary conditions, we obtain

$$A = \frac{P}{2(K + \mu/3)}\frac{\varpi_1^2}{(\varpi_2^2 - \varpi_1^2)}, \qquad B = \frac{P}{2\mu}\frac{\varpi_1^2\varpi_2^2}{(\varpi_2^2 - \varpi_1^2)}. \tag{11.85}$$

The only nonvanishing components of the strain then work out to be

$$S_{\varpi\varpi} = \frac{\partial\xi_{\varpi}}{\partial\varpi} = A - \frac{B}{\varpi^2}, \quad S_{\phi\phi} = \frac{\xi_{\varpi}}{\varpi} = A + \frac{B}{\varpi^2}. \tag{11.86}$$

This strain is maximal at the inner wall of the pipe; expressing it in terms of the ratio $\zeta \equiv \varpi_2/\varpi_1$ of the outer to the inner pipe radius and using the values of $K = 180$ GPa and $\mu = 81$ GPa for steel, we bring this maximum strain into the form

$$S_{\varpi\varpi} \simeq -\frac{P}{\mu}\frac{5\zeta^2 - 2}{10(\zeta^2 - 1)}, \quad S_{\phi\phi} \simeq \frac{P}{\mu}\frac{5\zeta^2 + 2}{10(\zeta^2 - 1)}. \tag{11.87}$$

Note that $|S_{\phi\phi}| > |S_{\varpi\varpi}|$.

The pipe will fracture at a strain $\sim 10^{-3}$; for safety it is best to keep the actual strain smaller than this by an order of magnitude, $|S_{ij}| \lesssim 10^{-4}$. A typical pressure for an oil pipeline is $P \simeq 10$ atmospheres ($\simeq 10^6$ Pa), compared to the shear modulus of steel $\mu = 81$ GPa, so $P/\mu \simeq 1.2 \times 10^{-5}$. Inserting this number into Eq. (11.87) with $|S_{\phi\phi}| \lesssim 10^{-4}$, we deduce that the ratio of the pipe's outer radius to its inner radius must be $\zeta = \varpi_2/\varpi_1 \gtrsim 1.04$. If the pipe has a diameter of a half meter, then its wall thickness should be about 1 cm or more. This is typical of oil pipelines.

criterion for safety against fracture

Exercise 11.25 presents a second fairly simple example of elastostatics in cylindrical coordinates: a computation of the period of a torsion pendulum.

EXERCISES

Exercise 11.24 *Derivation and Practice: Fracture of a Pipe* **T2**
Fill in the details of the text's analysis of the deformation of a pipe carrying a high-pressure fluid, and the wall thickness required to protect the pipe against fracture. (See Fig. 11.16.)

Exercise 11.25 *Practice: Torsion Pendulum* **T2**
A torsion pendulum is a very useful tool for testing the equivalence principle (Sec. 25.2), for seeking evidence for hypothetical fifth (not to mention sixth!) forces, and for searching for deviations from gravity's inverse-square law on submillimeter scales, which could arise from gravity being influenced by macroscopic higher spatial dimensions. (See, e.g., Kapner et al., 2008; Wagner et al., 2012.) It would be advantageous to design a torsion pendulum with a 1-day period (Fig. 11.17). Here we estimate whether this is possible. The pendulum consists of a thin cylindrical wire of length ℓ and radius a. At the bottom of the wire are suspended three masses at the corners of an equilateral triangle at a distance b from the wire.

(a) Show that the longitudinal strain is

$$\xi_{z;z} = \frac{3mg}{\pi a^2 E}. \tag{11.88a}$$

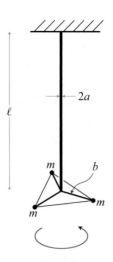

FIGURE 11.17 Torsion pendulum.

(b) What component of shear is responsible for the restoring force in the wire, which causes the torsion pendulum to oscillate?

(c) Show that the pendulum undergoes torsional oscillations with period

$$P = 2\pi \left(\frac{\ell}{g}\right)^{1/2} \left(\frac{2b^2 E \xi_{z;z}}{a^2 \mu}\right)^{1/2}. \tag{11.88b}$$

(d) Do you think you could design a pendulum that attains the goal of a 1-day period?

11.9.2

11.9.2 Separation of Variables and Green's Functions: Thermoelastic Noise in Mirrors T2

In complicated situations that have moderate amounts of symmetry, the elastostatic equations can be solved by the same kinds of sophisticated mathematical techniques as one uses in electrostatics: separation of variables, Green's functions, complex potentials, or integral transform methods (see, e.g., Gladwell, 1980). We provide an example in this section, focusing on separation of variables and Green's functions.

MOTIVATION

Our example is motivated by an important issue in high-precision measurements with light, including, among others, gravitational-wave detectors and quantum-optics experiments in which photons and atoms are put into entangled nonclassical states by coupling them to one another inside Fabry-Perot cavities.

In these situations, noise due to thermal motions of the mirror is a serious issue. It can hide a gravitational wave, and it can cause decoherence of the atom/photon quantum states. In Sec. 6.8.2, we formulated a generalized fluctuation-dissipation theorem by which this mirror thermal noise can be computed (Levin, 1998).

Specifically, in a thought experiment one applies to the mirror face a force F_o that oscillates at some frequency f at which one wants to evaluate the thermal noise. This force has the same transverse pressure distribution as the light beam—say, for concreteness, a Gaussian distribution:

$$T_{zz}^{\text{applied}} = \frac{e^{-\varpi^2/\varpi_o^2}}{\pi \varpi_o^2} F_o \cos(2\pi f t). \qquad (11.89)$$

This applied pressure induces a strain distribution $\mathbf{S}$ inside the mirror, and that oscillating strain interacts with imperfections to dissipate energy at some rate $W_{\text{diss}}(f)$. The fluctuation-dissipation theorem states that in the real experiment, where the light beam bounces off the mirror, the reflected light will encode a noisy transverse-averaged position q for the mirror face, and the noise spectral density for q will be

$$S_q(f) = \frac{8 W_{\text{diss}}(f) k_B T}{(2\pi f)^2 F_o^2} \qquad (11.90)$$

[Eq. (6.88b)].

Even if one could make the mirror perfect (no dislocations or impurities), so there is no dissipation due to imperfections, there will remain one other source of dissipation in this thought experiment: the applied pressure (11.89) will produce a spatially inhomogeneous expansion $\Theta(\mathbf{x}, t)$ inside the mirror, which in turn will produce the thermoelastic temperature change $\Delta T / T = -[3\alpha K / (\rho c_V)]\Theta$ [Eq. (11.29)]. The gradient of this temperature will induce heat flow, with a thermal energy flux $\mathbf{F}_{\text{th}} = -\kappa \boldsymbol{\nabla} \Delta T$, where κ is the thermal conductivity. When an amount Q of this thermal energy flows from a region with temperature T to a region of lower temperature $T - dT$, it produces an entropy increase $dS = Q/(T - dT) - Q/T = Q dT/T^2$; and correspondingly, there is a rate of entropy increase per unit volume given by $d^2 S/dV dt = -\mathbf{F}_{\text{th}} \cdot \boldsymbol{\nabla} \Delta T / T^2 = \kappa (\boldsymbol{\nabla} \Delta T)^2 / T^2$. This entropy increase has an accompanying energy dissipation rate $W_{\text{diss}} = \int T(d^2 S/dt dV) dV = \int (\kappa/T)(\boldsymbol{\nabla} \Delta T)^2 dV$. Expressing ΔT in terms of the expansion that drives it via $\Delta T / T = -[3\alpha K / (\rho c_V)]\Theta$ and inserting that into Eq. (11.90) and using the third of Eqs. (11.39), we obtain the thermal noise spectral density that the experimenters must contend with:

$$S_q(f) = \frac{8\kappa E^2 \alpha^2 k_B T^2}{(1 - 2\nu)^2 c_V^2 \rho^2 F_o^2 (2\pi f)^2} \left\langle \int (\boldsymbol{\nabla}\Theta)^2 \varpi \, d\phi \, d\varpi \, dz \right\rangle. \qquad (11.91)$$

Here $\langle \cdot \rangle$ means average over time as Θ oscillates due to the oscillation of the driving force $F_o \cos(2\pi f t)$. Because the dissipation producing this noise is due to heat flowing down a thermoelastic temperature gradient, it is called *thermoelastic noise*.

This is the motivation for an elasticity problem that we shall solve to illustrate separation of variables: to evaluate this thermoelastic noise, we must compute the expansion $\Theta(\mathbf{x}, t)$ inside a mirror, produced by the oscillating pressure (11.89) on the mirror face; and we must then perform the integral (11.91).

The frequencies f at which we wish to evaluate the thermal noise are low compared to the inverse sound travel time across the mirror, so when computing Θ we can regard the force as oscillating very slowly (i.e., we can use our elastostatic equations rather than dynamical equations of the next chapter). Also, the size of the light spot on the mirror is usually small compared to the mirror's transverse size and thickness, so we can idealize the mirror as being infinitely large and thick—a homogeneous half-space of isotropic, elastic material.

Because the applied stress is axially symmetric, the induced strain and expansion will also be axially symmetric and are thus computed most easily using cylindrical coordinates. Our challenge, then, is to solve the Navier-Cauchy equation $\mathbf{f} = (K + \frac{1}{3}\mu)\mathbf{\nabla}(\mathbf{\nabla} \cdot \boldsymbol{\xi}) + \mu\nabla^2\boldsymbol{\xi} = 0$ for the cylindrical components $\xi_\varpi(z, \varpi)$ and $\xi_z(z, \varpi)$ of the displacement, and then evaluate the expansion $\Theta = \mathbf{\nabla} \cdot \boldsymbol{\xi}$. (The component ξ_ϕ vanishes by symmetry.)

Equations of elasticity in cylindrical coordinates, and their homogeneous solution
It is straightforward, using the techniques of Sec. 11.8, to compute the cylindrical components of $\mathbf{f}$. Reexpressing the bulk K and shear μ moduli in terms of Young's modulus E and Poisson's ratio ν [Eqs. (11.39)] and setting the internal forces to zero, we obtain

elastostatic force balance in cylindrical coordinates

$$f_\varpi = \frac{E}{2(1+\nu)(1-2\nu)}\left[2(1-\nu)\left(\frac{\partial^2\xi_\varpi}{\partial\varpi^2} + \frac{1}{\varpi}\frac{\partial\xi_\varpi}{\partial\varpi} - \frac{\xi_\varpi}{\varpi^2}\right)\right.$$
$$\left. + (1-2\nu)\frac{\partial^2\xi_\varpi}{\partial z^2} + \frac{\partial^2\xi_z}{\partial z\partial\varpi}\right] = 0, \quad (11.92a)$$

$$f_z = \frac{E}{2(1+\nu)(1-2\nu)}\left[(1-2\nu)\left(\frac{\partial^2\xi_z}{\partial\varpi^2} + \frac{1}{\varpi}\frac{\partial\xi_z}{\partial\varpi}\right)\right.$$
$$\left. + 2(1-\nu)\frac{\partial^2\xi_z}{\partial z^2} + \frac{\partial^2\xi_\varpi}{\partial z\partial\varpi} + \frac{1}{\varpi}\frac{\partial\xi_\varpi}{\partial z}\right] = 0. \quad (11.92b)$$

These are two coupled, linear, second-order differential equations for the two unknown components of the displacement vector. As with the analogous equations of electrostatics and magnetostatics, these can be solved by separation of variables, that is, by setting $\xi_\varpi = R_\varpi(\varpi)Z_\varpi(z)$ and $\xi_z = R_z(\varpi)Z_z(z)$, and inserting into Eq. (11.92a). We seek the general solution that dies out at large ϖ and z. The general solution of this sort, to the complicated-looking Eqs. (11.92), turns out to be (really!!)

separation-of-variables solution of force-balance equation $f_j = 0$

$$\xi_\varpi = \int_0^\infty [\alpha(k) - (2 - 2\nu - kz)\beta(k)]\, e^{-kz} J_1(k\varpi)\, dk,$$

$$\xi_z = \int_0^\infty [\alpha(k) + (1 - 2\nu + kz)\beta(k)]\, e^{-kz} J_0(k\varpi)\, dk. \quad (11.93)$$

Here J_0 and J_1 are Bessel functions of order 0 and 1, and $\alpha(k)$ and $\beta(k)$ are arbitrary functions.

Boundary conditions

The functions $\alpha(k)$ and $\beta(k)$ are determined by boundary conditions on the face of the test mass. The force per unit area exerted across the face by the strained test-mass material, T_{zj} at $z = 0$ with $j = \{\varpi, \phi, z\}$, must be balanced by the applied force per unit area, T_{zj}^{applied} [Eq. (11.89)]. The (shear) forces in the ϕ direction, $T_{z\phi}$ and $T_{z\phi}^{\text{applied}}$, vanish because of cylindrical symmetry and thus provide no useful boundary condition. The (shear) force in the ϖ direction, which must vanish at $z = 0$ since $T_{z\varpi}^{\text{applied}} = 0$, is given by [cf. Eq. (2) in Box 11.4]

$$T_{z\varpi}(z = 0) = -2\mu \Sigma_{z\varpi} = -\mu \left(\frac{\partial \xi_z}{\partial \varpi} + \frac{\partial \xi_\varpi}{\partial z} \right)$$

$$= -\mu \int_0^\infty [\beta(k) - \alpha(k)] J_1(kz) k \, dk = 0, \quad (11.94)$$

shear-force boundary condition at $z = 0$

which implies that $\beta(k) = \alpha(k)$. The (normal) force in the z direction, which must balance the applied pressure (11.89), is $T_{zz} = -K\Theta - 2\mu \Sigma_{zz}$; using Eq. (2) in Box 11.4 and Eqs. (11.39), (11.76), (11.93) and (11.89), this reduces to

$$T_{zz}(z = 0) = 2\mu \int_0^\infty \alpha(k) J_0(k\varpi) k \, dk = T_{zz}^{\text{applied}} = \frac{e^{-\varpi^2/\varpi_o^2}}{\pi \varpi_o^2} F_o \cos(2\pi f t),$$

longitudinal-force boundary condition at $z = 0$

$$(11.95)$$

which can be inverted[11] to give

$$\alpha(k) = \beta(k) = \frac{1}{4\pi\mu} e^{-k^2 \varpi_o^2/4} F_o \cos(2\pi f t). \quad (11.96)$$

solution for expansion coefficients

Inserting this equation into the Eqs. (11.93) for the displacement and then evaluating the expansion $\Theta = \nabla \cdot \boldsymbol{\xi} = \xi_{z,z} + \varpi^{-1}(\varpi \xi_\varpi)_{,\varpi}$, we obtain

$$\Theta = 2(1 - 2\nu) \int_0^\infty \alpha(k) e^{-kz} J_0(k\varpi) k \, dk. \quad (11.97)$$

As in electrostatics and magnetostatics, so also in elasticity theory, one can solve an elastostatics problem using Green's functions instead of separation of variables. We explore this option for our applied Gaussian force in Ex. 11.27. For greater detail on

---

11. The inversion and the subsequent evaluation of the integral of $(\nabla \Theta)^2$ are aided by the following expressions for the Dirac delta function:

$$\delta(k - k') = k \int_0^\infty J_0(k\varpi) J_0(k'\varpi) \varpi \, d\varpi = k \int_0^\infty J_1(k\varpi) J_1(k'\varpi) \varpi \, d\varpi.$$

Green's functions in elastostatics and their applications from an engineer's viewpoint, see Johnson (1985). For other commonly used solution techniques, see Box 11.3.

THERMOELASTIC NOISE SPECTRAL DENSITY

Let us return to the mirror-noise problem that motivated our calculation. It is straightforward to compute the gradient of the expansion (11.97), and square and integrate it to get the spectral density $S_q(f)$ [Eq. (11.91)]. The result is (Braginsky, Gorodetsky, and Vyatchanin, 1999; Liu and Thorne, 2000)

$$S_q(f) = \frac{8(1+\nu)^2 \kappa \alpha^2 k_B T^2}{\sqrt{2\pi} c_V^2 \rho^2 \varpi_o^3 (2\pi f)^2}.$$ (11.98)

Early plans for advanced LIGO gravitational wave detectors (Sec. 9.5; LIGO Scientific Collaboration, 2015) called for mirrors made of high-reflectivity dielectric coatings on sapphire crystal substrates. Sapphire was chosen because it can be grown in giant crystals with very low impurities and dislocations, resulting in low thermal noise. However, the thermoelastic noise (11.98) in sapphire turns out to be uncomfortably high. Using sapphire's $\nu = 0.29$, $\kappa = 40$ W m^{-1} K^{-1}, $\alpha = 5.0 \times 10^{-6}$ K^{-1}, $c_V = 790$ J kg^{-1} K^{-1}, $\rho = 4{,}000$ kg m^{-3}, and a light-beam radius $\varpi_o = 4$ cm and room temperature $T = 300$ K, Eq. (11.98) gives the following for the noise in a bandwidth equal to frequency:

$$\sqrt{f S_q(f)} = 5 \times 10^{-20} \text{ m} \sqrt{\frac{100 \text{ Hz}}{f}}.$$ (11.99)

Because this was uncomfortably high at low frequencies, $f \sim 10$ Hz, and because of the birefringence of sapphire, which could cause technical problems, a decision was made to switch to fused silica for the advanced LIGO mirrors.

EXERCISES

Exercise 11.26 *Derivation and Practice: Evaluation of Elastostatic Force in Cylindrical Coordinates* T2

Derive Eqs. (11.92) for the cylindrical components of the internal elastostatic force per unit volume $\mathbf{f} = (K + \frac{1}{3}\mu)\nabla(\nabla \cdot \boldsymbol{\xi}) + \mu \nabla^2 \boldsymbol{\xi}$ in a cylindrically symmetric situation.

Exercise 11.27 **Example: Green's Function for Normal Force on a Half-Infinite Body* T2

Suppose that a stress $T_{zj}^{\text{applied}}(\mathbf{x}_o)$ is applied on the face $z = 0$ of a half-infinite elastic body (one that fills the region $z > 0$). Then by virtue of the linearity of the elastostatics equation $\mathbf{f} = (K + \frac{1}{3}\mu)\nabla(\nabla \cdot \boldsymbol{\xi}) + \mu \nabla^2 \boldsymbol{\xi} = 0$ and the linearity of its boundary conditions, $T_{zj}^{\text{internal}} = T_{zj}^{\text{applied}}$, there must be a Green's function $G_{jk}(\mathbf{x} - \mathbf{x}_o)$ such that the body's internal displacement $\boldsymbol{\xi}(\mathbf{x})$ is given by

$$\xi_j(\mathbf{x}) = \int G_{jk}(\mathbf{x} - \mathbf{x}_0) T_{kz}^{\text{applied}}(\mathbf{x}_o) d^2 x_o.$$ (11.100)

Here the integral is over all points $\mathbf{x}_o$ on the face of the body ($z = 0$), and $\mathbf{x}$ can be anywhere inside the body, $z \geq 0$.

(a) Show that if a force F_j is applied on the body's surface at a single point (the origin of coordinates), then the displacement inside the body is

$$\xi_j(\mathbf{x}) = G_{jk}(\mathbf{x}) F_k. \tag{11.101}$$

Thus, the Green's function can be thought of as the body's response to a point force on its surface.

(b) As a special case, consider a point force F_z directed perpendicularly into the body. The resulting displacement turns out to have cylindrical components[12]

$$\xi_z = G_{zz}(\varpi, z) F_z = \frac{(1+\nu)}{2\pi E} \left[\frac{2(1-\nu)}{r} + \frac{z^2}{r^3} \right] F_z,$$

$$\xi_\varpi = G_{\varpi z}(\varpi, z) F_z = -\frac{(1+\nu)}{2\pi E} \left[\frac{(1-2\nu)\varpi}{r(r+z)} - \frac{\varpi z}{r^3} \right] F_z, \tag{11.102}$$

where $r = \sqrt{\varpi^2 + z^2}$. It is straightforward to show that this displacement does satisfy the elastostatics equations (11.92). Show that it also satisfies the required boundary condition $T_{z\varpi}(z = 0) = -2\mu\Sigma_{z\varpi} = 0$.

(c) Show that for this displacement [Eq. (11.102)], $T_{zz} = -K\Theta - 2\mu\Sigma_{zz}$ vanishes everywhere on the body's surface $z = 0$ except at the origin $\varpi = 0$ and is infinite there. Show that the integral of this normal stress over the surface is F_z, and therefore, $T_{zz}(z = 0) = F_z\delta_2(\mathbf{x})$, where δ_2 is the 2-dimensional Dirac delta function on the surface. This is the second required boundary condition.

(d) Plot the integral curves of the displacement vector $\boldsymbol{\xi}$ (i.e., the curves to which $\boldsymbol{\xi}$ is parallel) for a reasonable choice of Poisson's ratio ν. Explain physically why the curves have the form you find.

(e) One can use the Green's function (11.102) to compute the displacement $\boldsymbol{\xi}$ induced by the Gaussian-shaped pressure (11.89) applied to the body's face, and to then evaluate the induced expansion and thence the thermoelastic noise; see Braginsky, Gorodetsky, and Vyatchanin (1999) and Liu and Thorne (2000). The results agree with Eqs. (11.97) and (11.98) deduced using separation of variables.

Bibliographic Note

Elasticity theory was developed in the eighteenth, nineteenth, and early twentieth centuries. The classic advanced textbook from that era is Love (1927). An outstanding, somewhat more modern advanced text is Landau and Lifshitz (1986)—originally

---

12. For the other components of the Green's function, written in Cartesian coordinates (since a non-normal applied force breaks the cylindrical symmetry), see Landau and Lifshitz [1986, Eqs. (8.18)].

written in the 1950s and revised in a third edition shortly before Lifshitz's death. This is among the most readable textbooks that Landau and Lifshitz wrote and is still widely used by physicists in the early twenty-first century.

Some significant new insights, both mathematical and physical, have been developed in recent decades, for example, catastrophe theory and its applications to bifurcations and stability, practical insights from numerical simulations, and practical applications based on new materials (e.g., carbon nanotubes). For a modern treatment that deals with these and much else from an engineering viewpoint, we strongly recommend Ugural and Fenster (2012). For a fairly brief and elementary modern treatment, we recommend Lautrup (2005, Part III). Other good texts that focus particularly on solving the equations for elastostatic equilibrium include Southwell (1941), Timoshenko and Goodier (1970), Gladwell (1980), Johnson (1985), Boresi and Chong (1999), and Slaughter (2002); see also the discussion and references in Box 11.3.

12

Elastodynamics

> . . . logarithmic plots are a device of the devil.
>
> CHARLES RICHTER (1980)

12.1 Overview

12.1

In the previous chapter we considered elastostatic equilibria, in which the forces acting on elements of an elastic solid were balanced, so the solid remained at rest. When this equilibrium is disturbed, the solid will undergo accelerations. This is the subject of this chapter—*elastodynamics*.

In Sec. 12.2, we derive the equations of motion for elastic media, paying particular attention to the underlying conservation laws and focusing especially on elastodynamic waves. We show that there are two distinct wave modes that propagate in a uniform, isotropic solid—longitudinal waves and shear waves—and both are non-dispersive (their phase speeds are independent of frequency).

A major use of elastodynamics is in structural engineering, where one encounters vibrations (usually standing waves) on the beams that support buildings and bridges. In Sec. 12.3, we discuss the types of waves that propagate on bars, rods, and beams and find that the boundary conditions at the free transverse surfaces make the waves dispersive. We also return briefly to the problem of bifurcation of equilibria (treated in Sec. 11.6) and show how, as the parameters controlling an equilibrium are changed so it passes through a bifurcation point, the equilibrium becomes unstable; the instability drives the system toward the new, stable equilibrium state.

A second application of elastodynamics is to seismology (Sec. 12.4). Earth is mostly a solid body through which waves can propagate. The waves can be excited naturally by earthquakes or artificially using human-made explosions. Understanding how waves propagate through Earth is important for locating the sources of earthquakes, for diagnosing the nature of an explosion (was it an illicit nuclear bomb test?), and for analyzing the structure of Earth. We briefly describe some of the wave modes that propagate through Earth and some of the inferences about Earth's structure that have been drawn from studying their propagation. In the process, we gain some experience in applying the tools of geometric optics to new types of waves, and we learn how rich can be the Green's function for elastodynamic waves, even when the medium is as simple as a homogeneous half-space. We briefly discuss how the methods by which geophysicists probe Earth's structure using seismic waves also find application in ultrasonic technology: imaging solid structures and the human body using high-frequency sound waves.

Finally, in Sec. 12.5, we return to physics to consider the quantum theory of elastodynamic waves. We compare the classical theory with the quantum theory, specializing to quantized vibrations in an elastic solid: phonons.

12.2   Basic Equations of Elastodynamics; Waves in a Homogeneous Medium

12.2

In this section, we derive a vectorial equation that governs the dynamical displacement $\boldsymbol{\xi}(\mathbf{x}, t)$ of a dynamically disturbed elastic medium. We then consider monochromatic plane wave solutions and explore wave propagation through inhomogeneous media in the geometric optics limit. This general approach to wave propagation in continuum mechanics will be taken up again in Chap. 16 for waves in fluids, Part VI for waves in plasmas, and Chap. 27 for general relativistic gravitational waves. We conclude this section with a discussion of the energy density and energy flux of these waves.

12.2.1   Equation of Motion for a Strained Elastic Medium

12.2.1

In Chap. 11, we learned that when an elastic medium undergoes a displacement $\boldsymbol{\xi}(\mathbf{x})$, it builds up a strain $S_{ij} = \frac{1}{2}(\xi_{i;j} + \xi_{j;i})$, which in turn produces an internal stress

$\mathbf{T} = -K\Theta\mathbf{g} - 2\mu\boldsymbol{\Sigma}$. Here $\Theta \equiv \boldsymbol{\nabla} \cdot \boldsymbol{\xi}$ is the expansion, and $\boldsymbol{\Sigma} \equiv$ (the trace-free part of $\mathbf{S}$) is the shear; see Eqs. (11.4) and (11.18). The stress $\mathbf{T}$ produces an elastic force per unit volume

$$\mathbf{f} = -\boldsymbol{\nabla} \cdot \mathbf{T} = \left(K + \frac{1}{3}\mu\right) \boldsymbol{\nabla}(\boldsymbol{\nabla} \cdot \boldsymbol{\xi}) + \mu\nabla^2\boldsymbol{\xi} \qquad (12.1)$$

[Eq. (11.30)], where K and μ are the bulk and shear moduli.

In Chap. 11, we restricted ourselves to elastic media that are in elastostatic equilibrium, so they are static. This equilibrium requires that the net force per unit volume acting on the medium vanish. If the only force is elastic, then $\mathbf{f}$ must vanish. If the pull of gravity is also significant, then $\mathbf{f} + \rho\mathbf{g}$ vanishes, where ρ is the medium's mass density and $\mathbf{g}$ the acceleration of gravity.

In this chapter, we focus on dynamical situations, in which an unbalanced force per unit volume causes the medium to move—with the motion taking the form of an elastodynamic wave. For simplicity, we assume unless otherwise stated that the only significant force is elastic (i.e., the gravitational force is negligible by comparison). In Ex. 12.1, we show that this is the case for elastodynamic waves in most media on Earth when the wave frequency $\omega/(2\pi)$ is higher than about 0.001 Hz (which is usually the case in practice). Stated more precisely, in a homogeneous medium we can ignore the gravitational force when the elastodynamic wave's angular frequency ω is much larger than g/C, where g is the acceleration of gravity, and C is the wave's propagation speed.

when gravity can be ignored

Consider, then, a dynamical, strained medium with elastic force per unit volume (12.1) and no other significant force (negligible gravity), and with velocity

$$\boxed{\mathbf{v} = \frac{\partial\boldsymbol{\xi}}{\partial t}.} \qquad (12.2a)$$

The law of momentum conservation states that the force per unit volume $\mathbf{f}$, if nonzero, must produce a rate of change of momentum per unit volume $\rho\mathbf{v}$ according to the equation[1]

$$\frac{\partial(\rho\mathbf{v})}{\partial t} = \mathbf{f} = -\boldsymbol{\nabla} \cdot \mathbf{T} = \left(K + \frac{1}{3}\mu\right) \boldsymbol{\nabla}(\boldsymbol{\nabla} \cdot \boldsymbol{\xi}) + \mu\nabla^2\boldsymbol{\xi}. \qquad (12.2b)$$

elastodynamic momentum conservation

Notice that, when rewritten in the form

$$\frac{\partial(\rho\mathbf{v})}{\partial t} + \boldsymbol{\nabla} \cdot \mathbf{T} = 0, \qquad (12.2b')$$

---

1. In Sec. 13.5, we learn that the motion of the medium produces a stress $\rho\mathbf{v} \otimes \mathbf{v}$ that must be included in this equation if the velocities are large. However, this subtle dynamical stress is always negligible in elastodynamic waves, because the displacements and hence velocities $\mathbf{v}$ are tiny and $\rho\mathbf{v} \otimes \mathbf{v}$ is second order in the displacement. For this reason, we delay studying this subtle nonlinear effect until Chap. 13. A similar remark applies to Eq. (12.2c). In general the conservation laws of continuum mechanics are nonlinear, as we discuss in Box 12.2.

this is the version of the law of momentum conservation discussed in Chap. 1 [Eq. (1.36)]. It has the standard form for a conservation law (time derivative of density of something plus divergence of flux of that something vanishes; see end of Sec. 1.8); $\rho\mathbf{v}$ is the density of momentum, and the stress tensor $\mathbf{T}$ is by definition the flux of momentum. Equations (12.2a) and (12.2b), together with the law of mass conservation [the obvious analog of Eqs. (1.30) for conservation of charge and particle number],

mass conservation

$$\boxed{\frac{\partial \rho}{\partial t} + \boldsymbol{\nabla} \cdot (\rho\mathbf{v}) = 0,}$$

(12.2c)

are a complete set of equations for the evolution of the displacement $\boldsymbol{\xi}(\mathbf{x}, t)$, the velocity $\mathbf{v}(\mathbf{x}, t)$, and the density $\rho(\mathbf{x}, t)$.

To derive a linear wave equation, we must find some small parameter in which to expand. The obvious choice in elastodynamics is the magnitude of the components of the strain $S_{ij} = \frac{1}{2}(\xi_{i;j} + \xi_{j;i})$, which are less than about 10^{-3} so as to remain below the proportionality limit (i.e., to remain in the realm where stress is proportional to strain; Sec. 11.3.2). Equally well, we can regard the displacement $\boldsymbol{\xi}$ itself as our small parameter (or more precisely, ξ/λ, the magnitude of $\boldsymbol{\xi}$ divided by the reduced wavelength of its perturbations).

If the medium's equilibrium state were homogeneous, the linearization would be trivial. However, we wish to be able to treat perturbations of inhomogeneous equilibria, such as seismic waves in Earth, or perturbations of slowly changing equilibria, such as vibrations of a pipe or mirror that is gradually changing temperature. In most practical situations the lengthscale $\mathcal{L}$ and timescale $\mathcal{T}$ on which the medium's equilibrium properties (ρ, K, μ) vary are extremely large compared to the lengthscale and timescale of the dynamical perturbations [their reduced wavelength $\lambda = \text{wavelength}/(2\pi)$ and $1/\omega = \text{period}/(2\pi)$]. This permits us to perform a two-lengthscale expansion (like the one that underlies geometric optics; Sec. 7.3) alongside our small-strain expansion.

When analyzing a dynamical perturbation of an equilibrium state, we use $\boldsymbol{\xi}(\mathbf{x}, t)$ to denote the dynamical displacement (i.e., we omit from it the equilibrium's static displacement, and similarly we omit from $\mathbf{S}$ the equilibrium strain). We write the density

density perturbation

as $\rho + \delta\rho$, where $\rho(\mathbf{x})$ is the equilibrium density distribution, and $\delta\rho(\mathbf{x}, t)$ is the dynamical density perturbation, which is first order in the dynamical displacement $\boldsymbol{\xi}$. Inserting these into the equation of mass conservation (12.2c), we obtain $\partial\delta\rho/\partial t + \boldsymbol{\nabla} \cdot [(\rho + \delta\rho)\mathbf{v}] = 0$. Because $\mathbf{v} = \partial\boldsymbol{\xi}/\partial t$ is first order, the term $(\delta\rho)\mathbf{v}$ is second order and can be dropped, resulting in the linearized equation $\partial\delta\rho/\partial t + \boldsymbol{\nabla} \cdot (\rho\mathbf{v}) = 0$. Because ρ varies on a much longer lengthscale than does $\mathbf{v}$ ($\mathcal{L}$ versus λ), we can pull ρ out of the derivative. Setting $\mathbf{v} = \partial\boldsymbol{\xi}/\partial t$ and interchanging the time derivative and divergence, we then obtain $\partial\delta\rho/\partial t + \rho\partial(\boldsymbol{\nabla} \cdot \boldsymbol{\xi})/\partial t = 0$. Noting that ρ varies on a

BOX 12.2. WAVE EQUATIONS IN CONTINUUM MECHANICS

Here we make an investment for future chapters by considering wave equations in some generality. Most wave equations arise as approximations to the full set of equations that govern a dynamical physical system. It is usually possible to arrange those full equations as a set of first-order partial differential equations that describe the dynamical evolution of a set of n physical quantities, V_A, with $A = 1, 2, \ldots, n$:

$$\frac{\partial V_A}{\partial t} + F_A(V_B) = 0. \tag{1}$$

[For elastodynamics there are $n = 7$ quantities V_A: $\{\rho, \rho v_x, \rho v_y, \rho v_z, \xi_x, \xi_y, \xi_z\}$ (in Cartesian coordinates); and the seven equations (1) are mass conservation, momentum conservation, and $\partial \xi_j / \partial t = v_j$; Eqs. (12.2).]

Most dynamical systems are intrinsically nonlinear (Maxwell's equations in a vacuum being a conspicuous exception), and it is usually quite hard to find nonlinear solutions. However, it is generally possible to make a perturbation expansion in some small physical quantity about a time-independent equilibrium and retain only those terms that are linear in this quantity. We then have a set of n linear partial differential equations that are much easier to solve than the nonlinear ones—and that usually turn out to have the character of wave equations (i.e., to be *hyperbolic*; see Arfken, Weber, and Harris, 2013). Of course, the solutions will only be a good approximation for small-amplitude waves. [In elastodynamics, we justify linearization by requiring that the strains be below the proportionality limit and linearize in the strain or displacement of the dynamical perturbation. The resulting linear wave equation is $\rho \partial^2 \boldsymbol{\xi} / \partial t^2 = (K + \frac{1}{3}\mu)\boldsymbol{\nabla}(\boldsymbol{\nabla} \cdot \boldsymbol{\xi}) + \mu \nabla^2 \boldsymbol{\xi}$; Eq. (12.4b).]

BOUNDARY CONDITIONS

In some problems (e.g., determining the normal modes of vibration of a building during an earthquake, or analyzing the sound from a violin or the vibrations of a finite-length rod), the boundary conditions are intricate and have to be incorporated as well as possible to have any hope of modeling the problem. The situation is rather similar to that familiar from elementary quantum mechanics. The waves are often localized in some region of space, like bound states, in such a way that the eigenfrequencies are discrete (e.g., standing-wave modes of a plucked string). In other problems the volume in which the wave propagates is essentially infinite (e.g., waves on the surface of the ocean, or seismic waves propagating through Earth), as happens with

(continued)

BOX 12.2. (continued)

unbound quantum states. Then the only boundary condition is essentially that the wave amplitude remain finite at large distances. In this case, the wave spectrum is usually continuous.

GEOMETRIC-OPTICS LIMIT AND DISPERSION RELATIONS

The solutions to the wave equation will reflect the properties of the medium through which the wave is propagating, as well as its boundaries. If the medium and boundaries have a finite number of discontinuities but are otherwise smoothly varying, there is a simple limiting case: waves of short enough wavelength and high enough frequency can be analyzed in the geometric-optics approximation (Chap. 7).

The key to geometric optics is the dispersion relation, which (as we learned in Sec. 7.3) acts as a hamiltonian for the propagation. Recall from Chap. 7 that although the medium may actually be inhomogeneous and might even be changing with time, when deriving the dispersion relation, we can approximate it as precisely homogeneous and time independent and can resolve the waves into plane-wave modes [i.e., modes in which the perturbations vary $\propto \exp i(\mathbf{k} \cdot \mathbf{x} - \omega t)$]. Here $\mathbf{k}$ is the wave vector and ω is the angular frequency. This allows us to remove all the temporal and spatial derivatives and converts our set of partial differential equations into a set of homogeneous, linear algebraic equations. When we do this, we say that our normal modes are *local*. If, instead, we were to go to the trouble of solving the partial differential wave equation with its attendant boundary conditions, the modes would be referred to as *global*.

The linear algebraic equations for a local problem can be written in the form $M_{AB}V_B = 0$, where V_A is the vector of n dependent variables, and the elements M_{AB} of the $n \times n$ matrix $||M_{AB}||$ depend on $\mathbf{k}$ and ω as well as on parameters p_α that describe the local conditions of the medium. [See, e.g., Eq. (12.6).] This set of equations can be solved in the usual manner by requiring that the determinant of $||M_{AB}||$ vanish. Carrying through this procedure yields a polynomial, usually of nth order, for $\omega(\mathbf{k}, p_\alpha)$. This polynomial is the dispersion relation. It can be solved (analytically in simple cases and numerically in general) to yield a number of complex values for ω (the eigenfrequencies), with $\mathbf{k}$ regarded as real. (Some problems involve complex $\mathbf{k}$ and real ω, but for concreteness, we shall take $\mathbf{k}$ to be real.) Armed with the complex eigenfrequencies, we can solve for the associated eigenvectors V_A. The eigenfrequencies and eigenvectors fully characterize the

(continued)

solution of the local problem and can be used to solve for the waves' temporal evolution from some given initial conditions in the usual manner. (As we shall see several times, especially when we discuss Landau damping in Sec. 22.3, there are some subtleties that can arise.)

What does a complex value of the angular frequency ω mean? We have posited that all small quantities vary $\propto \exp[i(\mathbf{k} \cdot \mathbf{x} - \omega t)]$. If ω has a positive imaginary part, then the small perturbation quantities will grow exponentially with time. Conversely, if it has a negative imaginary part, they will decay. Now, polynomial equations with real coefficients have complex conjugate solutions. Therefore, if there is a decaying mode there must also be a growing mode. Growing modes correspond to instability, a topic that we shall encounter often.

timescale $\mathcal{T}$ long compared to the $1/\omega$ of $\boldsymbol{\xi}$ and $\delta\rho$, we can integrate this to obtain the linear relation

$$\frac{\delta\rho}{\rho} = -\nabla \cdot \boldsymbol{\xi}. \qquad (12.3)$$

This linearized equation for the fractional perturbation of density could equally well have been derived by considering a small volume V of the medium that contains mass $M = \rho V$ and noting that the dynamical perturbations lead to a volume change $\delta V/V = \Theta = \nabla \cdot \boldsymbol{\xi}$ [Eq. (11.7)]. Then conservation of mass requires $0 = \delta M = \delta(\rho V) = V\delta\rho + \rho\delta V = V\delta\rho + \rho V\nabla \cdot \boldsymbol{\xi}$, which implies $\delta\rho/\rho = -\nabla \cdot \boldsymbol{\xi}$. This is the same as Eq. (12.3).

The equation of momentum conservation (12.2b) can be handled similarly. By setting $\rho \to \rho(\mathbf{x}) + \delta\rho(\mathbf{x}, t)$, then linearizing (i.e., dropping the $\delta\rho\,\mathbf{v}$ term) and pulling the slowly varying $\rho(\mathbf{x})$ out from the time derivative, we convert $\partial(\rho\mathbf{v})/\partial t$ into $\rho\partial\mathbf{v}/\partial t = \rho\partial^2\boldsymbol{\xi}/\partial t^2$. Inserting this into Eq. (12.2b), we obtain the linear wave equation

$$\rho\frac{\partial^2\boldsymbol{\xi}}{\partial t^2} = -\nabla \cdot \mathbf{T}_{\text{el}}, \qquad (12.4a)$$

where $\mathbf{T}_{\text{el}}$ is the elastic contribution to the stress tensor. Expanding it, we obtain

$$\rho\frac{\partial^2\boldsymbol{\xi}}{\partial t^2} = \left(K + \frac{1}{3}\mu\right)\nabla(\nabla \cdot \boldsymbol{\xi}) + \mu\nabla^2\boldsymbol{\xi}. \qquad (12.4b)$$

elastodynamic wave
equation

In this equation, terms involving a derivative of K or μ have been omitted, because the two-lengthscale assumption $\mathcal{L} \gg \lambda$ makes them negligible compared to the terms we have kept.

Equation (12.4b) is the first of many wave equations we shall encounter in elastodynamics, fluid mechanics, plasma physics, and general relativity.

12.2.2 Elastodynamic Waves

Continuing to follow our general procedure for deriving and analyzing wave equations as outlined in Box 12.2, we next derive dispersion relations for two types of waves (longitudinal and transverse) that are jointly incorporated into the general elastodynamic wave equation (12.4b).

Recall from Sec. 7.3.1 that, although a dispersion relation can be used as a hamiltonian for computing wave propagation through an inhomogeneous medium, one can derive the dispersion relation most easily by specializing to monochromatic plane waves propagating through a medium that is precisely homogeneous. Therefore, we seek a plane-wave solution, that is, a solution of the form

$$\boldsymbol{\xi}(\mathbf{x}, t) \propto e^{i(\mathbf{k}\cdot\mathbf{x} - \omega t)}, \tag{12.5}$$

to the wave equation (12.4b) with ρ, K, and μ regarded as homogeneous (constant). (To deal with more complicated perturbations of a homogeneous medium, we can think of this wave as being an individual Fourier component and linearly superpose many such waves as a Fourier integral.) Since our wave is planar and monochromatic, we can remove the derivatives in Eq. (12.4b) by making the substitutions $\nabla \to i\mathbf{k}$ and $\partial/\partial t \to -i\omega$ (the first of which implies $\nabla^2 \to -k^2$, $\nabla \cdot \to i\mathbf{k}\cdot$, $\nabla \times \to i\mathbf{k}\times$). We thereby reduce the partial differential equation (12.4b) to a vectorial algebraic equation:

elastodynamic
eigenequation for plane
waves

$$\rho\omega^2\boldsymbol{\xi} = \left(K + \frac{1}{3}\mu\right)\mathbf{k}(\mathbf{k}\cdot\boldsymbol{\xi}) + \mu k^2\boldsymbol{\xi}. \tag{12.6}$$

[This reduction is only possible because the medium is uniform, or in the geometric-optics limit of near uniformity; otherwise, we must solve the second-order partial differential equation (12.4b).]

How do we solve this equation? The sure way is to write it as a 3×3 matrix equation $M_{ij}\xi_j = 0$ for the vector $\boldsymbol{\xi}$ and set the determinant of M_{ij} to zero (Box 12.2 and Ex. 12.3). This is not hard for small or sparse matrices. However, some wave equations are more complicated, and it often pays to think about the waves in a geometric, coordinate-independent way before resorting to brute force.

The quantity that oscillates in the elastodynamic waves of Eq. (12.6) is the vector field $\boldsymbol{\xi}$. The nature of its oscillations is influenced by the scalar constants ρ, μ, K, and ω and by just one quantity that has directionality: the constant vector $\mathbf{k}$. It seems reasonable to expect the description (12.6) of the oscillations to simplify, then, if we resolve the oscillations into a "longitudinal" component (or "mode") along $\mathbf{k}$ and a "transverse" component (or "mode") perpendicular to $\mathbf{k}$, as shown in Fig. 12.1:

decomposition into
longitudinal and
transverse modes

$$\boxed{\boldsymbol{\xi} = \boldsymbol{\xi}_L + \boldsymbol{\xi}_T, \quad \boldsymbol{\xi}_L = \xi_L\hat{\mathbf{k}}, \quad \boldsymbol{\xi}_T \cdot \hat{\mathbf{k}} = 0.} \tag{12.7a}$$

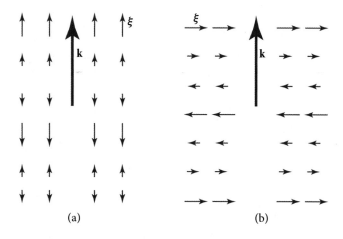

FIGURE 12.1 Displacements in an isotropic, elastic solid, perturbed by (a) a longitudinal mode and (b) a transverse mode.

Here $\hat{\mathbf{k}} \equiv \mathbf{k}/k$ is the unit vector along the propagation direction. It is easy to see that the longitudinal mode $\boldsymbol{\xi}_L$ has nonzero expansion $\Theta \equiv \boldsymbol{\nabla} \cdot \boldsymbol{\xi}_L \neq 0$ but vanishing rotation $\boldsymbol{\phi} = \frac{1}{2}\boldsymbol{\nabla} \times \boldsymbol{\xi}_L = 0$. It can therefore be written as the gradient of a scalar potential:

$$\boldsymbol{\xi}_L = \boldsymbol{\nabla}\psi . \tag{12.7b}$$

By contrast, the transverse mode has zero expansion but nonzero rotation and can thus be written as the curl of a vector potential:

$$\boldsymbol{\xi}_T = \boldsymbol{\nabla} \times \mathbf{A}; \tag{12.7c}$$

cf. Ex. 12.2.

12.2.3 Longitudinal Sound Waves

12.2.3

For the longitudinal mode the algebraic wave equation (12.6) reduces to the following simple relation [as one can easily see by inserting $\boldsymbol{\xi} \equiv \boldsymbol{\xi}_L = \xi_L\hat{\mathbf{k}}$ into Eq. (12.6), or, alternatively, by taking the divergence of Eq. (12.6), which is equivalent to taking the scalar product with $\mathbf{k}$]:

$$\omega^2 = \frac{K + \frac{4}{3}\mu}{\rho}k^2; \quad \text{or} \quad \boxed{\omega = \Omega(\mathbf{k}) = \left(\frac{K + \frac{4}{3}\mu}{\rho}\right)^{1/2} k.} \tag{12.8}$$

longitudinal dispersion relation

Here $k = |\mathbf{k}|$ is the wave number (the magnitude of $\mathbf{k}$). This relation between ω and $\mathbf{k}$ is the longitudinal mode's dispersion relation.

From the geometric-optics analysis in Sec. 7.3 we infer that if K, μ, and ρ vary spatially on an inhomogeneity lengthscale $\mathcal{L}$ large compared to $1/k = \lambdabar$, and vary temporally on a timescale $\mathcal{T}$ large compared to $1/\omega$, then the dispersion relation (12.8)—with Ω now depending on $\mathbf{x}$ and t through K, μ, and ρ—serves as a hamiltonian for the wave propagation. In Sec. 12.4 and Fig. 12.5 below we use this hamiltonian to deduce details of the propagation of seismic waves through Earth's inhomogeneous interior.

As discussed in great detail in Sec. 7.2, associated with any wave mode is its phase velocity $\mathbf{V}_{ph} = (\omega/k)\hat{\mathbf{k}}$ and its phase speed $V_{ph} = \omega/k$. The dispersion relation (12.8) implies that for longitudinal elastodynamic modes, the phase speed is

longitudinal wave speed

$$C_L = \frac{\omega}{k} = \left(\frac{K + \frac{4}{3}\mu}{\rho} \right)^{1/2}. \tag{12.9a}$$

As Eq. (12.9a) does not depend on the wave number $k \equiv |\mathbf{k}|$, the mode is non-dispersive. As it does not depend on the direction $\hat{\mathbf{k}}$ of propagation through the medium, the phase speed is also isotropic, naturally enough, and the group velocity $V_{gj} = \partial\Omega/\partial k_j$ is equal to the phase velocity:

$$\mathbf{V}_g = \mathbf{V}_{ph} = C_L\hat{\mathbf{k}}. \tag{12.9b}$$

Elastodynamic longitudinal modes are similar to sound waves in a fluid. However, in a fluid, as we can infer from Eq. (16.48), the sound waves travel with phase speed $V_{ph} = (K/\rho)^{1/2}$ [the limit of Eq. (12.9a) when the shear modulus vanishes].[2] This fluid sound speed is lower than the C_L of a solid with the same bulk modulus, because the longitudinal displacement necessarily entails shear (note that in Fig. 12.1a the motions are not an isotropic expansion), and in a solid there is a restoring shear stress (proportional to μ) that is absent in a fluid.

Because the longitudinal phase velocity is independent of frequency, we can write down general planar longitudinal-wave solutions to the elastodynamic wave equation (12.4b) in the following form:

general longitudinal plane wave

$$\boldsymbol{\xi} = \xi_L\hat{\mathbf{k}} = F(\hat{\mathbf{k}} \cdot \mathbf{x} - C_Lt)\hat{\mathbf{k}}, \tag{12.10}$$

where $F(x)$ is an arbitrary function. This describes a wave propagating in the (arbitrary) direction $\hat{\mathbf{k}}$ with an arbitrary profile determined by the function F.

12.2.4 Transverse Shear Waves

12.2.4

To derive the dispersion relation for a transverse wave we can simply make use of the transversality condition $\mathbf{k} \cdot \boldsymbol{\xi}_T = 0$ in Eq. (12.6); or, equally well, we can take the curl of Eq. (12.6) (multiply it by $i\mathbf{k}\times$), thereby projecting out the transverse piece, since

---

2. Equation (16.48) below gives the fluid sound speed as $C = \sqrt{(\partial P/\partial\rho)_s}$ (i.e., the square root of the derivative of the fluid pressure with respect to density at fixed entropy). In the language of elasticity theory, the fractional change of density is related to the expansion Θ by $\delta\rho/\rho = -\Theta$ [Eq. (12.3)], and the accompanying change of pressure is $\delta P = -K\Theta$ [sentence preceding Eq. (11.18)], so $\delta P = K(\delta\rho/\rho)$. Therefore, the fluid mechanical sound speed is $C = \sqrt{\delta P/\delta\rho} = \sqrt{K/\rho}$ [see passage following Eq. (12.43)].

the longitudinal part of $\boldsymbol{\xi}$ has vanishing curl. The result is

$$\omega^2 = \frac{\mu}{\rho} k^2, \quad \text{or} \quad \boxed{\omega = \Omega(\mathbf{k}) \equiv \left(\frac{\mu}{\rho}\right)^{1/2} k.} \tag{12.11}$$

transverse dispersion relation

This dispersion relation $\omega = \Omega(\mathbf{k})$ serves as a geometric-optics hamiltonian for wave propagation when μ and ρ vary slowly with $\mathbf{x}$ and/or t. It also implies that the transverse waves propagate with a phase speed C_T and phase and group velocities given by

$$\boxed{C_T = \left(\frac{\mu}{\rho}\right)^{1/2};} \tag{12.12a}$$

transverse wave speed

$$\boxed{\mathbf{V}_{\mathrm{ph}} = \mathbf{V}_{\mathrm{g}} = C_T \hat{\mathbf{k}}.} \tag{12.12b}$$

As $K > 0$, the shear wave speed C_T is always less than the speed C_L of longitudinal waves [Eq. (12.9a)].

These transverse modes are known as *shear waves*, because they are driven by the shear stress; cf. Fig. 12.1b. There is no expansion and therefore no change in volume associated with shear waves. They do not exist in fluids, but they are close analogs of the transverse vibrations of a string.

Longitudinal waves can be thought of as scalar waves, since they are fully describable by a single component ξ_L of the displacement $\boldsymbol{\xi}$: that along $\hat{\mathbf{k}}$. Shear waves, by contrast, are inherently vectorial. Their displacement $\boldsymbol{\xi}_T$ can point in any direction orthogonal to $\mathbf{k}$. Since the directions orthogonal to $\mathbf{k}$ form a 2-dimensional space, once $\mathbf{k}$ has been chosen, there are two independent states of polarization for the shear wave. These two polarization states, together with the single one for the scalar, longitudinal wave, make up the three independent degrees of freedom in the displacement $\boldsymbol{\xi}$.

polarization

In Ex. 12.3 we deduce these properties of $\boldsymbol{\xi}$ using matrix techniques.

EXERCISES

Exercise 12.1 **Problem: Influence of Gravity on Wave Speed*
Modify the wave equation (12.4b) to include the effect of gravity. Assume that the medium is homogeneous and the gravitational field is constant. By comparing the orders of magnitude of the terms in the wave equation, verify that the gravitational terms can be ignored for high-enough frequency elastodynamic modes: $\omega \gg g/C_{L,T}$. For wave speeds ~ 3 km s^{-1}, this gives $\omega/(2\pi) \gg 0.0005$ Hz. Seismic waves are mostly in this regime.

Exercise 12.2 *Example: Scalar and Vector Potentials for Elastic Waves in a Homogeneous Solid*
Just as in electromagnetic theory, it is sometimes useful to write the displacement $\boldsymbol{\xi}$ in terms of scalar and vector potentials:

$$\boldsymbol{\xi} = \nabla \psi + \nabla \times \mathbf{A}. \tag{12.13}$$

(The vector potential $\mathbf{A}$ is, as usual, only defined up to a gauge transformation, $\mathbf{A} \to \mathbf{A} + \nabla\varphi$, where φ is an arbitrary scalar field.) By inserting Eq. (12.13) into the general elastodynamic wave equation (12.4b), show that the scalar and vector potentials satisfy the following wave equations in a homogeneous solid:

$$\frac{\partial^2 \psi}{\partial t^2} = C_L^2 \nabla^2 \psi, \qquad \frac{\partial^2 \mathbf{A}}{\partial t^2} = C_T^2 \nabla^2 \mathbf{A}. \tag{12.14}$$

Thus, the scalar potential ψ generates longitudinal waves, while the vector potential $\mathbf{A}$ generates transverse waves.

Exercise 12.3 *Example: Solving the Algebraic Wave Equation by Matrix Techniques*
By using the matrix techniques discussed in the next-to-the-last paragraph of Box 12.2, deduce that the general solution to the algebraic wave equation (12.6) is the sum of a longitudinal mode with the properties deduced in Sec. 12.2.3 and two transverse modes with the properties deduced in Sec. 12.2.4. [Note: This matrix technique is necessary and powerful when the algebraic dispersion relation is complicated, such as for plasma waves; Secs. 21.4.1 and 21.5.1. Elastodynamic waves are simple enough that we did not need this matrix technique in the text.] Specifically, do the following.

(a) Rewrite the algebraic wave equation in the matrix form $M_{ij}\xi_j = 0$, obtaining thereby an explicit form for the matrix $||M_{ij}||$ in terms of ρ, K, μ, ω and the components of $\mathbf{k}$.

(b) This matrix equation from part (a) has a solution if and only if the determinant of the matrix $||M_{ij}||$ vanishes. (Why?) Show that $\det ||M_{ij}|| = 0$ is a cubic equation for ω^2 in terms of k^2, and that one root of this cubic equation is $\omega = C_L k$, while the other two roots are $\omega = C_T k$ with C_L and C_T given by Eqs. (12.9a) and (12.12a).

(c) Orient Cartesian axes so that $\mathbf{k}$ points in the z direction. Then show that, when $\omega = C_L k$, the solution to $M_{ij}\xi_j = 0$ is a longitudinal wave (i.e., a wave with $\boldsymbol{\xi}$ pointing in the z direction, the same direction as $\mathbf{k}$).

(d) Show that, when $\omega = C_T k$, there are two linearly independent solutions to $M_{ij}\xi_j = 0$, one with $\boldsymbol{\xi}$ pointing in the x direction (transverse to $\mathbf{k}$) and the other in the y direction (also transverse to $\mathbf{k}$).

12.2.5

12.2.5 Energy of Elastodynamic Waves

Elastodynamic waves transport energy, just like waves on a string. The waves' kinetic energy density is obviously $\frac{1}{2}\rho v^2 = \frac{1}{2}\rho\dot{\boldsymbol{\xi}}^2$, where the dot means $\partial/\partial t$. The elastic energy density is given by Eq. (11.24), so the total energy density is

$$U = \frac{1}{2}\rho\dot{\boldsymbol{\xi}}^2 + \frac{1}{2}K\Theta^2 + \mu\Sigma_{ij}\Sigma_{ij}. \tag{12.15a}$$

In Ex. 12.4 we show that (as one might expect) the elastodynamic wave equation (12.4b) can be derived from an action whose lagrangian density is the kinetic energy density minus the elastic energy density. We also show that associated with the waves is an energy flux **F** (not to be confused with a force for which we use the same notation) given by

$$F_i = -K \Theta \dot{\xi}_i - 2\mu \Sigma_{ij} \dot{\xi}_j = T_{ij} \dot{\xi}_j. \qquad (12.15b)$$

energy density and flux for general elastodynamic wave in isotropic medium

As the waves propagate, energy sloshes back and forth between the kinetic and the elastic parts, with the time-averaged kinetic energy being equal to the time-averaged elastic energy (equipartition of energy). For the planar, monochromatic, longitudinal mode, the time-averaged energy density and flux are

$$\boxed{U_L = \rho \langle \dot{\xi}_L^2 \rangle, \qquad \mathbf{F}_L = U_L C_L \hat{\mathbf{k}},} \qquad (12.16)$$

energy density and flux for plane, monochromatic elastodynamic wave

where $\langle \cdot \rangle$ denotes an average over one period or wavelength of the wave. Similarly, for the planar, monochromatic, transverse mode, the time-averaged density and flux of energy are

$$\boxed{U_T = \rho \langle \dot{\boldsymbol{\xi}}_T^2 \rangle, \quad \mathbf{F}_T = U_T C_T \hat{\mathbf{k}}} \qquad (12.17)$$

(Ex. 12.4). Thus, elastodynamic waves transport energy at the same speed $C_{L,T}$ as the waves propagate, and in the same direction $\hat{\mathbf{k}}$. This is the same behavior as electromagnetic waves in vacuum, whose Poynting flux and energy density are related by $\mathbf{F}_{\mathrm{EM}} = U_{\mathrm{EM}} c \hat{\mathbf{k}}$ with c the speed of light, and the same as all forms of dispersion-free scalar waves [e.g., sound waves in a medium; cf. Eq. (7.31)]. Actually, this is the dispersion-free limit of the more general result that the energy of any wave, in the geometric-optics limit, is transported with the wave's group velocity $\mathbf{V}_{\mathrm{g}}$; see Sec. 7.2.2.

Exercise 12.4 *Example: Lagrangian and Energy for Elastodynamic Waves*
Derive the energy-density, energy-flux, and lagrangian properties of elastodynamic waves given in Sec. 12.2.5. Specifically, do the following.

(a) For ease of calculation (and for greater generality), consider an elastodynamic wave in a possibly anisotropic medium, for which

$$T_{ij} = -Y_{ijkl} \xi_{k;l}, \qquad (12.18)$$

with Y_{ijkl} the tensorial modulus of elasticity, which is symmetric under interchange (i) of the first two indices ij, (ii) of the last two indices kl, and (iii) of the first pair ij with the last pair kl [Eq. (11.17) and associated discussion]. Show that for an isotropic medium

$$Y_{ijkl} = \left(K - \frac{2}{3} \mu \right) g_{ij} g_{kl} + \mu (g_{ik} g_{jl} + g_{il} g_{jk}). \qquad (12.19)$$

(Recall that in the orthonormal bases to which we have confined ourselves, the components of the metric are $g_{ij} = \delta_{ij}$, i.e., the Kronecker delta.)

(b) For these waves the elastic energy density is $\frac{1}{2}Y_{ijkl}\xi_{i;j}\xi_{k;l}$ [Eq. (11.25)]. Show that the kinetic energy density minus the elastic energy density,

$$\mathcal{L} = \frac{1}{2}\rho\,\dot{\xi}_i\dot{\xi}_i - \frac{1}{2}Y_{ijkl}\xi_{i;j}\xi_{k;l},\tag{12.20}$$

is a lagrangian density for the waves; that is, show that the vanishing of its variational derivative $\delta\mathcal{L}/\delta\xi_j \equiv \partial\mathcal{L}/\partial\xi_j - (\partial/\partial t)(\partial\mathcal{L}/\partial\dot{\xi}_j) = 0$ is equivalent to the elastodynamic equations $\rho\ddot{\boldsymbol{\xi}} = -\nabla\cdot\mathbf{T}$.

(c) The waves' energy density and flux can be constructed by the vector-wave analog of the canonical procedure of Eq. (7.36c):

$$U = \frac{\partial\mathcal{L}}{\partial\dot{\xi}_i}\dot{\xi}_i - \mathcal{L} = \frac{1}{2}\rho\,\dot{\xi}_i\dot{\xi}_i + \frac{1}{2}Y_{ijkl}\xi_{i;j}\xi_{k;l},$$

$$F_j = \frac{\partial\mathcal{L}}{\partial\xi_{i;j}}\dot{\xi}_i = -Y_{ijkl}\dot{\xi}_i\xi_{k;l}.\tag{12.21}$$

Verify that these density and flux values satisfy the energy conservation law, $\partial U/\partial t + \nabla\cdot\mathbf{F} = 0$. Using Eq. (12.19), verify that for an isotropic medium, expressions (12.21) for the energy density and flux become the expressions (12.15) given in the text.

(d) Show that in general (for an arbitrary mixture of wave modes), the time average of the total kinetic energy in some huge volume is equal to that of the total elastic energy. Show further that for an individual longitudinal or transverse, planar, monochromatic mode, the time-averaged kinetic energy density and time-averaged elastic energy density are both independent of spatial location. Combining these results, infer that for a single mode, the time-averaged kinetic and elastic energy densities are equal, and therefore the time-averaged total energy density is equal to twice the time-averaged kinetic energy density. Show that this total time-averaged energy density is given by the first of Eqs. (12.16) and (12.17).

(e) Show that the time average of the energy flux (12.15b) for the longitudinal and transverse modes is given by the second of Eqs. (12.16) and (12.17), so the energy propagates with the same speed and direction as the waves themselves.

12.3

12.3 Waves in Rods, Strings, and Beams

Before exploring applications (Sec. 12.4) of the longitudinal and transverse waves just described, we discuss how the wave equations and wave speeds are modified when the medium through which they propagate is bounded. Despite this situation being formally "global" in the sense of Box 12.2, elementary considerations enable us to derive the relevant dispersion relations without much effort.

12.3.1 Compression Waves in a Rod

First consider a longitudinal wave propagating along a light (negligible gravity), thin, unstressed rod. We shall call this a *compression wave*. Introduce a Cartesian coordinate system with the x-axis parallel to the rod. When there is a small displacement ξ_x independent of y and z, the restoring stress is given by $T_{xx} = -E\partial\xi_x/\partial x$, where E is Young's modulus [cf. Eq. (11.37)]. Hence the restoring force density $\mathbf{f} = -\nabla \cdot \mathbf{T}$ is $f_x = E\partial^2\xi_x/\partial x^2$. The wave equation then becomes

$$\frac{\partial^2\xi_x}{\partial t^2} = \left(\frac{E}{\rho}\right)\frac{\partial^2\xi_x}{\partial x^2}, \tag{12.22}$$

and so the sound speed for compression waves in a long straight rod is

speed of compression wave in a rod

$$\boxed{C_C = \left(\frac{E}{\rho}\right)^{\frac{1}{2}}.} \tag{12.23}$$

Referring to Table 11.1 we see that a typical value of Young's modulus in a solid is ~ 100 GPa. If we adopt a typical density of $\sim 3 \times 10^3$ kg m^{-3}, then we estimate the compressional sound speed to be ~ 5 km s^{-1}. This is roughly 15 times the sound speed in air.

As this compressional wave propagates, in regions where the rod is compressed longitudinally, it bulges laterally by an amount given by Poisson's ratio; where it is expanded longitudinally, the rod shrinks laterally. By contrast, for a longitudinal wave in a homogeneous medium, transverse forces do not cause lateral bulging and shrinking. This accounts for the different propagation speeds, $C_C \neq C_L$; see Ex. 12.6.

12.3.2 Torsion Waves in a Rod

Next consider a rod with circular cross section of radius a, subjected to a twisting force (Fig. 12.2). Let us introduce an angular displacement $\Delta\phi \equiv \varphi(x)$ that depends on distance x down the rod. The only nonzero component of the displacement vector is $\xi_\phi = \varpi\varphi(x)$, where $\varpi = \sqrt{y^2 + z^2}$ is the cylindrical radius. We can calculate the total torque, exerted by the portion of the rod to the left of x on the portion to the right, by integrating over a circular cross section. For small twists, there is no expansion, and the only components of the shear tensor are

$$\Sigma_{\phi x} = \Sigma_{x\phi} = \frac{1}{2}\xi_{\phi,x} = \frac{\varpi}{2}\frac{\partial\varphi}{\partial x}. \tag{12.24}$$

The torque contributed by an annular ring of radius ϖ and thickness $d\varpi$ is $\varpi \cdot T_{\phi x} \cdot 2\pi\varpi d\varpi$, and we substitute $T_{\phi x} = -2\mu\Sigma_{\phi x}$ to obtain the total torque of the rod to the left of x on that to the right:

$$N = -\int_0^a 2\pi\mu\varpi^3 d\varpi\,\frac{\partial\varphi}{\partial x} = -\frac{\mu}{\rho}I\frac{\partial\varphi}{\partial x}, \quad \text{where} \quad I = \frac{\pi}{2}\rho a^4 \tag{12.25}$$

is the moment of inertia per unit length.

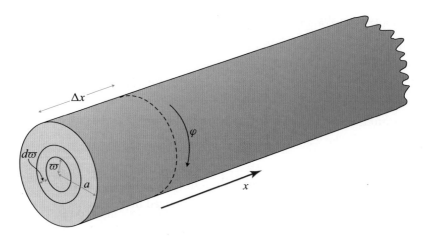

FIGURE 12.2 When a rod with circular cross section is twisted, there will be a restoring torque.

Equating the net torque on a segment with length Δx to the rate of change of its angular momentum, we obtain

$$-\frac{\partial N}{\partial x}\Delta x = I\frac{\partial^2 \varphi}{\partial t^2}\Delta x,\tag{12.26}$$

which, using Eq. (12.25), becomes the wave equation for torsional waves:

$$\left(\frac{\mu}{\rho}\right)\frac{\partial^2 \varphi}{\partial x^2} = \frac{\partial^2 \varphi}{\partial t^2}.\tag{12.27}$$

The speed of torsional waves is thus

speed of torsion wave in a rod

$$\boxed{C_T = \left(\frac{\mu}{\rho}\right)^{\frac{1}{2}}.}\tag{12.28}$$

Note that this is the same speed as that of transverse shear waves in a uniform medium. This might have been anticipated, as there is no change in volume in a torsional oscillation and so only the shear stress acts to produce a restoring force.

12.3.3 Waves on Strings

This example is surely all too familiar. When a string under a tensile force T (*not* force per unit area) is plucked, there will be a restoring force proportional to the curvature of the string. If $\xi_z \equiv \eta$ is the transverse displacement (in the same notation as we used for rods in Secs. 11.5 and 11.6), then the wave equation will be

$$T\frac{\partial^2 \eta}{\partial x^2} = \Lambda\frac{\partial^2 \eta}{\partial t^2},\tag{12.29}$$

where Λ is the mass per unit length. The wave speed is thus

$$c_S = \left(\frac{T}{\Lambda}\right)^{1/2}.$$

(12.30)

speed of wave in a string under tension

12.3.4 Flexural Waves on a Beam

Now consider the small-amplitude, transverse displacement of a horizontal rod or beam that can be flexed. In Sec. 11.5 we showed that such a flexural displacement produces a net elastic restoring force per unit length given by $D\partial^4\eta/\partial x^4$, and we considered a situation where that force was balanced by the beam's weight per unit length, $W = \Lambda g$ [Eq. (11.45)]. Here

$$D = \frac{1}{12}Ewh^3$$

(12.31)

is the flexural rigidity [Eq. (11.41b)], h is the beam's thickness in the direction of bend, w is its width, $\eta = \xi_z$ is the transverse displacement of the neutral surface from the horizontal, Λ is the mass per unit length, and g is Earth's acceleration of gravity. The solution of the resulting force-balance equation, $-D\partial^4\eta/\partial x^4 = W = \Lambda g$, is the quartic (11.46a), which describes the equilibrium beam shape.

When gravity is absent and the beam is allowed to bend dynamically, the acceleration of gravity g gets replaced by the beam's dynamical acceleration, $\partial^2\eta/\partial t^2$. The result is a wave equation for flexural waves on the beam:

$$-D\frac{\partial^4\eta}{\partial x^4} = \Lambda\frac{\partial^2\eta}{\partial t^2}.$$

(12.32)

(This derivation of the wave equation is an elementary illustration of the principle of equivalence—the equivalence of gravitational and inertial forces, or gravitational and inertial accelerations—which underlies Einstein's general relativity theory; see Sec. 25.2.)

The wave equations we have encountered so far in this chapter have all described nondispersive waves, for which the wave speed is independent of the frequency. Flexural waves, by contrast, are dispersive. We can see this by assuming that $\eta \propto \exp[i(kx - \omega t)]$ and thereby deducing from Eq. (12.32) the dispersion relation:

dispersion relation for flexural waves on a beam

$$\omega = \sqrt{D/\Lambda}\, k^2 = \sqrt{Eh^2/(12\rho)}\, k^2.$$

(12.33)

Here we have used Eq. (12.31) for the flexural rigidity D and $\Lambda = \rho wh$ for the mass per unit length.

Before considering the implications of this dispersion relation, we complicate the equilibrium a little. Let us suppose that, in addition to the net shearing force per

dispersion relation for
flexural waves on a
stretched beam

unit length $-D\partial^4\eta/\partial x^4$, the beam is also held under a tensile force T. We can then combine the two wave equations (12.29) and (12.32) to obtain

$$-D\frac{\partial^4\eta}{\partial x^4} + T\frac{\partial^2\eta}{\partial x^2} = \Lambda\frac{\partial^2\eta}{\partial t^2}, \tag{12.34}$$

for which the dispersion relation is

$$\omega^2 = C_S^2 k^2\left(1 + \frac{k^2}{k_c^2}\right), \tag{12.35}$$

where $C_S = \sqrt{T/\Lambda}$ is the wave speed when the flexural rigidity D is negligible so the beam is string-like, and

$$k_c = \sqrt{T/D} \tag{12.36}$$

is a critical wave number. If the average strain induced by the tension is $\epsilon = T/(Ewh)$, then $k_c = (12\epsilon)^{1/2}h^{-1}$, where h is the thickness of the beam, w is its width, and we have used Eq. (12.31). [Notice that k_c is equal to 1/(the lengthscale on which a pendulum's support wire—analog of our beam—bends), as discussed in Ex. 11.11.] For short wavelengths $k \gg k_c$, the shearing force dominates, and the beam behaves like a tension-free beam; for long wavelengths $k \ll k_c$, it behaves like a string.

phase and group velocities
for flexural waves on a
stretched beam

A consequence of dispersion is that waves with different wave numbers k propagate with different speeds, and correspondingly, the group velocity $V_g = d\omega/dk$ with which wave packets propagate differs from the phase velocity $V_{ph} = \omega/k$ with which a wave's crests and troughs move (see Sec. 7.2.2). For the dispersion relation (12.35), the phase and group velocities are

$$V_{ph} \equiv \omega/k = C_S(1 + k^2/k_c^2)^{1/2},$$

$$V_g \equiv d\omega/dk = C_S(1 + 2k^2/k_c^2)(1 + k^2/k_c^2)^{-1/2}. \tag{12.37}$$

As we discussed in detail in Sec. 7.2.2 and Ex. 7.2, for dispersive waves such as this one, the fact that different Fourier components in the wave packet propagate with different speeds causes the packet to gradually spread; we explore this quantitatively for longitudinal waves on a beam in Ex. 12.5.

EXERCISES

Exercise 12.5 *Derivation: Dispersion of Flexural Waves*
Verify Eqs. (12.35) and (12.37). Sketch the dispersion-induced evolution of a Gaussian wave packet as it propagates along a stretched beam.

Exercise 12.6 *Problem: Speeds of Elastic Waves*
Show that the sound speeds for the following types of elastic waves in an isotropic material are in the ratios $1 : (1 - \nu^2)^{-1/2} : \left(\frac{1-\nu}{(1+\nu)(1-2\nu)}\right)^{1/2} : [2(1 + \nu)]^{-1/2} : [2(1 + \nu)]^{-1/2}$. The elastic waves are (i) longitudinal waves along a rod, (ii) longitudinal waves along a sheet, (iii) longitudinal waves along a rod embedded in an

incompressible fluid, (iv) shear waves in an extended solid, and (v) torsional waves along a rod. [Note: Here and elsewhere in this book, if you encounter grungy algebra (e.g., frequent conversions from $\{K, \mu\}$ to $\{E, \nu\}$), do not hesitate to use Mathematica, Matlab, Maple, or other symbolic manipulation software to do the algebra!]

Exercise 12.7 *Problem: Xylophones*
Consider a beam of length ℓ, whose weight is negligible in the elasticity equations, supported freely at both ends (so the slope of the beam is unconstrained at the ends). Show that the frequencies of standing flexural waves satisfy

$$\omega = \left(\frac{n\pi}{\ell}\right)^2 \left(\frac{D}{\rho A}\right)^{1/2},$$

where A is the cross sectional area, and n is an integer. Now repeat the exercise when the ends are clamped. Based on your result, explain why xylophones don't have clamped ends.

12.3.5 Bifurcation of Equilibria and Buckling (Once More)

We conclude this discussion by returning to the problem of buckling, which we introduced in Sec. 11.6. The example we discussed there was a playing card compressed until it wants to buckle. We can analyze small dynamical perturbations of the card, $\eta(x, t)$, by treating the tension T of the previous section as negative, $T = -F$, where F is the compressional force applied to the card's two ends in Fig. 11.11. Then the equation of motion (12.34) becomes

$$-D\frac{\partial^4 \eta}{\partial x^4} - F\frac{\partial^2 \eta}{\partial x^2} = \Lambda \frac{\partial^2 \eta}{\partial t^2}. \tag{12.38}$$

We seek solutions for which the ends of the playing card are held fixed (as shown in Fig. 11.11): $\eta = 0$ at $x = 0$ and at $x = \ell$. Solving Eq. (12.38) by separation of variables, we see that

$$\eta = \eta_n \sin\left(\frac{n\pi}{\ell}x\right) e^{-i\omega_n t}. \tag{12.39}$$

Here $n = 1, 2, 3, \ldots$ labels the card's modes of oscillation, $n - 1$ is the number of nodes in the card's sinusoidal shape for mode n, η_n is the amplitude of deformation for mode n, and the mode's eigenfrequency ω_n (of course) satisfies the same dispersion relation (12.35) as for waves on a long, stretched beam, with $T \to -F$ and $k \to n\pi/\ell$:

eigenfrequencies for normal modes of a compressed beam

$$\omega_n^2 = \frac{1}{\Lambda}\left(\frac{n\pi}{\ell}\right)^2 \left[\left(\frac{n\pi}{\ell}\right)^2 D - F\right] = \frac{1}{\Lambda}\left(\frac{n\pi}{\ell}\right)^2 \left(n^2 F_{\text{crit}} - F\right), \tag{12.40}$$

where $F_{\text{crit}} = \pi^2 D/\ell^2$ is the critical force that we introduced in Chap. 11 [Eq. (11.54)].

Consider the lowest normal mode, $n = 1$, for which the playing card is bent in the single-arch manner of Fig. 11.11 as it oscillates. When the compressional force F is small, ω_1^2 is positive, so ω_1 is real and the normal mode oscillates sinusoidally and stably. But for $F > F_{\text{crit}} = \pi^2 D/\ell^2$, ω_1^2 is negative, so ω_1 is imaginary and there are two normal-mode solutions, one decaying exponentially with time, $\eta \propto \exp(-|\omega_1|t)$, and the other increasing exponentially with time, $\eta \propto \exp(+|\omega_1|t)$, signifying an instability against buckling.

onset of buckling for a compressed beam

Notice that the onset of instability occurs at the same compressional force, $F = F_{\text{crit}}$, as the bifurcation of equilibria [Eq. (11.54)], where a new (bent) equilibrium state for the playing card comes into existence. Notice, moreover, that the card's $n = 1$ normal mode has zero frequency, $\omega_1 = 0$, at this onset of instability and bifurcation of equilibria; the card can bend by an amount that grows linearly in time, $\eta = A \sin(\pi x/\ell)\, t$, with no restoring force or exponential growth. This zero-frequency motion leads the card from its original, straight equilibrium shape, to its new, bent equilibrium shape.

This is an example of a very general phenomenon, which we shall meet again in fluid mechanics (Sec. 15.6.1). For mechanical systems without dissipation (no energy losses to friction, viscosity, radiation, or anything else), as one gradually changes some "control parameter" (in this case the compressional force F), there can occur bifurcation of equilibria. At each bifurcation point, a normal mode of the original equilibrium becomes unstable, and at its onset of instability the mode has zero frequency and represents a motion from the original equilibrium (which is becoming unstable) to the new, stable equilibrium.

zero-frequency mode at bifurcation of equilibria

In our simple playing-card example, we see this phenomenon repeated again and again as the control parameter F is increased. One after another, at $F = n^2 F_{\text{crit}}$, the modes $n = 1$, $n = 2$, $n = 3$, ... become unstable. At each onset of instability, ω_n vanishes, and the zero-frequency mode (with $n - 1$ nodes in its eigenfunction) leads from the original, straight-card equilibrium to the new, stable, $(n - 1)$-noded, bent equilibrium.

12.4 Body Waves and Surface Waves—Seismology and Ultrasound

In Sec. 12.2, we derived the dispersion relations $\omega = C_L k$ and $\omega = C_T k$ for longitudinal and transverse elastodynamic waves in uniform media. We now consider how the waves are modified in an inhomogeneous, finite body: Earth. Earth is well approximated as a sphere of radius $R \sim 6{,}000\,\text{km}$ and mean density $\bar{\rho} \sim 6{,}000\,\text{kg m}^{-3}$. The outer crust, extending down to 5–10 km below the ocean floor and comprising rocks of high tensile strength, rests on a more malleable mantle, the two regions being separated by the famous Mohorovičić (or Moho for short) discontinuity. Underlying the mantle is an outer core mainly composed of liquid iron, which itself surrounds a denser, solid inner core of mostly iron; see Table 12.1 and Fig. 12.5 below.

The pressure in Earth's interior is much larger than atmospheric, and the rocks are therefore quite compressed. Their atomic structure cannot be regarded as a small

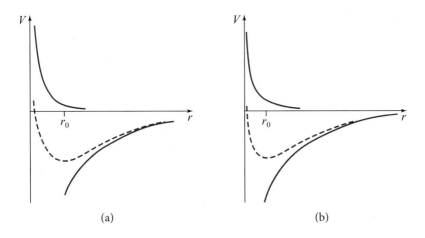

(a)   (b)

FIGURE 12.3 Potential energy curves (dashed) for nearest neighbors in a crystal lattice. (a) At atmospheric (effectively zero) pressure, the equilibrium spacing is set by the minimum in the potential energy, which is a combination of hard electrostatic repulsion by the nearest neighbors (upper solid curve) and a softer overall attraction associated with all the nearby ions (lower solid curve). (b) At much higher pressure, the softer, attractive component is moved inward, and the equilibrium spacing is greatly reduced. The bulk modulus is proportional to the curvature of the potential energy curve at its minimum and so is considerably increased.

perturbation from their structure in a vacuum. Nevertheless, we can still use linear elasticity theory to discuss small perturbations about their equilibrium. This is because the crystal lattice has had plenty of time to establish a new equilibrium with a much smaller lattice spacing than at atmospheric pressure (Fig. 12.3). The density of lattice defects and dislocations will probably not differ appreciably from those at atmospheric pressure, so the proportionality limit and yield stress should be about the same as for rocks near Earth's surface. The linear stress-strain relation will still apply below the proportionality limit, although the elastic moduli are much greater than those measured at atmospheric pressure.

We can estimate the magnitude of the pressure P in Earth's interior by idealizing the planet as an isotropic medium with negligible shear stress, so its stress tensor is like that of a fluid, $\mathbf{T} = P\mathbf{g}$ (where $\mathbf{g}$ is the metric tensor). Then the equation of static equilibrium takes the form

$$\frac{dP}{dr} = -g\rho, \qquad (12.41)$$

where ρ is density, and $g(r)$ is the acceleration of gravity at radius r. This equation can be approximated by

$$P \sim \bar{\rho} g R \sim 300\,\mathrm{GPa} \sim 3 \times 10^6\,\text{atmospheres}, \qquad (12.42)$$

pressure at Earth's center

where g is now the acceleration of gravity at Earth's surface $r = R$, and $\bar{\rho}$ is Earth's mean density. This agrees well numerically with the accurate value of 360 GPa at Earth's center. The bulk modulus produces the isotropic pressure $P = -K\Theta$ [Eq. (11.18)]; and since $\Theta = -\delta\rho/\rho$ [cf. Eq. (12.3)], the bulk modulus can be expressed as

$$K = \frac{dP}{d\ln\rho}.$$ (12.43)

[Strictly speaking, we should distinguish between adiabatic and isothermal variations in Eq. (12.43), but the distinction is small for solids; see Sec. 11.3.5. It is significant for gases.] Typically, the bulk modulus inside Earth is 4 to 5 times the pressure, and the shear modulus in the crust and mantle is about half the bulk modulus.

12.4.1 Body Waves

Virtually all our direct information about the internal structure of Earth comes from measurements of the propagation times of elastic waves that are generated by earthquakes and propagate through Earth's interior (*body waves*). There are two fundamental kinds of body waves: the longitudinal and transverse modes of Sec. 12.2. These are known in seismology as P-modes (P for pressure) and S-modes (S for shear), respectively. The two polarizations of the transverse shear waves are designated SH and SV, where H and V stand for "horizontal" and "vertical" displacements (i.e., displacements orthogonal to $\mathbf{k}$ that are fully horizontal, or that are obtained by projecting the vertical direction $\mathbf{e}_z$ orthogonal to $\hat{\mathbf{k}}$).

We shall first be concerned with what seismologists call high-frequency (of order 1 Hz) modes, which leads to three related simplifications. As typical wave speeds lie in the range 3–14 km s^{-1}, the wavelengths lie in the range $\sim$1 to 10 km, which is generally small compared with the distance over which gravity causes the pressure to change significantly—the *pressure scale height*. It turns out that we then can ignore the effects of gravity on the propagation of small perturbations. In addition, we can regard the medium locally as effectively homogeneous and so use the local dispersion relations $\omega = C_{L,T}k$. Finally, as the wavelengths are short, we can trace rays through Earth using geometrical optics (Sec. 7.3).

Now, Earth is quite inhomogeneous globally, and the sound speeds therefore vary significantly with radius; see Table 12.1. To a fair approximation, Earth is horizontally stratified below the outer crust (whose thickness is irregular). Two types of variation can be distinguished, the abrupt and the gradual. There are several abrupt changes in the crust and mantle (including the Moho discontinuity at the interface between crust and mantle), and also at the transitions between mantle and outer core, and between outer core and inner core. At these *surfaces of discontinuity,* the density and elastic constants apparently change over distances short compared with a wavelength. Seismic waves incident on these discontinuities behave like light incident on the surface of a glass plate; they can be reflected and refracted. In addition, as there are now two different waves with different phase speeds, it is possible to generate SV waves from pure P waves and vice versa at a discontinuity (Fig. 12.4). However, this

TABLE 12.1: Typical outer radii (R), densities (ρ), bulk moduli (K), shear moduli (μ), P-wave speeds (C_P), and S-wave speeds (C_S) in different zones of Earth

Zone	R (10^3 km)	ρ (10^3 kg m^{-3})	K (GPa)	μ (GPa)	C_P (km s^{-1})	C_S (km s^{-1})
Inner core	1.2	13	1,400	160	11	2
Outer core	3.5	10–12	600–1,300	—	8–10	—
Mantle	6.35	3–5	100–600	70–250	8–14	5–7
Crust	6.37	3	50	30	6–7	3–4
Ocean	6.37	1	2	—	1.5	—

Notes: Note the absence of shear waves (denoted by a —) in the fluid regions. Adapted from Stacey (1977).

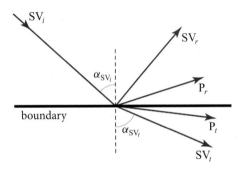

FIGURE 12.4 An incident shear wave polarized in the vertical direction (SV$_i$), incident from above on a surface of discontinuity, produces both a longitudinal (P) wave and an SV wave in reflection and in transmission. If the wave speeds increase across the boundary (the case shown), then the transmitted waves, SV$_t$, P$_t$, will be refracted away from the vertical. A shear mode, SV$_r$, will be reflected at the same angle as the incident wave. However, the reflected P mode, P$_r$, will be reflected at a greater angle to the vertical, as it has greater speed than the incident wave.

wave-wave mixing is confined to SV and P; the SH waves do not mix with SV or P; see Ex. 12.9.

The junction conditions that control this wave mixing and all other details of the waves' behavior at a surface of discontinuity are: (i) the displacement $\boldsymbol{\xi}$ must be continuous across the discontinuity (otherwise infinite strain and infinite stress would develop there); and (ii) the net force acting on an element of surface must be zero (otherwise the surface, having no mass, would have infinite acceleration), so the force per unit area acting from the front face of the discontinuity to the back must be balanced by that acting from the back to the front. If we take the unit normal to the horizontal discontinuity to be $\mathbf{e}_z$, then these boundary conditions become

$$[\xi_j] = [T_{jz}] = 0,$$

(12.44)

wave boundary conditions at interfaces

where the notation $[X]$ signifies the difference in X across the boundary, and the j is a vector index. (For an alternative, more formal derivation of $[T_{jz}] = 0$, see Ex. 12.8.)

One consequence of these boundary conditions is Snell's law for the directions of propagation of the waves. Since these continuity conditions must be satisfied all along the surface of discontinuity and at all times, the phase $\phi = \mathbf{k} \cdot \mathbf{x} - \omega t$ of the wave must be continuous across the surface at all locations $\mathbf{x}$ on it and all times, which means that the phase ϕ must be the same on the surface for all transmitted waves and all reflected waves as for the incident waves. This is possible only if the frequency ω, the horizontal wave number $k_H = k \sin \alpha$, and the horizontal phase speed $C_H = \omega / k_H = \omega/(k \sin \alpha)$ are the same for all the waves. (Here $k_H = k \sin \alpha$ is the magnitude of the horizontal component of a wave's propagation vector, and α is the angle between its propagation direction and the vertical; cf. Fig. 12.4.) Thus we arrive at Snell's law: for every reflected or transmitted wave J, the horizontal phase speed must be the same as for the incident wave:

Snell's law for waves at interfaces

$$\boxed{\frac{C_J}{\sin \alpha_J} = C_H \text{ is the same for all } J.}$$

(12.45)

It is straightforward though tedious to compute the reflection and transmission coefficients (e.g., the strength of a transmitted P-wave produced by an incident SV wave) for the general case using the boundary conditions (12.44) and (12.45) (see, e.g., Eringen and Suhubi, 1975, Sec. 7.7). For the very simplest of examples, see Ex. 12.10.

In the regions between the discontinuities, the pressures (and consequently, the elastic moduli) increase steadily, over many wavelengths, with depth. The elastic moduli generally increase more rapidly than the density does, so the wave speeds generally also increase with depth (i.e., $dC/dr < 0$). This radial variation in C causes the rays along which the waves propagate to bend. The details of this bending are governed by Hamilton's equations, with the hamiltonian $\Omega(\mathbf{x}, \mathbf{k})$ determined by the simple non-dispersive dispersion relation $\Omega = C(\mathbf{x})k$ (Sec. 7.3.1). Hamilton's equations in this case reduce to the simple ray equation (7.48), which (since the index of refraction is $\propto 1/C$) can be rewritten as

geometric optics ray equation for elastodynamic waves in inhomogeneous elastic medium

$$\frac{d}{ds}\left(\frac{1}{C}\frac{d\mathbf{x}}{ds}\right) = \nabla\left(\frac{1}{C}\right).$$

(12.46)

Here s is distance along the ray, so $d\mathbf{x}/ds = \mathbf{n}$ is the unit vector tangent to the ray. This ray equation can be reexpressed in the following form:

$$d\mathbf{n}/ds = -(\nabla \ln C)_\perp,$$

(12.47)

where the subscript $\perp$ means "projected perpendicular to the ray;" and this in turn means that the ray bends *away* from the direction in which C increases (i.e., it bends upward inside Earth, since C increases downward) with the radius of curvature of the bend given by

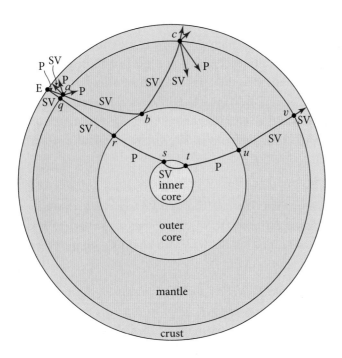

FIGURE 12.5 Seismic wave propagation in a schematic Earth model. A P wave made by an earthquake, E, propagates to the crust-mantle boundary at a where it generates two transmitted waves (SV and P) and two reflected waves (SV and P). The transmitted SV wave propagates along a ray that bends upward a bit (geometric-optics bending) and hits the mantle–outer-core boundary at b. There can be no transmitted SV wave at b, because the outer core is fluid; there can be no transmitted or reflected P wave, because the angle of incidence of the SV wave is too great. So the SV wave is perfectly reflected. It then travels along an upward curving ray, to the crust-mantle interface at c, where it generates four waves, two of which hit Earth's surface. The earthquake E also generates an SV wave traveling almost radially inward, through the crust-mantle interface at q, to the mantle–outer-core interface at r. Because the outer core is liquid, it cannot support an SV wave, so only a P wave is transmitted into the outer core at r. That P wave propagates to the interface with the inner core at s, where it regenerates an SV wave along with the transmitted and reflected P waves (not shown). The SV wave refracts back upward in the inner core and generates a P wave at the interface with the outer core t. That P wave propagates through the liquid outer core to u, where it generates an SV wave along with its transmitted and reflected P waves (not shown); that SV wave travels nearly radially outward, through v and to Earth's surface.

$$\mathcal{R} = \frac{1}{|(\boldsymbol{\nabla} \ln C)_{\perp}|} = \frac{1}{|(d \ln C/dr) \sin \alpha|}. \tag{12.48}$$

Here α is the angle between the ray and the radial direction.

Figure 12.5 shows schematically the propagation of seismic waves through Earth. At each discontinuity in Earth's material, Snell's law governs the directions of the reflected and transmitted waves. As an example, note from Eq. (12.45) that an SV mode incident on a boundary cannot generate any P mode when its angle of incidence exceeds $\sin^{-1}(C_{Ti}/C_{Lt})$. (Here we use the standard notation C_T for the phase speed of an S wave and C_L for that of a P wave.) This is what happens at point b in Fig. 12.5.

seismic wave propagation

Exercise 12.8 *Derivation: Junction Condition at a Discontinuity*

Derive the junction condition $[T_{jz}] = 0$ at a horizontal discontinuity between two media by the same method as one uses in electrodynamics to show that the normal component of the magnetic field must be continuous: Integrate the equation of motion $\rho d\mathbf{v}/dt = -\nabla \cdot \mathbf{T}$ over the volume of an infinitesimally thin pill box centered on the boundary (see Fig. 11.7), and convert the volume integral to a surface integral via Gauss's theorem.

Exercise 12.9 *Derivation: Wave Mixing at a Surface of Discontinuity*

Using the boundary conditions (12.44), show that at a surface of discontinuity inside Earth, SV and P waves mix, but SH waves do not mix with the other waves.

Exercise 12.10 *Example: Reflection and Transmission of Normal, Longitudinal Waves at a Boundary*

Consider a longitudinal elastic wave incident normally on the boundary between two media, labeled 1 and 2. By matching the displacement and the normal component of stress at the boundary, show that the ratio of the transmitted wave amplitude to the incident amplitude is given by

$$t = \frac{2Z_1}{Z_1 + Z_2},$$

where $Z_{1,2} = [\rho_{1,2}(K_{1,2} + 4\mu_{1,2}/3)]^{1/2}$ is known as the *acoustic impedance*. (The impedance is independent of frequency and is just a characteristic of the material.) Likewise, evaluate the amplitude reflection coefficient, and verify that wave energy flux is conserved.

12.4.2

12.4.2 Edge Waves

One phenomenon that is important in seismology (and also occurs in plasmas; Ex. 21.17) but is absent for many other types of wave motion is *edge waves,* i.e., waves that propagate along a discontinuity in the elastic medium. An important example is *surface waves*, which propagate along the surface of a medium (e.g., Earth's surface) and that decay exponentially with depth. Waves with such exponential decay are sometimes called *evanescent*.

edge waves and surface waves

The simplest type of surface wave is the *Rayleigh wave*, which propagates along the surface of an elastic medium. We analyze Rayleigh waves for the idealization of a *plane semi-infinite solid*—also sometimes called a *homogeneous half-space*. When it is applied to Earth, this discussion must be modified to allow for both the density stratification and (if the wavelength is sufficiently long) the surface curvature. However, the qualitative character of the mode is unchanged.

Rayleigh wave

Rayleigh waves are an intertwined mixture of P and SV waves. When analyzing them, it is useful to resolve their displacement vector $\boldsymbol{\xi}$ into a sum of a (longitudinal) P-wave component $\boldsymbol{\xi}^L$ and a (transverse) SV-wave component $\boldsymbol{\xi}^T$.

Consider a semi-infinite elastic medium, and introduce a local Cartesian coordinate system with $\mathbf{e}_z$ normal to the surface, $\mathbf{e}_x$ lying in the surface, and the propagation vector $\mathbf{k}$ in the $\mathbf{e}_z$-$\mathbf{e}_x$ plane. The propagation vector has a real component along the horizontal ($\mathbf{e}_x$) direction, corresponding to true propagation, and an imaginary component along the $\mathbf{e}_z$ direction, corresponding to an exponential decay of the amplitude as one goes down into the medium. For the longitudinal (P-wave) and transverse (SV-wave) parts of the wave to remain in phase with each other as they propagate along the boundary, they must have the same values of the frequency ω and horizontal wave number k_x. However, there is no reason why their vertical e-folding lengths should be the same (i.e., why their imaginary k_z values should be the same). We therefore denote their imaginary k_zs by $-iq_L$ for the longitudinal (P-wave) component and $-iq_T$ for the transverse (S-wave) component, and we denote their common k_x by k.

First focus attention on the longitudinal part of the wave. Its displacement can be written as the gradient of a scalar [Eq. (12.7b)], $\boldsymbol{\xi}_L = \boldsymbol{\nabla}(\psi_0 e^{q_L z + i(kx - \omega t)})$:

$$\xi_x^L = ik\psi_0 e^{q_L z + i(kx - \omega t)}, \quad \xi_z^L = q_L \psi_0 e^{q_L z + i(kx - \omega t)}, \quad z \leq 0. \tag{12.49}$$

Substituting into the general dispersion relation $\omega^2 = C_L^2 \mathbf{k}^2$ for longitudinal waves, we obtain

$$q_L = \left(k^2 - \frac{\omega^2}{C_L^2} \right)^{1/2}. \tag{12.50}$$

Because the transverse part of the wave is divergence free, the wave's expansion comes entirely from the longitudinal part, $\Theta = \boldsymbol{\nabla} \cdot \boldsymbol{\xi}_L$, and is given by

$$\Theta = (q_L^2 - k^2)\psi_0 e^{q_L z + i(kx - \omega t)}. \tag{12.51}$$

The transverse (SV) part of the wave can be written as the curl of a vector potential [Eq. (12.7c)], which we can take to point in the y direction, $\boldsymbol{\xi}_T = \boldsymbol{\nabla} \times (A_0 \mathbf{e}_y e^{q_T z + i(kx - \omega t)})$:

$$\xi_x^T = -q_T A_0 e^{q_T z + i(kx - \omega t)}, \quad \xi_z^T = ik A_0 e^{q_T z + i(kx - \omega t)}, \quad z \leq 0. \tag{12.52}$$

The dispersion relation $\omega^2 = C_T^2 \mathbf{k}^2$ for transverse waves tells us that

$$q_T = \left(k^2 - \frac{\omega^2}{C_T^2} \right)^{1/2}. \tag{12.53}$$

We next impose boundary conditions at the surface. Since the surface is free, there will be no force acting on it:

$$\mathbf{T} \cdot \mathbf{e}_z|_{z=0} = 0, \tag{12.54}$$

which is a special case of the general boundary condition (12.44). (Note that we can evaluate the stress at the unperturbed surface rather than at the displaced surface as we are only working to linear order.) The tangential stress is $T_{xz} = -2\mu(\xi_{z,x} + \xi_{x,z}) = 0$, so its vanishing is equivalent to

$$\xi_{z,x} + \xi_{x,z} = 0 \quad \text{at } z = 0. \tag{12.55}$$

Inserting $\xi = \xi_L + \xi_T$ and Eqs. (12.49) and (12.52) for the components and solving for the ratio of the transverse amplitude to the longitudinal amplitude, we obtain

$$\frac{A_0}{\psi_0} = \frac{2ikq_L}{k^2 + q_T^2}. \tag{12.56}$$

The normal stress is $T_{zz} = -K\Theta - 2\mu(\xi_{z,z} - \frac{1}{3}\Theta) = 0$, so from the values $C_L^2 = (K + \frac{4}{3}\mu)/\rho$ and $C_T^2 = \mu/\rho$ of the longitudinal and transverse sound speeds, which imply $K/\mu = (C_L/C_T)^2 - \frac{4}{3}$, we deduce that vanishing T_{zz} is equivalent to

$$(1 - 2\kappa)\Theta + 2\kappa\xi_{z,z} = 0 \quad \text{at } z = 0, \quad \text{where} \quad \kappa \equiv \frac{C_T^2}{C_L^2} = \frac{1 - 2\nu}{2(1 - \nu)}. \tag{12.57}$$

Here we have used Eqs. (11.39) to express the speed ratio in terms of Poisson's ratio ν. Inserting Θ from Eq. (12.51) and $\xi_z = \xi_z^L + \xi_z^T$ with components from Eqs. (12.49) and (12.52), using the relation $q_L^2 - k^2 + 2\kappa k^2 = \kappa(q_T^2 + k^2)$ [which follows from Eqs. (12.50), (12.53), and (12.57)], and solving for the ratio of amplitudes, we obtain

$$\frac{A_0}{\psi_0} = \frac{k^2 + q_T^2}{-2ikq_T}. \tag{12.58}$$

The dispersion relation for Rayleigh waves is obtained by equating the two expressions (12.56) and (12.58) for the amplitude ratio:

$$4k^2 q_T q_L = (k^2 + q_T^2)^2. \tag{12.59}$$

We can express this dispersion relation more explicitly in terms of the ratio of the Rayleigh-wave speed $C_R = \omega/k$ to the transverse-wave speed C_T:

$$\zeta = \left(\frac{\omega}{C_T k}\right)^2 = \left(\frac{C_R}{C_T}\right)^2. \tag{12.60}$$

By inserting $q_T = k\sqrt{1 - \zeta}$ [from Eqs. (12.53) and (12.60)] and $q_L = k\sqrt{1 - \kappa\zeta}$ [from Eqs. (12.50), (12.57), and (12.60)], and expressing κ in terms of Poisson's ratio via Eq. (12.57), we bring Eq. (12.59) into the explicit form

dispersion relation for Rayleigh wave

$$\zeta^3 - 8\zeta^2 + 8\left(\frac{2 - \nu}{1 - \nu}\right)\zeta - \frac{8}{(1 - \nu)} = 0. \tag{12.61}$$

This dispersion relation is a third-order polynomial in $\zeta \propto \omega^2$ with just one positive real root $\zeta(\nu)$, which we plot in Fig. 12.6. From that plot, we see that for a Poisson's

phase speed of Rayleigh wave

ratio characteristic of rocks, $0.2 \lesssim \nu \lesssim 0.3$, *the phase speed of a Rayleigh wave is roughly 0.9 times the speed of a pure shear wave*; cf. Fig. 12.6.

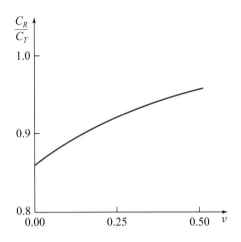

FIGURE 12.6 Solution $C_R/C_T = \sqrt{\zeta}$ of the dispersion relation (12.61) as a function of Poisson's ratio ν.

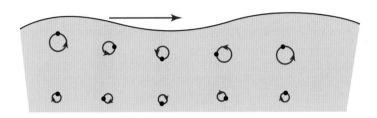

FIGURE 12.7 Rayleigh waves in a semi-infinite elastic medium.

The displacement vector for Rayleigh waves is the sum of Eqs. (12.49) and (12.52) with the amplitudes related by (12.56):

$$\xi_x = ik\psi_o \left[e^{q_L z} - \frac{2q_L q_T}{k^2 + q_T^2} e^{q_T z} \right] e^{i(kx - \omega t)},$$

material displacement in Rayleigh wave

$$\xi_z = q_L \psi_o \left[e^{q_L z} - \frac{2k^2}{k^2 + q_T^2} e^{q_T z} \right] e^{i(kx - \omega t)}. \tag{12.62}$$

Equation (12.62) represents a backward rotating, elliptical motion for each fluid element near the surface (as depicted in Fig. 12.7), reversing to a forward rotation at depths where the sign of ξ_x has flipped.

Rayleigh waves propagate around the surface of Earth rather than penetrate its interior. However, our treatment is inadequate, because their wavelengths—typically 1–10 km if generated by an earthquake—are not necessarily small compared with the scale heights in the outer crust over which C_S and C_T vary. Our wave equation has to be modified to include these vertical gradients.

This vertical stratification has an important additional consequence. Ignoring these gradients, if we attempt to find an orthogonal surface mode just involving SH waves, we find that we cannot simultaneously satisfy the surface boundary conditions on displacement and stress with a single evanescent wave. We need two modes to do this. However, when we allow for stratification, the strong refraction allows an SH surface wave to propagate. This is known as a *Love wave*. The reason for its practical importance is that seismic waves are also created by underground nuclear explosions, and it is important to be able to distinguish explosion-generated waves from earthquake waves. An earthquake is usually caused by the transverse slippage of two blocks of crust across a fault line. It is therefore an efficient generator of shear modes, including Love waves. By contrast, explosions involve radial motions away from the point of explosion and are inefficient emitters of Love waves. This allows these two sources of seismic disturbance to be distinguished.

Love waves (margin note)

EXERCISES

Exercise 12.11 *Example: Earthquakes*
The magnitude M of an earthquake, on modern variants of the *Richter scale*, is a quantitative measure of the strength of the seismic waves it creates. The earthquake's seismic-wave energy release can be estimated using a rough semi-empirical formula due to Båth (1966):

earthquake magnitude (margin note)

$$E = 10^{5.24+1.44M} \text{J}. \qquad (12.63)$$

The largest earthquakes have magnitude ~ 9.5.

One type of earthquake is caused by slippage along a fault deep in the crust. Suppose that most of the seismic power in an earthquake with $M \sim 8.5$ is emitted at frequencies ~ 1 Hz and that the quake lasts for a time $T \sim 100$ s. If C is an average wave speed, then it is believed that the stress is relieved over an area of fault of length $\sim CT$ and a depth of order one wavelength (Fig. 12.8). By comparing the stored elastic energy with the measured energy release, make an estimate of the minimum strain prior to the earthquake. Is your answer reasonable? Hence estimate the typical displacement during the earthquake in the vicinity of the fault. Make an order-of-magnitude estimate of the acceleration measurable by a seismometer in the next state and in the next continent. (Ignore the effects of density stratification, which are actually quite significant.)

12.4.3 (margin note)

12.4.3 Green's Function for a Homogeneous Half-Space

To gain insight into the combination of waves generated by a localized source, such as an explosion or earthquake, it is useful to examine the Green's function for excitations in a homogeneous half-space. Physicists define the Green's function $G_{jk}(\mathbf{x}, t; \mathbf{x}', t')$ to be the displacement response $\xi_j(\mathbf{x}, t)$ to a unit delta-function force in the $\mathbf{e}_k$ direction at location $\mathbf{x}'$ and time t': $\mathbf{F} = \delta(\mathbf{x} - \mathbf{x}')\delta(t - t')\mathbf{e}_k$. Geophysicists sometimes find

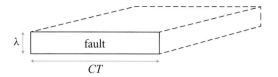

FIGURE 12.8 Earthquake: The region of a fault that slips (solid rectangle), and the volume over which the strain is relieved, on one side of the fault (dashed region).

it useful to work, instead, with the "Heaviside Green's function," $G^H_{jk}(\mathbf{x}, t; \mathbf{x}', t')$, which is the displacement response $\xi_j(\mathbf{x}, t)$ to a unit step-function force (one that turns on to unit strength and remains forever constant afterward) at $\mathbf{x}'$ and t': $\mathbf{F} = \delta(\mathbf{x} - \mathbf{x}')H(t - t')\mathbf{e}_k$. Because $\delta(t - t')$ is the time derivative of the Heaviside step function $H(t - t')$, *the Heaviside Green's function is the time integral of the physicists' Green's function*. The Heaviside Green's function has the advantage that one can easily see the size of the step functions it contains, by contrast with the size of the delta functions contained in the physicists' Green's function.

It is a rather complicated task to compute the Heaviside Green's function, and geophysicists have devoted much effort to doing so. We shall not give the details of such computations, but merely show the function graphically in Fig. 12.9 for an instructive situation: the displacement produced by a step-function force in a homogeneous half-space with the observer at the surface and the force at two different locations: a point nearly beneath the observer (Fig. 12.9a), and a point close to the surface and some distance away in the x direction (Fig. 12.9b).

Several features of this Heaviside Green's function deserve note:

- Because of their relative propagation speeds, the P waves arrive at the observer first, then (about twice as long after the excitation) the S waves, and shortly thereafter the Rayleigh waves. From the time interval ΔT between the first P waves and first S waves, one can estimate the distance to the source: $\ell \simeq (C_P - C_S)\Delta T \sim 3(\Delta T/\text{s})$ km.

- For the source nearly beneath the observer (Fig. 12.9a), there is no sign of any Rayleigh wave, whereas for the source close to the surface, the Rayleigh wave is the strongest feature in the x and z (longitudinal and vertical) displacements but is absent from the y (transverse) displacement. From this, one can infer the waves' propagation direction.

- The y (transverse) component of force produces a transverse displacement that is strongly concentrated in the S wave.

- The x and z (longitudinal and vertical) components of force produce x and z displacements that include P waves, S waves, and (for the source near the surface) Rayleigh waves.

Heaviside Green's function for elastodynamic waves

Properties of Heaviside Green's function

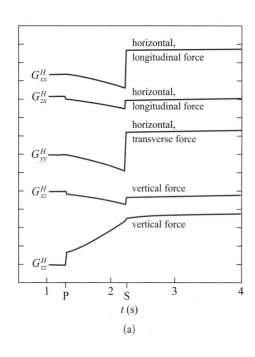

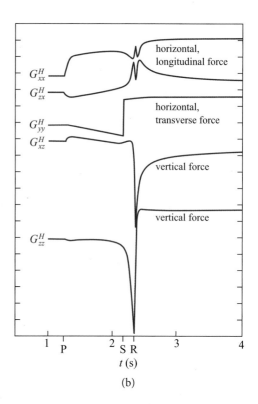

(a) (b)

FIGURE 12.9 The Heaviside Green's function (displacement response to a step-function force) in a homogeneous half-space. The observer is at the surface. The force is applied at a point in the x-z plane, with a direction stated in words and also given by the second index of G^H; the displacement direction is given by the first index of G^H. The longitudinal and transverse speeds are $C_L = 8.00$ km s^{-1} and $C_S = 4.62$ km s^{-1}, and the density is 3.30×10^3 kg m^{-3}. For a force of 1 Newton, a division on the vertical scale is 10^{-16} m. The moments of arrival of the P wave, S wave, and Rayleigh (R) wave from the moment the force is turned on are indicated on the horizontal axis. (a) The source is nearly directly beneath the observer, so the waves propagate nearly vertically upward. More specifically, the source is at 10 km depth and is 2 km distant along the horizontal x direction. (b) The source is close to the surface, and the waves propagate nearly horizontally (in the x direction). More specifically, the source is at 2 km depth and is 10 km distant along the horizontal x direction. Adapted from Johnson (1974, Figs. 2, 4).

- The gradually changing displacements that occur between the arrival of the turn-on P wave and the turn-on S wave are due to P waves that hit the surface some distance from the observer, and from there diffract to the observer as a mixture of P and S waves. Similarly for gradual changes of displacement after the turn-on S wave.

The complexity of seismic waves arises in part from the richness of features in this homogeneous-half-space Heaviside Green's function, in part from the influences of Earth's inhomogeneities, and in part from the complexity of an earthquake's or explosion's forces.

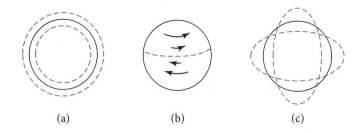

FIGURE 12.10 Displacements associated with three types of global modes for an elastic sphere, such as Earth. The polar axis points upward. (a) A radial mode shown on a cross section through the sphere's center; matter originally on the solid circle moves in and out between the two dashed circles. (b) An $l = 1$, $m = 0$ torsional mode; the arrows are proportional to the displacement vector on the sphere's surface at one moment of time. (c) An $l = 2$, $m = 0$ spheroidal mode shown on a cross section through the sphere's center; matter originally on the solid circle moves between the two ellipses.

12.4.4 Free Oscillations of Solid Bodies

12.4.4

When computing the dispersion relations for body (P- and S-wave) modes and surface (Rayleigh-wave) modes, we have assumed that the wavelength is small compared with Earth's radius; therefore the modes have a continuous frequency spectrum. However, it is also possible to excite global modes in which the whole Earth "rings" with a discrete spectrum. If we approximate Earth as spherically symmetric and ignore its rotation (whose period is long compared to a normal-mode period), then we can isolate three types of global modes: *radial, torsional,* and *spheroidal.*

types of normal modes of an elastic sphere

To compute the details of these modes, we can solve by separation of variables the equations of elastodynamics for the displacement vector in spherical polar coordinates. This is much like solving the Schrödinger equation for a central potential. See Ex. 12.12 for a relatively simple example. In general (as in that example), each of the three types of modes has a displacement vector $\boldsymbol{\xi}$ characterized by its own type of spherical harmonic.

The *radial modes* have spherically symmetric, purely radial displacements $\xi_r(r)$ and so have spherical harmonic order $l = 0$; see Fig. 12.10a and Ex. 12.12a.

radial modes

The *torsional modes* have vanishing radial displacements, and their nonradial displacements are proportional to the vector spherical harmonic $\hat{\mathbf{L}} Y_l^m(\theta, \phi)$, where θ, ϕ are spherical coordinates, Y_l^m is the scalar spherical harmonic, and $\hat{\mathbf{L}} = \mathbf{x} \times \nabla$ is the angular momentum operator (aside from an omitted factor $\hbar/i$), which plays a major role in quantum theory. Spherical symmetry guarantees that these modes' eigenfrequencies are independent of m (because by reorienting the polar axis, the various m are mixed among one another), so $m = 0$ is representative. In this case, $\nabla Y_l^0 \propto \nabla P_l(\cos\theta)$ is in the θ direction, so $\hat{\mathbf{L}} Y_l^0$ is in the ϕ direction. Hence the only nonzero component of the displacement vector is

torsional modes

$$\xi_\phi \propto \partial P_l(\cos\theta)/\partial\theta = \sin\theta P_l'(\cos\theta). \tag{12.64}$$

(Here P_l is the Legendre polynomial and P_l' is the derivative of P_l with respect to its argument.) Therefore, in these modes alternate zones of different latitude oscillate in opposite directions (clockwise or counterclockwise at some chosen moment of time) in such a way as to conserve total angular momentum. See Fig. 12.10b and Ex. 12.12b. In the high-frequency limit, the torsional modes become SH waves (since their displacements are horizontal).

spheroidal modes

The *spheroidal modes* have radial displacements proportional to $Y_l^m(\theta, \phi)\mathbf{e}_r$, and they have nonradial components proportional to ∇Y_l^m. These two displacements can combine into a single mode, because they have the same parity (and opposite parity from the torsional modes) as well as the same spherical-harmonic orders l and m. The eigenfrequencies again are independent of m and thus can be studied by specializing to $m = 0$, in which case the displacements become

$$\xi_r \propto P_l(\cos\theta), \qquad \xi_\theta \propto \sin\theta\, P_l'(\cos\theta). \tag{12.65}$$

These displacements deform the sphere in a spheroidal manner for the special case $l = 2$ (Fig. 12.10c and Ex. 12.12c), whence their name "spheroidal." The radial modes are the special case $l = 0$ of these spheroidal modes. It is sometimes mistakenly asserted that there are no $l = 1$ spheroidal modes because of conservation of momentum. In fact, $l = 1$ modes do exist: for example, the central regions of the sphere can move up, while the outer regions move down. For Earth, the lowest-frequency $l = 2$ spheroidal mode has a period of 53 minutes and can ring for about 1,000 periods (i.e., its quality factor is $Q \sim 1,000$). This is typical for solid planets. In the high-frequency limit, the spheroidal modes become a mixture of P and SV waves.

solving elastodynamic equations for modes

When one writes the displacement vector $\boldsymbol{\xi}$ for a general vibration of Earth as a sum over these various types of global modes and inserts that sum into the wave equation (12.4b) (augmented, for greater realism, by gravitational forces), spherical symmetry of unperturbed Earth guarantees that the various modes will separate from one another. For each mode the wave equation will give a radial wave equation analogous to that for a hydrogen atom in quantum mechanics. The boundary condition $\mathbf{T} \cdot \mathbf{e}_r = 0$

mode spectra

at Earth's surface constrains the solutions of the radial wave equation for each mode to be a discrete set, which one can label by the number n of radial nodes that the mode possesses (just as for the hydrogen atom). The frequencies of the modes increase with both n and l. See Ex. 12.12 for details in a relatively simple case.

For small values of the quantum numbers l and n, the modes are quite sensitive to the model assumed for Earth's structure. For example, they are sensitive to whether one correctly includes the gravitational restoring force in the wave equation. However, for large l and n, the spheroidal and toroidal modes become standing combinations of P, SV, SH, Rayleigh, and Love waves, and therefore they are rather insensitive to the effects of gravity.

12.4.5 Seismic Tomography

Observations of all of these types of seismic waves clearly code much information about Earth's structure. Inverting the measurements to infer this structure has become a highly sophisticated and numerically intensive branch of geophysics—and also of oil exploration! The travel times of the P and S body waves can be measured between various pairs of points over Earth's surface and essentially allow C_L and C_T (and hence K/ρ and μ/ρ) to be determined as functions of radius inside Earth. Travel times are $\lesssim 1$ hour. Using this type of analysis, seismologists can infer the presence of hot and cold regions in the mantle and then infer how the rocks are circulating under the crust.

It is also possible to combine the observed travel times with Earth's equation of elastostatic equilibrium

$$\frac{dP}{dr} = -g(r)\rho, \quad g(r) = \frac{4\pi G}{r^2} \int_0^r r'^2 \rho(r')dr', \tag{12.66}$$

elastostatic equilibrium equation for gravitating, elastic sphere (e.g., Earth)

where $g(r)$ is the gravitational acceleration, to determine the distributions of density, pressure, and elastic constants. Measurements of Rayleigh and Love waves can be used to probe the surface layers. The results of this procedure are then input to obtain free oscillation frequencies (global mode frequencies), which compare well with the observations. The damping rates for the free oscillations furnish information on the interior viscosity.

12.4.6 Ultrasound; Shock Waves in Solids

Sound waves at frequencies above 20,000 Hz (where human hearing ends) are widely used in modern technology, especially for imaging and tomography. This is much like exploring Earth with seismic waves.

Just as seismic waves can travel from Earth's solid mantle through its liquid outer core and into its solid inner core, so these *ultrasonic waves* can travel through both solids and liquids, with reflection, refraction, and transmission similar to those in Earth.

Applications of ultrasound include, among others, medical imaging (e.g., of structures in the eye and of a human fetus during pregnancy), inspection of materials (e.g., of cracks, voids, and welding joints), acoustic microscopy at frequencies up to ~3 GHz (e.g., by scanning acoustic microscopes), ultrasonic cleaning (e.g., of jewelry), and ultrasonic welding (with sonic vibrations creating heat at the interface of the materials to be joined).

When an ultrasonic wave (or any other sound wave) in an elastic medium reaches a sufficiently high amplitude, its high-amplitude regions propagate faster than its low-amplitude regions. This causes it to develop a *shock front,* in which the compression increases almost discontinuously. This nonlinear mechanism for shock formation in a solid is similar to that in a fluid, where we shall study the mechanism in detail (Sec. 17.4.1). The theory of the elastodynamic shock itself, especially *jump conditions across the shock,* is also similar to that in a fluid (Sec. 17.5.1). For details of the theory in solids, including effects of plastic flow at sufficiently high compressions, see Davison (2010). Ultrasonic shock waves are used to break up kidney stones in the human body. In laboratories, shocks from the impact of a rapidly moving projectile on a material specimen, or from intense pulsed laser-induced ablation (e.g., at the U.S. National Ignition Facility, Sec. 10.2.2), are used to compress the specimen to pressures similar to those in Earth's core or higher and thereby explore the specimen's high-density properties. Strong shock waves from meteorite impacts alter the properties of the rock through which they travel and can be used to infer aspects of the equation of state for planetary interiors.

EXERCISES

Exercise 12.12 **Example: Normal Modes of a Homogeneous, Elastic Sphere*[3]
Show that, for frequency-ω oscillations of a homogeneous elastic sphere with negligible gravity, the displacement vector can everywhere be written as $\boldsymbol{\xi} = \boldsymbol{\nabla}\psi + \boldsymbol{\nabla} \times \mathbf{A}$, where ψ is a scalar potential that satisfies the longitudinal-wave Helmholtz equation, and $\mathbf{A}$ is a divergence-free vector potential that satisfies the transverse-wave Helmholtz equation:

$$\boldsymbol{\xi} = \boldsymbol{\nabla}\psi + \boldsymbol{\nabla} \times \mathbf{A}, \quad (\nabla^2 + k_L^2)\psi = 0,$$

$$(\nabla^2 + k_T^2)\mathbf{A} = 0, \quad \boldsymbol{\nabla} \cdot \mathbf{A} = 0; \quad k_L = \frac{\omega}{C_L}, \quad k_T = \frac{\omega}{C_T}. \tag{12.67}$$

[Hint: See Ex. 12.2, and make use of gauge freedom in the vector potential.]

(a) *Radial Modes.* The radial modes are purely longitudinal; their scalar potential [general solution of $(\nabla^2 + k_L^2)\psi = 0$ that is regular at the origin] is the spherical Bessel function of order zero: $\psi = j_0(k_L r) = \sin(k_L r)/(k_L r)$. The corresponding displacement vector $\boldsymbol{\xi} = \boldsymbol{\nabla}\psi$ has as its only nonzero component

$$\xi_r = j_0'(k_L r), \tag{12.68}$$

where the prime on j_0 and on any other function in this exercise means the derivative with respect to its argument. (Here we have dropped a multiplicative factor k_L.) Explain why the only boundary condition that need be imposed is $T_{rr} = 0$ at the sphere's surface. By computing T_{rr} for the displacement vector of

---

3. For a detailed and extensive treatment of this problem and many references, see Eringen and Suhubi (1975, Secs. 8.13, 8.14). Our approach to the mathematics is patterned after that of Ashby and Dreitlein (1975, Sec. III), and our numerical evaluations are for the same cases as Love (1927, Secs. 195, 196) and as the classic paper on this topic, Lamb (1882). It is interesting to compare the mathematics of our analysis with the nineteenth-century mathematics of Lamb and Love, which uses radial functions $\psi_l(r) \sim r^{-l} j_l(r)$ and *solid harmonics* that are sums over m of $r^l Y_l^m(\theta, \phi)$ and satisfy Laplace's equation. Here j_l is the spherical Bessel function of order l, and Y_l^m is the spherical harmonic. Solid harmonics can be written as $\mathcal{F}_{ij...q} x_i x_j \ldots x_q$, with $\mathcal{F}$ a symmetric, trace-free tensor of order l. A variant of them is widely used today in multipolar expansions of gravitational radiation (see, e.g., Thorne, 1980, Sec. II).

Eq. (12.68) and setting it to zero, deduce the following eigenequation for the wave numbers k_L and thence frequencies $\omega = C_L k_L$ of the radial modes:

$$\frac{\tan x_L}{x_L} = \frac{1}{1 - (x_T/2)^2},$$
(12.69)

where

$$x_L \equiv k_L R = \frac{\omega R}{C_L} = \omega R \sqrt{\frac{\rho}{K + 4\mu/3}}, \quad x_T \equiv k_T R = \frac{\omega R}{C_T} = \omega R \sqrt{\frac{\rho}{\mu}},$$
(12.70)

with R the radius of the sphere. For $(x_T/x_L)^2 = 3$, which corresponds to a Poisson's ratio $\nu = 1/4$ (about that of glass and typical rock; see Table 11.1), numerical solution of this eigenequation gives the spectrum for modes $\{l = 0;$ $n = 0, 1, 2, 3, \ldots\}$:

$$\frac{x_L}{\pi} = \frac{\omega}{\pi C_L/R} = \{0.8160, 1.9285, 2.9539, 3.9658, \ldots\}.$$
(12.71)

Note that these eigenvalues get closer and closer to integers as one ascends the spectrum; this also will be true for any other physically reasonable value of x_T/x_L. Explain why.

(b) **T2** *Torsional Modes.* The general solution to the scalar Helmholtz equation $(\nabla^2 + k^2)\psi$ that is regular at the origin is a sum over eigenfunctions of the form $j_l(kr)Y_l^m(\theta, \phi)$. Show that the angular momentum operator $\hat{\mathbf{L}} = \mathbf{x} \times \nabla$ commutes with the laplacian ∇^2, and thence infer that $\hat{\mathbf{L}}[j_l Y_l^m]$ satisfies the vector Helmholtz equation. Verify further that it is divergence free, which means it must be expressible as the curl of a vector potential that is also divergence free and satisfies the vector Helmholtz equation. This means that $\boldsymbol{\xi} = \hat{\mathbf{L}}[j_l(k_T r)Y_l^m(\theta, \phi)]$ is a displacement vector that satisfies the elastodynamic equation for transverse waves. Since $\hat{\mathbf{L}}$ differentiates only transversely (in the θ and ϕ directions), we can rewrite this expression as $\boldsymbol{\xi} = j_l(k_T r)\hat{\mathbf{L}} Y_l^m(\theta, \phi)$. To simplify computing the eigenfrequencies (which are independent of m), specialize to $m = 0$, and show that the only nonzero component of the displacement is

$$\xi_\phi \propto j_l(k_T r) \sin \theta P_l'(\theta).$$
(12.72)

Equation (12.72), for our homogeneous sphere, is the torsional mode discussed in the text [Eq. (12.64)]. Show that the only boundary condition that must be imposed on this displacement function is $T_{\phi r} = 0$ at the sphere's surface, $r = R$. Compute this component of the stress tensor (with the aid of Box 11.4 for the shear), and by setting it to zero, derive the following eigenequation for the torsional-mode frequencies:

$$x_T \, j_l'(x_T) - j_l(x_T) = 0.$$
(12.73)

For $l = 1$ (the case illustrated in Fig. 12.10b), this eigenequation reduces to $(3 - x_T^2) \tan x_T = 3x_T$, and the lowest few eigenfrequencies are

$$\frac{x_T}{\pi} = \frac{\omega}{\pi C_T/R} = \{1.8346, 2.8950, 3.9225, 4.9385, \ldots\}. \tag{12.74}$$

As for the radial modes, these eigenvalues get closer and closer to integers as one ascends the spectrum. Explain why.

(c) ◻T2 *Ellipsoidal Modes.* The displacement vector for the ellipsoidal modes is the sum of a longitudinal piece $\nabla \psi$ [with $\psi = j_l(k_L r) Y_l^m(\theta, \phi)$ satisfying the longitudinal wave Helmholtz equation] and a transverse piece $\nabla \times \mathbf{A}$ [with $\mathbf{A} = j_l(k_T r) \hat{\mathbf{L}} Y_l^m(\theta, \phi)$, which as we saw in part (b) satisfies the transverse wave Helmholtz equation and is divergence free]. Specializing to $m = 0$ to derive the eigenequation (which is independent of m), show that the components of the displacement vector are

$$\xi_r = \left[\frac{\alpha}{k_L} j_l'(k_L r) + \frac{\beta}{k_T} l(l+1) \frac{j_l(k_T r)}{k_T r} \right] P_l(\cos \theta),$$

$$\xi_\theta = \left[-\frac{\alpha}{k_L} \frac{j_l(k_L r)}{k_L r} + \frac{\beta}{k_T} \left(j'(k_T r) + \frac{j(k_T r)}{k_T r} \right) \right] \sin \theta \, P_l'(\cos \theta), \tag{12.75}$$

where α/k_L and β/k_T are constants that determine the weightings of the longitudinal and transverse pieces, and we have included the k_L and k_T to simplify the stress tensor derived below. (To get the β term in ξ_r, you will need to use a differential equation satisfied by P_l.) Show that the boundary conditions we must impose on these eigenfunctions are $T_{rr} = 0$ and $T_{r\theta} = 0$ at the sphere's surface, $r = R$. By evaluating these [using the shear components in Box 11.4, the differential equation satisfied by $j_l(x)$, and $(K + \frac{4}{3}\mu)/\mu = (x_T/x_L)^2$], obtain the following expressions for these components of the stress tensor at the surface:

$$T_{rr} = -\mu P_l(\cos \theta) \left[\alpha \left\{ 2 j_l''(x_L) - [(x_T/x_L)^2 - 2] j_l(x_L) \right\} + \beta \, 2l(l+1) f_1(x_T) \right] = 0,$$

$$T_{r\theta} = \mu \sin \theta \, P_l'(\cos \theta) \left[\alpha \, 2 f_1(x_L) + \beta \left\{ j_l''(x_T) + [l(l+1) - 2] f_0(x_T) \right\} \right] = 0,$$

where $f_0(x) \equiv j_l(x)/x^2$ and $f_1(x) \equiv (j_l(x)/x)'$. $\tag{12.76}$

These simultaneous equations for the ratio α/β have a solution if and only if the determinant of their coefficients vanishes:

$$\left\{ 2 j_l''(x_L) - [(x_T/x_L)^2 - 2] j_l(x_L) \right\} \left\{ j_l''(x_T) + (l+2)(l-1) f_0(x_T) \right\}$$

$$- 4l(l+1) f_1(x_L) f_1(x_T) = 0. \tag{12.77}$$

This is the eigenequation for the ellipsoidal modes. For $l = 2$ and $(x_T/x_L)^2 = 3$, it predicts for the lowest four ellipsoidal eigenfrequencies

$$\frac{x_T}{\pi} = \frac{\omega}{\pi C_T/R} = \{0.8403, 1.5487, 2.6513, 3.1131, \ldots\}. \tag{12.78}$$

Notice that these are significantly smaller than the torsional frequencies. Show that, in the limit of an incompressible sphere ($K \to \infty$) and for $l = 2$, the eigenequation becomes $(4 - x_T^2)[j_2''(x_T) + 4f_0(x_T)] - 24f_1(x_T) = 0$, which predicts for the lowest four eigenfrequencies

$$\frac{x_T}{\pi} = \frac{\omega}{\pi C_T / R} = \{0.8485, 1.7421, 2.8257, 3.8709, \ldots\}. \tag{12.79}$$

These are modestly larger than the compressible case, Eq. (12.78).

12.5 The Relationship of Classical Waves to Quantum Mechanical Excitations [T2]

In the previous chapter, we identified the effects of atomic structure on the continuum approximation for elastic solids. Specifically, we showed that atomic structure accounts for the magnitude of the elastic moduli and explains why most solids yield under comparatively small strain. A quite different connection of the continuum theory to atomic structure is provided by the normal modes of vibration of a finite solid body (e.g., the sphere treated in Sec. 12.4.4 and Ex. 12.12).

For any such body, one can solve the vector wave equation (12.4b) [subject to the vanishing-surface-force boundary condition $\mathbf{T} \cdot \mathbf{n} = 0$, Eq. (11.33)] to find the body's normal modes, as we did in Ex. 12.12 for the sphere. In this section, we label the normal modes by a single index N (encompassing $\{l, m, n\}$ in the case of a sphere) and denote the eigenfrequency of mode N by ω_N and its (typically complex) eigenfunction by $\boldsymbol{\xi}_N$. Then any general, small-amplitude disturbance in the body can be decomposed into a linear superposition of these normal modes:

normal modes of a general elastic body

$$\boxed{\boldsymbol{\xi}(\mathbf{x}, t) = \Re \sum_N a_N(t) \boldsymbol{\xi}_N(\mathbf{x}), \qquad a_N = A_N \exp(-i\omega_N t).} \tag{12.80}$$

displacement expanded in normal modes

Here $\Re$ means to take the real part, a_N is the *complex generalized coordinate* of mode N, and A_N is its complex amplitude. It is convenient to normalize the eigenfunctions so that

$$\int \rho |\boldsymbol{\xi}_N|^2 dV = M, \tag{12.81}$$

where M is the mass of the body; A_N then measures the mean physical displacement in mode N.

Classical electromagnetic waves in a vacuum are described by linear Maxwell equations; so after they have been excited, they will essentially propagate forever. This is not so for elastic waves, where the linear wave equation is only an approximation. Nonlinearities—and most especially impurities and defects in the structure of the body's material—will cause the different modes to interact weakly and also damp, so that their complex amplitudes A_N change slowly with time according to a damped simple harmonic oscillator differential equation of the form

$$\ddot{a}_N + (2/\tau_N)\dot{a}_N + \omega_N^2 a_N = F'_N/M. \tag{12.82}$$

Here the second term on the left-hand side is a damping term (due to frictional heating and weak coupling with other modes) that will cause the mode to decay as long as $\tau_N > 0$; F'_N is a fluctuating or *stochastic* force on mode N (also caused by weak coupling to the other modes). Equation (12.82) is the Langevin equation that we studied in Sec. 6.8.1. The spectral density of the fluctuating force F'_N is proportional to $1/\tau_N$ and is determined by the fluctuation-dissipation theorem, Eqs. (6.74) or (6.86). If the modes are thermalized at temperature T, then the fluctuating forces maintain an average energy of kT in each mode.

What happens quantum mechanically? The ions and electrons in an elastic solid interact so strongly that it is difficult to analyze them directly. A quantum mechanical treatment is much easier if one makes a canonical transformation from the coordinates and momenta of the individual ions or atoms to new, generalized coordinates $\hat{x}_N$ and momenta $\hat{p}_N$ that represent weakly interacting normal modes. These coordinates and momenta are Hermitian operators, and they are related to the quantum mechanical complex generalized coordinate $\hat{a}_N$ by

$$\hat{x}_N = \frac{1}{2}(\hat{a}_N + \hat{a}_N^\dagger), \tag{12.83a}$$

$$\hat{p}_N = \frac{M\omega_N}{2i}(\hat{a}_N - \hat{a}_N^\dagger), \tag{12.83b}$$

where the dagger denotes the Hermitian adjoint. We can transform back to obtain an expression for the displacement of the ith ion:

$$\hat{\mathbf{x}}_i = \frac{1}{2}\Sigma_N[\hat{a}_N\xi_N(\mathbf{x}_i) + \hat{a}_N^\dagger\xi_N^*(\mathbf{x}_i)] \tag{12.84}$$

[a quantum version of Eq. (12.80)].

The hamiltonian can be written in terms of these coordinates as

$$\hat{H} = \Sigma_N\left(\frac{\hat{p}_N^2}{2M} + \frac{1}{2}M\omega_N^2\hat{x}_N^2\right) + \hat{H}_{\text{int}}, \tag{12.85}$$

where the first term is a sum of simple harmonic oscillator hamiltonians for individual modes; and $\hat{H}_{\text{int}}$ is the perturbative interaction hamiltonian, which takes the place of the combined damping and stochastic forcing terms in the classical Langevin equation (12.82). When the various modes are thermalized, the mean energy in mode N takes on the standard Bose-Einstein form:

$$\bar{E}_N = \hbar\omega_N \left[\frac{1}{2} + \frac{1}{\exp[\hbar\omega_N/(k_B T)] - 1}\right]$$

(12.86)

thermalized normal modes

[Eq. (4.28b) with vanishing chemical potential and augmented by a "zero-point energy" of $\frac{1}{2}\hbar\omega$], which reduces to $k_B T$ in the classical limit $\hbar \to 0$.

As the unperturbed hamiltonian for each mode is identical to that for a particle in a harmonic oscillator potential well, it is sensible to think of each wave mode as analogous to such a particle-in-well. Just as the particle-in-well can reside in any one of a series of discrete energy levels lying above the zero-point energy of $\hbar\omega/2$ and separated by $\hbar\omega$, so each wave mode with frequency ω_N must have an energy $(n + 1/2)\hbar\omega_N$, where n is an integer. The operator that causes the energy of the mode to decrease by $\hbar\omega_N$ is the *annihilation operator* for mode n:

$$\hat{\alpha}_N = \left(\frac{M\omega_N}{2\hbar}\right)^{1/2} \hat{a}_N,$$

(12.87)

creation and annihilation operators for normal-mode quanta (phonons)

the operator that causes an increase in the energy by $\hbar\omega_N$ is its Hermitian conjugate, the *creation operator* $\hat{\alpha}_N^\dagger$, and their commutator is $[\hat{\alpha}_N, \hat{\alpha}_N^\dagger] = 1$, corresponding to $[\hat{x}_N, \hat{p}_N] = i\hbar$; see for example Chap. 5 of Cohen-Tannoudji, Diu, and Laloë (1977). It is useful to think of each increase or decrease of a mode's energy as the creation or annihilation of an individual quantum or "particle" of energy, so that when the energy in mode N is $(n + 1/2)\hbar\omega_N$, there are n quanta (particles) present. These particles are called *phonons*. Because phonons can coexist in the same state (the same mode), they are bosons. They have individual energies and momenta which must be conserved in their interactions with one another and with other types of particles (e.g., electrons). This conservation law shows up classically as resonance conditions in mode-mode mixing (cf. the discussion in nonlinear optics, Sec. 10.6.1).

phonons are bosons

The important question is, given an elastic solid at finite temperature, do we think of its thermal agitation as a superposition of classical modes, or do we regard it as a gas of quanta? The answer depends on what we want to do. From a purely fundamental viewpoint, the quantum mechanical description takes precedence. However, for many problems where the number of phonons per mode (the mode's mean occupation number) $\eta_N \sim k_B T/(\hbar\omega_N)$ is large compared to one, the classical description is amply adequate and much easier to handle. We do not need a quantum treatment when computing the normal modes of a vibrating building excited by an earthquake or when trying to understand how to improve the sound quality of a violin. Here the difficulty often is in accommodating the boundary conditions so as to determine the normal modes. All this was expected. What comes as more of a surprise is that often, for purely classical problems (where $\hbar$ is quantitatively irrelevant), the fastest way to

relation of classical and quantum theories for normal modes

analyze a practical problem formally is to follow the quantum route and then take the limit $\hbar \rightarrow 0$. We shall see this graphically demonstrated when we discuss nonlinear plasma physics in Chap. 23.

Bibliographic Note

The classic textbook treatments of elastodynamics from a physicist's viewpoint are Landau and Lifshitz (1986) and—in nineteenth-century language—Love (1927). For a lovely and very readable introduction to the basic concepts, with a focus on elasto-dynamic waves, see Kolsky (1963). Our favorite advanced textbook and treatise is Eringen and Suhubi (1975).

By contrast with elastostatics, where there are a number of good, twenty-first-century engineering-oriented textbooks at the elementary and intermediate levels, there are none that we know of for elastodynamics.

However, we do know two good, advanced engineering-oriented books on methods to solve the elastodynamic equations in nontrivial situations: Poruchikov, Khokhryakov, and Groshev (1993) and Kausel (2006). And for a compendium of practical engineering lore about vibrations of engineering structures, from building foundations to bell towers to suspension bridges, see Bachman (1994).

For seismic waves and their geophysical applications, we recommend the textbooks by Stein and Wysession (2003) and by Shearer (2009).

PART V

FLUID DYNAMICS

Having studied elasticity theory, we now turn to a second branch of continuum mechanics: *fluid dynamics*. Three of the four states of matter (gases, liquids, and plasmas) can be regarded as fluids, so it is not surprising that interesting fluid phenomena surround us in our everyday lives. Fluid dynamics is an experimental discipline; much of our current understanding has come in response to laboratory investigations. Fluid dynamics finds experimental application in engineering, physics, biophysics, chemistry, and many other fields. The observational sciences of oceanography, meteorology, astrophysics, and geophysics, in which experiments are less frequently performed, also rely heavily on fluid dynamics. Many of these fields have enhanced our appreciation of fluid dynamics by presenting flows under conditions that are inaccessible to laboratory study.

Despite this rich diversity, the fundamental principles are common to all these applications. The key assumption that underlies the equations governing the motion of a fluid is that the length- and timescales associated with the flow are long compared with the corresponding microscopic scales, so the continuum approximation can be invoked.

The fundamental equations of fluid dynamics are, in some respects, simpler than the corresponding laws of elastodynamics. However, as with particle dynamics, simplicity of equations does not imply the solutions are simple. Indeed, they are not! One reason is that fluid displacements are usually not small (by contrast with elastodynamics, where the elastic limit keeps them small), so most fluid phenomena are immediately nonlinear.

Relatively few problems in fluid dynamics admit complete, closed-form, analytic solutions, so progress in describing fluid flows has usually come from introducing clever physical models and using judicious mathematical approximations. Semi-empirical scaling laws are also common, especially for engineering applications. In more recent years, numerical fluid dynamics has come of age, and in many areas of fluid dynamics, computer simulations are complementing and even supplanting laboratory experiments and measurements. For example, most design work for airplanes and automobiles is now computational.

In fluid dynamics, considerable insight accrues from visualizing the flow. This is true of fluid experiments, where much technical skill is devoted to marking the fluid so it can be imaged; it is also true of numerical simulations, where frequently more time is devoted to computer graphics than to solving the underlying partial differential equations. Indeed, obtaining an analytic solution to the equations of fluid dynamics is not the same as understanding the flow; as a tool for understanding, at the very least it is usually a good idea to sketch the flow pattern.

We present the fundamental concepts of fluid dynamics in Chap. 13, focusing particularly on the underlying physical principles and the conservation laws for mass, momentum, and energy. We explain why, when flow velocities are very subsonic, a fluid's density changes very little (i.e., it is effectively incompressible), and we specialize the fundamental principles and equations to incompressible flows.

Vorticity plays major roles in fluid dynamics. In Chap. 14, we focus on those roles for incompressible flows, both in the fundamental equations of fluid dynamics and in applications. Our applications include, among others, tornados and whirlpools, boundary layers abutting solid bodies, the influence of boundary layers on bulk flows, and how wind drives ocean waves and is ultimately responsible for deep-ocean currents.

Viscosity has a remarkably strong influence on fluid flows, even when the viscosity is weak. When strong, it keeps a flow laminar (smooth); when weak, it controls details of the turbulence that pervades the bulk flow (the flow away from boundary layers). In Chap. 15, we describe turbulence, a phenomenon so difficult to handle theoretically that semiquantitative ideas and techniques pervade its theoretical description, even in the incompressible approximation (to which we adhere). The onset of turbulence is especially intriguing. We illuminate it by exploring a closely related phenomenon: chaotic behavior in mathematical maps.

In Chap. 16, we focus on waves in fluids, beginning with waves on the surface of water, where we shall see, for shallow water, how nonlinear effects and dispersion together give rise to "solitary waves" (solitons) that hold themselves together as they propagate. In this chapter, we abandon the incompressible approximation, which has permeated Part V thus far, to study sound waves. Radiation reaction in sound generation is much simpler than in, for example, electrodynamics, so we use sound waves to elucidate the physical origin of radiation reaction and the nonsensical nature of pre-acceleration.

In Chap. 17, we turn to transonic and supersonic flows, in which density changes are of major importance. Here we meet some beautiful and powerful mathematical tools: characteristics and their associated Riemann invariants. We focus especially on flow through rocket nozzles and other constrictions, and on shock fronts, with applications to explosions (bombs and supernovae).

Convection is another phenomenon in which density changes are crucial—though here the density changes are induced by thermal expansion rather than by physical compression. We study convection in Chap. 18, paying attention to the (usually small but sometimes large) influence of diffusive heat conduction and the diffusion of chemical constituents (e.g., salt).

When a fluid is electrically conducting and has an embedded magnetic field, the exchange of momentum between the field and the fluid can produce remarkable phenomena (e.g., dynamos that amplify a seed magnetic field, a multitude of instabilities, and Alfvén waves and other magnetosonic waves). This is the realm of magnetohydrodynamics, which we explore in Chap. 19. The most common application of magnetohydrodynamics is to a highly ionized plasma, the topic of Part VI of this book, so Chap. 19 serves as a transition from fluid dynamics (Part V) to plasma physics (Part VI).

13

Foundations of Fluid Dynamics

$\epsilon \H{\upsilon}\rho\eta\kappa\alpha$

ARCHIMEDES (CA. 250 BC)

13.1 Overview

13.1

In this chapter, we develop the fundamental concepts and equations of fluid dynamics, first in the flat-space venue of Newtonian physics (Track One) and then in the Minkowski spacetime venue of special relativity (Track Two). Our relativistic treatment is rather brief. This chapter contains a large amount of terminology that may be unfamiliar to readers. A glossary of terminology is given in Box 13.5, near the end of the chapter.

We begin in Sec. 13.2 with a discussion of the physical nature of a fluid: the possibility of describing it by a piecewise continuous density, velocity, and pressure and the relationship between density changes and pressure changes. Then in Sec. 13.3, we discuss hydrostatics (density and pressure distributions of a static fluid in a static gravitational field); this parallels our discussion of elastostatics in Chap. 11. After explaining the physical basis of Archimedes' law, we discuss stars, planets, Earth's atmosphere, and other applications.

Our foundation for moving from hydrostatics to hydrodynamics will be conservation laws for mass, momentum, and energy. To facilitate that transition, in Sec. 13.4, we examine in some depth the physical and mathematical origins of these conservation laws in Newtonian physics.

The stress tensor associated with most fluids can be decomposed into an isotropic pressure and a viscous term linear in the rate of shear (i.e., in the velocity gradient). Under many conditions the viscous stress can be neglected over most of the flow, and diffusive heat conductivity is negligible. The fluid is then called *ideal* or *perfect*.[1]
We study the laws governing ideal fluids in Sec. 13.5. After deriving the relevant

ideal fluid (perfect fluid)

---

1. An ideal fluid (also called a *perfect fluid*) should not be confused with an ideal or perfect gas—one whose pressure is due solely to kinetic motions of particles and thus is given by $P = nk_BT$, with n the particle number density, k_B Boltzmann's constant, and T temperature, and that may (ideal gas) or may not (perfect gas) have excited internal molecular degrees of freedom; see Box 13.2.

675

conservation laws and equation of motion (the Euler equation), we derive and discuss Bernoulli's theorem for an ideal fluid and explain how it can simplify the description of many flows. In flows for which the fluid velocities are much smaller than the speed of sound and gravity is too weak to compress the fluid much, the fractional changes in fluid density are small. It can then be a good approximation to treat the fluid as *incompressible,* which leads to considerable simplification, also addressed in Sec. 13.5. As we shall see in Sec. 13.6, incompressibility can be a good approximation not just for liquids (which tend to have large bulk moduli) but also, more surprisingly, for gases. It is so widely applicable that we restrict ourselves to the incompressible approximation throughout Chaps. 14 and 15.

In Sec. 13.7, we augment our basic equations with terms describing viscous stresses and also heat conduction. This allows us to derive the famous Navier-Stokes equation and illustrate its use by analyzing the flow of a fluid through a pipe, and then use this to make a crude model of blood flowing through an artery. Much of our study of fluids in future chapters will focus on this Navier-Stokes equation.

In our study of fluids we often deal with the influence of a uniform gravitational field (e.g., Earth's, on lengthscales small compared to Earth's radius). Occasionally, however, we consider inhomogeneous gravitational fields produced by the fluid whose motion we study. For such situations it is useful to introduce gravitational contributions to the stress tensor and energy density and flux. We present and discuss these in Box 13.4, where they will not impede the flow of the main stream of ideas.

We conclude this chapter in Sec. 13.8 with a brief, Track-Two overview of relativistic fluid mechanics for a perfect (ideal) fluid. As an important application, we explore the structure of a relativistic astrophysical jet: the conversion of internal thermal energy into the energy of organized bulk flow as the jet travels outward from the nucleus of a galaxy into intergalactic space and widens. We also explore how the fundamental equations of Newtonian fluid mechanics arise as low-velocity limits of the relativistic equations.

13.2 The Macroscopic Nature of a Fluid: Density, Pressure, Flow Velocity; Liquids versus Gases

The macroscopic nature of a fluid follows from two simple observations. The first is that in most flows the macroscopic continuum approximation is valid. Because the molecular mean free paths in a fluid are small compared to macroscopic lengthscales, we can define a mean local velocity $\mathbf{v}(\mathbf{x}, t)$ of the fluid's molecules, which varies smoothly both spatially and temporally; we call this the fluid's velocity. For the same reason, other quantities that characterize the fluid [e.g., the density $\rho(\mathbf{x}, t)$] also vary smoothly on macroscopic scales. This need not be the case everywhere in the flow. An important exception is a shock front, which we study in Chap. 17; there the flow varies rapidly, over a length of order the molecules' mean free path for collisions. In this case, the continuum approximation is only piecewise valid, and we must perform a matching at the shock front. One might think that a second exception is a turbulent flow, where it seems plausible that the average molecular velocity will vary rapidly on whatever lengthscale we choose to study, all the way down to intermolecular distances, so averaging becomes problematic. As we shall see in Chap. 15, this is not the case; turbulent flows generally have a lengthscale far larger than intermolecular distances, below which the flow varies smoothly.

The second observation is that fluids do not oppose a steady shear strain. This is easy to understand on microscopic grounds, as there is no lattice to deform, and the molecular velocity distribution remains locally isotropic in the presence of a static shear. By kinetic theory considerations (Chap. 3), we therefore expect that a fluid's stress tensor $\mathbf{T}$ will be isotropic in the local rest frame of the fluid (i.e., in a frame where $\mathbf{v} = 0$). Because of viscosity, this is not quite true when the shear varies with time. However, we neglect viscosity as well as diffusive heat flow until Sec. 13.7 (i.e., we restrict ourselves to ideal fluids). This assumption allows us to write $\mathbf{T} = P\mathbf{g}$ in the local rest frame, where P is the fluid's pressure, and $\mathbf{g}$ is the metric (with Kronecker delta components, $g_{ij} = \delta_{ij}$, in Cartesian coordinates).

fluid

The laws of fluid mechanics as we develop them are equally valid for liquids, gases, and (under many circumstances) plasmas. In a liquid, as in a solid, the molecules are packed side by side (but can slide over one another easily). In a gas or plasma, the molecules are separated by distances large compared to their sizes. This difference leads to different behaviors under compression.

For a liquid (e.g., water in a lake), the molecules resist strongly even a small compression; as a result, it is useful to characterize the pressure increase by a bulk modulus K, as in an elastic solid (Chap. 11):

pressure changes in a liquid

$$\delta P = -K\Theta = K\frac{\delta\rho}{\rho} \quad \text{for a liquid.} \tag{13.1}$$

(Here we have used the fact that the expansion Θ is the fractional increase in volume, or equivalently by mass conservation, it is the fractional decrease in density.) The bulk modulus for water is about 2.2 GPa, so as one goes downward in a lake far enough to double the pressure from one atmosphere (10^5 Pa to 2×10^5 Pa), the fractional change in density is only $\delta\rho/\rho = (2 \times 10^5/2.2 \times 10^9) \simeq 1$ part in 10,000.

By contrast, gases and plasmas are much less resistant to compression. Due to the large distance between molecules, a doubling of the pressure requires, in order of magnitude, a doubling of the density:

pressure changes in a gas or plasma

$$\frac{\delta P}{P} = \Gamma\frac{\delta\rho}{\rho} \quad \text{for a gas,} \tag{13.2}$$

where Γ is a proportionality factor of order unity. The numerical value of Γ depends on the physical situation. If the gas is ideal [so $P = \rho k_B T/(\mu m_p)$ in the notation of Box 13.2, Eq. (4)] and the temperature T is being held fixed by thermal contact with some heat source as the density changes (*isothermal process*), then $\delta P \propto \delta\rho$, and $\Gamma = 1$. Alternatively, and much more commonly, a fluid element's entropy may remain constant, because no significant heat can flow in or out of it during the density change. In this case Γ is called the *adiabatic index,* and (continuing to assume ideality), it can be shown using the laws of thermodynamics that

adiabatic index

$$\Gamma = \gamma \equiv c_P/c_V \quad \text{for adiabatic processes in an ideal gas.} \tag{13.3}$$

specific heats per unit mass

Here c_P and c_V are the specific heats at constant pressure and volume; see Ex. 5.4 in Chap. 5.[2]

---

2. In fluid dynamics, our specific heats, and other extensive variables, such as energy, entropy, and enthalpy, are defined per unit mass and are denoted by lowercased letters. So $c_P = T(\partial s/\partial T)_P$ is the amount of heat that must be added to a unit mass of the fluid to increase its temperature by one unit, and similarly for $c_V = T(\partial s/\partial T)_\rho$. By contrast, in statistical thermodynamics (Chap. 5) our extensive variables are defined for some chosen sample of material and are denoted by capital letters, e.g., $C_P = T(\partial S/\partial T)_P$.

BOX 13.2. THERMODYNAMIC CONSIDERATIONS

One feature of fluid dynamics (especially gas dynamics) that distinguishes it from elastodynamics is that the thermodynamic properties of the fluid are often important, so we must treat energy conservation explicitly. In this box we review, from Chap. 5 or any book on thermodynamics (e.g., Kittel and Kroemer, 1980), the thermodynamic concepts needed in our study of fluids. We have no need for partition functions, ensembles, and other statistical aspects of thermodynamics. Instead, we only need elementary thermodynamics.

We begin with the nonrelativistic first law of thermodynamics (5.8) for a sample of fluid with energy E, entropy S, volume V, number N_I of molecules of species I, temperature T, pressure P, and chemical potential μ_I for species I:

$$dE = TdS - PdV + \sum_I \mu_I dN_I. \tag{1}$$

Almost everywhere in our treatment of fluid dynamics (and throughout this chapter), we assume that the term $\sum_I \mu_I dN_I$ vanishes. Physically this holds because (i) all relevant nuclear reactions are frozen (occur on timescales τ_{react} far longer than the dynamical timescales τ_{dyn} of interest to us), so $dN_I = 0$, and (ii) each chemical reaction is either frozen $dN_I = 0$ or goes so rapidly ($\tau_{\text{react}} \ll \tau_{\text{dyn}}$) that it and its inverse are in local thermodynamic equilibrium: $\sum_I \mu_I dN_I = 0$ for those species involved in the reactions. In the rare intermediate situation, where some relevant reaction has $\tau_{\text{react}} \sim \tau_{\text{dyn}}$, we would have to carefully keep track of the relative abundances of the chemical or nuclear species and their chemical potentials.

Consider a small fluid element with mass Δm, internal energy per unit mass u, entropy per unit mass s, and volume per unit mass $1/\rho$. Then inserting $E = u\Delta m$, $S = s\Delta m$, and $V = \Delta m/\rho$ into the first law $dE = TdS - PdV$, we obtain the form of the first law that we use in almost all of our fluid-dynamics studies:

$$\boxed{du = Tds - Pd\left(\frac{1}{\rho}\right).} \tag{2}$$

The internal energy (per unit mass) u comprises the random translational energy of the molecules that make up the fluid, together with the energy associated with their internal degrees of freedom (rotation, vibration, etc.) and with their intermolecular forces. The term Tds represents some amount of heat (per unit mass) that may be injected into a fluid element (e.g., by

(continued)

BOX 13.2. (continued)

viscous heating; Sec. 13.7) or may be removed (e.g., by radiative cooling). The term $-Pd(1/\rho)$ represents work done on the fluid.

In fluid mechanics it is useful to introduce the enthalpy $H = E + PV$ of a fluid element (cf. Ex. 5.5) and the corresponding enthalpy per unit mass $h = u + P/\rho$. Inserting $u = h - P/\rho$ into the left-hand side of the first law (2), we obtain the first law in the enthalpy representation [Eq. (5.47)]:

$$dh = T ds + \frac{dP}{\rho}. \tag{3}$$

Because we assume that all reactions are frozen or are in local thermodynamic equilibrium, the relative abundances of the various nuclear and chemical species are fully determined by a fluid element's density ρ and temperature T (or by any two other variables in the set ρ, T, s, and P). Correspondingly, the thermodynamic state of a fluid element is completely determined by any two of these variables. To calculate all features of that state from two variables, we must know the relevant equations of state, such as $P(\rho, T)$ and $s(\rho, T)$; $P(\rho, s)$ and $T(\rho, s)$; or the fluid's fundamental thermodynamic potential (Table 5.1), from which follow the equations of state.

We often deal with ideal gases (i.e., gases in which intermolecular forces and the volume occupied by the molecules are treated as totally negligible). For any ideal gas, the pressure arises solely from the kinetic motions of the molecules, and so the equation of state $P(\rho, T)$ is

$$P = \frac{\rho k_B T}{\mu m_p}. \tag{4}$$

Here μ is the *mean molecular weight*, and m_p is the proton mass [cf. Eqs. (3.39b), with the number density of particles n reexpressed as $\rho/\mu m_p$]. The mean molecular weight μ is the mean mass per gas molecule in units of the proton mass (e.g., $\mu = 1$ for hydrogen, $\mu = 32$ for O_2, and $\mu = 28.8$ for air). This μ should not be confused with the chemical potential of species I, μ_I (which will rarely if ever be used in our fluid dynamics analyses). [The concept of an ideal gas must not be confused with an ideal fluid; see footnote 1.]

An idealization that is often accurate in fluid dynamics is that the fluid is *adiabatic*: no heating or cooling from dissipative processes, such as viscosity,

(continued)

BOX 13.2. (continued)

thermal conductivity, or the emission and absorption of radiation. When this is a good approximation, the entropy per unit mass s of a fluid element is constant.

In an adiabatic flow, there is only one thermodynamic degree of freedom, so we can write $P = P(\rho, s) = P(\rho)$. Of course, this function will be different for fluid elements that have different s. In the case of an ideal gas, a standard thermodynamic argument (Ex. 5.4) shows that the pressure in an adiabatically expanding or contracting fluid element varies with density as $\delta P / P = \gamma \delta \rho / \rho$, where $\gamma = c_P / c_V$ is the adiabatic index [Eqs. (13.2) and (13.3)]. If, as is often the case, the adiabatic index remains constant over several doublings of the pressure and density, then we can integrate this expression to obtain the equation of state

$$P = K(s)\rho^\gamma, \tag{5}$$

where $K(s)$ is some function of the entropy. Equation (5) is sometimes called the *polytropic* equation of state, and a *polytropic index n* (not to be confused with number density of particles!) is defined by $\gamma = 1 + 1/n$. See the discussion of stars and planets in Sec. 13.3.2 and Ex. 13.4. A special case of adiabatic flow is *isentropic* flow. In this case, the entropy is constant everywhere, not just inside individual fluid elements.

When the pressure can be regarded as a function of the density alone (the same function everywhere), the fluid is called *barotropic*.

From Eqs. (13.1) and (13.2), we see that $K = \Gamma P$; so why use K for liquids and Γ for gases and plasmas? Because in a liquid K remains nearly constant when P changes by large fractional amounts $\delta P / P \gtrsim 1$, while in a gas or plasma it is Γ that remains nearly constant.

For other thermodynamic aspects of fluid dynamics that will be relevant as we proceed, see Box 13.2.

13.3 Hydrostatics

Just as we began our discussion of elasticity with a treatment of elastostatics, so we introduce fluid mechanics by discussing hydrostatic equilibrium.

The *equation of hydrostatic equilibrium* for a fluid at rest in a gravitational field $\mathbf{g}$ is the same as the equation of elastostatic equilibrium with a vanishing shear stress, so $\mathbf{T} = P\mathbf{g}$:

equation of hydrostatic equilibrium

$$\mathbf{\nabla} \cdot \mathbf{T} = \mathbf{\nabla} P = \rho \mathbf{g} \tag{13.4}$$

[Eq. (11.14) with $\mathbf{f} = -\nabla \cdot \mathbf{T}$]. Here $\mathbf{g}$ is the acceleration of gravity (which need not be constant; e.g., it varies from location to location inside the Sun). It is often useful to express $\mathbf{g}$ as the gradient of the Newtonian gravitational potential Φ:

Newtonian gravitational potential

$$\boxed{\mathbf{g} = -\nabla\Phi.}$$

(13.5)

Note our sign convention: Φ is negative near a gravitating body and zero far from all bodies, and it is determined by Newton's field equation for gravity:

Newtonian field equation for gravity

$$\boxed{\nabla^2\Phi = -\nabla \cdot \mathbf{g} = 4\pi G\rho.}$$

(13.6)

We can draw some immediate and important inferences from Eq. (13.4). Take the curl of Eq. (13.4) and use Eq. (13.5) to obtain

$$\nabla\Phi \times \nabla\rho = 0.$$

(13.7)

theorems about hydrostatic structure

Equation (13.7) tells us that in hydrostatic equilibrium, the contours of constant density (isochores) coincide with the equipotential surfaces: $\rho = \rho(\Phi)$. Equation (13.4) itself, with (13.5), then tells us that, as we move from point to point in the fluid, the changes in P and Φ are related by $dP/d\Phi = -\rho(\Phi)$. This, in turn, implies that the difference in pressure between two equipotential surfaces Φ_1 and Φ_2 is given by

$$\Delta P = -\int_{\Phi_1}^{\Phi_2} \rho(\Phi)d\Phi.$$

(13.8)

Moreover, as $\nabla P \propto \nabla\Phi$, the surfaces of constant pressure (the *isobars*) coincide with the gravitational equipotentials. This is all true whether $\mathbf{g}$ varies inside the fluid or is constant.

pressure is weight of overlying fluid

The gravitational acceleration $\mathbf{g}$ is actually constant to high accuracy in most non-astrophysical applications of fluid dynamics, for example, on the surface of Earth. In this case the pressure at a point in a fluid is, from Eq. (13.8), equal to the total weight of fluid per unit area above the point,

$$P(z) = g\int_z^{\infty} \rho dz,$$

(13.9)

where the integral is performed by integrating upward in the gravitational field (cf. Fig. 13.1). For example, the deepest point in the world's oceans is the bottom of the Mariana Trench in the Pacific, 11 km below sea level. Adopting a density $\rho \simeq 10^3 \, \mathrm{kg\,m^{-3}}$ for water and $g \simeq 10 \, \mathrm{m\,s^{-2}}$, we obtain a pressure of $\simeq 10^8$ Pa or $\simeq 10^3$ atmospheres. This is comparable with the yield stress of the strongest materials. It should therefore come as no surprise that the record for the deepest dive ever recorded by a submersible—a depth of 10.91 km (just $\sim$100 m shy of the lowest point in the trench) achieved by the *Trieste* in 1960—remained unbroken for more than half a century. Only in 2012 was that last 100 m conquered and the trench's bottom reached, by the filmmaker James Cameron in the *Deep Sea Challenger*. Since the bulk modulus

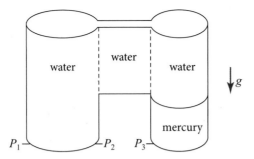

FIGURE 13.1 Elementary demonstration of the principles of hydrostatic equilibrium. Water and mercury, two immiscible fluids of different density, are introduced into a container with two connected chambers as shown. In each chamber, isobars (surfaces of constant pressure) coincide with surfaces of constant $\Phi = -gz$, and so are horizontal. The pressure at each point on the flat bottom of a container is equal to the weight per unit area of the overlying fluid [Eq. (13.9)]. The pressures P_1 and P_2 at the bottom of the left chamber are equal, but because of the density difference between mercury and water, they differ from the pressure P_3 at the bottom of the right chamber.

of sea water is $K = 2.3\,\mathrm{GPa}$, at the bottom of the trench the water is compressed by $\delta\rho/\rho = P/K \simeq 0.05$.

Exercise 13.1 *Example: Earth's Atmosphere*

As mountaineers know, it gets cooler as you climb. However, the rate at which the temperature falls with altitude depends on the thermal properties of air. Consider two limiting cases.

(a) In the lower stratosphere (Fig. 13.2), the air is isothermal. Use the equation of hydrostatic equilibrium (13.4) to show that the pressure decreases exponentially with height z:

$$P \propto \exp(-z/H), \tag{13.10a}$$

where the scale height H is given by

$$H = \frac{k_B T}{\mu m_p g}, \tag{13.10b}$$

with μ the mean molecular weight of air and m_p the proton mass. Evaluate this numerically for the lower stratosphere, and compare with the stratosphere's thickness. By how much does P drop between the bottom and top of the isothermal region?

(b) Suppose that the air is isentropic, so that $P \propto \rho^\gamma$ [Eq. (5) of Box 13.2], where γ is the specific heat ratio. (For diatomic gases like nitrogen and oxygen, $\gamma \sim 1.4$; see Ex. 17.1.) Show that the temperature gradient satisfies

$$\frac{dT}{dz} = -\left(\frac{\gamma - 1}{\gamma}\right)\frac{g\mu m_p}{k}. \tag{13.10c}$$

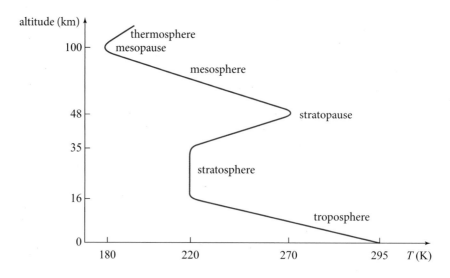

FIGURE 13.2 Actual temperature variation of Earth's mean atmosphere at temperate latitudes.

Note that the temperature gradient vanishes when $\gamma \rightarrow 1$. Evaluate the temperature gradient, also known at low altitudes as the *lapse rate*. The average lapse rate is measured to be ~ 6 K km^{-1} (Fig. 13.2). Show that this is intermediate between the two limiting cases of isentropic and isothermal lapse rates.

13.3.1

Archimedes' law

13.3.1 Archimedes' Law

The law of Archimedes states that, *when a solid body is totally or partially immersed in a fluid in a uniform gravitational field* $\mathbf{g} = -g\mathbf{e}_z$, *the total buoyant upward force of the fluid on the body is equal to the weight of the displaced fluid.*

Proof of Archimedes' Law. A formal proof can be made as follows (Fig. 13.3). The fluid, pressing inward on the body across a small element of the body's surface $d\mathbf{\Sigma}$, exerts a force $d\mathbf{F}^{\text{buoy}} = \mathbf{T}(\_\_, -d\mathbf{\Sigma})$ [Eq. (1.32)], where $\mathbf{T}$ is the fluid's stress tensor, and the minus sign is because by convention, $d\mathbf{\Sigma}$ points out of the body rather than into it. Converting to index notation and integrating over the body's surface ∂V, we obtain for the net buoyant force

$$F_i^{\text{buoy}} = -\int_{\partial V} T_{ij} d\Sigma_j. \tag{13.11a}$$

Now, imagine removing the body and replacing it by fluid that has the same pressure $P(z)$ and density $\rho(z)$, at each height z as the surrounding fluid; this is the fluid that was originally displaced by the body. Since the fluid stress on ∂V has not changed, the buoyant force will be unchanged. Use Gauss's law to convert the surface integral (13.11a) into a volume integral over the interior fluid (the originally displaced fluid):

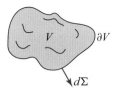

FIGURE 13.3 Derivation of Archimedes' law.

$$F_i^{\text{buoy}} = - \int_{\mathcal{V}} T_{ij;j} dV. \qquad (13.11b)$$

The displaced fluid obviously is in hydrostatic equilibrium with the surrounding fluid, and its equation of hydrostatic equilibrium $T_{ij;j} = \rho g_i$ [Eq. (13.4)], when inserted into Eq. (13.11b), implies that

$$\boxed{\mathbf{F}^{\text{buoy}} = -\mathbf{g} \int_{\mathcal{V}} \rho dV = -M\mathbf{g},} \qquad (13.12)$$

where M is the mass of the displaced fluid. Thus, the upward buoyant force on the original body is equal in magnitude to the weight Mg of the displaced fluid. Clearly, if the body has a higher density than the fluid, then the downward gravitational force on it (its weight) will exceed the weight of the displaced fluid and thus exceed the buoyant force it feels, and the body will fall. If the body's density is less than that of the fluid, the buoyant force will exceed its weight and it will be pushed upward.

∎

A key piece of physics underlying Archimedes' law is the fact that the inter-molecular forces acting in a fluid, like those in a solid (cf. Sec. 11.3.6), are short range. If, instead, the forces were long range, Archimedes' law could fail. For example, consider a fluid that is electrically conducting, with currents flowing through it that produce a magnetic field and resulting long-range magnetic forces (the magnetohy-drodynamic situation studied in Chap. 19). If we then substitute an insulating solid for some region $\mathcal{V}$ of the conducting fluid, the force that acts on the solid will be different from the force that acted on the displaced fluid.

short-range versus long-range forces

Exercise 13.2 *Practice: Weight in Vacuum*
How much more would you weigh in a vacuum?

Exercise 13.3 *Problem: Stability of Boats*
Use Archimedes' law to explain qualitatively the conditions under which a boat float-ing in still water will be stable to small rolling motions from side to side. [Hint: You might want to define and introduce a *center of buoyancy* and a *center of gravity* inside

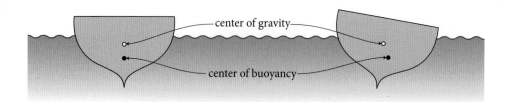

FIGURE 13.4 The center of gravity and center of buoyancy of a boat when it is upright (left) and tilted (right).

the boat, and pay attention to the change in the center of buoyancy when the boat tilts. See Fig. 13.4.]

13.3.2 Nonrotating Stars and Planets

Stars and massive planets—if we ignore their rotation—are self-gravitating fluid spheres. We can model the structure of such a nonrotating, spherical, self-gravitating fluid body by combining the equation of hydrostatic equilibrium (13.4) in spherical polar coordinates,

$$\frac{dP}{dr} = -\rho \frac{d\Phi}{dr}, \tag{13.13}$$

with Poisson's equation,

$$\nabla^2 \Phi = \frac{1}{r^2}\frac{d}{dr}\left(r^2 \frac{d\Phi}{dr}\right) = 4\pi G \rho, \tag{13.14}$$

to obtain

$$\frac{1}{r^2}\frac{d}{dr}\left(\frac{r^2}{\rho}\frac{dP}{dr}\right) = -4\pi G \rho. \tag{13.15}$$

Equation (13.15) can be integrated once radially with the aid of the boundary condition $dP/dr = 0$ at $r = 0$ (pressure cannot have a cusp-like singularity) to obtain

equations of structure for nonrotating star or planet

$$\boxed{\frac{dP}{dr} = -\rho \frac{Gm}{r^2},} \tag{13.16a}$$

where

$$\boxed{m = m(r) \equiv \int_0^r 4\pi \rho r^2 dr} \tag{13.16b}$$

is the total mass inside radius r. Equation (13.16a) is an alternative form of the equation of hydrostatic equilibrium (13.13) at radius r inside the body: Gm/r^2 is the gravitational acceleration g at r, $\rho(Gm/r^2) = \rho g$ is the downward gravitational force per unit volume on the fluid, and dP/dr is the upward buoyant force per unit volume.

Equations (13.13)–(13.16b) are a good approximation for solid planets (e.g., Earth) as well as for stars and fluid planets (e.g., Jupiter) because at the enormous stresses encountered in the interior of a solid planet, the strains are so large that plastic flow occurs. In other words, the shear stresses are much smaller than the isotropic part of the stress tensor.

Let us make an order-of-magnitude estimate of the interior pressure in a star or planet of mass M and radius R. We use the equation of hydrostatic equilibrium (13.4) or (13.16a), approximating m by M, the density ρ by M/R^3, and the gravitational acceleration by GM/R^2; the result is

$$P \sim \frac{GM^2}{R^4}. \tag{13.17}$$

To improve this estimate, we must solve Eq. (13.15). For that, we need a prescription relating the pressure to the density (i.e., an equation of state). A common idealization is the polytropic relation:

$$P \propto \rho^{1+1/n}, \tag{13.18}$$

polytropic equation of state

where n is the polytropic index (cf. last part of Box 13.2). [This finesses the issue of the generation and flow of heat in stellar interiors, which determines the temperature $T(r)$ and thence the pressure $P(\rho, T)$.] Low-mass white-dwarf stars are well approximated as $n = 1.5$ polytropes [Eq. (3.53c)], and red-giant stars are somewhat similar in structure to $n = 3$ polytropes. The giant planets, Jupiter and Saturn, are mainly composed of an H-He fluid that can be approximated by an $n = 1$ polytrope, and the density of a small planet like Mercury is roughly constant ($n = 0$).

To solve Eqs. (13.16), we also need boundary conditions. We can choose some density ρ_c and corresponding pressure $P_c = P(\rho_c)$ at the star's center $r = 0$, then integrate Eqs. (13.16) outward until the pressure P drops to zero, which will be the star's (or planet's) surface. The values of r and m there will be the star's radius R and mass M. For mathematical details of polytropic stellar models constructed in this manner, see Ex. 13.4. This exercise is particularly important as an example of the power of converting to dimensionless variables, a procedure we use frequently in this part of the book.

boundary conditions for stellar/planetary structure

We can easily solve the equations of hydrostatic equilibrium (13.16) for a planet with constant density ($n = 0$) to obtain $m = (4\pi/3)\rho r^3$ and

$$P = P_0 \left(1 - \frac{r^2}{R^2}\right), \tag{13.19}$$

where the central pressure is

$$P_0 = \left(\frac{3}{8\pi}\right) \frac{GM^2}{R^4}, \tag{13.20}$$

pressure in constant-density planet

consistent with our order-of-magnitude estimate (13.17).

Exercise 13.4 **Example: Polytropes—The Power of Dimensionless Variables*
When dealing with differential equations describing a physical system, it is often helpful to convert to dimensionless variables. *Polytropes* (nonrotating, spherical fluid bodies with the polytropic equation of state $P = K\rho^{1+1/n}$) are a nice example.

(a) Combine the two equations of stellar structure (13.16) to obtain a single second-order differential equation for P and ρ as functions of r.

(b) In the equation from part (a) set $P = K\rho^{1+1/n}$ to obtain a nonlinear, second-order differential equation for $\rho(r)$.

(c) It is helpful to change dependent variables from $\rho(r)$ to some other variable, call it $\theta(r)$, so chosen that the quantity being differentiated is linear in θ and the only θ nonlinearity is in the driving term. Show that choosing $\rho \propto \theta^n$ achieves this.

(d) It is helpful to choose the proportionality constant in $\rho \propto \theta^n$ in such a way that θ is dimensionless and takes the value 1 at the polytrope's center and 0 at its surface. This is achieved by setting

$$\rho = \rho_c \theta^n, \tag{13.21a}$$

where ρ_c is the polytrope's (possibly unknown) central density.

(e) Similarly, it is helpful to make the independent variable r dimensionless by setting $r = a\xi$, where a is a constant with dimensions of length. The value of a should be chosen wisely, so as to simplify the differential equation as much as possible. Show that the choice

$$r = a\xi, \quad \text{where } a = \left[\frac{(n+1)K\rho_c^{(1/n-1)}}{4\pi G}\right]^{1/2}, \tag{13.21b}$$

brings the differential equation into the form

$$\frac{1}{\xi^2}\frac{d}{d\xi}\xi^2\frac{d\theta}{d\xi} = -\theta^n. \tag{13.22}$$

Equation (13.22) is called the *Lane-Emden equation of stellar structure*, after Jonathan Homer Lane and Jacob Robert Emden, who introduced and explored it near the end of the nineteenth century. There is an extensive literature on solutions of the Lane-Emden equation (see, e.g., Chandrasekhar, 1939, Chap. 4; Shapiro and Teukolsky, 1983, Sec. 3.3).

(f) Explain why the Lane-Emden equation (13.22) must be solved subject to the following boundary conditions (where $\theta' \equiv d\theta/d\xi$):

$$\theta = 1 \quad \text{and} \quad \theta' = 0 \quad \text{at } \xi = 0. \tag{13.23}$$

(g) One can integrate the Lane-Emden equation, numerically or analytically, outward from $\xi = 0$ until some radius ξ_1 at which θ (and thus also ρ and P) goes to zero. That is the polytrope's surface. Its physical radius is then $R = a\xi_1$, and its mass is $M = \int_0^R 4\pi\rho r^2 dr$, which is readily shown to be $M = 4\pi a^3 \rho_c \xi_1^2 |\theta'(\xi_1)|$.

Then, using the value of a given in Eq. (13.21b), we have:

$$R = \left[\frac{(n+1)K}{4\pi G} \right]^{1/2} \rho_c^{(1-n)/2n} \xi_1,$$

$$M = 4\pi \left[\frac{(n+1)K}{4\pi G} \right]^{3/2} \rho_c^{(3-n)/2n} \xi_1^2 |\theta'(\xi_1)|, \qquad (13.24a)$$

whence

$$M = 4\pi R^{(3-n)/(1-n)} \left[\frac{(n+1)K}{4\pi G} \right]^{n/(n-1)} \xi_1^{(n+1)/(n-1)} |\theta'(\xi_1)|. \qquad (13.24b)$$

(h) When one converts a problem into dimensionless variables that satisfy some differential or algebraic equation(s) and then expresses physical quantities in terms of the dimensionless variables, the resulting expressions describe how the physical quantities scale with one another. As an example, Jupiter and Saturn are both made up of an H-He fluid that is well approximated by a polytrope of index $n = 1$, $P = K\rho^2$, with the same constant K. Use the information that $M_J = 2 \times 10^{27}$ kg, $R_J = 7 \times 10^4$ km, and $M_S = 6 \times 10^{26}$ kg to estimate the radius of Saturn. For $n = 1$, the Lane-Emden equation has a simple analytical solution: $\theta = \sin \xi / \xi$. Compute the central densities of Jupiter and Saturn.

13.3.3 Rotating Fluids

The equation of hydrostatic equilibrium (13.4) and the applications of it discussed above are valid only when the fluid is static in a nonrotating reference frame. However, they are readily extended to bodies that rotate rigidly with some uniform angular velocity $\boldsymbol{\Omega}$ relative to an inertial frame. In a frame that corotates with the body, the fluid will have vanishing velocity $\mathbf{v}$ (i.e., will be static), and the equation of hydrostatic equilibrium (13.4) will be changed only by the addition of the centrifugal force per unit volume:

$$\boxed{\nabla P = \rho(\mathbf{g} + \mathbf{g}_{\text{cen}}) = -\rho\nabla(\Phi + \Phi_{\text{cen}}).} \qquad (13.25)$$

hydrostatic equilibrium in corotating reference frame

Here

$$\boxed{\mathbf{g}_{\text{cen}} = -\boldsymbol{\Omega} \times (\boldsymbol{\Omega} \times \mathbf{r}) = -\nabla\Phi_{\text{cen}}} \qquad (13.26)$$

is the centrifugal acceleration, $\rho\mathbf{g}_{\text{cen}}$ is the centrifugal force per unit volume, and

$$\boxed{\Phi_{\text{cen}} = -\frac{1}{2}(\boldsymbol{\Omega} \times \mathbf{r})^2} \qquad (13.27)$$

is a *centrifugal potential* whose gradient is equal to the centrifugal acceleration when $\boldsymbol{\Omega}$ is constant. The centrifugal potential can be regarded as an augmentation of the gravitational potential Φ. Indeed, *in the presence of uniform rotation, all hydrostatic theorems [e.g., Eqs. (13.7) and (13.8)] remain valid in the corotating reference frame with Φ replaced by $\Phi + \Phi_{\text{cen}}$.*

hydrostatic theorems for rotating body

We can illustrate this by considering the shape of a spinning fluid planet. Let us suppose that almost all the planet's mass is concentrated in its core, so the gravitational potential $\Phi = -GM/r$ is unaffected by the rotation. The surface of the planet must be an equipotential of $\Phi + \Phi_{\text{cen}}$ [coinciding with the zero-pressure isobar; cf. the sentence following Eq. (13.8), with $\Phi \to \Phi + \Phi_{\text{cen}}$]. The contribution of the centrifugal potential at the equator is $-\Omega^2 R_e^2/2$, and at the pole it is zero. The difference in the gravitational potential Φ between the equator and the pole is $\simeq g(R_e - R_p)$ where R_e and R_p are the equatorial and polar radii, respectively, and g is the gravitational acceleration at the planet's surface. Therefore, adopting this centralized-mass model and requiring that $\Phi + \Phi_{\text{cen}}$ be the same at the equator as at the pole, we estimate the difference between the polar and equatorial radii to be

centrifugal flattening

$$R_e - R_p \simeq \frac{\Omega^2 R^2}{2g}. \qquad (13.28a)$$

Earth, although not a fluid, is unable to withstand large shear stresses, because its shear strain cannot exceed the yield strain of rock, ~ 0.001; see Sec. 11.3.2 and Table 11.1. Since the heights of the tallest mountains are also governed by the yield strain, Earth's surface will not deviate from its equipotential by more than the maximum height of a mountain, $\simeq 9$ km.

If, for Earth, we substitute $g \simeq 10 \text{ m s}^{-2}$, $R \simeq 6 \times 10^6 \text{ m}$, and $\Omega \simeq 7 \times 10^{-5} \text{ rad s}^{-1}$ into Eq. (13.28a), we obtain $R_e - R_p \simeq 10$ km, about half the correct value of 21 km. The reason for this discrepancy lies in our assumption that all the planet's mass resides at its center. In fact, the mass is distributed fairly uniformly in radius and, in particular, some mass is found in the equatorial bulge. This deforms the gravitational equipotential surfaces from spheres to ellipsoids, which accentuates the flattening. If, following Newton (in his *Principia Mathematica*, published in 1687), we assume that Earth has uniform density, then the flattening estimate is 2.5 times larger than our centralized-mass estimate (Ex. 13.5) (i.e., $R_e - R_p \simeq 25$ km), in fairly good agreement with Earth's actual shape.

EXERCISES

Exercise 13.5 *Example: Shape of a Constant-Density, Spinning Planet*

(a) Show that the spatially variable part of the gravitational potential for a uniform-density, nonrotating planet can be written as $\Phi = 2\pi G\rho r^2/3$, where ρ is the density.

(b) Hence argue that the gravitational potential for a slowly spinning planet can be written in the form

$$\Phi = \frac{2\pi G\rho r^2}{3} + Ar^2 P_2(\mu),$$

where A is a constant, and P_2 is the Legendre polynomial with argument $\mu = \sin(\text{latitude})$. What happens to the P_1 term?

(c) Give an equivalent expansion for the potential outside the planet.

(d) Now transform into a frame spinning with the planet, and add the centrifugal potential to give a total potential.

(e) By equating the potential and its gradient at the planet's surface, show that the difference between the polar and the equatorial radii is given by

$$R_e - R_p \simeq \frac{5\Omega^2 R^2}{4g}, \qquad (13.28b)$$

where g is the gravitational acceleration at the surface. Note that this is 5/2 times the answer for a planet whose mass is concentrated at its center [Eq. (13.28a)].

Exercise 13.6 *Problem: Shapes of Stars in a Tidally Locked Binary System*
Consider two stars with the same mass M orbiting each other in a circular orbit with diameter (separation between the stars' centers) a. Kepler's laws tell us that the stars' orbital angular velocity is $\Omega = \sqrt{2GM/a^3}$. Assume that each star's mass is concentrated near its center, so that everywhere except near a star's center the gravitational potential, in an inertial frame, is $\Phi = -GM/r_1 - GM/r_2$ with r_1 and r_2 the distances of the observation point from the center of star 1 and star 2. Suppose that the two stars are "tidally locked": tidal gravitational forces have driven them each to rotate with rotational angular velocity equal to the orbital angular velocity Ω. (The Moon is tidally locked to Earth, which is why it always keeps the same face toward Earth.) Then in a reference frame that rotates with angular velocity Ω, each star's gas will be at rest, $\mathbf{v} = 0$.

(a) Write down the total potential $\Phi + \Phi_{\text{cen}}$ for this binary system in the rotating frame.

(b) Using Mathematica, Maple, Matlab, or some other computer software, plot the equipotentials $\Phi + \Phi_{\text{cen}} = $ const for this binary in its orbital plane, and use these equipotentials to describe the shapes that these stars will take if they expand to larger and larger radii (with a and M held fixed). You should obtain a sequence in which the stars, when compact, are well separated and nearly round. As they grow, tidal gravity elongates them ultimately into tear-drop shapes, followed by merger into a single, highly distorted star. With further expansion, the merged star starts flinging mass off into the surrounding space (a process not included in this hydrostatic analysis).

13.4 Conservation Laws

As a foundation for the transition from hydrostatics to hydrodynamics [i.e., to situations with nonzero fluid velocity $\mathbf{v}(\mathbf{x}, t)$], we give a general discussion of Newtonian conservation laws, focusing especially on the conservation of mass and of linear momentum.

We begin with the differential law of mass conservation,

$$\frac{\partial \rho}{\partial t} + \boldsymbol{\nabla} \cdot (\rho \mathbf{v}) = 0, \tag{13.29}$$

mass conservation: differential form

which we met and used in our study of elastic media [Eq. (12.2c)]. Eq. (13.29) is the obvious analog of the laws of conservation of charge, $\partial \rho_e / \partial t + \boldsymbol{\nabla} \cdot \mathbf{j} = 0$, and of particles, $\partial n / \partial t + \boldsymbol{\nabla} \cdot \mathbf{S} = 0$, which we met in Chap. 1 [Eqs. (1.30)]. In each case the law has the form $(\partial / \partial t)$(density of something) $+ \boldsymbol{\nabla} \cdot$ (flux of that something) $= 0$. In fact, this form is universal for a differential conservation law.

Each Newtonian differential conservation law has a corresponding integral conservation law (Sec. 1.8), which we obtain by integrating the differential law over some arbitrary 3-dimensional volume $\mathcal{V}$, for example, the volume used in Fig. 13.3 to discuss Archimedes' law: $(d/dt) \int_{\mathcal{V}} \rho dV = \int_{\mathcal{V}} (\partial \rho / \partial t) dV = - \int_{\mathcal{V}} \boldsymbol{\nabla} \cdot (\rho \mathbf{v}) dV$. Applying Gauss's law to the last integral, we obtain

mass conservation: integral form

$$\frac{d}{dt} \int_{\mathcal{V}} \rho dV = - \int_{\partial \mathcal{V}} \rho \mathbf{v} \cdot d\boldsymbol{\Sigma}, \tag{13.30}$$

where $\partial \mathcal{V}$ is the closed surface bounding $\mathcal{V}$. The left-hand side is the rate of change of mass inside the region $\mathcal{V}$. The right-hand side is the rate at which mass flows into $\mathcal{V}$ through $\partial \mathcal{V}$ (since $\rho \mathbf{v}$ is the mass flux, and the inward pointing surface element is $-d\boldsymbol{\Sigma}$). This is the same argument, connecting differential to integral conservation laws, as we gave when deriving Eqs. (1.29) and (1.30) for electric charge and for particles, but going in the opposite direction. And this argument depends in no way on whether the flowing material is a fluid or not. The mass conservation laws (13.29) and (13.30) are valid for any kind of material.

Writing the differential conservation law in the form (13.29), where we monitor the changing density at a given location in space rather than moving with the material, is called the *Eulerian* approach. There is an alternative *Lagrangian* approach to mass conservation, in which we focus on changes of density as measured by somebody who moves, locally, with the material (i.e., with velocity $\mathbf{v}$). We obtain this approach by differentiating the product $\rho \mathbf{v}$ in Eq. (13.29) to obtain

mass conservation: alternative differential form

$$\frac{d\rho}{dt} = -\rho \boldsymbol{\nabla} \cdot \mathbf{v}, \tag{13.31}$$

where

convective (advective) time derivative

$$\frac{d}{dt} \equiv \frac{\partial}{\partial t} + \mathbf{v} \cdot \boldsymbol{\nabla}. \tag{13.32}$$

The operator d/dt is known as the *convective time derivative* (or *advective time derivative*) and crops up often in continuum mechanics. Its physical interpretation is very simple. Consider first the partial derivative $(\partial/\partial t)_{\mathbf{x}}$. This is the rate of change of some quantity [the density ρ in Eq. (13.31)] at a fixed point $\mathbf{x}$ in space in some reference frame. In other words, if there is motion, $\partial/\partial t$ compares this quantity at the same point $\mathcal{P}$ in space for two different points in the material: one that is at $\mathcal{P}$ at time $t + dt$; the other that was at $\mathcal{P}$ at the earlier time t. By contrast, the convective time derivative d/dt follows the motion, taking the difference in the value of the quantity at successive times at the same point in the moving matter. It is the time derivative for the Lagrangian approach.

For a fluid, the Lagrangian approach can also be expressed in terms of fluid elements. Consider a small fluid element with a bounding surface attached to the fluid, and denote its volume by V. The mass inside the fluid element is $M = \rho V$. As the fluid flows, this mass must be conserved, so $dM/dt = (d\rho/dt)V + \rho(dV/dt) = 0$, which we can rewrite as

$$\frac{d\rho}{dt} = -\rho \frac{dV/dt}{V}.$$

(13.33)

Lagrangian formulation of mass conservation

Comparing with Eq. (13.31), we see that

$$\boxed{\nabla \cdot \mathbf{v} = \frac{dV/dt}{V}.}$$

(13.34)

Thus, the divergence of $\mathbf{v}$ is the fractional rate of increase of a fluid element's volume. Notice that this is just the time derivative of our elastostatic equation $\Delta V/V = \nabla \cdot \boldsymbol{\xi} = \Theta$ [Eq. (11.7)] (since $\mathbf{v} = d\boldsymbol{\xi}/dt$). Correspondingly we denote

$$\boxed{\nabla \cdot \mathbf{v} \equiv \theta = d\Theta/dt,}$$

(13.35)

rate of expansion of fluid

and call it the fluid's *rate of expansion*.

Equation (13.29), $\partial\rho/\partial t + \nabla \cdot (\rho\mathbf{v}) = 0$, is our model for Newtonian conservation laws. It says that there is a quantity (in this case mass) with a certain density (ρ), and a certain flux ($\rho\mathbf{v}$), and this quantity is neither created nor destroyed. The temporal derivative of the density (at a fixed point in space) added to the divergence of the flux must vanish. Of course, not all physical quantities have to be conserved. If there were sources or sinks of mass, then these would be added to the right-hand side of Eq. (13.29).

Now turn to momentum conservation. The (Newtonian) law of momentum conservation must take the standard conservation-law form $(\partial/\partial t)(\text{momentum density}) + \nabla \cdot (\text{momentum flux}) = 0$.

If we just consider the *mechanical momentum* associated with the motion of mass, its density is the vector field $\rho\mathbf{v}$. There can also be other forms of momentum density (e.g., electromagnetic), but these do not enter into Newtonian fluid mechanics. For fluids, as for an elastic medium (Chap. 12), the momentum density is simply $\rho\mathbf{v}$.

The momentum flux is more interesting and rich. Quite generally it is, by definition, the stress tensor **T**, and the differential conservation law states

momentum conservation: Eulerian formulation

$$\boxed{\frac{\partial(\rho\mathbf{v})}{\partial t} + \nabla \cdot \mathbf{T} = 0}$$

(13.36)

[Eq. (1.36)]. For an elastic medium, $\mathbf{T} = -K\Theta\mathbf{g} - 2\mu\boldsymbol{\Sigma}$ [Eq. (11.18)], and the conservation law (13.36) gives rise to the elastodynamic phenomena that we explored in Chap. 12. For a fluid we build up **T** piece by piece.

We begin with the rate $d\mathbf{p}/dt$ that mechanical momentum flows through a small element of surface area $d\boldsymbol{\Sigma}$, from its back side to its front. The rate that mass flows through is $\rho\mathbf{v} \cdot d\boldsymbol{\Sigma}$, and we multiply that mass by its velocity **v** to get the momentum flow rate: $d\mathbf{p}/dt = (\rho\mathbf{v})(\mathbf{v} \cdot d\boldsymbol{\Sigma})$. This rate of flow of momentum is the same thing as a force $\mathbf{F} = d\mathbf{p}/dt$ acting across $d\boldsymbol{\Sigma}$; so it can be computed by inserting $d\boldsymbol{\Sigma}$ into the second slot of a "mechanical" stress tensor $\mathbf{T}_m$: $d\mathbf{p}/dt = \mathbf{T}_m(\_\_, d\boldsymbol{\Sigma})$ [cf. the definition (1.32) of the stress tensor]. By writing these two expressions for the momentum flow in index notation, $dp_i/dt = (\rho v_i)v_j d\Sigma_j = T_{mij}d\Sigma_j$, we read off the mechanical stress tensor: $T_{mij} = \rho v_i v_j$:

mechanical stress tensor

$$\boxed{\mathbf{T}_m = \rho\mathbf{v} \otimes \mathbf{v}.}$$

(13.37)

This tensor is symmetric (as any stress tensor must be; Sec. 1.9), and it obviously is the flux of mechanical momentum, since it has the form (momentum density) $\otimes$ (velocity).

force f per unit volume acting on fluid

Let us denote by **f** the net force per unit volume that acts on the fluid. Then instead of writing momentum conservation in the usual Eulerian differential form (13.36), we can write it as

$$\frac{\partial(\rho\mathbf{v})}{\partial t} + \nabla \cdot \mathbf{T}_m = \mathbf{f}$$

(13.38)

(conservation law with a source on the right-hand side). Inserting $\mathbf{T}_m = \rho\mathbf{v} \otimes \mathbf{v}$ into this equation, converting to index notation, using the rule for differentiating products, and combining with the law of mass conservation, we obtain the *Lagrangian law*

momentum conservation: Lagrangian formulation

$$\boxed{\rho\frac{d\mathbf{v}}{dt} = \mathbf{f}.}$$

(13.39)

Here $d/dt = \partial/\partial t + \mathbf{v} \cdot \nabla$ is the convective time derivative (i.e., the time derivative moving with the fluid); so this equation is just Newton's $\mathbf{F} = m\mathbf{a}$, per unit volume. For the equivalent equations (13.38) and (13.39) of momentum conservation to also be equivalent to the Eulerian formulation (13.36), there must be a stress tensor $\mathbf{T}_f$ such that

stress tensor for the force f

$$\boxed{\mathbf{f} = -\nabla \cdot \mathbf{T}_f; \quad \text{and} \quad \mathbf{T} = \mathbf{T}_m + \mathbf{T}_f.}$$

(13.40)

Then Eq. (13.38) becomes the Eulerian conservation law (13.36).

Evidently, a knowledge of the stress tensor $\mathbf{T}_f$ for some material is equivalent to a knowledge of the force density $\mathbf{f}$ that acts on the material. It often turns out to be much easier to figure out the form of the stress tensor, for a given situation, than the form of the force. Correspondingly, as we add new pieces of physics to our fluid analysis (e.g., isotropic pressure, viscosity, gravity, magnetic forces), an efficient way to proceed at each stage is to insert the relevant physics into the stress tensor $\mathbf{T}_f$ and then evaluate the resulting contribution $\mathbf{f} = -\nabla \cdot \mathbf{T}_f$ to the force density and thence to the Lagrangian law of force balance (13.39). At each step, we get out in the form $\mathbf{f} = -\nabla \cdot \mathbf{T}_f$ the physics that we put into $\mathbf{T}_f$.

stress tensor for force density f

force density as divergence of stress tensor

There may seem something tautological about the procedure (13.40) by which we went from the Lagrangian $\mathbf{F} = m\mathbf{a}$ equation (13.39) to the Eulerian conservation law (13.36), (13.40). The $\mathbf{F} = m\mathbf{a}$ equation makes it look like mechanical momentum is not conserved in the presence of the force density $\mathbf{f}$. But we make it be conserved by introducing the momentum flux $\mathbf{T}_f$. It is almost as if we regard conservation of momentum as a principle to be preserved at all costs, and so every time there appears to be a momentum deficit, we simply define it as a bit of the momentum flux. However, this is not the whole story. What is important is that the force density $\mathbf{f}$ can always be expressed as the divergence of a stress tensor; that fact is central to the nature of force and of momentum conservation. An erroneous formulation of the force would not necessarily have this property, and so no differential conservation law could be formulated. Therefore, the fact that we can create elastostatic, thermodynamic, viscous, electromagnetic, gravitational, etc. contributions to some grand stress tensor (that go to zero outside the regions occupied by the relevant matter or fields)—as we shall do in the coming chapters—is significant. It affirms that our physical model is complete at the level of approximation to which we are working.

We can proceed in the same way with energy conservation as we have with momentum. There is an energy density $U(\mathbf{x}, t)$ for a fluid and an energy flux $\mathbf{F}(\mathbf{x}, t)$, and they obey a conservation law with the standard form

energy density U, energy flux F, and energy conservation

$$\frac{\partial U}{\partial t} + \nabla \cdot \mathbf{F} = 0. \tag{13.41}$$

At each stage in our buildup of fluid dynamics (adding, one by one, the influences of compressional energy, viscosity, gravity, and magnetism), we can identify the relevant contributions to U and $\mathbf{F}$ and then grind out the resulting conservation law (13.41). At each stage we get out the physics that we put into U and $\mathbf{F}$.

13.5 The Dynamics of an Ideal Fluid

13.5

We now use the general conservation laws of the previous section to derive the fundamental equations of fluid dynamics. We do so in several stages. In this section and Sec. 13.6, we confine our attention to ideal fluids—flows for which it is safe to ignore dissipative processes (viscosity and thermal conductivity) and thus for which the entropy of a fluid element remains constant with time. In Sec. 13.7, we introduce the effects of viscosity and diffusive heat flow.

13.5.1

13.5.1 Mass Conservation

As we have seen, mass conservation takes the (Eulerian) form $\partial\rho/\partial t + \mathbf{\nabla} \cdot (\rho\mathbf{v}) = 0$ [Eq. (13.29)], or equivalently the (Lagrangian) form $d\rho/dt = -\rho\mathbf{\nabla} \cdot \mathbf{v}$ [Eq. (13.31)], where $d/dt = \partial/\partial t + \mathbf{v} \cdot \mathbf{\nabla}$ is the convective time derivative [i.e., the time derivative moving with the fluid; Eq. (13.32)].

As we shall see in Sec. 13.6, when flow speeds are small compared to the speed of sound and the effects of gravity are sufficiently modest, the density of a fluid element remains nearly constant: $|(1/\rho)d\rho/dt| = |\mathbf{\nabla} \cdot \mathbf{v}| \ll 1/\tau$, where τ is the fluid flow's **incompressible approximation** characteristic timescale. It is then a good approximation to rewrite the law of mass conservation as $\mathbf{\nabla} \cdot \mathbf{v} = 0$, which is the *incompressible approximation*.

13.5.2

13.5.2 Momentum Conservation

For an ideal fluid, the only forces that can act are those of gravity and of the fluid's isotropic pressure P. We have already met and discussed the contribution of P to the stress tensor, $\mathbf{T} = P\mathbf{g}$, when dealing with elastic media (Chap. 11) and in hydrostatics (Sec. 13.3). The gravitational force density $\rho\mathbf{g}$ is so familiar that it is easier to write it down than the corresponding gravitational contribution to the stress. Correspondingly, we can most easily write momentum conservation in the form

momentum conservation for an ideal fluid

$$\frac{\partial(\rho\mathbf{v})}{\partial t} + \mathbf{\nabla} \cdot \mathbf{T} = \rho\mathbf{g}; \quad \text{or} \quad \frac{\partial(\rho\mathbf{v})}{\partial t} + \mathbf{\nabla} \cdot (\rho\mathbf{v} \otimes \mathbf{v} + P\mathbf{g}) = \rho\mathbf{g}, \quad (13.42)$$

where the stress tensor is given by

stress tensor for an ideal fluid

$$\boxed{\mathbf{T} = \rho\mathbf{v} \otimes \mathbf{v} + P\mathbf{g} \quad \text{for an ideal fluid}} \quad (13.43)$$

[cf. Eqs. (13.37), (13.38), and (13.4)]. The first term, $\rho\mathbf{v} \otimes \mathbf{v}$, is the mechanical momentum flux (also called the *kinetic* stress), and the second, $P\mathbf{g}$, is that associated with the fluid's pressure.

In most of our Newtonian applications, the gravitational field $\mathbf{g}$ will be externally imposed (i.e., it will be produced by some object, e.g., Earth that is different from the fluid we are studying). However, the law of momentum conservation remains the same [Eq. (13.42)], independently of what produces gravity—the fluid, or an external body, or both. And independently of its source, one can write the stress tensor $\mathbf{T}_g$ for the gravitational field $\mathbf{g}$ in a form presented and discussed in the Track-Two Box 13.4 later in the chapter—a form that has the required property $-\mathbf{\nabla} \cdot \mathbf{T}_g = \rho\mathbf{g} = $ (the gravitational force density).

13.5.3 Euler Equation

The Euler equation is the equation of motion that one gets out of the momentum conservation law (13.42) for an ideal fluid by performing the differentiations and invoking mass conservation (13.29):[3]

$$\frac{d\mathbf{v}}{dt} \equiv \frac{\partial \mathbf{v}}{\partial t} + (\mathbf{v} \cdot \nabla)\mathbf{v} = -\frac{\nabla P}{\rho} + \mathbf{g} \quad \text{for an ideal fluid.} \qquad (13.44)$$

Euler equation (momentum conservation) for an ideal fluid

The Euler equation has a simple physical interpretation: $d\mathbf{v}/dt$ is the convective derivative of the velocity (i.e., the derivative moving with the fluid), which means it is the acceleration felt by the fluid. This acceleration has two causes: gravity, $\mathbf{g}$, and the pressure gradient, ∇P. In a hydrostatic situation, $\mathbf{v} = 0$, the Euler equation reduces to the equation of hydrostatic equilibrium: $\nabla P = \rho \mathbf{g}$ [Eq. (13.4)].

In Cartesian coordinates, the Euler equation (13.44) and mass conservation $d\rho/dt + \rho \nabla \cdot \mathbf{v} = 0$ [Eq. (13.31)] comprise four equations in five unknowns, ρ, P, v_x, v_y, and v_z. The remaining fifth equation gives P as a function of ρ. For an ideal fluid, this equation comes from the fact that the entropy of each fluid element is conserved (because there is no mechanism for dissipation):

$$\frac{ds}{dt} = 0, \qquad (13.45)$$

entropy conservation

together with an equation of state for the pressure in terms of the density and the entropy: $P = P(\rho, s)$. In practice, the equation of state is often well approximated by incompressibility, $\rho = \text{const}$, or by a polytropic relation, $P = K(s)\rho^{1+1/n}$ [Eq. (13.18)].

13.5.4 Bernoulli's Theorem

Bernoulli's theorem is well known. Less well appreciated are the conditions under which it is true. To deduce these, we must first introduce a kinematic quantity known as the *vorticity*:

$$\boldsymbol{\omega} \equiv \nabla \times \mathbf{v}. \qquad (13.46)$$

vorticity

The physical interpretation of vorticity is simple. Consider a small fluid element. As it moves and deforms over a tiny period of time δt, each bit of fluid inside it undergoes a tiny displacement $\boldsymbol{\xi} = \mathbf{v}\delta t$. The gradient of that displacement field can be decomposed into an expansion, rotation, and shear (as we discussed in the context of an elastic medium in Sec. 11.2.2). The vectorial angle of the rotation is $\boldsymbol{\phi} = \frac{1}{2}\nabla \times \boldsymbol{\xi}$ [Eq. (11.9b)]. The time derivative of that vectorial angle, $d\boldsymbol{\phi}/dt = \frac{1}{2}d\boldsymbol{\xi}/dt = \frac{1}{2}\nabla \times \mathbf{v}$, is obviously the fluid element's rotational angular velocity; hence *the vorticity $\boldsymbol{\omega} = \nabla \times \mathbf{v}$ is twice*

angular velocity of a fluid element

---

3. This equation was first derived in 1757 by the Swiss mathematician and physicist Leonhard Euler—the same Euler who formulated the theory of buckling of a compressed beam (Sec. 11.6.1).

the angular velocity of rotation of a fluid element. Vorticity plays a major role in fluid mechanics, as we shall see in Chap. 14.

To derive Bernoulli's theorem (with the aid of vorticity), we begin with the Euler equation for an ideal fluid, $d\mathbf{v}/dt = -(1/\rho)\nabla P + \mathbf{g}$. We express $\mathbf{g}$ as $-\nabla\Phi$ and convert the convective derivative of velocity (i.e., the acceleration) into its two parts $d\mathbf{v}/dt = \partial\mathbf{v}/\partial t + (\mathbf{v} \cdot \nabla)\mathbf{v}$. Then we rewrite $(\mathbf{v} \cdot \nabla)\mathbf{v}$ using the vector identity

$$\mathbf{v} \times \boldsymbol{\omega} \equiv \mathbf{v} \times (\nabla \times \mathbf{v}) = \frac{1}{2}\nabla v^2 - (\mathbf{v} \cdot \nabla)\mathbf{v}. \tag{13.47}$$

The result is

$$\boxed{\frac{\partial\mathbf{v}}{\partial t} + \nabla\left(\frac{1}{2}v^2 + \Phi\right) + \frac{\nabla P}{\rho} - \mathbf{v} \times \boldsymbol{\omega} = 0.} \tag{13.48}$$

This is just the Euler equation written in a new form, but it is also *the most general version of Bernoulli's theorem*—valid for any ideal fluid. Two special cases are of interest.

BERNOULLI'S THEOREM FOR STEADY FLOW OF AN IDEAL FLUID

Since the fluid is ideal, dissipation (due to viscosity and heat flow) can be ignored, so the entropy is constant following the flow: $ds/dt = (\mathbf{v} \cdot \nabla)s = 0$. When, in addition, the flow is steady, meaning $\partial(\text{everything})/\partial t = 0$, the thermodynamic identity $dh = T\,ds + dP/\rho$ [Eq. (3) of Box 13.2] combined with $ds/dt = 0$ implies

$$(\mathbf{v} \cdot \nabla)P = \rho(\mathbf{v} \cdot \nabla)\,h. \tag{13.49}$$

Dotting the velocity $\mathbf{v}$ into the most general Bernoulli theorem (13.48) and invoking Eq. (13.49) and $\partial\mathbf{v}/\partial t = 0$, we obtain

$$\frac{dB}{dt} = (\mathbf{v} \cdot \nabla)B = 0, \tag{13.50}$$

where

$$\boxed{B \equiv \frac{1}{2}v^2 + h + \Phi.} \tag{13.51}$$

Equation (13.50) states that *in a steady flow of an ideal fluid, the Bernoulli function B, like the entropy, is conserved moving with a fluid element.* This is the most elementary form of the Bernoulli theorem.

Let us define *streamlines*, analogous to lines of force of a magnetic field, by the differential equations

$$\frac{dx}{v_x} = \frac{dy}{v_y} = \frac{dz}{v_z}. \tag{13.52}$$

In the language of Sec. 1.5, these are just the integral curves of the (steady) velocity field; they are also the spatial world lines of fluid elements. Equation (13.50) states: *In a steady flow of an ideal fluid, the Bernoulli function B is constant along streamlines.*

BOX 13.3. FLOW VISUALIZATION

Various methods are used to visualize fluid flows. One way is via *streamlines,* which are the integral curves of the velocity field **v** at a given time [Eq. (13.52)]. Streamlines are the analog of magnetic field lines. They coincide with the *paths* of individual fluid elements if the flow is steady, but not if the flow is time dependent. In general, the paths are the solutions of the equation $d\mathbf{x}/dt = \mathbf{v}(\mathbf{x}, t)$. These paths are the analog of particle trajectories in mechanics.

Another type of flow line is a *streak*. Monitoring streaks is a common way of visualizing a flow experimentally. Streaks are usually produced by introducing some colored or fluorescent tracer into the flow continuously at some fixed release point $\mathbf{x}_r$, and observing the locus of the tracer at some fixed time, say, t_0. Each point on the streak can be parameterized by the common release point $\mathbf{x}_r$, the common time of observation t_0, and the time t_r at which its marker was released, $\mathbf{x}(\mathbf{x}_r, t_r; t_0)$; so the streak is the parameterized curve $\mathbf{x}(t_r) = \mathbf{x}(\mathbf{x}_r, t_r; t_0)$.

Examples of streamlines, paths, and streaks are sketched below.

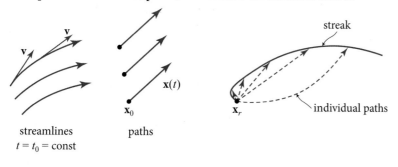

Streamlines are a powerful way to visualize fluid flows. There are other ways, sketched in Box 13.3.

The Bernoulli function $B = \frac{1}{2}v^2 + h + \Phi = \frac{1}{2}v^2 + u + P/\rho + \Phi$ has a simple physical meaning. It is the fluid's total energy density (kinetic plus internal plus potential) per unit mass, plus the work $P(1/\rho)$ that must be done to inject a unit mass of fluid (with volume $1/\rho$) into surrounding fluid that has pressure P. This goes hand in hand with the enthalpy $h = u + P/\rho$ being the *injection energy* (per unit mass) in the absence of kinetic and potential energy; see the last part of Ex. 5.5. This meaning of B leads to the following physical interpretation of the constancy of B for a stationary, ideal flow.

In a steady flow of an ideal fluid, consider a *stream tube* made of a bundle of streamlines (Fig. 13.5). A fluid element with unit mass occupies region $\mathcal{A}$ of the stream

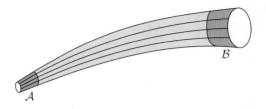

FIGURE 13.5 A stream tube used to explain the Bernoulli theorem for stationary flow of an ideal fluid.

tube at some early time and has moved into region $\mathcal{B}$ at some later time. When it vacates region $\mathcal{A}$, the fluid element carries an energy B [including the energy $P(1/\rho)$ it acquires by being squeezed out of $\mathcal{A}$ by the pressure of the surrounding fluid]. When it moves into region $\mathcal{B}$, it similarly carries the total injection energy B. Because the flow is steady, the energy it extracts from $\mathcal{A}$ must be precisely what it needs to occupy $\mathcal{B}$ (i.e., B must be constant along the stream tube, and hence also along each streamline in the tube).

The most immediate consequence of Bernoulli's theorem for steady flow is that, if gravity is having no significant effect, then the enthalpy falls when the speed increases, and conversely. This is just the conversion of internal (injection) energy into bulk kinetic energy, and conversely. For our ideal fluid, entropy must be conserved moving with a fluid element, so the first law of thermodynamics says $dh = T\,ds + dP/\rho = dP/\rho$. Therefore, as the speed increases and h decreases, P will also decrease.

Pitot tube

This behavior is the foundation for the *Pitot tube*, a simple device used to measure the air speed of an aircraft (Fig. 13.6). The Pitot tube extends out from the side of the aircraft, all the way through a boundary layer of slow-moving air and into the bulk flow. There it bends into the flow. The Pitot tube is actually two tubes: (i) an outer tube with several orifices along its sides, past which the air flows with its incoming speed V and pressure P, so the pressure inside that tube is also P; and (ii) an inner tube with a small orifice at its end, where the flowing air is brought essentially to rest (to *stagnation*). At this stagnation point and inside the orifice's inner tube, the pressure, by Bernoulli's theorem with Φ and ρ both essentially constant, is

stagnation pressure

the *stagnation pressure*: $P_{\text{stag}} = P + \frac{1}{2}\rho V^2$. The pressure difference $\Delta P = P_{\text{stag}} - P$ between the two tubes is measured by an instrument called a *manometer*, from which the air speed is computed as $V = (2\Delta P/\rho)^{1/2}$. If $V \sim 100\ \text{m s}^{-1}$ and $\rho \sim 1\ \text{kg m}^{-3}$, then $\Delta P \sim 5{,}000\ \text{N m}^{-3} \sim 0.05$ atmospheres.

In this book, we shall meet many other applications of the Bernoulli theorem for steady, ideal flows.

BERNOULLI'S THEOREM FOR IRROTATIONAL FLOW OF AN IDEAL, ISENTROPIC FLUID

An even more specialized type of flow is one that is *isentropic* (so s is the same everywhere) and *irrotational* (meaning its vorticity vanishes everywhere), as well as ideal.

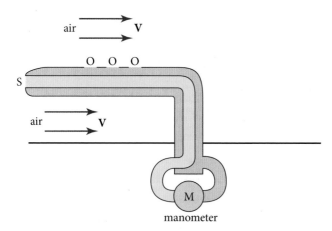

FIGURE 13.6 Schematic illustration of a Pitot tube used to measure air speed. The manometer M measures the pressure difference between the stagnation point S and the orifices O.

(In Sec. 14.2, we shall learn that if an incompressible flow initially is irrotational and it encounters no walls and experiences no significant viscous stresses, then it remains irrotational.) As $\boldsymbol{\omega} = \boldsymbol{\nabla} \times \mathbf{v}$ vanishes, we can follow the electrostatic precedent and introduce a *velocity potential* $\psi(\mathbf{x}, t)$, so that at any time,

$$\mathbf{v} = \boldsymbol{\nabla}\psi \quad \text{for an irrotational flow.} \tag{13.53}$$

velocity potential for irrotational (vorticity-free) flow

The first law of thermodynamics [Eq. (3) of Box 13.2] implies that $\boldsymbol{\nabla}h = T\boldsymbol{\nabla}s + (1/\rho)\boldsymbol{\nabla}P$. Therefore, in an isentropic flow, $\boldsymbol{\nabla}P = \rho\boldsymbol{\nabla}h$. Imposing these conditions on Eq. (13.48), we obtain, for a (possibly unsteady) isentropic, irrotational flow:

$$\boldsymbol{\nabla}\left(\frac{\partial\psi}{\partial t} + B\right) = 0. \tag{13.54}$$

Bernoulli's theorem for isentropic, irrotational flow of an ideal fluid

Thus *in an isentropic, irrotational flow of an ideal fluid, the quantity $\partial\psi/\partial t + B$ is constant everywhere.* (If $\partial\psi/\partial t + B$ is a function of time, we can absorb that function into ψ without affecting $\mathbf{v}$, leaving it constant in time as well as in space.) Of course, if the flow is steady, so $\partial(\text{everything})/\partial t = 0$, then B itself is constant.

Exercise 13.7 *Problem: A Hole in My Bucket*
There's a hole in my bucket. How long will it take to empty? (Try an experiment, and if the time does not agree with the estimate, explain why not.)

Exercise 13.8 *Problem: Rotating Planets, Stars, and Disks*
Consider a stationary, axisymmetric planet, star, or disk differentially rotating under the action of a gravitational field. In other words, the motion is purely in the azimuthal direction.

(a) Suppose that the fluid has a *barotropic* equation of state $P = P(\rho)$. Write down the equations of hydrostatic equilibrium, including the centrifugal force, in cylindrical polar coordinates. Hence show that the angular velocity must be constant on surfaces of constant cylindrical radius. This is called *von Zeipel's theorem*. (As an application, Jupiter is differentially rotating and therefore might be expected to have similar rotation periods at the same latitudes in the north and the south. This is only roughly true, suggesting that the equation of state is not completely barotropic.)

(b) Now suppose that the structure is such that the surfaces of constant entropy per unit mass and angular momentum per unit mass coincide. (This state of affairs can arise if slow convection is present.) Show that the Bernoulli function (13.51) is also constant on these surfaces. [Hint: Evaluate ∇B.]

Exercise 13.9 **Problem: Crocco's Theorem*

(a) Consider steady flow of an ideal fluid. The Bernoulli function (13.51) is conserved along streamlines. Show that the variation of B across streamlines is given by Crocco's theorem:

$$\nabla B = T \nabla s + \mathbf{v} \times \boldsymbol{\omega}. \tag{13.55}$$

(b) As an example, consider the air in a tornado. In the tornado's core, the velocity vanishes; it also vanishes beyond the tornado's outer edge. Use Crocco's theorem to show that the pressure in the core is substantially different from that at the outer edge. Is it lower, or is it higher? How does this explain the ability of a tornado to make the walls of a house explode? For more detail, see Ex. 14.5.

Exercise 13.10 *Problem: Cavitation (Suggested by P. Goldreich)*
A hydrofoil moves with speed V at a depth $D = 3$ m below the surface of a lake; see Fig. 13.7. Estimate how fast V must be to make the water next to the hydrofoil boil. [This boiling, which is called *cavitation*, results from the pressure P trying to go negative (see, e.g., Batchelor, 2000, Sec. 6.12; Potter, Wiggert, and Ramadan, 2012, Sec. 8.3.4).] [Note: For a more accurate value of the speed V that triggers cavitation,

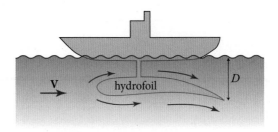

FIGURE 13.7 Water flowing past a hydrofoil as seen in the hydrofoil's rest frame.

one would have to compute the velocity field $\mathbf{v}(\mathbf{x})$ around the hydrofoil—for example, using the method of Ex. 14.17—and identify the maximum value of $v = |\mathbf{v}|$ near the hydrofoil's surface.]

Exercise 13.11 *Example: Collapse of a Bubble*
Suppose that a spherical bubble has just been created in the water above the hydrofoil in the previous exercise. Here we analyze its collapse—the decrease of the bubble's radius $R(t)$ from its value R_o at creation, using the incompressible approximation (which is rather good in this situation). This analysis is an exercise in solving the Euler equation.

(a) Introduce spherical polar coordinates with origin at the center of the bubble, so the collapse entails only radial fluid motion, $\mathbf{v} = v(r, t)\mathbf{e}_r$. Show that the incompressibility approximation $\nabla \cdot \mathbf{v} = 0$ implies that the radial velocity can be written in the form $v = w(t)/r^2$. Then use the radial component of the Euler equation (13.44) to show that

$$\frac{1}{r^2}\frac{dw}{dt} + v\frac{\partial v}{\partial r} + \frac{1}{\rho}\frac{\partial P}{\partial r} = 0.$$

At fixed time t, integrate this outward from the bubble surface at radius $R(t)$ to a large enough radius that the bubble's influence is no longer felt. Thereby obtain

$$\frac{-1}{R}\frac{dw}{dt} + \frac{1}{2}\dot{R}^2(R) = \frac{P_0}{\rho},$$

where P_0 is the ambient pressure and $-\dot{R}(R)$ is the speed of collapse of the bubble's surface when its radius is R. Assuming vanishing collapse speed when the bubble is created, $\dot{R}(R_o) = 0$, show that

$$\dot{R}(R) = -\left(\frac{2P_0}{3\rho}\right)^{1/2}\left[\left(\frac{R_0}{R}\right)^3 - 1\right]^{1/2},$$

which can be integrated to get $R(t)$.

(b) Suppose that bubbles formed near the pressure minimum on the surface of the hydrofoil are swept back onto a part of the surface where the pressure is much larger. By what factor R_o/R must the bubbles collapse if they are to create stresses that inflict damage on the hydrofoil?

Pistol shrimp can create collapsing bubbles and use the shock waves to stun their prey. A modification of this solution is important in interpreting the fascinating phenomenon of *sonoluminescence* (Brenner, Hilgenfeldt, and Lohse, 2002), which arises when fluids are subjected to high-frequency acoustic waves that create oscillating bubbles. The temperatures inside these bubbles can get so large that the air becomes ionized and radiates.

13.5.5 Conservation of Energy

As well as imposing conservation of mass and momentum, we must also address energy conservation in its general form (by contrast with the specialized version of energy conservation inherent in Bernoulli's theorem for a stationary, ideal flow).

In general, energy conservation is needed for determining the temperature T of a fluid, which in turn is needed to compute the pressure $P(\rho, T)$. So far in our treatment of fluid dynamics, we have finessed this issue by either postulating some relationship between the pressure P and the density ρ (e.g., the polytropic relation $P = K\rho^\gamma$) or by focusing on the flow of ideal fluids, where the absence of dissipation guarantees the entropy is constant moving with the flow, so that $P = P(\rho, s)$ with constant s. In more general situations, one cannot avoid confronting energy conservation. Moreover, even for ideal fluids, understanding how energy is conserved is often useful for gaining physical insight—as we have seen in our discussion of Bernoulli's theorem.

The most fundamental formulation of the law of energy conservation is Eq. (13.41): $\partial U/\partial t + \nabla \cdot \mathbf{F} = 0$. To explore its consequences for an ideal fluid, we must insert the appropriate ideal-fluid forms of the energy density U and energy flux $\mathbf{F}$.

When (for simplicity) the fluid is in an externally produced gravitational field Φ, its energy density is obviously

energy density for ideal fluid with external gravity

$$U = \rho \left(\frac{1}{2}v^2 + u + \Phi \right) \quad \text{for ideal fluid with external gravity.} \tag{13.56}$$

Here the three terms are kinetic, internal, and gravitational energy. When the fluid participates in producing gravity and one includes the energy of the gravitational field itself, the energy density is a bit more subtle; see the Track-Two Box 13.4.

In an external gravitational field, one might expect the energy flux to be $\mathbf{F} = U\mathbf{v}$, but this is not quite correct. Consider a bit of surface area dA orthogonal to the direction in which the fluid is moving (i.e., orthogonal to $\mathbf{v}$). The fluid element that crosses dA during time dt moves through a distance $dl = v dt$, and as it moves, the fluid behind this element exerts a force PdA on it. That force, acting through the distance dl, feeds an energy $dE = (PdA)dl = Pv dA dt$ across dA; the corresponding energy flux across dA has magnitude $dE/dA dt = Pv$ and points in the $\mathbf{v}$ direction, so it contributes $P\mathbf{v}$ to the energy flux $\mathbf{F}$. This contribution is missing from our initial guess $\mathbf{F} = U\mathbf{v}$. We explore its importance at the end of this subsection. When it is added to our guess, we obtain for the total energy flux

energy flux for ideal fluid with external gravity

$$\mathbf{F} = \rho \mathbf{v} \left(\frac{1}{2}v^2 + h + \Phi \right) \quad \text{for ideal fluid with external gravity.} \tag{13.57}$$

Here $h = u + P/\rho$ is the enthalpy per unit mass (cf. Box 13.2). Inserting Eqs. (13.56) and (13.57) into the law of energy conservation (13.41), and requiring that the external

BOX 13.4. SELF-GRAVITY T2

In the text, we mostly treat the gravitational field as externally imposed and independent of the fluid. This approximation is usually a good one. However, it is inadequate for planets and stars, whose self-gravity is crucial. It is easiest to discuss the modifications due to the fluid's self-gravitational effects by amending the conservation laws.

As long as we work in the domain of Newtonian physics, the mass conservation equation (13.29) is unaffected by self-gravity. However, we included the gravitational force per unit volume $\rho \mathbf{g}$ as a source of momentum in the momentum conservation law (13.42). It would fit much more neatly in our formalism if we could express it as the divergence of a gravitational stress tensor $\mathbf{T}_g$. To see that this is indeed possible, use Poisson's equation $\nabla \cdot \mathbf{g} = -4\pi G\rho$ (which embodies self-gravity) to write

$$\nabla \cdot \mathbf{T}_g = -\rho \mathbf{g} = \frac{(\nabla \cdot \mathbf{g})\mathbf{g}}{4\pi G} = \frac{\nabla \cdot [\mathbf{g} \otimes \mathbf{g} - \frac{1}{2}g^2\mathbf{g}]}{4\pi G},$$

so

$$\boxed{\mathbf{T}_g = \frac{\mathbf{g} \otimes \mathbf{g} - \frac{1}{2}g^2\mathbf{g}}{4\pi G}.} \tag{1}$$

Readers familiar with classical electromagnetic theory will notice an obvious and understandable similarity to the Maxwell stress tensor [Eqs. (1.38) and (2.80)], whose divergence equals the Lorentz force density.

What of the gravitational momentum density? We expect that it can be related to the gravitational energy density using a Lorentz transformation. That is to say, it is $O(v/c^2)$ times the gravitational energy density, where v is some characteristic speed. However, in the Newtonian approximation, the speed of light c is regarded as infinite, and so we should expect the gravitational momentum density to be identically zero in Newtonian theory—and indeed it is. We therefore can write the full equation of motion (13.42), including gravity, as a conservation law:

$$\frac{\partial(\rho \mathbf{v})}{\partial t} + \nabla \cdot \mathbf{T}_{\text{total}} = 0, \tag{2}$$

where $\mathbf{T}_{\text{total}}$ includes $\mathbf{T}_g$.

Now consider energy conservation. We have seen in the text that in a constant, external gravitational field, the fluid's total energy density U and flux $\mathbf{F}$ are given by Eqs. (13.56) and (13.57), respectively. In a general

(continued)

BOX 13.4. (continued)

situation, we must add to these some field energy density and flux. On dimensional grounds, these must be $U_{\text{field}} \propto \mathbf{g}^2/G$ and $\mathbf{F}_{\text{field}} \propto \Phi_{,t}\mathbf{g}/G$ (where $\mathbf{g} = -\nabla\Phi$). The proportionality constants can be deduced by demanding that for an ideal fluid in the presence of gravity, the law of energy conservation when combined with mass conservation, momentum conservation, and the first law of thermodynamics, lead to $ds/dt = 0$ (no dissipation in, so no dissipation out); see Eq. (13.59) and associated discussion. The result (Ex. 13.13) is

$$U = \rho\left(\frac{1}{2}v^2 + u + \Phi\right) + \frac{g^2}{8\pi G}, \tag{3}$$

$$\mathbf{F} = \rho\mathbf{v}\left(\frac{1}{2}v^2 + h + \Phi\right) + \frac{1}{4\pi G}\frac{\partial\Phi}{\partial t}\mathbf{g}. \tag{4}$$

Actually, there is an ambiguity in how the gravitational energy is localized. This ambiguity arises physically because one can transform away the gravitational acceleration $\mathbf{g}$, at any point in space, by transforming to a reference frame that falls freely there. Correspondingly, it turns out, one can transform away the gravitational energy density at any desired point in space. This possibility is embodied mathematically in the possibility of adding to the energy flux $\mathbf{F}$ the time derivative of $\alpha\Phi\nabla\Phi/(4\pi G)$ and adding to the energy density U minus the divergence of this quantity (where α is an arbitrary constant), while preserving energy conservation $\partial U/\partial t + \nabla\cdot\mathbf{F} = 0$. Thus the following choice of energy density and flux is just as good as Eqs. (3) and (4); both satisfy energy conservation:

$$U = \rho\left(\frac{1}{2}v^2 + u + \Phi\right) + \frac{g^2}{8\pi G} - \alpha\nabla\cdot\left(\frac{\Phi\nabla\Phi}{4\pi G}\right)$$

$$= \rho\left[\frac{1}{2}v^2 + u + (1-\alpha)\Phi\right] + (1 - 2\alpha)\frac{g^2}{8\pi G}, \tag{5}$$

$$\mathbf{F} = \rho\mathbf{v}\left(\frac{1}{2}v^2 + h + \Phi\right) + \frac{1}{4\pi G}\frac{\partial\Phi}{\partial t}\mathbf{g} + \alpha\frac{\partial}{\partial t}\left(\frac{\Phi\nabla\Phi}{4\pi G}\right)$$

$$= \rho\mathbf{v}\left(\frac{1}{2}v^2 + h + \Phi\right) + (1-\alpha)\frac{1}{4\pi G}\frac{\partial\Phi}{\partial t}\mathbf{g} - \frac{\alpha}{4\pi G}\Phi\frac{\partial\mathbf{g}}{\partial t}. \tag{6}$$

[Here we have used the gravitational field equation $\nabla^2\Phi = 4\pi G\rho$ and $\mathbf{g} = -\nabla\Phi$.] Note that the choice $\alpha = 1/2$ puts all the energy density into the

(continued)

$\rho\Phi$ term, while the choice $\alpha = 1$ puts all the energy density into the field term $\mathbf{g}^2$. In Ex. 13.14 it is shown that the total gravitational energy of an isolated system is independent of the arbitrary parameter α, as it must be on physical grounds.

A full understanding of the nature and limitations of the concept of gravitational energy requires the general theory of relativity (Part VII). The relativistic analog of the arbitrariness of Newtonian energy localization is an arbitrariness in the gravitational "stress-energy pseudotensor" (see, e.g., Misner, Thorne, and Wheeler, 1973, Sec. 20.3).

gravity be static (time independent), so the work it does on the fluid is conservative, we obtain the following ideal-fluid equation of energy balance:

$$\frac{\partial}{\partial t}\left[\rho\left(\frac{1}{2}v^2 + u + \Phi\right)\right] + \nabla\cdot\left[\rho\mathbf{v}\left(\frac{1}{2}v^2 + h + \Phi\right)\right] = 0$$

energy conservation for ideal fluid with external gravity

for an ideal fluid and static external gravity. (13.58)

When the gravitational field is dynamical or is being generated by the fluid itself (or both), we must use a more complete gravitational energy density and stress; see Box 13.4.

By combining the law of energy conservation (13.58) with the corresponding laws of momentum (13.29) and mass conservation (13.42), and using the first law of thermodynamics $dh = T ds + (1/\rho)dP$, we obtain the remarkable result that the entropy per unit mass is conserved moving with the fluid:

$$\boxed{\frac{ds}{dt} = 0 \quad \text{for an ideal fluid.}}$$

(13.59)

entropy conservation for ideal fluid

The same conclusion can be obtained when the gravitational field is dynamical and not external (cf. Box 13.4 and Ex. 13.13), so no statement about gravity is included with this equation. This entropy conservation should not be surprising. If we put no dissipative processes into the energy density, energy flux, or stress tensor, then we get no dissipation out. Moreover, the calculation that leads to Eq. (13.59) ensures that, *so long as we take full account of mass and momentum conservation, then the full and sole content of the law of energy conservation for an ideal fluid is $ds/dt = 0$*.

Table 13.1 summarizes our formulas for the density and flux of mass, momentum, and energy in an ideal fluid with externally produced gravity.

TABLE 13.1: Densities and fluxes of mass, momentum, and energy for an ideal fluid in an externally produced gravitational field

Quantity	Density	Flux
Mass	ρ	$\rho\mathbf{v}$
Momentum	$\rho\mathbf{v}$	$\mathbf{T} = P\mathbf{g} + \rho\mathbf{v} \otimes \mathbf{v}$
Energy	$U = (\frac{1}{2}v^2 + u + \Phi)\rho$	$\mathbf{F} = (\frac{1}{2}v^2 + h + \Phi)\rho\mathbf{v}$

EXERCISES

Exercise 13.12 *Example: Joule-Kelvin Cooling*

A good illustration of the importance of the Pv term in the energy flux is provided by the *Joule-Kelvin method* commonly used to cool gases (Fig. 13.8). Gas is driven from a high-pressure chamber 1 through a nozzle or porous plug into a low-pressure chamber 2, where it expands and cools.

(a) Using the energy flux (13.57), including the Pv term contained in h, show that a mass ΔM ejected through the nozzle carries a total energy ΔE_1 that is equal to the enthalpy ΔH_1 that this mass had while in chamber 1.

(b) This ejected gas expands and crashes into the gas of chamber 2, temporarily going out of statistical (thermodynamic) equilibrium. Explain why, after it has settled down into statistical equilibrium as part of the chamber-2 gas, the total energy it has deposited into chamber 2 is its equilibrium enthalpy ΔH_2. Thereby conclude that the enthalpy per unit mass is the same in the two chambers, $h_1 = h_2$.

(c) From $h_1 = h_2$, show that the temperature drop between the two chambers is

$$\Delta T = \int_{P_1}^{P_2} \mu_{\mathrm{JK}} dP, \tag{13.60}$$

where $\mu_{\mathrm{JK}} \equiv (\partial T/\partial P)_h$ is the Joule-Kelvin coefficient (also called Joule-Thomson). A straightforward thermodynamic calculation yields the identity

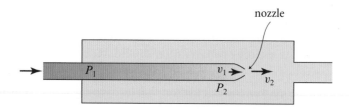

FIGURE 13.8 Schematic illustration of Joule-Kelvin cooling of a gas. Gas flows steadily through a nozzle from a chamber at high pressure P_1 to one at low pressure P_2. The flow proceeds at constant enthalpy. Work done against attractive intermolecular forces leads to cooling. The efficiency of cooling can be enhanced by exchanging heat between the two chambers. Gases can also be liquified in this manner.

$$\mu_{JK} \equiv \left(\frac{\partial T}{\partial P}\right)_h = -\frac{1}{\rho^2 c_p}\left(\frac{\partial(\rho T)}{\partial T}\right)_P. \tag{13.61}$$

(d) Show that the Joule-Kelvin coefficient of an ideal gas vanishes. Therefore, the cooling must arise because of the attractive forces (van der Waals forces; Sec. 5.3.2) between the molecules, which are absent in an ideal gas. When a real gas expands, work is done against these forces, and the gas therefore cools.

Exercise 13.13 *Derivation: No Dissipation "in" Means No Dissipation "out" and Verification of the Claimed Gravitational Energy Density and Flux* T2

Consider an ideal fluid interacting with a (possibly dynamical) gravitational field that the fluid itself generates via $\nabla^2\Phi = 4\pi G\rho$. For this fluid, take the law of energy conservation, $\partial U/\partial t + \mathbf{\nabla} \cdot \mathbf{F} = 0$, and from it subtract the scalar product of $\mathbf{v}$ with the law of momentum conservation, $\mathbf{v} \cdot [\partial(\rho\mathbf{v})/\partial t + \mathbf{\nabla} \cdot \mathbf{T})]$; then simplify using the law of mass conservation and the first law of thermodynamics, to obtain $\rho ds/dt = 0$. In your computation, use for U and $\mathbf{F}$ the expressions given in Eqs. (3) and (4) of Box 13.4. This calculation tells us two things. (i) The law of energy conservation for an ideal fluid reduces simply to conservation of entropy moving with the fluid; we have put no dissipative physics into the fluxes of momentum and energy, so we get no dissipation out. (ii) The gravitational energy density and flux contained in Eqs. (3) and (4) of Box 13.4 must be correct, since they guarantee that gravity does not alter this "no dissipation in, no dissipation out" result.

Exercise 13.14 *Example: Gravitational Energy* T2

Integrate the energy density U of Eq. (5) of Box 13.4 over the interior and surroundings of an isolated gravitating system to obtain the system's total energy. Show that the gravitational contribution to this total energy (i) is independent of the arbitrariness (parameter α) in the energy's localization, and (ii) can be written in the following forms:

$$E_g = \int dV \frac{1}{2}\rho\Phi = -\frac{1}{8\pi G}\int dV g^2 = \frac{-G}{2}\int\int dV dV'\frac{\rho(\mathbf{x})\rho(\mathbf{x}')}{|\mathbf{x} - \mathbf{x}'|}. \tag{13.62}$$

Interpret each of these expressions physically.

13.6 Incompressible Flows

A common assumption made when discussing the fluid dynamics of highly subsonic flows is that the density is constant—that the fluid is incompressible. This is a natural approximation to make when dealing with a liquid like water, which has a very large bulk modulus. It is a bit of a surprise that it is also useful for flows of gases, which are far more compressible under static conditions.

To see its validity, suppose that we have a flow in which the characteristic length L over which the fluid variables P, ρ, v, and so forth vary is related to the characteristic timescale T over which they vary by $L \lesssim vT$. In this case, we can compare the magnitude of the various terms in the Euler equation (13.44) to obtain an estimate of the magnitude of the pressure variation:

$$\underbrace{\frac{\partial \mathbf{v}}{\partial t}}_{v/T} + \underbrace{(\mathbf{v} \cdot \nabla)\mathbf{v}}_{v^2/L} = - \underbrace{\frac{\nabla P}{\rho}}_{\delta P/\rho L} - \underbrace{\nabla \Phi}_{\delta \Phi/L} .$$

(13.63)

Multiplying through by L and using $L/T \lesssim v$, we obtain $\delta P/\rho \sim v^2 + |\delta \Phi|$. The variation in pressure will be related to the variation in density by $\delta P \sim C^2 \delta \rho$, where $C = \sqrt{(\partial P/\partial \rho)_s}$ is the sound speed (Sec. 16.5), and we drop constants of order unity when making these estimates. Inserting this into our expression for δP, we obtain the estimate for the fractional density fluctuation:

$$\boxed{\frac{\delta \rho}{\rho} \sim \frac{v^2}{C^2} + \frac{\delta \Phi}{C^2}.}$$

(13.64)

incompressible approximation

Therefore, *if the fluid speeds are highly subsonic ($v \ll C$) and the gravitational potential does not vary greatly along flow lines ($|\delta \Phi| \ll C^2$), then we can ignore the density variations moving with the fluid when solving for the velocity field.* More specifically, since $\rho^{-1} d\rho/dt = -\nabla \cdot \mathbf{v} = -\theta$ [Eq. (13.35)], we can make the incompressible approximation

$$\nabla \cdot \mathbf{v} \simeq 0$$

(13.65)

(which means that the velocity field is *solenoidal,* like a magnetic field; i.e., is expressible as the curl of some potential). This argument breaks down when we are dealing with sound waves for which $L \sim CT$.

For air at atmospheric temperature, the speed of sound is $C \sim 300$ m/s, which is very fast compared to most flow speeds one encounters, so most flows are incompressible.

It should be emphasized, though, that the incompressible approximation for the velocity field, $\nabla \cdot \mathbf{v} \simeq 0$, does not imply that the density variation can be neglected in all other contexts. A particularly good example is provided by convection flows, which are driven by buoyancy, as we shall discuss in Chap. 18.

Incompressibility is a weaker condition than requiring the density to be constant everywhere; for example, the density varies substantially from Earth's center to its surface, but if the material inside Earth were moving more or less on surfaces of constant radius, the flow would be incompressible.

We restrict ourselves to incompressible flows throughout the next two chapters and then abandon incompressibility in subsequent chapters on fluid dynamics.

13.7

13.7 Viscous Flows with Heat Conduction

13.7.1

13.7.1 Decomposition of the Velocity Gradient into Expansion, Vorticity, and Shear

It is an observational fact that many fluids, when they flow, develop a shear stress (also called a *viscous stress*). Honey pouring off a spoon is a nice example. Most fluids, however, appear to flow quite freely; for example, a cup of tea appears to offer little resistance to stirring other than the inertia of the water. In such cases, it might be thought that viscous effects only produce a negligible correction to the flow's details. However, this is not so.

One of the main reasons is that most flows touch solid bodies, at whose surfaces the velocity must vanish. This leads to the formation of boundary layers, whose thickness and behavior are controlled by viscous forces. The boundary layers in turn can exert a controlling influence on the bulk flow (where the viscosity is negligible); for example, they can trigger the development of turbulence in the bulk flow—witness the stirred tea. For details see Chaps. 14 and 15.

We must therefore augment our equations of fluid dynamics to include viscous stresses. Our formal development proceeds in parallel to that used in elasticity, with the velocity field $\mathbf{v} = d\boldsymbol{\xi}/dt$ replacing the displacement field $\boldsymbol{\xi}$. We decompose the velocity gradient tensor $\boldsymbol{\nabla}\mathbf{v}$ into its irreducible tensorial parts: a *rate of expansion* θ; a symmetric, trace-free *rate of shear* tensor $\boldsymbol{\sigma}$; and an antisymmetric *rate of rotation* tensor $\mathbf{r}$:

$$\boldsymbol{\nabla}\mathbf{v} = \frac{1}{3}\theta\mathbf{g} + \boldsymbol{\sigma} + \mathbf{r}. \tag{13.66}$$

Note that we use lowercased symbols to distinguish the fluid case from its elastic counterpart: $\theta = d\Theta/dt$, $\boldsymbol{\sigma} = d\boldsymbol{\Sigma}/dt$, $\mathbf{r} = d\mathbf{R}/dt$. Proceeding directly in parallel to the treatment in Sec. 11.2.2 and Box 11.2, we can invert Eq. (13.66) to obtain

$$\theta = \boldsymbol{\nabla}\cdot\mathbf{v}, \tag{13.67a}$$

$$\sigma_{ij} = \frac{1}{2}(v_{i;j} + v_{j;i}) - \frac{1}{3}\theta g_{ij}, \tag{13.67b}$$

$$r_{ij} = \frac{1}{2}(v_{i;j} - v_{j;i}) = -\frac{1}{2}\epsilon_{ijk}\omega^k, \tag{13.67c}$$

rates of expansion, shear, and rotation

where $\boldsymbol{\omega} = 2d\boldsymbol{\phi}/dt$ is the vorticity, which we introduced and discussed in Sec. 13.5.4.

EXERCISES

Exercise 13.15 **Example: Kinematic Interpretation of Vorticity*
Consider a velocity field with nonvanishing curl. Define a locally orthonormal basis at a point in the velocity field, so that one basis vector, $\mathbf{e}_x$, is parallel to the vorticity. Now imagine the remaining two basis vectors as being frozen into the fluid. Show that they will both rotate about the axis defined by $\mathbf{e}_x$ and that the vorticity will be the sum of their angular velocities (i.e., twice the average of their angular velocities).

13.7.2 Navier-Stokes Equation

13.7.2

Although, as we have emphasized, a fluid at rest does not exert a shear stress, and this distinguishes it from an elastic solid, a fluid in motion can resist shear in the

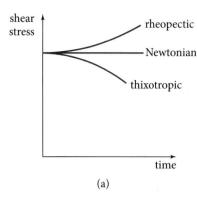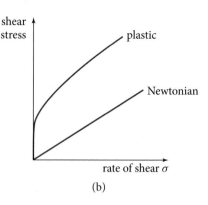

(a) (b)

FIGURE 13.9 Some examples of non-Newtonian behavior in fluids. (a) In a Newtonian fluid the shear stress is proportional to the rate of shear σ and does not vary with time when σ is constant as here. However, some substances, such as paint, flow more freely with time and are said to be *thixotropic*. Microscopically, the long, thin paint molecules gradually become aligned with the flow, which reduces their resistance to shear. The opposite behavior is exhibited by *rheopectic* substances such as ink and some lubricants. (b) An alternative type of non-Newtonian behavior is exhibited by various plastics, where a threshold stress is needed before flow will commence.

velocity field. It has been found experimentally that in most fluids the magnitude of this shear stress is linearly related to the velocity gradient. This law, due to Hooke's contemporary, Isaac Newton, is the analog of the linear relation between stress and strain that we used in our discussion of elasticity. Fluids that obey this law are known as *Newtonian*. (Some examples of non-Newtonian fluid behavior are shown in Fig. 13.9.) We analyze only Newtonian fluids in this book.

Fluids are usually isotropic. (Important exceptions include *smectic* liquid crystals.) In this book we restrict ourselves to isotropic fluids. By analogy with the theory of elasticity, we describe the linear relation between stress and rate of strain using

coefficients of bulk and shear viscosity

two constants called the coefficients of *bulk* and *shear* viscosity, denoted ζ and η, respectively. We write the viscous contribution to the stress tensor as

viscous stress tensor

$$\boxed{\mathbf{T}_{\mathrm{vis}} = -\zeta\theta\mathbf{g} - 2\eta\boldsymbol{\sigma},} \tag{13.68}$$

by analogy to Eq. (11.18), $\mathbf{T}_{\mathrm{elas}} = -K\Theta\mathbf{g} - 2\mu\boldsymbol{\Sigma}$, for an elastic solid. Here as there, shear-free rotation about a point does not produce a resistive stress.

If we include this viscous contribution in the stress tensor, then the law of momentum conservation $\partial(\rho\mathbf{v})/\partial t + \nabla \cdot \mathbf{T} = \rho\mathbf{g}$ gives the following generalization of Euler's equation (13.44):

Navier-Stokes equation: general form

$$\rho\frac{d\mathbf{v}}{dt} = -\nabla P + \rho\mathbf{g} + \nabla(\zeta\theta) + 2\nabla \cdot (\eta\boldsymbol{\sigma}). \tag{13.69}$$

This is the *Navier-Stokes equation,* and the last two terms are the viscous force density.

For incompressible flows (e.g., when the flow is highly subsonic; Sec. 13.6), θ can be approximated as zero, so the bulk viscosity can be ignored. The viscosity coefficient

TABLE 13.2: Approximate kinematic viscosity for common fluids

Quantity	Kinematic viscosity ν (m² s⁻¹)
Water	10^{-6}
Air	10^{-5}
Glycerine	10^{-3}
Blood	3×10^{-6}

η generally varies in space far more slowly than the shear $\boldsymbol{\sigma}$, and so can be taken outside the divergence. In this case, Eq. (13.69) simplifies to

$$\frac{d\mathbf{v}}{dt} = -\frac{\nabla P}{\rho} + \mathbf{g} + \nu\nabla^2\mathbf{v},$$ (13.70)

Navier-Stokes equation for incompressible flow

where

$$\nu \equiv \frac{\eta}{\rho}$$ (13.71)

kinematic shear viscosity coefficient

is known as the *kinematic viscosity,* by contrast to η, which is often called the *dynamic viscosity*. Equation (13.70) is the commonly quoted form of the Navier-Stokes equation; it is the form that we shall almost always use. Approximate values of the kinematic viscosity for common fluids are given in Table 13.2.

13.7.3 Molecular Origin of Viscosity

13.7.3

We can distinguish gases from liquids microscopically. In a gas, a molecule of mass m travels a distance of order its *mean free path* λ before it collides. If there is a shear in the fluid (Fig. 13.10), then the molecule, traveling in the y direction, on average

estimate of shear viscosity

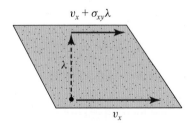

FIGURE 13.10 Molecular origin of viscosity in a gas. A molecule travels a distance λ in the y direction between collisions. Its mean x velocity at its point of origin is that of the fluid there, v_x, which differs from the mean x velocity at its next collision by $-\sigma_{xy}\lambda$. As a result, it transports a momentum $-m\sigma_{xy}\lambda$ to the location of its next collision.

will transfer an x momentum of about $-m\lambda\sigma_{xy}$ between collision points. If there are n molecules per unit volume traveling with mean thermal speeds v_{th}, then the transferred momentum crossing a unit area in unit time is $T_{xy} \sim -nmv_{\text{th}}\lambda\sigma_{xy}$, from which, by comparison with Eq. (13.68), we can extract an estimate of the coefficient of shear viscosity:

$$\eta \simeq \frac{1}{3}\rho v_{\text{th}}\lambda. \tag{13.72}$$

Here the numerical coefficient of 1/3 (which arises from averaging over molecular directions and speeds) has been inserted to agree with a proper kinetic-theory calculation; see Ex. 3.19 in Chap. 3. Note from Eq. (13.72) that in a gas, where the mean thermal kinetic energy $\frac{3}{2}k_B T$ is $\sim mv_{\text{th}}^2$, the coefficient of viscosity will increase with temperature as $\nu \propto T^{1/2}$.

In a liquid, where the molecules are less mobile, it is the close intermolecular attraction that produces the shear stress. The ability of molecules to slide past one another increases rapidly with their thermal activation, causing typical liquid viscosity coefficients to fall dramatically with rising temperature.

EXERCISES

Exercise 13.16 *Problem: Mean Free Path*
Estimate the collision mean free path of the air molecules around you. Hence verify the estimate for the kinematic viscosity of air given in Table 13.2.

13.7.4

13.7.4 Energy Conservation and Entropy Production

The viscous stress tensor represents an additional momentum flux that can do work on the fluid at a rate $\mathbf{T}_{\text{vis}} \cdot \mathbf{v}$ per unit area. Therefore a contribution

viscous energy flux

$$\mathbf{F}_{\text{vis}} = \mathbf{T}_{\text{vis}} \cdot \mathbf{v} \tag{13.73}$$

is made to the energy flux, just like the term $P\mathbf{v}$ appearing (as part of the $\rho\mathbf{v}h$) in Eq. (13.57). Diffusive heat flow (thermal conductivity) can also contribute to the energy flux; its contribution is [Eq. (3.70b)]

energy flux for heat conduction

$$\mathbf{F}_{\text{cond}} = -\kappa\nabla T, \tag{13.74}$$

coefficient of thermal conductivity κ

where κ is the coefficient of thermal conductivity. The molecules or particles that produce the viscosity and the heat flow also carry energy, but their energy density already is included in u, the total internal energy per unit mass, and their energy flux in $\rho\mathbf{v}h$. The total energy flux, including these contributions, is shown in Table 13.3, along with the energy density and the density and flux of momentum.

We see most clearly the influence of the dissipative viscous forces and heat conduction on energy conservation by inserting the energy density and flux from Table 13.3 into the law of energy conservation $\partial U/\partial t + \nabla \cdot \mathbf{F} = 0$, subtracting $\mathbf{v} \cdot [\partial(\rho\mathbf{v})/\partial t +$

TABLE 13.3: Densities and fluxes of mass, momentum, and energy for a dissipative fluid in an externally produced gravitational field

Quantity	Density	Flux
Mass	ρ	$\rho\mathbf{v}$
Momentum	$\rho\mathbf{v}$	$\mathbf{T} = \rho\mathbf{v} \otimes \mathbf{v} + P\mathbf{g} - \zeta\theta\mathbf{g} - 2\eta\boldsymbol{\sigma}$
Energy	$U = (\frac{1}{2}v^2 + u + \Phi)\rho$	$\mathbf{F} = (\frac{1}{2}v^2 + h + \Phi)\rho\mathbf{v} - \zeta\theta\mathbf{v} - 2\eta\boldsymbol{\sigma} \cdot \mathbf{v} - \kappa\nabla T$

Note: For self-gravitating systems, see Box 13.4.

$\nabla \cdot \mathbf{T} = \rho\mathbf{g}$] (**v** dotted into momentum conservation), and simplifying using mass conservation and the first law of thermodynamics. The result (Ex. 13.17) is the following equation for the evolution of entropy:

$$T\left[\rho\left(\frac{ds}{dt}\right) + \nabla \cdot \left(\frac{\mathbf{F}_{\text{cond}}}{T}\right)\right] = \zeta\theta^2 + 2\eta\boldsymbol{\sigma} : \boldsymbol{\sigma} + \frac{\kappa}{T}(\nabla T)^2, \qquad (13.75)$$

Lagrangian equation for entropy evolution, for viscous, heat-conducting fluid

where $\boldsymbol{\sigma} : \boldsymbol{\sigma}$ is the double scalar product $\sigma_{ij}\sigma_{ij}$. The term in square brackets on the left-hand side represents an increase of entropy per unit volume moving with the fluid due to dissipation (the total increase minus that due to heat flowing conductively into a unit volume); multiplied by T, this is the dissipative increase in entropy density. This increase of random, thermal energy is being produced, on the right-hand side, by viscous heating (first two terms), and by the flow of heat $\mathbf{F}_{\text{cond}} = -\kappa\nabla T$ down a temperature gradient $-\nabla T$ (third term).

The dissipation equation (13.75) is the full content of the law of energy conservation for a dissipative fluid, when one takes account of mass conservation, momentum conservation, and the first law of thermodynamics.

We can combine this Lagrangian rate of viscous dissipation with the equation of mass conservation (13.29) to obtain an Eulerian differential equation for the entropy increase:

$$\frac{\partial(\rho s)}{\partial t} + \nabla \cdot (\rho s \mathbf{v} - \kappa\nabla\ln T) = \frac{1}{T}\left(\zeta\theta^2 + 2\eta\boldsymbol{\sigma} : \boldsymbol{\sigma} + \frac{\kappa}{T}(\nabla T)^2\right). \qquad (13.76)$$

Eulerian equation for entropy evolution

The left-hand side of this equation describes the rate of change of entropy density plus the divergence of entropy flux. The right-hand side is therefore the rate of production of entropy per unit volume. Invoking the second law of thermodynamics, this quantity must be positive definite. Therefore the two coefficients of viscosity, like the bulk and shear moduli, must be positive, as must the coefficient of thermal conductivity κ (heat must flow from hotter regions to cooler ones).

In most laboratory and geophysical flows, thermal conductivity is unimportant, so we largely ignore it until our discussion of convection in Chap. 18.

Exercise 13.17 *Derivation: Entropy Increase*

(a) Derive the Lagrangian equation (13.75) for the rate of increase of entropy in a dissipative fluid by carrying out the steps in the sentence preceding that equation. [Hint: If you have already done the analogous problem (Ex. 13.13) for an ideal fluid, then you need only compute the new terms that arise from the dissipative momentum flux $\mathbf{T}_{\text{vis}} = -\zeta\theta\mathbf{g} - 2\eta\boldsymbol{\sigma}$ and dissipative energy fluxes $\mathbf{F}_{\text{vis}} = \mathbf{T}_{\text{vis}} \cdot \mathbf{v}$ and $\mathbf{F}_{\text{cond}} = -\kappa\nabla T$. The sum of these new contributions, when you subtract $\mathbf{v} \cdot$ (momentum conservation) from energy conservation, is $\nabla \cdot \mathbf{F}_{\text{cond}} + \nabla \cdot (\mathbf{T}_{\text{vis}} \cdot \mathbf{v}) - \mathbf{v} \cdot (\nabla \cdot \mathbf{T}_{\text{vis}})$; this must be added to the left-hand side of the result $\rho T\, ds/dt = 0$, Eq. (13.59), for an ideal fluid. When doing the algebra, it may be useful to decompose the gradient of the velocity into its irreducible tensorial parts, Eq. (13.66).]

(b) From the Lagrangian equation of entropy increase (13.75) derive the corresponding Eulerian equation (13.76).

13.7.5

13.7.5 Reynolds Number

The kinematic viscosity ν has dimensions length^2/time. This suggests that we quantify the importance of viscosity in a fluid flow by comparing ν with the product of the flow's characteristic velocity V and its characteristic lengthscale L. The dimensionless combination

Reynolds number

$$\boxed{\text{Re} = \frac{LV}{\nu}} \qquad (13.77)$$

is known as the *Reynolds number* and is the first of many dimensionless numbers we shall encounter in our study of fluid mechanics. Flows with Reynolds number much less than unity are dominated by viscosity. Large Reynolds number flows can also be strongly influenced by viscosity (as we shall see in later chapters), especially when the viscosity acts near boundaries—even though the viscous stresses are negligible over most of the flow's volume.

13.7.6

13.7.6 Pipe Flow

Let us now consider a simple example of viscous stresses at work, namely, the steady-state flow of blood down an artery.[4] We model the artery as a cylindrical pipe of radius a, through which the blood is forced by a time-independent pressure gradient. This is an example of what is called *pipe flow*.

Because gravity is unimportant and the flow is time independent, the Navier-Stokes equation (13.70) reduces to

---

4. We approximate the blood as a Newtonian fluid although, in reality, its shear viscosity η decreases at high rates of shear σ.

$$(\mathbf{v} \cdot \nabla)\mathbf{v} = -\frac{\nabla P}{\rho} + \nu \nabla^2 \mathbf{v}. \tag{13.78}$$

We assume that the flow is *laminar* (smooth, as is usually the case for blood in arteries), so **v** points solely along the z direction and is only a function of cylindrical radius ϖ. (This restriction is very important. As we discuss in Chap. 15, in other types of pipe flow, e.g., in crude oil pipelines, it often fails because the flow becomes turbulent, which has a major impact on the flow. In arteries, turbulence occasionally occurs and can lead to blood clots and a stroke!) Writing Eq. (13.78) in cylindrical coordinates, and denoting by $v(\varpi)$ the z component of velocity (the only nonvanishing component), we deduce that the nonlinear $\mathbf{v} \cdot \nabla \mathbf{v}$ term vanishes, and the pressure P is a function of z only and not of ϖ:

laminar flow contrasted with turbulent flow

$$\frac{1}{\varpi}\frac{d}{d\varpi}\left(\varpi\frac{dv}{d\varpi}\right) = \frac{1}{\eta}\frac{dP}{dz}. \tag{13.79}$$

Here dP/dz (which is negative) is the pressure gradient along the pipe, and $\eta = \nu\rho$ is the dynamic viscosity. This differential equation must be solved subject to the boundary conditions that the velocity gradient vanish at the center of the pipe and the velocity vanish at its walls. The solution is

velocity profile for laminar pipe flow

$$v(\varpi) = -\frac{dP}{dz}\frac{a^2 - \varpi^2}{4\eta}. \tag{13.80}$$

Using this velocity field, we can evaluate the pipe's total *flow rate*—volume per unit time—for (incompressible) blood volume:

Poiseuille's law for pipe flow

$$\mathcal{F} = \int_0^a v 2\pi \varpi \, d\varpi = -\frac{dP}{dz}\frac{\pi a^4}{8\eta}. \tag{13.81}$$

This relation is known as *Poiseuille's law* and because of the parabolic shape of the velocity profile (13.80), this pipe flow is sometimes called *parabolic Poiseuille flow*.

Now let us apply this result to a human body. The healthy adult heart, beating at about 60 beats per minute, pumps $\mathcal{F} \sim 5\,\mathrm{L\,min^{-1}}$ (liters per minute) of blood into a circulatory system of many branching arteries that reach into all parts of the body and then return. This circulatory system can be thought of as like an electric circuit, and Poiseuille's law (13.81) is like the current-voltage relation for a small segment of wire in the circuit. The flow rate $\mathcal{F}$ in an arterial segment plays the role of the electric current I in the wire segment, the pressure gradient dP/dz is the voltage drop per unit length dV/dz, and $dR/dz \equiv 8\eta/\pi a^4$ is the resistance per unit length. Thus $-dP/dz = \mathcal{F}\,dR/dz$ [Eq. (13.81)] is equivalent to the voltage-current relation $-dV/dz = I\,dR/dz$. Moreover, just as the total current is conserved at a circuit branch point (sum of currents in equals sum of currents out), so also the total blood flow rate is conserved at an arterial branch point. These identical conservation laws and identical pressure-and-voltage-drop equations imply that the analysis of pressure changes and flow distributions in the body's many-branched circulatory system is the same as that of voltage changes and current distributions in an equivalent many-branched electrical circuit.

blood flow in human body

Because of the heart's periodic pumping, blood flow is *pulsatile* (pulsed; periodic) in the great vessels leaving the heart (the aorta and its branches); see Ex. 13.19. However, as the vessels divide into smaller and smaller arterial branches, the pulsatility becomes lost, so the flow is steady in the smallest vessels, the *arterioles*. Since a vessel's resistance per unit length scales with its radius as $dR/dz \propto 1/a^4$, and hence its pressure drop per unit length $-dP/dz = \mathcal{F}dR/dz$ also scales as $1/a^4$ [Eq. (13.81)], it should not be surprising that in a healthy human the circulatory system's pressure drop occurs primarily in the tiny arterioles, which have radii $a \sim 5$–$50\ \mu$m.

The walls of these arterioles have circumferentially oriented smooth muscle structures, which are capable of changing the vessel radius a by as much as a factor ~ 2 or 3 in response to various stimuli (exercise, cold, stress, etc.). Note that a factor 3 radius increase means a factor $3^4 \sim 100$ decrease in pressure gradient at fixed flow rate! Accordingly, drugs designed to lower blood pressure do so by triggering radius changes in the arterioles. And anything you can do to keep your arteries from hardening, narrowing, or becoming blocked will help keep your blood pressure down. Eat salads!

EXERCISES

Exercise 13.18 *Problem: Steady Flow between Two Plates*
A viscous fluid flows steadily (no time dependence) in the z direction, with the flow confined between two plates that are parallel to the x-z plane and are separated by a distance $2a$. Show that the flow's velocity field is

$$v_z = -\frac{dP}{dz}\frac{a^2}{2\eta}\left[1 - \left(\frac{y}{a}\right)^2\right], \qquad (13.82a)$$

and the mass flow rate (the discharge) per unit width of the plates is

$$\frac{dm}{dt\,dx} = -\frac{dP}{dz}\frac{2\rho a^3}{3\eta}. \qquad (13.82b)$$

Here dP/dz (which is negative) is the pressure gradient along the direction of flow. (In Sec. 19.4 we return to this problem, augmented by a magnetic field and electric current, and discover great added richness.)

Exercise 13.19 *Example: Pulsatile Blood Flow*
Consider the pulsatile flow of blood through one of the body's larger arteries. The pressure gradient $dP/dz = P'(t)$ consists of a steady term plus a term that is periodic, with the period of the heart's beat.

(a) Assuming laminar flow with $\mathbf{v}$ pointing in the z direction and being a function of radius and time, $\mathbf{v} = v(\varpi, t)\mathbf{e}_z$, show that the Navier-Stokes equation reduces to $\partial v/\partial t = -P'/\rho + \nu\nabla^2 v$.

(b) Explain why $v(\varpi, t)$ is the sum of a steady term produced by the steady (time-independent) part of P', plus terms at angular frequencies $\omega_0, 2\omega_0, \ldots$, produced by parts of P' that have these frequencies. Here $\omega_0 \equiv 2\pi/(\text{heart's beat period})$.

(c) Focus on the component with angular frequency $\omega = n\omega_0$ for some integer n. For what range of ω do you expect the ϖ dependence of v to be approximately Poiseuille [Eq. (13.80)], and what ϖ dependence do you expect in the opposite extreme, and why?

(d) By solving the Navier-Stokes equation for the frequency-ω component, which is driven by the pressure-gradient term $dP/dz = \Re(P'_\omega e^{-i\omega t})$, and by imposing appropriate boundary conditions at $\varpi = 0$ and $\varpi = a$, show that

$$v = \Re\left[\frac{P'_\omega e^{-i\omega t}}{i\omega\rho}\left(1 - \frac{J_0(\sqrt{i}\, W\varpi/a)}{J_0(\sqrt{i}\, W)}\right)\right]. \tag{13.83}$$

Here $\Re$ means take the real part, a is the artery's radius, J_0 is the Bessel function, i is $\sqrt{-1}$, and $W \equiv \sqrt{\omega a^2/\nu}$ is called the (dimensionless) *Womersley number*.

(e) Plot the pieces of this $v(\varpi)$ that are in phase and out of phase with the driving pressure gradient. Compare with the prediction you made in part (b). Explain the phasing physically. Notice that in the extreme non-Poiseuille regime, there is a boundary layer attached to the artery's wall, with sharply changing flow velocity. What is its thickness in terms of a and the Womersley number? We study boundary layers like this one in Sec. 14.4 and especially Ex. 14.18.

13.8 Relativistic Dynamics of a Perfect Fluid [T2]

When a fluid's speed $v = |\mathbf{v}|$ becomes comparable to the speed of light c, or $P/(\rho c^2)$ or u/c^2 become of order unity, Newtonian fluid mechanics breaks down and must be replaced by a relativistic treatment. In this section, we briefly sketch the resulting laws of relativistic fluid dynamics for an ideal (perfect) fluid. For the extension to a fluid with dissipation (viscosity and heat conductivity), see, e.g., Misner, Thorne, and Wheeler (1973, Ex. 22.7).

Our treatment takes off from the brief description of an ideal, relativistic fluid in Secs. 2.12.3 and 2.13.3. As done there, we shall use geometrized units in which the speed of light is set to unity: $c = 1$ (see Sec. 1.10).

13.8.1 Stress-Energy Tensor and Equations of Relativistic Fluid Mechanics [T2]

For relativistic fluids we use ρ to denote the total density of mass-energy (including rest mass and internal energy) in the fluid's local rest frame; it is sometimes written as

$$\rho = \rho_o(1 + u), \quad \text{where} \quad \rho_o = \bar{m}_B n \tag{13.84}$$

is the density of rest mass, $\bar{m}_B$ is some standard mean rest mass per baryon, n is the number density of baryons, ρ_o is the density of rest mass, and u is the specific internal energy (Sec. 2.12.3).

The stress-energy tensor $T^{\alpha\beta}$ for a relativistic, ideal fluid takes the form [Eq. (2.74b)]

$$T^{\alpha\beta} = (\rho + P)u^{\alpha}u^{\beta} + Pg^{\alpha\beta}, \tag{13.85}$$

where P is the fluid's pressure (as measured in its local rest frame), u^{α} is its 4-velocity, and $g^{\alpha\beta}$ is the spacetime metric. In the fluid's local rest frame, where $u^0 = 1$ and $u^j = 0$, the components of this stress-energy tensor are, of course, $T^{00} = \rho$, $T^{j0} = T^{0j} = 0$, and $T^{jk} = Pg^{jk} = P\delta^{jk}$.

The dynamics of our relativistic, ideal fluid are governed by five equations. The first equation is the law of *rest-mass conservation*, $(\rho_o u^{\alpha})_{;\alpha} = 0$, which can be rewritten in the form [Eqs. (2.64) and (2.65) of Ex. 2.24]

$$\frac{d\rho_o}{d\tau} = -\rho_o \vec{\nabla} \cdot \vec{u}, \quad \text{or} \quad \frac{d(\rho_o V)}{d\tau} = 0, \tag{13.86a}$$

where $d/d\tau = \vec{u} \cdot \vec{\nabla}$ is the derivative with respect to proper time moving with the fluid, $\vec{\nabla} \cdot \vec{u} = (1/V)(dV/d\tau)$ is the divergence of the fluid's 4-velocity, and V is the volume of a fluid element. The second equation is *energy conservation*, in the form of the vanishing divergence of the stress-energy tensor projected onto the fluid 4-velocity, $u_{\alpha} T^{\alpha\beta}_{\;;\beta} = 0$, which, when combined with the law of rest-mass conservation, reduces to [Eqs. (2.76)]

$$\frac{d\rho}{d\tau} = -(\rho + P)\vec{\nabla} \cdot \vec{u}, \quad \text{or} \quad \frac{d(\rho V)}{d\tau} = -P\frac{dV}{d\tau}. \tag{13.86b}$$

The third equation follows from the first law of thermodynamics moving with the fluid, $d(\rho V)/d\tau = -PdV/d\tau + Td(\rho_o Vs)/d\tau$, combined with rest-mass conservation (13.86a) and energy conservation (13.86b), to yield *conservation of the entropy per unit rest mass s* (adiabaticity of the flow):

$$\frac{ds}{d\tau} = 0. \tag{13.86c}$$

As in Newtonian theory, the ultimate source of this adiabaticity is our restriction to an ideal fluid (i.e., one without any dissipation). The fourth equation is *momentum conservation*, which we obtain by projecting the vanishing divergence of the stress-energy tensor orthogonally to the fluid's 4-velocity, $P_{\alpha\mu} T^{\mu\nu}_{\;;\nu} = 0$, resulting in [Eq. (2.76c)]

$$(\rho + P)\frac{du^{\alpha}}{d\tau} = -P^{\alpha\mu} P_{;\mu}, \quad \text{where } P^{\alpha\mu} = g^{\alpha\mu} + u^{\alpha}u^{\mu}. \tag{13.86d}$$

This is the *relativistic Euler equation*, and $P^{\alpha\mu}$ (not to be confused with the fluid pressure P or its gradient $P_{;\mu}$) is the tensor that projects orthogonally to $\vec{u}$. Note that the inertial mass per unit volume is $\rho + P$ (Ex. 2.27), and that the pressure gradient produces a force that is orthogonal to $\vec{u}$. The fifth equation is an *equation of state*, for example, in the form

$$P = P(\rho_o, s). \tag{13.86e}$$

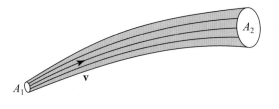

FIGURE 13.11 A tube generated by streamlines for a stationary flow. The tube ends are orthogonal to the streamlines and have areas A_1 and A_2.

Equations (13.86) are four independent scalar equations and one vector equation for the four scalars ρ_o, ρ, P, and s, and one vector $\vec{u}$.

As an example of an equation of state, one that we use below, consider a fluid so hot that its pressure and energy density are dominated by thermal motions of relativistic particles (photons, electron-positron pairs, etc.), so $P = \rho/3$ [Eqs. (3.54a)]. Then from the first law of thermodynamics for a fluid element, $d(\rho V) = -P dV$, and the law of rest-mass conservation, $d(\rho_o V) = 0$, one can deduce the relativistic polytropic equation of state

$$P = \frac{1}{3}\rho = K(s)\rho_o^{4/3}. \tag{13.87}$$

13.8.2 Relativistic Bernoulli Equation and Ultrarelativistic Astrophysical Jets T2

When the relativistic flow is steady (independent of time t in some chosen inertial frame), the law of energy conservation in that frame, $T^{0\mu}{}_{;\mu} = 0$, and the law of mass conservation, $(\rho_o u^\mu)_{;\mu} = 0$, together imply that the relativistic Bernoulli function B is conserved along flow lines; specifically:

$$\frac{dB}{d\tau} = \gamma v_j \frac{\partial B}{\partial x^j} = 0, \quad \text{where } B = \frac{(\rho + P)\gamma}{\rho_o}. \tag{13.88}$$

Bernoulli theorem for steady flow of relativistic, ideal fluid

Here $\gamma = u^0 = 1/\sqrt{1 - \mathbf{v}^2}$. A direct proof is left as an exercise (Ex. 13.20). The following more indirect geometric proof provides useful insight.

Consider a narrow tube in space, whose walls are generated by steady flow lines (streamlines), which are tangent to the steady velocity field $\mathbf{v}$ (Fig. 13.11). Denote the tube's interior by $\mathcal{V}$ and its boundary by $\partial\mathcal{V}$. Because the flow is steady, the law of mass conservation, $(\rho_o u^\alpha)_{;\alpha} = 0$, reduces to the vanishing spatial divergence: $(\rho_o u^j)_{;j} = (\rho_o \gamma v^j)_{;j} = 0$. Integrate this equation over the tube's interior, and apply Gauss's law to obtain $\int_{\partial\mathcal{V}} \rho_o \gamma v^j d\Sigma_j = 0$. Because the walls of the tube are parallel to v^j, they make no contribution. The contribution from the ends is $(\rho_o \gamma v)_2 A_2 - (\rho_o \gamma v)_1 A_1 = 0$. In other words, *the product $\rho_o \gamma v A$—the discharge—is constant along the stream tube.*

Similarly, for our steady flow the law of energy conservation, $T^{0\alpha}{}_{;\alpha} = 0$, reduces to the vanishing spatial divergence, $T^{0j}{}_{;j} = [(\rho + P)\gamma^2 v^j]_{;j} = 0$, which, when integrated over the tube's interior and converted to a surface integral using Gauss's theorem, implies that *the product* $(\rho + P)\gamma^2 vA$—*the* power—*is constant along the stream tube.*

The ratio of these two constants, $(\rho + P)\gamma/\rho_o = B$, *must also be constant along the stream tube;* equivalently (since it is independent of the area of the narrow tube), *B must be constant along a streamline,* which is Bernoulli's theorem.

relativistic jets

Important venues for relativistic fluid mechanics are the early universe (Secs. 28.4 and 28.5), and also the narrow relativistic jets that emerge from some galactic nuclei and gamma ray bursts. The flow velocities in some of these jets are measured to be so close to the speed of light that $\gamma \sim 1,000$. For the moment, ignore dissipation and electromagnetic contributions to the stress-energy tensor (which are quite important in practice), and assume that the gas pressure P and mass-energy density ρ are dominated by relativistic particles, so the equation of state is $P = \rho/3 = K\rho_o^{4/3}$ [Eq. (13.87)]. Then we can use the above, italicized conservation laws for the discharge, power, and Bernoulli function to learn how ρ, P, and γ evolve along such a jet.

From the relativistic Bernoulli theorem (the ratio B of the two constants, discharge and power) and the equation of state, we deduce that $\gamma \propto B\rho_o/(\rho + P) \propto \rho_o/\rho_o^{4/3} \propto \rho_o^{-1/3} \propto \rho^{-1/4}$. This describes a conversion of the jet's internal mass-energy ($\sim\rho$) into bulk flow energy ($\sim\gamma$): as the internal energy (energy of random thermal motions of the jet's particles) goes down, the bulk flow energy (energy of organized flow) goes up.

This energy exchange is actually driven by changes in the jet's cross sectional area, which (in some jets) is controlled by competition between the inward pressure of surrounding, slowly moving gas and outward pressure from the jet itself. Since $\rho_o \gamma vA \simeq \rho_o \gamma A$ is constant along the jet, we have $A \propto 1/(\rho_o \gamma) \propto \gamma^2 \propto \rho^{-1/2}$. Therefore, as the jet's cross sectional area A increases, internal energy ($\rho \propto 1/A^2$) goes down, and the energy of bulk flow ($\gamma \propto A^{1/2}$) goes up. Another way to think about this is in terms of the relativistic distribution function of the constituent particles. As the volume of configuration space dV_x that they occupy increases, their volume of momentum space dV_p decreases (Secs. 3.2.2 and 3.6).

We explore the origin of relativistic jets in Sec. 26.5.

EXERCISES

Exercise 13.20 *Derivation: Relativistic Bernoulli Theorem* T2
By manipulating the differential forms of the law of rest-mass conservation and the law of energy conservation, derive the constancy of $B = (\rho + P)\gamma/\rho_o$ along steady flow lines, Eq. (13.88).

Exercise 13.21 *Example: Relativistic Momentum Conservation* T2
Give an expression for the change in the *thrust*—the momentum crossing a surface perpendicular to the tube per unit time—along a slender stream tube when the discharge and power are conserved. Explain why the momentum has to change.

It is instructive to evaluate the nonrelativistic limit of the perfect-fluid stress-energy tensor, $T^{\alpha\beta} = (\rho + P)u^\alpha u^\beta + Pg^{\alpha\beta}$, and verify that it has the form we deduced in our study of nonrelativistic fluid mechanics (see Table 13.1) with vanishing gravitational potential $\Phi = 0$.

In the nonrelativistic limit, the fluid is nearly at rest in the chosen Lorentz reference frame. It moves with ordinary velocity $\mathbf{v} = d\mathbf{x}/dt$ that is small compared to the speed of light, so the temporal part of its 4-velocity $u^0 = 1/\sqrt{1 - v^2}$ and the spatial part $\mathbf{u} = u^0\mathbf{v}$ can be approximated as

$$u^0 \simeq 1 + \frac{1}{2}v^2, \qquad \mathbf{u} \simeq \left(1 + \frac{1}{2}v^2\right)\mathbf{v}. \tag{13.89a}$$

nonrelativistic limit of 4-velocity

We write $\rho = \rho_o(1 + u)$ [Eq. (13.84)], where u is the specific internal energy (not to be confused with the fluid 4-velocity $\vec{u}$ or its spatial part $\mathbf{u}$). In our chosen Lorentz frame the volume of each fluid element is Lorentz contracted by the factor $\sqrt{1 - v^2}$, and therefore the rest-mass density is increased from ρ_o to $\rho_o/\sqrt{1 - v^2} = \rho_o u^0$. Correspondingly the rest-mass flux is increased from $\rho_o\mathbf{v}$ to $\rho_o u^0\mathbf{v} = \rho_o\mathbf{u}$ [Eq. (2.62)], and the law of rest-mass conservation becomes $\partial(\rho_o u^0)/\partial t + \partial(\rho_o u^j)/\partial x^j = 0$. When taking the Newtonian limit, we should identify the Newtonian mass ρ_N with the low-velocity limit of this Lorentz-contracted rest-mass density:

$$\rho_N = \rho_o u^0 \simeq \rho_o\left(1 + \frac{1}{2}v^2\right). \tag{13.89b}$$

nonrelativistic limit of Lorentz-contracted mass density

In the nonrelativistic limit the specific internal energy u, the kinetic energy per unit mass $\frac{1}{2}v^2$, and the ratio of pressure to rest-mass density P/ρ_o are of the same order of smallness:

$$u \sim \frac{1}{2}v^2 \sim \frac{P}{\rho_o} \ll 1, \tag{13.90}$$

and the momentum density T^{j0} is accurate to first order in $v \equiv |\mathbf{v}|$, the momentum flux (stress) T^{jk} and the energy density T^{00} are both accurate to second order in v, and the energy flux T^{0j} is accurate to third order in v. To these accuracies, the perfect-fluid stress-energy tensor (13.85), when combined with Eqs. (13.84) and (13.89), takes the following form:

nonrelativistic limit of stress-energy tensor for ideal (perfect) fluid

$$T^{j0} = \rho_N v^j, \quad T^{jk} = Pg^{jk} + \rho_N v^j v^k,$$

$$T^{00} = \rho_N + \frac{1}{2}\rho_N v^2 + \rho_N u, \quad T^{0j} = \rho_N v^j + \left(\frac{1}{2}v^2 + u + \frac{P}{\rho_N}\right)\rho_N v^j; \tag{13.91}$$

see Ex. 13.22. These are precisely the same as the nonrelativistic momentum density, momentum flux, energy density, and energy flux in Table 13.1, aside from (i) the notational change $\rho \to \rho_N$ from there to here, (ii) including the rest mass-energy, $\rho_N = \rho_N c^2$, in T_{00} here but not there, and (iii) including the rest-mass-energy flux $\rho_N v^j$ in T^{0j} here but not there.

EXERCISES

Exercise 13.22 *Derivation: Nonrelativistic Limit of Perfect-Fluid Stress-Energy Tensor* T2

(a) Show that in the nonrelativistic limit, the components of the perfect-fluid stress-energy tensor (13.85) take on the forms (13.91), and verify that these agree with the densities and fluxes of energy and momentum that are used in nonrelativistic fluid mechanics (Table 13.1).

(b) Show that the contribution of the pressure P to the relativistic density of inertial mass causes the term $(P/\rho_N)\rho_N \mathbf{v} = P\mathbf{v}$ to appear in the nonrelativistic energy flux.

BOX 13.5.   TERMINOLOGY USED IN CHAPTER 13

This chapter introduces a large amount of terminology. We list much of it here.

adiabatic A process in which each fluid element conserves its entropy.

adiabatic index The parameter Γ that relates pressure and density changes, $\delta P/P = \Gamma \delta\rho/\rho$, in an adiabatic process. For an ideal gas, it is the ratio of specific heats: $\Gamma = \gamma \equiv C_P/C_V$.

advective time derivative The time derivative moving with the fluid: $d/dt = \partial/\partial t + \mathbf{v} \cdot \nabla$. Also called the convective time derivative.

barotropic A process or equation in which pressure can be regarded as a function solely of density: $P = P(\rho)$.

Bernoulli function, also sometimes called Bernoulli constant: $B = \rho(\frac{1}{2}v^2 + h + \Phi)$.

bulk viscosity, coefficient of The proportionality constant ζ relating rate of expansion to viscous stress: $\mathbf{T}_{\mathrm{vis}} = -\zeta\theta\mathbf{g}$.

convective time derivative See advective time derivative.

dissipation A process that increases the entropy. Viscosity and diffusive heat flow (heat conduction) are forms of dissipation.

dynamic viscosity The coefficient of shear viscosity, η.

(continued)

BOX 13.5.   (continued)

equation of state In this chapter, where chemical and nuclear reactions do not occur: relations of the form $u(\rho, s)$, $P(\rho, s)$ or $u(\rho, T)$, $P(\rho, T)$.

Euler equation Newton's second law for an ideal fluid: $\rho d\mathbf{v}/dt = -\nabla P + \rho\mathbf{g}$.

Eulerian changes Changes in a quantity measured at fixed locations in space; cf. Lagrangian changes.

expansion, rate of Fractional rate of increase of a fluid element's volume: $\theta = \nabla \cdot \mathbf{v}$.

gas A fluid in which the separations between molecules are large compared to the molecular sizes, and no long-range forces act among molecules except gravity; cf. liquid.

ideal flow A flow in which there is no dissipation.

ideal fluid A fluid in which there are no dissipative processes (also called "perfect fluid").

ideal gas A gas in which the sizes of the molecules and (nongravitational) forces among them are neglected, so the pressure is due solely to the molecules' kinetic motions: $P = nk_BT = [\rho/(\mu m_p)]k_BT$.

incompressible A process or fluid in which the fractional changes of density are small, $\delta\rho/\rho \ll 1$, so the velocity can be approximated as divergence free: $\nabla \cdot \mathbf{v} = 0$.

inviscid Having negligible viscosity.

irrotational A flow or fluid with vanishing vorticity.

isentropic A process or fluid in which the entropy per unit rest mass s is the same everywhere.

isobar A surface of constant pressure.

isothermal A process or fluid in which the temperature is the same everywhere.

kinematic viscosity The ratio of the coefficient of shear viscosity to the density: $\nu \equiv \eta/\rho$.

Lagrangian changes Changes measured moving with the fluid; cf. Eulerian changes.

laminar flow A nonturbulent flow.

(continued)

BOX 13.5.   (continued)

liquid A fluid in which the molecules are packed side by side (e.g., water); contrast this with a gas.

mean molecular weight The average mass of a molecule in a fluid, divided by the mass of a proton: μ.

Navier-Stokes equation Newton's second law for a viscous, incompressible fluid: $d\mathbf{v}/dt = -(1/\rho)\boldsymbol{\nabla} P + \nu\nabla^2\mathbf{v} + \mathbf{g}$.

Newtonian fluid A (i) nonrelativistic fluid, or (ii) a fluid in which the shear-stress tensor is proportional to the rate of shear $\boldsymbol{\sigma}$ and is time-independent when $\boldsymbol{\sigma}$ is constant.

perfect fluid See ideal fluid.

perfect gas An ideal gas (with $P = [\rho/(\mu m_p)]k_B T$) that has negligible excitation of internal molecular degrees of freedom.

polytropic A barotropic pressure-density relation of the form $P \propto \rho^{1+1/n}$ for some constant n called the *polytropic index*. The proportionality constant is usually a function of entropy.

Reynolds number The ratio $\mathrm{Re} = LV/\nu$, where L is the characteristic lengthscale of a flow, V is the characteristic velocity, and ν is the kinematic viscosity. In order of magnitude it is the ratio of inertial acceleration $(\mathbf{v} \cdot \boldsymbol{\nabla})\mathbf{v}$ to viscous acceleration $\nu\nabla^2\mathbf{v}$ in the Navier-Stokes equation.

rotation, rate of Antisymmetric part of the gradient of velocity; vorticity converted into an antisymmetric tensor using the Levi-Civita tensor.

shear, rate of Symmetric, trace-free part of the gradient of velocity: $\boldsymbol{\sigma}$.

shear viscosity, coefficient of The proportionality constant η relating rate of shear to viscous stress: $\mathbf{T}_{\mathrm{vis}} = -\eta\boldsymbol{\sigma}$.

steady flow Flow that is independent of time in some chosen reference frame.

turbulent flow Flow characterized by chaotic fluid motions.

vorticity The curl of the velocity field: $\boldsymbol{\omega} = \boldsymbol{\nabla} \times \mathbf{v}$.

Bibliographic Note

There are many good texts on fluid mechanics. Among those with a physicist's perspective, we particularly like Acheson (1990) and Lautrup (2005) at an elementary level, and Lighthill (1986) and Batchelor (2000) at a more advanced level. Landau and Lifshitz (1959), as always, is terse but good for physicists who already have some knowledge of the subject. Tritton (1987) takes an especially physical approach to the subject with lots of useful diagrams and photographs of fluid flows. Faber (1995) gives a more deductive yet still physical approach, including a broader range of flows. A general graduate text covering many topics that we discuss (including blood flow) is Kundu, Cohen, and Dowling (2012). For relativistic fluid mechanics we recommend Rezzolla and Zanotti (2013).

Given the importance of fluids to modern engineering and technology, it should not be surprising that there are many more texts with an engineering perspective than with a physics one. Those we particularly like include Potter, Wiggert, and Ramadan (2012), which has large numbers of useful examples, illustrations, and exercises; also recommended are Munson, Young, and Okiishi (2006) and White (2008).

Physical intuition is very important in fluid mechanics and is best developed with the aid of visualizations—both movies and photographs. In recent years many visualizations have been made available on the web. Movies that we have found especially useful are those produced by the National Committee for Fluid Mechanics Films (Shapiro 1961a) and those produced by Hunter Rouse (1963a–f).

The numerical solution of the equations of fluid dynamics on computers (computational fluid dynamics, or CFD) is a mature field of science in its own right. CFD simulations are widely used in engineering, geophysics, astrophysics, and the movie industry. We do not treat CFD in this book. For an elementary introduction, see Ferziger and Peric (2001), Lautrup (2005, Chap. 21) and Kundu, Cohen, and Dowling (2012, Chap. 11). A text that emphasizs fundamental principles and formal development is Hosking and Dewar (2016). For more thorough pedagogical treatments see, for example, Fletcher (1991), Toro (2010), and Canuto et al. (2014); in the relativistic domain, see Rezzolla and Zanotti (2013).

CHAPTER FOURTEEN

Vorticity

The flow of wet water

RICHARD FEYNMAN (1964, VOLUME 2, CHAP. 41)

14.1 Overview

In the last chapter, we introduced an important quantity called *vorticity*, which is the principal subject of the present chapter. Although the most mathematically simple flows are potential, with velocity $\mathbf{v} = \nabla\psi$ for some ψ so the vorticity $\boldsymbol{\omega} = \nabla \times \mathbf{v}$ vanishes, most naturally occurring flows are *vortical,* with $\boldsymbol{\omega} \neq 0$. By studying vorticity, we shall develop an intuitive understanding of how flows evolve. We shall also see that computing the vorticity can be a powerful step along the path to determining a flow's full velocity field.

We all think we can recognize a vortex. The most hackneyed example is water disappearing down a drainhole in a bathtub or shower. The angular velocity around the drain increases inward, because the angular momentum per unit mass is conserved when the water moves radially slowly, in addition to rotating. Remarkably, angular momentum conservation means that the product of the circular velocity v_ϕ and the radius ϖ is independent of radius, which in turn implies that $\nabla \times \mathbf{v} = 0$. So this is a vortex without vorticity! (Except, as we shall see, a delta-function spike of vorticity right at the drainhole's center; see Sec. 14.2 and Ex. 14.24.) Vorticity is a precise physical quantity defined by $\boldsymbol{\omega} = \nabla \times \mathbf{v}$, not just any vaguely circulatory motion.

In Sec. 14.2, we introduce two tools for analyzing and using vorticity: vortex lines and circulation. Vorticity is a vector field and therefore has integral curves obtained by solving $d\mathbf{x}/d\lambda = \boldsymbol{\omega}$ for some parameter λ. These integral curves are the *vortex lines*; they are analogous to magnetic field lines. The flux of vorticity $\int_{\mathcal{S}} \boldsymbol{\omega} \cdot d\boldsymbol{\Sigma}$ across a surface $\mathcal{S}$ is equal to the integral of the velocity field, $\Gamma \equiv \int_{\partial\mathcal{S}} \mathbf{v} \cdot d\mathbf{x}$, around the surface's boundary $\partial\mathcal{S}$ (by Stokes' theorem). We call this Γ the *circulation* around $\partial\mathcal{S}$; it is analogous to magnetic-field flux. In fact, the analogy with magnetic fields turns out to be extremely useful. Vorticity, like a magnetic field, automatically has vanishing divergence, which means that the vortex lines are continuous, just like magnetic field lines. Vorticity, again like a magnetic field, is an axial vector and thus can be written as the curl of a polar vector potential, the velocity $\mathbf{v}$.[1] Vorticity has the interesting property that it evolves in a perfect fluid (ideal fluid) in such a manner that the flow carries the vortex lines along with it; we say that the vortex lines are "frozen into the fluid." When viscous stresses make the fluid imperfect, then the vortex lines diffuse through the moving fluid with a diffusion coefficient that is equal to the kinematic viscosity ν.

In Sec. 14.3, we study a classical problem that illustrates both the action and the propagation of vorticity: the *creeping* flow of a low-Reynolds-number fluid around a sphere. (Low-Reynolds-number flow arises when the magnitude of the viscous-acceleration term in the equation of motion is much larger than the magnitude of the inertial acceleration.) The solution to this problem finds contemporary application in the sedimentation rates of soot particles in the atmosphere.

In Sec. 14.4, we turn to high-Reynolds-number flows, in which the viscous stress is quantitatively weak over most of the fluid. Here, the action of vorticity can be concentrated in relatively thin *boundary layers* in which the vorticity, created at a wall, diffuses away into the main body of the flow. Boundary layers arise because in real fluids, intermolecular attraction requires that the component of the fluid velocity parallel to the boundary (not just the normal component) vanish. The vanishing of both components of velocity distinguishes real fluid flow at high Reynolds number (i.e., low viscosity) from the solutions obtained assuming no vorticity. Nevertheless, it is often (but sometimes not) a good approximation to seek a solution to the equations of fluid dynamics in which vortex-free fluid slips freely past the solid and then match it to a boundary-layer solution near the solid.

Stirred water in a tea cup and Earth's oceans and atmosphere rotate nearly rigidly, so they are most nicely analyzed in a co-rotating reference frame. In Sec. 14.5, we use

---

1. Pursuing the electromagnetic analogy further, we can ask the question, "Given a specified vorticity field $\boldsymbol{\omega}(\mathbf{x}, t)$, can I solve uniquely for the velocity $\mathbf{v}(\mathbf{x}, t)$?" The answer, of course, is "No." There is gauge freedom, so many solutions exist. Interestingly, if we specify that the flow be incompressible $\nabla \cdot \mathbf{v} = 0$ (i.e., be the analog of the Coulomb gauge), then $\mathbf{v}(\mathbf{x}, t)$ is unique up to an additive constant.

In the 1960s, Asher Shapiro and the National Committee for Fluid Mechanics Films and Hunter Rouse at the University of Iowa produced movies that are pedagogically powerful and still fully relevant a half-century later. Those most germane to this chapter are:

- Shapiro (1961b), relevant to the entire chapter;
- Taylor (1964), relevant to Sec. 14.3;
- Abernathy (1968), the portion dealing with (nonturbulent) laminar boundary layers, relevant to Sec. 14.4;
- Rouse (1963e), relevant to Sec. 14.4;
- Fultz (1969), relevant to Sec. 14.5; and
- Taylor (1968), relevant to Sec. 14.5.4.

Also relevant are many segments of the movies produced at the University of Iowa, such as Rouse (1963a–f).

such an analysis to discover novel phenomena produced by Coriolis forces—including winds around pressure depressions; Taylor columns of fluid that hang together like a rigid body; Ekman boundary layers with spiral-shaped velocity fields; gyres (humps of water, e.g., the Sargasso Sea), around which ocean currents (e.g., the Gulf Stream) circulate; and tea leaves that accumulate at the bottom center of a tea cup.

When a flow has a large amount of shear, Nature often finds ways to tap the relative kinetic energy of neighboring stream tubes. In Sec. 14.6 we explore the resulting instabilities, focusing primarily on horizontally stratified fluids with relative horizontal velocities, which have vorticity concentrated in regions where the velocity changes sharply. The instabilities we encounter show up, in Nature, as (among other things) billow clouds and clear-air turbulence in the stratosphere. These phenomena provide motivation for the principal topic of the next chapter: turbulence.

Physical insight into the phenomena of this chapter is greatly aided by movies of fluid flows. The reader is urged to view relevant movies in parallel with reading this chapter; see Box 14.2.

14.2 Vorticity, Circulation, and Their Evolution

14.2

In Sec. 13.5.4, we defined the vorticity as the curl of the velocity field, $\boldsymbol{\omega} = \nabla \times \mathbf{v}$, analogous to defining the magnetic field as the curl of a vector potential. To get insight into vorticity, consider the three simple 2-dimensional flows shown in Fig. 14.1.

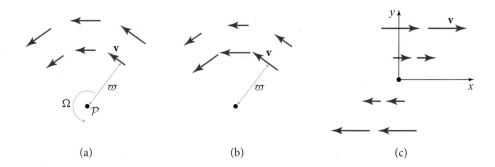

(a) (b) (c)

FIGURE 14.1 Vorticity in three 2-dimensional flows. The vorticity vector points in the z direction (orthogonal to the plane of the flow) and so can be thought of as a scalar ($\omega = \omega_z$). (a) Constant angular velocity Ω. If we measure radius ϖ from the center $\mathcal{P}$, the circular velocity satisfies $v = \Omega\varpi$. This flow has vorticity $\omega = 2\Omega$ everywhere. (b) Constant angular momentum per unit mass j, with $v = j/\varpi$. This flow has zero vorticity except at its center, $\omega = 2\pi j\delta(\mathbf{x})$. (c) Shearing flow in a laminar boundary layer, $v_x = -\omega y$ with $\omega < 0$. The vorticity is $\omega = -v_x/y$, and the rate of shear is $\sigma_{xy} = \sigma_{yx} = -\frac{1}{2}\omega$.

Figure 14.1a shows *uniform (rigid) rotation*, with constant angular velocity $\boldsymbol{\Omega} = \Omega\mathbf{e}_z$. The velocity field is $\mathbf{v} = \boldsymbol{\Omega} \times \mathbf{x}$, where $\mathbf{x}$ is measured from the rotation axis. Taking its curl, we discover that $\boldsymbol{\omega} = 2\boldsymbol{\Omega}$ everywhere.

Figure 14.1b shows a flow in which the angular momentum per unit mass $\mathbf{j} = j\mathbf{e}_z$ is constant, because it was approximately conserved as the fluid gradually drifted inward to create this flow. In this case the rotation is *differential* (radially changing angular velocity), with $\mathbf{v} = \mathbf{j} \times \mathbf{x}/\varpi^2$ (where $\varpi = |\mathbf{x}|$ and $\mathbf{j} = $ const). This is the kind of flow that occurs around a bathtub vortex and around a tornado—but outside the vortex's or tornado's core. The vorticity is $\boldsymbol{\omega} = 2\pi j\delta(\mathbf{x})$ (where $\delta(\mathbf{x})$ is the 2-dimensional Dirac delta function in the plane of the flow), so it vanishes everywhere except at the center, $\mathbf{x} = 0$ (or, more precisely, except in the vortex's or tornado's core). Anywhere in the flow, two neighboring fluid elements, separated tangentially, rotate about each other with an angular velocity $+\mathbf{j}/\varpi^2$, but when the two elements are separated radially, their relative angular velocity is $-\mathbf{j}/\varpi^2$; see Ex. 14.1. The average of these two angular velocities vanishes, which seems reasonable, since the vorticity vanishes.

The vanishing vorticity in this case is an illustration of a simple geometrical description of vorticity in any 2-dimensional flow (Ex. 13.15): If we orient the $\mathbf{e}_z$ axis of a Cartesian coordinate system along the vorticity, then

$$\omega = \left(\frac{\partial v_y}{\partial x} - \frac{\partial v_x}{\partial y}\right)\mathbf{e}_z. \tag{14.1}$$

This expression implies that the vorticity at a point is the sum of the angular velocities of a pair of mutually perpendicular, infinitesimal lines passing through that point (one along the x direction, the other along the y direction) and moving with the fluid; for example, these lines could be thin straws suspended in the fluid. If we float a little

vortex without vorticity except at its center

vorticity measured by a vane with orthogonal fins

vane with orthogonal fins in the flow, with the vane parallel to $\boldsymbol{\omega}$, then the vane will rotate with an angular velocity that is the average of the flow's angular velocities at its fins, which is half the vorticity. Equivalently, the vorticity is twice the rotation rate of the vane. In the case of constant-angular-momentum flow in Fig. 14.1b, the average of the two angular velocities is zero, the vane doesn't rotate, and the vorticity vanishes.

Figure 14.1c shows the flow in a plane-parallel shear layer. In this case, a line in the flow along the x direction does not rotate, while a line along the y direction rotates with angular velocity ω. The sum of these two angular velocities, $0 + \omega = \omega$, is the vorticity. Evidently, curved streamlines are not a necessary condition for vorticity.

Exercise 14.1 *Practice: Constant-Angular-Momentum Flow—Relative Motion of Fluid Elements*

Verify that for the constant-angular-momentum flow of Fig. 14.1b, with $\mathbf{v} = \mathbf{j} \times \mathbf{x}/\varpi^2$, two neighboring fluid elements move around each other with angular velocity $+j/\varpi^2$ when separated tangentially and $-j/\varpi^2$ when separated radially. [Hint: If the fluid elements' separation vector is $\boldsymbol{\xi}$, then their relative velocity is $\nabla_{\boldsymbol{\xi}} \mathbf{v} = \boldsymbol{\xi} \cdot \nabla \mathbf{v}$. Why?]

Exercise 14.2 *Practice: Vorticity and Incompressibility*

Sketch the streamlines for the following stationary 2-dimensional flows, determine whether the flow is compressible, and evaluate its vorticity. The coordinates are Cartesian in parts (a) and (b), and are circular polar with orthonormal bases $\{\mathbf{e}_{\varpi}, \mathbf{e}_{\phi}\}$ in (c) and (d).

(a) $v_x = 2xy$, $v_y = x^2$,

(b) $v_x = x^2$, $v_y = -2xy$,

(c) $v_{\varpi} = 0$, $v_{\phi} = \varpi$,

(d) $v_{\varpi} = 0$, $v_{\phi} = \varpi^{-1}$.

Exercise 14.3 **Example: Rotating Superfluids*

At low temperatures certain fluids undergo a phase transition to a superfluid state. A good example is ^{4}He, for which the transition temperature is 2.2 K. As a superfluid has no viscosity, it cannot develop vorticity. How then can it rotate? The answer (e.g., Feynman 1972, Chap. 11) is that not all the fluid is in a superfluid state; some of it is normal and can have vorticity. When the fluid rotates, all the vorticity is concentrated in microscopic vortex cores of normal fluid that are parallel to the rotation axis and have quantized circulations $\Gamma = h/m$, where m is the mass of the atoms and h is Planck's constant. The fluid external to these vortex cores is irrotational (has vanishing vorticity). These normal fluid vortices may be pinned at the walls of the container.

(a) Explain, using a diagram, how the vorticity of the macroscopic velocity field, averaged over many vortex cores, is twice the mean angular velocity of the fluid.

(b) Make an order-of-magnitude estimate of the spacing between these vortex cores in a beaker of superfluid helium on a turntable rotating at 10 rpm.

(c) Repeat this estimate for a neutron star, which mostly comprises superfluid neutron pairs at the density of nuclear matter and spins with a period of order a millisecond. (The mass of the star is roughly 3×10^{30} kg.)

14.2.1

14.2.1 Vorticity Evolution

vortex lines

By analogy with magnetic field lines, we define a flow's *vortex lines* to be parallel to the vorticity vector $\boldsymbol{\omega}$ and to have a line density proportional to $\omega = |\boldsymbol{\omega}|$. These vortex lines are always continuous throughout the fluid, because the vorticity field, like the magnetic field, is a curl and therefore is necessarily solenoidal ($\nabla \cdot \boldsymbol{\omega} = 0$). However, vortex lines can begin and end on solid surfaces, as the equations of fluid dynamics no longer apply there. Figure 14.2 shows an example: vortex lines that emerge from the wingtip of a flying airplane.

Vorticity and its vortex lines depend on the velocity field at a particular instant and evolve with time as the velocity field evolves. We can determine how by manipulating the Navier-Stokes equation.

In this chapter and the next one, we restrict ourselves to flows that are incompressible in the sense that $\nabla \cdot \mathbf{v} = 0$. As we saw in Sec. 13.6, this is the case when the flow is substantially subsonic and gravitational potential differences are not too extreme. We also require (as is almost always the case) that the shear viscosity vary spatially far

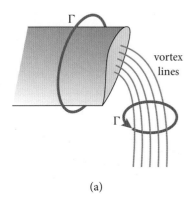

(a)

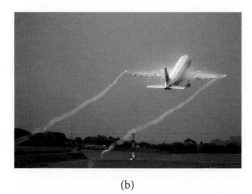

(b)

(c)

FIGURE 14.2 (a) Sketch of the wing of a flying airplane and the vortex lines that emerge from the wing tip and sweep backward behind the plane. The lines are concentrated in a region with small cross section, a vortex of whirling air. The closed red curves encircle the wing and the vortex; the integral of the velocity field around these curves, $\Gamma = \int \mathbf{v} \cdot d\mathbf{x}$, is the circulation contained in the wing and its boundary layers, and in the vortex; see Sec. 14.2.4 and especially Ex. 14.8. (b) Photograph of the two vortices emerging from the wingtips of an Airbus, made visible by light scattering off water droplets in the vortex cores (Ex. 14.6). Photo © Daniel Umaña. (c) Sketch of vortex lines (dashed) in the wingtip vortices of a flying bird and the flow lines (solid) of air around them. Sketch from Vogel, 1994, Fig. 12.7c, reprinted by permission.

more slowly than the shear itself. These restrictions allow us to write the Navier-Stokes equation in its simplest form:

$$\frac{d\mathbf{v}}{dt} \equiv \frac{\partial \mathbf{v}}{\partial t} + (\mathbf{v} \cdot \nabla)\mathbf{v} = -\frac{\nabla P}{\rho} - \nabla \Phi + \nu \nabla^2 \mathbf{v} \qquad (14.2)$$

[Eq. (13.70) with $\mathbf{g} = -\nabla \Phi$].

To derive the desired evolution equation for vorticity, we take the curl of Eq. (14.2) and use the vector identity $(\mathbf{v} \cdot \nabla)\mathbf{v} = \nabla(v^2)/2 - \mathbf{v} \times \boldsymbol{\omega}$ (easily derivable using the Levi-Civita tensor and index notation) to obtain

$$\frac{\partial \boldsymbol{\omega}}{\partial t} = \nabla \times (\mathbf{v} \times \boldsymbol{\omega}) - \frac{\nabla P \times \nabla \rho}{\rho^2} + \nu \nabla^2 \boldsymbol{\omega}. \qquad (14.3)$$

Although the flow is assumed incompressible, $\nabla \cdot \mathbf{v} = 0$, the density can vary spatially due to a varying chemical composition (e.g., some regions might be oil and others water) or varying temperature and associated thermal expansion. Therefore, we must not omit the $\nabla P \times \nabla \rho$ term.

It is convenient to rewrite the vorticity evolution equation (14.3) with the aid of the relation (again derivable using the Levi-Civita tensor)

$$\nabla \times (\mathbf{v} \times \boldsymbol{\omega}) = (\boldsymbol{\omega} \cdot \nabla)\mathbf{v} + \mathbf{v}(\nabla \cdot \boldsymbol{\omega}) - \boldsymbol{\omega}(\nabla \cdot \mathbf{v}) - (\mathbf{v} \cdot \nabla)\boldsymbol{\omega}. \qquad (14.4)$$

Inserting this into Eq. (14.3), using $\nabla \cdot \boldsymbol{\omega} = 0$ and $\nabla \cdot \mathbf{v} = 0$, and introducing a new type of time derivative[2]

$$\frac{D\boldsymbol{\omega}}{Dt} \equiv \frac{\partial \boldsymbol{\omega}}{\partial t} + (\mathbf{v} \cdot \nabla)\boldsymbol{\omega} - (\boldsymbol{\omega} \cdot \nabla)\mathbf{v} = \frac{d\boldsymbol{\omega}}{dt} - (\boldsymbol{\omega} \cdot \nabla)\mathbf{v}, \qquad (14.5)$$

we bring Eq. (14.3) into the following form:

$$\frac{D\boldsymbol{\omega}}{Dt} = -\frac{\nabla P \times \nabla \rho}{\rho^2} + \nu \nabla^2 \boldsymbol{\omega}. \qquad (14.6)$$

vorticity evolution equation for incompressible flow

This is our favorite form for the vorticity evolution equation for an incompressible flow, $\nabla \cdot \mathbf{v} = 0$. If there are additional accelerations acting on the fluid, then their curls must be added to the right-hand side. The most important examples are the Coriolis acceleration $-2\boldsymbol{\Omega} \times \mathbf{v}$ in a reference frame that rotates rigidly with angular velocity $\boldsymbol{\Omega}$ (Sec. 14.5.1), and the Lorentz-force acceleration $\mathbf{j} \times \mathbf{B}/\rho$ when the fluid has an internal electric current density $\mathbf{j}$ and an immersed magnetic field $\mathbf{B}$ (Sec. 19.2.1); then Eq. (14.6) becomes

$$\frac{D\boldsymbol{\omega}}{Dt} = -\frac{\nabla P \times \nabla \rho}{\rho^2} + \nu \nabla^2 \boldsymbol{\omega} - 2\nabla \times (\boldsymbol{\Omega} \times \mathbf{v}) + \nabla \times (\mathbf{j} \times \mathbf{B}/\rho). \qquad (14.7)$$

influence of Coriolis and Lorentz forces on vorticity

---

2. The combination of spatial derivatives appearing here is called the *Lie derivative* and is denoted $\mathcal{L}_\mathbf{v}\boldsymbol{\omega} \equiv (\mathbf{v} \cdot \nabla)\boldsymbol{\omega} - (\boldsymbol{\omega} \cdot \nabla)\mathbf{v}$; it is also the *commutator* of $\mathbf{v}$ and $\boldsymbol{\omega}$ and is denoted $[\mathbf{v}, \boldsymbol{\omega}]$. It is often encountered in differential geometry.

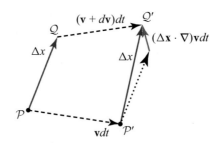

FIGURE 14.3 Equation of motion for an infinitesimal vector $\Delta \mathbf{x}$ connecting two fluid elements. As the fluid elements at $\mathcal{P}$ and $\mathcal{Q}$ move to $\mathcal{P}'$ and $\mathcal{Q}'$ in a time interval dt, the vector changes by $(\Delta \mathbf{x} \cdot \nabla)\mathbf{v}\,dt$.

In the remainder of Sec. 14.2, we explore the predictions of our favorite form [Eq. (14.6)] of the vorticity evolution equation.

The operator D/Dt [defined by Eq. (14.5) when acting on a vector and by $D/Dt = d/dt$ when acting on a scalar] is called the *fluid derivative*. (Warning: The notation D/Dt is used in some older texts for the convective derivative d/dt.) The geometrical meaning of the fluid derivative can be understood from Fig. 14.3. Denote by $\Delta \mathbf{x}(t)$ the vector connecting two points $\mathcal{P}$ and $\mathcal{Q}$ that are moving with the fluid. Then the figure shows that the convective derivative $d\Delta \mathbf{x}/dt$ is the relative velocity of these two points, namely $(\Delta \mathbf{x} \cdot \nabla)\mathbf{v}$. Therefore, by the second equality in Eq. (14.5), the fluid derivative of $\Delta \mathbf{x}$ vanishes:

$$\frac{D\Delta \mathbf{x}}{Dt} = 0. \tag{14.8}$$

meaning of the fluid derivative D/Dt

Correspondingly, *the fluid derivative of any vector is its rate of change relative to a vector, such as $\Delta \mathbf{x}$, whose tail and head move with the fluid.*

14.2.2

14.2.2 Barotropic, Inviscid, Compressible Flows: Vortex Lines Frozen into Fluid

To understand the vorticity evolution law (14.6) physically, we explore various special cases in this and the next few subsections.

Here we specialize to a barotropic [$P = P(\rho)$], inviscid ($\nu = 0$) fluid flow. (This kind of flow often occurs in Earth's atmosphere and oceans, well away from solid boundaries.) Then the right-hand side of Eq. (14.6) vanishes, leaving $D\boldsymbol{\omega}/Dt = 0$.

For generality, we temporarily (this subsection only) abandon our restriction to incompressible flow, $\nabla \cdot \mathbf{v} = 0$, but keep the flow barotropic and inviscid. Then it is straightforward to deduce, from the curl of the Euler equation (13.44), that

$$\frac{D\boldsymbol{\omega}}{Dt} = -\boldsymbol{\omega}\nabla \cdot \mathbf{v} \tag{14.9}$$

(Ex. 14.4). This equation shows that the vorticity has a fluid derivative parallel to itself: the fluid slides along its vortex lines, or equivalently, the vortex lines are frozen into

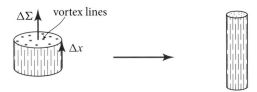

FIGURE 14.4 Simple demonstration of the kinematics of vorticity propagation in a compressible, barotropic, inviscid flow. A short, thick cylindrical fluid element with generators parallel to the local vorticity is deformed, by the flow, into a long, slender cylinder. By virtue of Eq. (14.10), we can think of the vortex lines as being convected with the fluid, with no creation of new lines or destruction of old ones, so that the number of vortex lines passing through the cylinder (through its end surface $\Delta\boldsymbol{\Sigma}$) remains constant.

(i.e., are carried by) the moving fluid. The wingtip vortex lines of Fig. 14.2 are an example. They are carried backward by the air that flowed over the wingtips, and they endow that air with vorticity that emerges from a wingtip.

We can actually make the fluid derivative vanish by substituting $\nabla \cdot \mathbf{v} = -\rho^{-1}d\rho/dt$ (the equation of mass conservation) into Eq. (14.9); the result is

$$\frac{D}{Dt}\left(\frac{\boldsymbol{\omega}}{\rho}\right) = 0 \quad \text{for barotropic, inviscid flow.} \tag{14.10}$$

Therefore, the quantity $\boldsymbol{\omega}/\rho$ evolves according to the same equation as the separation $\Delta\mathbf{x}$ of two points in the fluid. To see what this implies, consider a small cylindrical fluid element whose symmetry axis is parallel to $\boldsymbol{\omega}$ (Fig. 14.4). Denote its vectorial length by $\Delta\mathbf{x}$, its vectorial cross sectional area by $\Delta\boldsymbol{\Sigma}$, and its conserved mass by $\Delta M = \rho \Delta\mathbf{x} \cdot \Delta\boldsymbol{\Sigma}$. Then, since $\boldsymbol{\omega}/\rho$ points along $\Delta\mathbf{x}$ and both are frozen into the fluid, it must be that $\boldsymbol{\omega}/\rho = \text{const} \times \Delta\mathbf{x}$. Therefore, the fluid element's conserved mass is $\Delta M = \rho \Delta\mathbf{x} \cdot \Delta\boldsymbol{\Sigma} = \text{const} \times \boldsymbol{\omega} \cdot \Delta\boldsymbol{\Sigma}$, so $\boldsymbol{\omega} \cdot \Delta\boldsymbol{\Sigma}$ is conserved as the cylindrical fluid element moves and deforms. We thereby conclude that the fluid's vortex lines, with number per unit area proportional to $|\boldsymbol{\omega}|$, are convected by our barotropic, inviscid fluid, without being created or destroyed.

> for barotropic, inviscid flow: vortex lines are convected (frozen into the fluid)

Now return to an incompressible flow, $\nabla \cdot \mathbf{v} = 0$ (which includes, of course, Earth's oceans and atmosphere), so the vorticity evolution equation becomes $D\boldsymbol{\omega}/Dt = 0$. Suppose that the flow is 2 dimensional (as it commonly is to moderate accuracy when averaged over transverse scales large compared to the thickness of the atmosphere and oceans), so $\mathbf{v}$ is in the x and y directions and is independent of z. Then $\boldsymbol{\omega} = \omega\mathbf{e}_z$, and we can regard the vorticity as the scalar ω. Then Eq. (14.5) with $(\boldsymbol{\omega} \cdot \nabla)\mathbf{v} = 0$ implies that the vorticity obeys the simple propagation law

$$\frac{d\omega}{dt} = 0. \tag{14.11}$$

Thus, *in a 2-dimensional, incompressible, barotropic, inviscid flow, the scalar vorticity is conserved when convected,* just like entropy per unit mass in an adiabatic fluid.

EXERCISES

Exercise 14.4 *Derivation: Vorticity Evolution in a Compressible, Barotropic, Inviscid Flow*
By taking the curl of the Euler equation (13.44), derive the vorticity evolution equation (14.9) for a compressible, barotropic, inviscid flow.

14.2.3

14.2.3 Tornados

A particularly graphic illustration of the behavior of vorticity is provided by a tornado. Tornados in North America are most commonly formed at a front where cold, dry air from the north meets warm, moist air from the south, and huge, cumulonimbus thunderclouds form. A strong updraft of the warm, moist air creates rotational motion about a horizontal axis, and updraft of the central part of the rotation axis itself makes it somewhat vertical. A low-pressure vortical core is created at the center of this spinning fluid (recall Crocco's theorem, Ex. 13.9), and the spinning region lengthens under the action of up- and downdrafts. Now, consider this process in the context of vorticity propagation. As the flow (to first approximation) is incompressible, a lengthening of the spinning region's vortex lines corresponds to a reduction in the cross section and a strengthening of the vorticity. This in turn corresponds to an increase in the tornado's circulatory speeds. (Speeds in excess of 450 km/hr have been reported.) If and when the tornado touches down to the ground and its very-low-pressure core passes over the walls and roof of a building, the far larger, normal atmospheric pressure inside the building can cause the building to explode. Further details are explored in Exs. 13.9 and 14.5.

EXERCISES

Exercise 14.5 *Problem: Tornado*
(a) Figure 14.5 shows photographs of two particularly destructive tornados and one waterspout (a tornado sucking water from the ocean). For the tornados the wind speeds near the ground are particularly high: about 450 km/hr. Estimate the wind speeds at the top, where the tornados merge with the clouds. For the water spout, the wind speed near the water is about 150 km/hr. Estimate the wind speed at the top.

(b) Estimate the air pressure in atmospheres in the cores of these tornados and water spout. (Hint: Use Crocco's theorem, Ex. 13.9.)

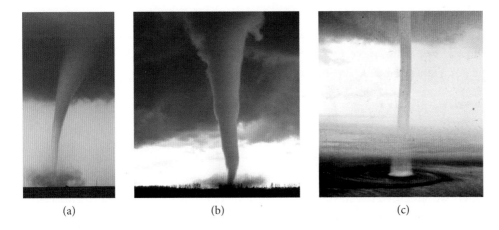

(a)   (b)   (c)

FIGURE 14.5 (a,b) Two destructive tornados and (c) a waterspout. (a) Eric Nguyen / Science Source; (b) Justin James Hobson, licensed under Creative Commons-ShareAlike 3.0 Unported (CC BY-SA 3.0); (c) NOAA; http://www.spc.noaa.gov/faq/tornado/wtrspout.htm.

Exercise 14.6 *Problem: Visualizing a Wingtip Vortex*

Explain why the pressure and temperature of the core of a wingtip vortex are significantly lower than the pressure and temperature of the ambient air. Under what circumstances will this lead to condensation of tiny water droplets in the vortex core, off which light can scatter, as in Fig. 14.2b?

14.2.4 Circulation and Kelvin's Theorem

14.2.4

circulation

Intimately related to vorticity is a quantity called *circulation* Γ; it is defined as the line integral of the velocity around a closed curve ∂S lying in the fluid:

$$\Gamma \equiv \int_{\partial S} \mathbf{v} \cdot d\mathbf{x}; \tag{14.12a}$$

it can be regarded as a property of the closed curve ∂S. We can invoke Stokes' theorem to convert this circulation into a surface integral of the vorticity passing through a surface S bounded by ∂S:

$$\Gamma = \int_{S} \boldsymbol{\omega} \cdot d\boldsymbol{\Sigma}. \tag{14.12b}$$

[Note, though, that Eq. (14.12b) is only valid if the area bounded by the contour is simply connected. If the area enclosed contains a solid body, this equation may fail.] Equation (14.12b) states that the circulation Γ around ∂S is the flux of vorticity through S, or equivalently, it is proportional to the number of vortex lines passing through S. Circulation is thus the fluid counterpart of magnetic flux.

circulation as the flux of vorticity

Kelvin's theorem tells us the rate of change of the circulation associated with a particular contour ∂S that is attached to the moving fluid. Let us evaluate this rate directly using the convective derivative of Γ. We do this by differentiating the two vector quantities inside the integral (14.12a):

$$\frac{d\Gamma}{dt} = \int_{\partial S} \frac{d\mathbf{v}}{dt} \cdot d\mathbf{x} + \int_{\partial S} \mathbf{v} \cdot d\left(\frac{d\mathbf{x}}{dt}\right)$$

$$= -\int_{\partial S} \frac{\nabla P}{\rho} \cdot d\mathbf{x} - \int_{\partial S} \nabla \Phi \cdot d\mathbf{x} + \nu \int_{\partial S} (\nabla^2 \mathbf{v}) \cdot d\mathbf{x} + \int_{\partial S} d\frac{1}{2}v^2, \quad (14.13)$$

where we have used the Navier-Stokes equation (14.2) with $\nu = \text{constant}$. The second and fourth terms on the right-hand side of Eq. (14.13) vanish around a closed curve, and the first can be rewritten in different notation to give

Kelvin's theorem for evolution of circulation

$$\boxed{\frac{d\Gamma}{dt} = -\int_{\partial S} \frac{dP}{\rho} + \nu \int_{\partial S} (\nabla^2 \mathbf{v}) \cdot d\mathbf{x}.} \quad (14.14)$$

This is Kelvin's theorem for the evolution of circulation. It is an integral version of our evolution equation (14.6) for vorticity. In a rotating reference frame it must be augmented by the integral of the Coriolis acceleration $-2\boldsymbol{\Omega} \times \mathbf{v}$ around the closed curve ∂S, and if the fluid is electrically conducting with current density $\mathbf{j}$ and possesses a magnetic field $\mathbf{B}$, it must be augmented by the integral of the Lorentz force per unit mass $(\mathbf{j} \times \mathbf{B})/\rho$ around ∂S:

$$\frac{d\Gamma}{dt} = -\int_{\partial S} \frac{dP}{\rho} + \nu \int_{\partial S} (\nabla^2 \mathbf{v}) \cdot d\mathbf{x} - 2\int_{\partial S} \boldsymbol{\Omega} \times \mathbf{v} \cdot d\mathbf{x} + \int_{\partial S} \frac{\mathbf{j} \times \mathbf{B}}{\rho} \cdot d\mathbf{x}. \quad (14.15)$$

This is the integral form of Eq. (14.7).

If the fluid is barotropic, $P = P(\rho)$, and the effects of viscosity are negligible (and the coordinates are inertial and there is no magnetic field and no electric current), then the right-hand sides of Eqs. (14.14) and (14.15) vanish, and Kelvin's theorem takes the simple form

conservation of circulation in a barotropic inviscid flow

$$\boxed{\frac{d\Gamma}{dt} = 0 \quad \text{for barotropic, inviscid flow.}} \quad (14.16)$$

Eq. (14.16) is the global version of our result that the circulation $\boldsymbol{\omega} \cdot \Delta\boldsymbol{\Sigma}$ of an infinitesimal fluid element is conserved.

The qualitative content of Kelvin's theorem is that vorticity in a fluid is long lived. A fluid's vorticity and circulation (or lack thereof) will persist, unchanged, unless and until viscosity or a $(\nabla P \times \nabla\rho)/\rho^2$ term (or a Coriolis or Lorentz force term) comes into play in the vorticity evolution equation (14.3). We now explore these sources and modifications of vorticity and circulation.

14.2.5 Diffusion of Vortex Lines

First consider the action of viscous stresses on an existing vorticity distribution. For an incompressible, barotropic fluid with nonnegligible viscosity, the vorticity evolution law (14.6) becomes

$$\boxed{\frac{D\boldsymbol{\omega}}{Dt} = \nu\nabla^2\boldsymbol{\omega} \quad \text{for an incompressible, barotropic fluid.}}$$ (14.17)

diffusion equation for vorticity

This is a *convective vectorial diffusion equation*: the viscous term $\nu\nabla^2\boldsymbol{\omega}$ causes the vortex lines to diffuse through the moving fluid, and the kinematic viscosity ν is the diffusion coefficient for the vorticity. When viscosity is negligible, the vortex lines are frozen into the flow. When viscosity is significant and no boundaries impede the vorticity's diffusion and the vorticity initially is concentrated in a thin vortex, then as time t passes the vorticity diffuses outward into a cross sectional area $\sim \nu t$ in the moving fluid (Ex. 14.7; i.e., the vortex expands). Thus the kinematic viscosity not only has the dimensions of a diffusion coefficient, it also actually controls the diffusion of vortex lines relative to the moving fluid.

diffusion of vortex lines

As a simple example of the spreading of vortex lines, consider an infinite plate moving parallel to itself relative to a fluid at rest. Transform to the rest frame of the plate (Fig. 14.6a) so the fluid moves past it. Suppose that at time $t = 0$ the velocity has

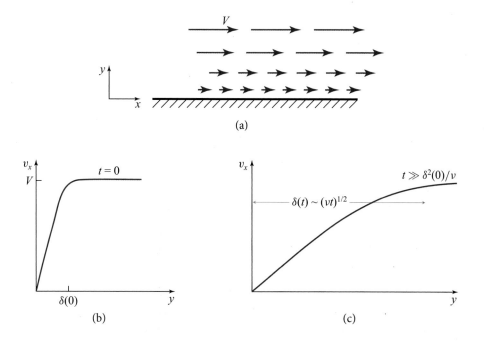

(a)

(b)

(c)

FIGURE 14.6 A simple shear layer that is translation invariant along the flow direction x. Vorticity diffuses away from the static plate at $y = 0$, under the action of viscous torques, in much the same way that heat diffuses away from a heated surface.

only a component v_x parallel to the plate, which depends solely on the distance y from the plate, so the flow is translation invariant in the x direction; then it will continue always to be translation invariant. Suppose further that, at $t = 0$, v_x is constant, $v_x = V$, except in a thin boundary layer near the plate, where it drops rapidly to 0 at $y = 0$ (as it must, because of the plate's no-slip boundary condition; Fig. 14.6b). As the flow is a function only of y (and t), and $\mathbf{v}$ and $\boldsymbol{\omega}$ point in directions orthogonal to $\mathbf{e}_y$, in this flow the fluid derivative D/Dt [Eq. (14.5)] reduces to $\partial/\partial t$, and the convective diffusion equation (14.17) becomes an ordinary scalar diffusion equation for the only nonzero component of the vorticity: $\partial\omega_z/\partial t = \nu\nabla^2\omega_z$. Let the initial thickness of the boundary layer at time $t = 0$ be $\delta(0)$. Then our experience with the diffusion equation (e.g., Exs. 3.17 and 6.3) tells us that the viscosity will diffuse through the fluid under the action of viscous stress, and as a result, the boundary-layer thickness will increase with time as

$$\delta(t) \sim (\nu t)^{\frac{1}{2}} \quad \text{for} \quad t \gtrsim \delta(0)^2/\nu; \tag{14.18}$$

see Fig. 14.6c. We compute the evolving structure $v_x(y, t)$ of the expanding boundary layer in Sec. 14.4.

EXERCISES

Exercise 14.7 **Example: Diffusive Expansion of a Vortex*
At time $t = 0$, a 2-dimensional barotropic flow has a velocity field, in circular polar coordinates, $\mathbf{v} = (j/\varpi)\mathbf{e}_\phi$ (Fig. 14.1b); correspondingly, its vorticity is $\boldsymbol{\omega} = 2\pi j\delta(x)\delta(y)\mathbf{e}_z$: it is a delta-function vortex. In this exercise you will solve for the full details of the subsequent evolution of the flow.

(a) Solve the vorticity evolution equation (14.6) to determine the vorticity as a function of time. From your solution, show that the area in which the vorticity is concentrated (the cross sectional area of the vortex) at time t is $A \sim \nu t$, and show that the vorticity is becoming smoothed out—it is evolving toward a state of uniform vorticity, where the viscosity will no longer have any influence.

(b) From your computed vorticity in part (a), plus circular symmetry, compute the velocity field as a function of time.

(c) From the Navier-Stokes equation (or equally well, from Crocco's theorem), compute the evolution of the pressure distribution $P(\varpi, t)$.

Remark: This exercise illustrates a frequent phenomenon in fluid mechanics: The pressure adjusts itself to whatever it needs to be to accommodate the flow. One can often solve for the pressure distribution in the end, after having worked out other details. This happens here because, when one takes the curl of the Navier-Stokes equation for a barotropic fluid (which we did to get the evolution equation for vorticity), the pressure drops out—it decouples from the evolution equation for vorticity and hence velocity.

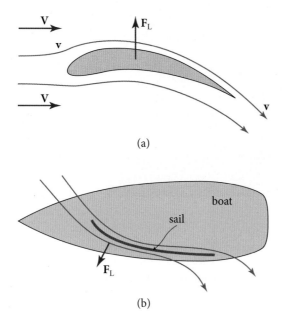

FIGURE 14.7 (a) Air flow around an airfoil viewed in cross section. The solid green lines are flow lines with velocity **v**; the incoming velocity is **V**. (b) Air flow around a sail. The shaded body represents the hull of a boat seen from above.

Exercise 14.8 **Example: The Lift on an Airplane Wing, Wingtip Vortices, Sailing Upwind, and Fish Locomotion*

When an appropriately curved airfoil (e.g., an airplane wing) is introduced into a steady flow of air, the air has to flow faster along the upper surface than along the lower surface, which can create a lifting force (Fig. 14.7a). In this situation, compressibility and gravity are usually unimportant for the flow.

(a) Show that the pressure difference across the airfoil is given approximately by $\Delta P = \frac{1}{2}\rho\Delta(v^2) = \rho v \Delta v$. Hence show that the lift exerted by the air on an airfoil of length L is given approximately by

$$\boxed{F_{\mathrm{L}} = L \int \Delta P \, dx = \rho V L \Gamma,} \tag{14.19}$$

where Γ is the circulation around the airfoil, and V is the air's incoming speed in the airfoil's rest frame. This is known as *Kutta-Joukowski's theorem*. Interpret this result in terms of the conservation of linear momentum, and sketch the overall flow pattern.

(b) Explain why the circulation around an airplane wing (left orange curve in Fig. 14.2a) is the same as that around the wingtip's vortex (right orange curve in

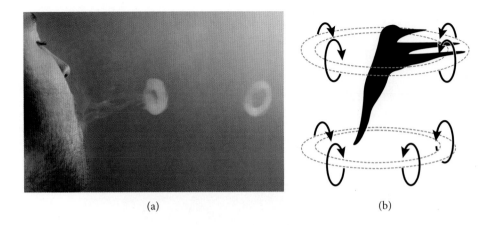

(a)   (b)

FIGURE 14.8 (a) Smoke rings blown by a man travel away from his mouth. Photo ©Andrew Vargas. (b) A ring-shaped wingtip vortex is produced by each half beat of the wings of a hovering hummingbird; the vortices travel downward. Sketch from Vogel, 1994, Fig. 12.7a, reprinted by permission.

Fig. 14.2a), and correspondingly explain why wingtip vortices are essential features of an airplane's flight. Without them, an airplane could not take off.

(c) How might birds' wingtip vortices (Fig. 14.2c) be related to the V-shaped configuration of birds in a flying flock? (For discussion, see Vogel, 1994, p. 288.)

(d) Explain how the same kind of lift as occurs on an airplane wing propels a sailboat forward when sailing upwind, as in Fig. 14.7b.

(e) Snakes, eels, and most fish undulate their bodies and/or fins as they swim. Draw pictures that explain how the same principle that propels a sailboat pushes these animals forward as well.

Exercise 14.9 *Problem: Vortex Rings*
Smoke rings (ring-shaped vortices) blown by a person (Fig. 14.8a) propagate away from him. Similarly, a hovering hummingbird produces ring-shaped vortices that propagate downward (Fig. 14.8b). Sketch the velocity field of such a vortex and explain how it propels itself through the ambient air. For the hovering hummingbird, discuss the role of the vortices in momentum conservation.

14.2.6   14.2.6 Sources of Vorticity

Having discussed how vorticity is conserved in simple inviscid flows and how it diffuses away under the action of viscosity, we now consider its sources. The most important source is a solid surface. When fluid suddenly encounters a solid surface, such as the leading edge of an airplane wing or a spoon stirring coffee, intermolecular forces act to decelerate the fluid rapidly in a thin boundary layer along the surface. This deceleration introduces circulation and consequently vorticity into the flow, where

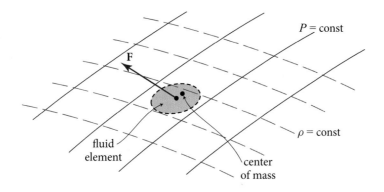

FIGURE 14.9 Mechanical explanation for the creation of vorticity in a nonbarotropic fluid. The net pressure gradient force **F**, acting at the geometric center of a small fluid element, is normal to the isobars (solid lines) and does not pass through the center of mass of the element; thereby a torque is produced.

none existed before; that vorticity then diffuses into the bulk flow, thickening the boundary layer (Sec. 14.4).

If the fluid is nonbarotropic (usually due to spatially variable chemical composition or spatially variable temperature), then pressure gradients can also create vorticity, as described by the first term on the right-hand side of the vorticity evolution law (14.6): $(-\boldsymbol{\nabla} P \times \boldsymbol{\nabla} \rho)/\rho^2$. Physically, when the surfaces of constant pressure (*isobars*) do not coincide with the surfaces of constant density (*isochors*), then the net pressure force on a small fluid element does not pass through its center of mass. The pressure therefore exerts a torque on the fluid element, introducing some rotational motion and vorticity (Fig. 14.9). Nonbarotropic pressure gradients can therefore create vorticity within the body of the fluid. Note that because the vortex lines must be continuous, any fresh ones that are created in the fluid must be created as loops that expand from a point or a line.

vorticity generated by nonbarotropic pressure gradients

There are three other common sources of vorticity in fluid dynamics:

1. Coriolis forces, when one's reference frame is rotating rigidly;

2. Lorentz forces, when the fluid is magnetized and electrically conducting [last two terms in Eqs. (14.7) and (14.15)]; and

3. curving shock fronts (when the fluid speed is supersonic).

vorticity generated by Coriolis forces, Lorentz forces, and curving shock fronts

We discuss these sources in Sec. 14.5 and Chaps. 19 and 17, respectively.

EXERCISES

Exercise 14.10 *Problem: Vortices Generated by a Spatula*
Fill a bathtub with water and sprinkle baby powder liberally over the water's surface to aid in viewing the motion of the surface water. Then take a spatula, insert it gently into the water, move it slowly and briefly perpendicular to its flat face, then extract it

gently from the water. Twin vortices will have been generated. Observe the vortices' motions. Explain (i) the generation of the vortices, and (ii) the sense in which the velocity field of each vortex convects the other vortex through the ambient water. Use your bathtub or a swimming pool to perform other experiments on the generation and propagation of vortices.

Exercise 14.11 *Example: Vorticity Generated by Heating*
Rooms are sometimes heated by radiators (hot surfaces) that have no associated blowers or fans. Suppose that, in a room whose air is perfectly still, a radiator is turned on to high temperature. The air will begin to circulate (convect), and that air motion contains vorticity. Explain how the vorticity is generated in terms of the $-\int dP/\rho$ term of Kelvin's theorem (14.14) and the $(-\nabla P \times \nabla\rho)/\rho^2$ term of the vorticity evolution equation (14.6).

14.3 Low-Reynolds-Number Flow—Stokes Flow and Sedimentation

Reynolds number as ratio of inertial acceleration to viscous acceleration

In the previous chapter, we defined the Reynolds number Re to be the product of the characteristic speed V and lengthscale a of a flow divided by its kinematic viscosity $\nu = \eta/\rho$: $\mathrm{Re} \equiv Va/\nu$. The significance of the Reynolds number follows from the fact that, in the Navier-Stokes equation (14.2), the ratio of the magnitude of the inertial acceleration $|(\mathbf{v} \cdot \nabla)\mathbf{v}|$ to the viscous acceleration $|\nu\nabla^2\mathbf{v}|$ is approximately equal to Re. Therefore, when $\mathrm{Re} \ll 1$, the inertial acceleration can often be ignored, and the velocity field is determined by balancing the pressure gradient against the viscous stress. The velocity then scales linearly with the magnitude of the pressure gradient and vanishes when the pressure gradient vanishes.

low-Re flow: pressure gradient balances viscous stress

low-Re flow is nearly reversible

This has the amusing consequence that a low-Reynolds-number flow driven by a solid object moving through a fluid at rest is effectively reversible. An example, depicted in the movie by Taylor (1964), is a rotating sphere. If it is rotated slowly in a viscous fluid for N revolutions in one direction, then rotated in reverse for N revolutions, the fluid elements will return almost to their original positions. This phenomenon is easily understood by examining the Navier-Stokes equation (14.2) with the inertial acceleration $d\mathbf{v}/dt$ neglected and gravity omitted: $\nabla P = \eta\nabla^2\mathbf{v}$; pressure gradient balances viscous force density. No time derivatives appear here, and the equation is linear. When the direction of the sphere's rotation is reversed, the pressure gradients reverse and the velocity field reverses, bringing the fluid back to its original state.

regimes of low-Re flow

From the magnitudes of viscosities of real fluids (Table 13.2), it follows that the low-Reynolds-number limit is appropriate either for small-scale flows (e.g., the motion of microorganisms in water; see Box 14.3) or for very viscous large scale fluids (e.g., Earth's mantle; Ex. 14.13).

Swimming provides insight into the differences between flows with low and high Reynolds numbers. In water, the flow around a swimming fish has Re $\sim 10^{+6}$, while that around a bacterium has Re $\sim 10^{-5}$. In the simplest variant of fish locomotion, the fish wags its tail fin back and forth nearly rigidly, pushing water backward and itself forward with each stroke [schematic drawing (a) below]. And a simple variant of bacterial propulsion is that of the *E. coli* bacterium: it has a rigid corkscrew-shaped tail (made from several flagella), which rotates, pushing water backward via friction (viscosity) and itself forward [schematic drawing (b), adapted from Nelson (2008)].

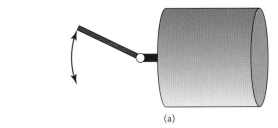

(a)

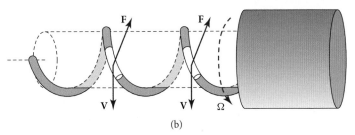

(b)

A fish's tail-wagging propulsion would fail at low Reynolds number, because the flow would be reversible: after each back-forth wagging cycle, the fluid would return to its original state—a consequence of the linear, no-time-derivatives balance of pressure and viscous forces: $\nabla P = \eta \nabla^2 \mathbf{v}$ (see second paragraph of Sec. 14.3). However, for the fish with high Reynolds number, aside from a very thin boundary layer near the fin's surface, viscosity is negligible, and so the flow is governed by Euler's equation: $d\mathbf{v}/dt = (\partial/\partial t + \mathbf{v} \cdot \nabla)\mathbf{v} = -\nabla P/\rho$ (and of course mass conservation). The time derivatives and nonlinearity make the fluid's motion nonreversible in this case. With each back-forth cycle of wag, the tail feeds substantial net backward momentum into the water, and the fish acquires net forward momentum (which will be counteracted by friction in boundary layers along the fish's body if the fish is moving fast enough).

(continued)

E. coli's corkscrew propulsion would fail at high Reynolds number, because its tail's flagella are so thin ($d \sim 20$ nm) that they could not push any noticeable amount of water inertially. However, at *E. coli's* low Reynolds number, the amount of water entrained by the tail's viscous friction is almost independent of the tail's thickness. To understand the resulting frictional propulsion, consider a segment of the tail shown white in drawing (b). It moves laterally with velocity **V**. If the segment's length ℓ is huge compared to its thickness d, then the water produces a drag force **F** on it of magnitude $F \sim 2\pi\eta V\ell/\ln(\nu/Vd)$ (Ex. 14.12) that points *not* opposite to **V** but rather somewhat forward of that direction, as shown in the drawing. The reason is that this segment of a thin rod has a drag that is larger (by about a factor two) when pulled perpendicular to its long axis than when pulled along its long axis. Hence the drag is a tensorial function of **V**, $F_i = H_{ij}V_j$, and is not parallel to **V**. In drawing (b) the transverse component of the drag cancels out when one integrates it along the winding tail, but the forward component adds coherently along the tail, giving the bacterium a net forward force.

For further discussion of fish as well as bacteria, see, for example, Vogel (1994) and Nelson (2008).

14.3.1 Motivation: Climate Change

An important example of a small-scale flow arises when we ask whether cooling of Earth due to volcanic explosions can mitigate global warming.

The context is concern about anthropogenic (human-made) climate change. Earth's atmosphere is a subtle and fragile protector of the environment that allows life to flourish. Especially worrisome is the increase in atmospheric carbon dioxide—by nearly 25% over the past 50 years to a mass of 3×10^{15} kg. As an important greenhouse gas, carbon dioxide traps solar radiation. Increases in its concentration are contributing to the observed increase in mean surface temperature, the rise of sea levels, and the release of oceanic carbon dioxide with potential runaway consequences.

These effects are partially mitigated by volcanos like Krakatoa, which exploded in 1883,[3] releasing roughly 250 megatons or $\sim 10^{18}$ J of energy and nearly 10^{14} kg of small particles (aerosols, ash, soot, etc.), of which $\sim 10^{12}$ kg was raised into the stratosphere,

---

3. A more recent example was the Pinatubo volcano in 1991, which released roughly a tenth the mass of Krakatoa. Studying the consequences of this explosion provided important calibration for models of greater catastrophes.

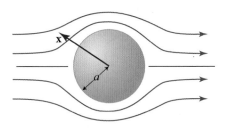

FIGURE 14.10 Flow lines for Stokes flow around a sphere.

where much of it remained for several years. These micron-sized particles absorb light with roughly their geometrical cross section. The area of Earth's surface is roughly 5×10^{14} m^2, and the density of the particles is roughly 2000 kg m^{-3}, so that 10^{12} kg of aerosols is sufficient to blot out the Sun. More specifically, the micron-sized particles absorb solar optical and ultraviolet radiation while remaining reasonably transparent to infrared radiation escaping from Earth's surface. The result is a noticeable global cooling of Earth for as long as the particles remain suspended in the atmosphere.[4]

A key issue in assessing how our environment is likely to change over the next century is how long particles of size 1–10 μm will remain in the atmosphere after volcanic explosions (i.e., their rate of sedimentation). This problem is one of low-Reynolds-number flow.

We model the sedimentation by computing the speed at which a spherical soot particle falls through quiescent air when the Reynolds number is small. The speed is governed by a balance between the downward force of gravity and the speed-dependent upward drag force of the air. We compute this sedimentation speed by first evaluating the force of the air on the moving particle ignoring gravity, and then, at the end of the calculation, inserting the influence of gravity.

14.3.2 Stokes Flow

We model the soot particle as a sphere with radius a. The low-Reynolds-number flow of a viscous fluid past such a sphere is known as *Stokes flow*. We calculate the flow's velocity field, and then from it, the force of the fluid on the sphere. (This calculation also finds application in the famous Millikan oil-drop experiment and is a prototype for many more complex calculations of low-Reynolds-number flow.)

SOLUTION FOR VELOCITY FIELD AND PRESSURE
It is easiest to tackle this problem in the frame of the sphere, where the flow is stationary (time-independent; Fig. 14.10). We seek a solution to the Navier-Stokes equation in which the flow velocity $\mathbf{v}(\mathbf{x})$ tends to a constant value $\mathbf{V}$ (the velocity of the sphere

---

4. Similar effects could follow the explosion of nuclear weapons in a major nuclear war according to Turco et al. (1986), a phenomenon they called *nuclear winter*.

through the fluid) at large distances from the sphere's center. We presume the asymptotic flow velocity $\mathbf{V}$ to be highly subsonic, so the flow is effectively incompressible: $\nabla \cdot \mathbf{v} = 0$.

insignificance of inertial acceleration
We define the Reynolds number for this flow by $\mathrm{Re} = \rho V a / \eta = V a / \nu$. As this is assumed to be small, in the Navier-Stokes equation (14.2) we can ignore the inertial term, which is $\mathrm{O}(V \Delta v / a)$, in comparison with the viscous term, which is $\mathrm{O}(\nu \Delta v / a^2)$; here $\Delta v \sim V$ is the total velocity variation. The time-independent Navier-Stokes equation (14.2) can thus be well approximated by $\nabla P = \rho \mathbf{g} + \eta \nabla^2 \mathbf{v}$. The uniform gravitational force density $\rho \mathbf{g}$ is obviously balanced by a uniform pressure gradient (hydrostatic equilibrium). Removing these uniform terms from both sides of the equation, we get

$$\nabla P = \eta \nabla^2 \mathbf{v}, \tag{14.20}$$

where ∇P is now just the nonuniform part of the pressure gradient required to balance the viscous force density. The full details of the flow are governed by this force-balance equation, the flow's incompressibility

$$\nabla \cdot \mathbf{v} = 0, \tag{14.21}$$

and the boundary conditions $\mathbf{v} = 0$ at $r = a$ and $\mathbf{v} \to \mathbf{V}$ at $r \to \infty$.

From force balance (14.20) we infer that in order of magnitude the difference between the fluid's pressure on the front of the sphere and that on the back is $\Delta P \sim \eta V / a$. We also expect a viscous drag stress along the sphere's sides of magnitude $T_{r\theta} \sim \eta V / a$, where V / a is the magnitude of the shear. These two stresses, acting on the sphere's surface area $\sim a^2$, will produce a net drag force $F \sim \eta V a$. Our goal is to verify this order-of-magnitude estimate, compute the force more accurately, then balance this force against gravity (adjusted for the much smaller Archimedes buoyancy force produced by the uniform part of the pressure gradient), and thereby infer the speed of fall V of a soot particle.

For a highly accurate analysis of the flow, we could write the full solution as a perturbation expansion in powers of the Reynolds number Re. We compute only the leading term in this expansion; the next term, which corrects for inertial effects, will be smaller than our solution by a factor $\mathrm{O}(\mathrm{Re})$.

some general solution ideas
Our solution to this classic problem is based on some general ideas that ought to be familiar from other areas of physics. First, we observe that the quantities in which we are interested are the pressure P, the velocity $\mathbf{v}$, and the vorticity $\boldsymbol{\omega}$—a scalar, a polar vector, and an axial vector, respectively—and to first order in the Reynolds number they should be linear in $\mathbf{V}$. The only scalar we can form that is linear in $\mathbf{V}$ is $\mathbf{V} \cdot \mathbf{x}$, so we expect the variable part of the pressure to be proportional to this combination. For the polar-vector velocity we have two choices, a part $\propto \mathbf{V}$ and a part $\propto (\mathbf{V} \cdot \mathbf{x})\mathbf{x}$; both terms are present. Finally, for the axial-vector vorticity, our only option is a term $\propto \mathbf{V} \times \mathbf{x}$. We use these combinations below.

Now take the divergence of Eq. (14.20), and conclude that the pressure must satisfy Laplace's equation: $\nabla^2 P = 0$. The solution should be axisymmetric about $\mathbf{V}$, and we know that axisymmetric solutions to Laplace's equation that decay as $r \to \infty$ can be expanded as a sum over Legendre polynomials, $\sum_{\ell=0}^{\infty} P_\ell(\mu)/r^{\ell+1}$, where μ is the cosine of the angle θ between $\mathbf{V}$ and $\mathbf{x}$, and r is $|\mathbf{x}|$. Since the variable part of P is proportional to $\mathbf{V} \cdot \mathbf{x}$, the dipolar ($\ell = 1$) term [for which $P_1(\mu) = \mu = \mathbf{V} \cdot \mathbf{x}/(Vr)$] is all we need; the higher-order polynomials will be higher-order in $\mathbf{V}$ and thus must arise at higher orders of the Reynolds-number expansion. We therefore write

$$P = P_\infty + \frac{k\eta(\mathbf{V} \cdot \mathbf{x})a}{r^3}. \tag{14.22}$$

Here k is a numerical constant that must be determined, we have introduced a factor η to make k dimensionless, and P_∞ is the pressure far from the sphere.

Next consider the vorticity. Since it is proportional to $\mathbf{V} \times \mathbf{x}$ and cannot depend in any other way on $\mathbf{V}$, it must be expressible as

$$\boldsymbol{\omega} = \frac{\mathbf{V} \times \mathbf{x}}{a^2} f(r/a). \tag{14.23}$$

The factor a appears in the denominator to make the unknown function f dimensionless. We determine this unknown function by rewriting Eq. (14.20) in the form

$$\nabla P = -\eta \nabla \times \boldsymbol{\omega} \tag{14.24}$$

[which we can do because $\nabla \times \boldsymbol{\omega} = -\nabla^2 \mathbf{v} + \nabla(\nabla \cdot \mathbf{v})$, and $\nabla \cdot \mathbf{v} = 0$], and then inserting Eqs. (14.22) and (14.23) into (14.24) to obtain $f(\xi) = k\xi^{-3}$; hence we have

$$\boldsymbol{\omega} = \frac{k(\mathbf{V} \times \mathbf{x})a}{r^3}. \tag{14.25}$$

Equation (14.25) for the vorticity looks familiar. It has the form of the Biot-Savart law for the magnetic field from a current element. We can therefore write down immediately a formula for its associated "vector potential," which in this case is the velocity:

electromagnetic analogy

$$\mathbf{v}(\mathbf{x}) = \frac{ka\mathbf{V}}{r} + \nabla\psi. \tag{14.26}$$

The addition of the $\nabla\psi$ term corresponds to the familiar gauge freedom in defining the vector potential. However, in the case of fluid dynamics, where the velocity is a directly observable quantity, the choice of the scalar ψ is fixed by the boundary conditions instead of being free. As ψ is a scalar linear in $\mathbf{V}$, it must be expressible in terms of a second dimensionless function $g(\xi)$ as

$$\psi = g(r/a)\mathbf{V} \cdot \mathbf{x}. \tag{14.27}$$

Next we recall that the flow is incompressible: $\mathbf{\nabla} \cdot \mathbf{v} = 0$. Substituting Eq. (14.27) into Eq. (14.26) and setting the divergence expressed in spherical polar coordinates to zero, we obtain an ordinary differential equation for g:

$$\frac{d^2 g}{d\xi^2} + \frac{4}{\xi} \frac{dg}{d\xi} - \frac{k}{\xi^3} = 0. \tag{14.28}$$

This has the solution

$$g(\xi) = A - \frac{k}{2\xi} + \frac{B}{\xi^3}, \tag{14.29}$$

where A and B are integration constants. As $\mathbf{v} \to \mathbf{V}$ far from the sphere, the constant $A = 1$. The constants B and k can be found by imposing the boundary condition $\mathbf{v} = 0$ for $r = a$. We thereby obtain $B = -1/4$ and $k = -3/2$. After substituting these values into Eq. (14.26), we obtain for the velocity field:

$$\mathbf{v} = \left[1 - \frac{3}{4} \left(\frac{a}{r} \right) - \frac{1}{4} \left(\frac{a}{r} \right)^3 \right] \mathbf{V} - \frac{3}{4} \left(\frac{a}{r} \right)^3 \left[1 - \left(\frac{a}{r} \right)^2 \right] \frac{(\mathbf{V} \cdot \mathbf{x}) \mathbf{x}}{a^2}. \tag{14.30}$$

velocity, pressure, and vorticity in Stokes flow

The associated pressure and vorticity, from Eqs. (14.22) and (14.25), are given by

$$P = P_\infty - \frac{3\eta a (\mathbf{V} \cdot \mathbf{x})}{2r^3},$$

$$\boldsymbol{\omega} = \frac{3a (\mathbf{x} \times \mathbf{V})}{2r^3}. \tag{14.31}$$

The pressure is seen to be largest on the upstream hemisphere, as expected. However, the vorticity, which points in the direction of $\mathbf{e}_\phi$, is seen to be symmetric between the front and the back of the sphere. This is because our low-Reynolds-number approximation neglects the advection of vorticity by the velocity field and only retains the diffusive term. Vorticity is generated on the front surface of the sphere and diffuses into the surrounding flow; then, after the flow passes the sphere's equator, the vorticity diffuses back inward and is absorbed onto the sphere's back face.

An analysis that includes higher orders in the Reynolds number would show that not all of the vorticity is reabsorbed; a small portion is left in the fluid downstream from the sphere.

We have been able to obtain a simple solution for low-Reynolds-number flow past a sphere. Although closed-form solutions like this are not common, the methods used to derive it are of widespread applicability. Let us recall them. First, we approximated the equation of motion by omitting the subdominant inertial term and invoked a symmetry argument. We used our knowledge of elementary electrostatics to write the pressure in the form of Eq. (14.22). We then invoked a second symmetry argument to solve for the vorticity and drew on another analogy with electromagnetic theory to derive a differential equation for the velocity field, which was solved subject to the no-slip boundary condition on the surface of the sphere.

Having obtained a solution for the velocity field and pressure, it is instructive to reexamine our approximations. The first point to notice is that the velocity perturbation, given by Eq. (14.30), dies off slowly—it is inversely proportional to distance r from the sphere. Thus for our solution to be valid, the region through which the sphere is moving must be much larger than the sphere; otherwise, the boundary conditions at $r \to \infty$ have to be modified. This is not a concern for a soot particle in the atmosphere. A second, related point is that, if we compare the sizes of the inertial term (which we neglected) and the pressure gradient (which we kept) in the full Navier-Stokes equation, we find that

$$|(\mathbf{v} \cdot \nabla)\mathbf{v}| \sim \frac{V^2 a}{r^2}, \quad \left|\frac{\nabla P}{\rho}\right| \sim \frac{\eta a V}{\rho r^3}. \tag{14.32}$$

At $r = a$ their ratio is $V a \rho / \eta = V a / \nu$, which is the (small) Reynolds number. However, at a distance $r \sim \eta / \rho V = a/\text{Re}$ from the sphere's center, the inertial term becomes comparable to the pressure term. Correspondingly, to improve on our zero-order solution, we must perform a second expansion at large r including inertial effects and then match it asymptotically to our near-zone expansion (see, e.g., Panton, 2005, Sec. 21.9). This technique of *matched asymptotic expansions* (Panton, 2005, Chap. 15) is a very powerful and general way of finding approximate solutions valid over a wide range of lengthscales, where the dominant physics changes from one scale to the next. We present an explicit example of such a matched asymptotic expansion in Sec. 16.5.3.

matched asymptotic expansions

DRAG FORCE

Let us return to the problem that motivated this calculation: computing the drag force on the sphere. It can be computed by integrating the stress tensor $\mathbf{T} = P\mathbf{g} - 2\eta\boldsymbol{\sigma}$ over the sphere's surface. If we introduce a local orthonormal basis $\{\mathbf{e}_r, \mathbf{e}_\theta, \mathbf{e}_\phi\}$ with polar axis ($\theta = 0$) along the flow direction $\mathbf{V}$, then we readily see that the only nonzero viscous contribution to the surface stress tensor is $T_{r\theta} = T_{\theta r} = \eta \partial v_\theta / \partial r$. The net resistive force along the direction of the velocity (drag force) is then given by

$$F = \int_{r=a} \frac{d\boldsymbol{\Sigma} \cdot \mathbf{T} \cdot \mathbf{V}}{V}$$

$$= \int_0^{2\pi} 2\pi a^2 \sin\theta \, d\theta \left[-P_\infty \cos\theta + \frac{3\eta V \cos^2\theta}{2a} + \frac{3\eta V \sin^2\theta}{2a} \right], \tag{14.33}$$

where the first two terms are from the fluid's pressure on the sphere and the third is from its viscous stress. The integrals are easy and give $F = 6\pi \eta a V$ for the force, in the direction of the flow. In the rest frame of the fluid, the sphere moves with velocity $\mathbf{V}_{\text{sphere}} = -\mathbf{V}$, so the drag force that the sphere experiences is

$$\boxed{\mathbf{F} = -6\pi \eta a \mathbf{V}_{\text{sphere}}.} \tag{14.34}$$

Stokes' law for drag force in low-Re flow

Eq. (14.34) is Stokes' law for the drag force in low-Reynolds-number flow. Two-thirds of the force comes from the viscous stress and one third from the pressure. When the influence of inertial forces at $r \gtrsim a/\text{Re}$ is taken into account via matched asymptotic expansions, one obtains a correction to the drag force:

$$\mathbf{F} = -6\pi\eta a \mathbf{V}_{\text{sphere}}\left(1 + \frac{3aV}{8\nu}\right) = -6\pi\eta a \mathbf{V}_{\text{sphere}}\left(1 + \frac{3\,\text{Re}_d}{16}\right), \qquad (14.35)$$

where (as is common) the Reynolds number Re_d is based on the sphere's diameter $d = 2a$ rather than its radius.

Exercise 14.12 *Problem: Stokes Flow around a Cylinder: Stokes' Paradox*
Consider low-Reynolds-number flow past an infinite cylinder whose axis coincides with the z-axis. Try to repeat the analysis we used for a sphere to obtain an order-of-magnitude estimate for the drag force per unit length. [Hint: You might find it useful to write $\mathbf{v} = \mathbf{\nabla} \times (\zeta\mathbf{e}_z)$, which guarantees $\mathbf{\nabla} \cdot \mathbf{v} = 0$ (cf. Box 14.4); then show that the scalar *stream function* $\zeta(\varpi, \phi)$ satisfies the biharmonic equation $\nabla^2\nabla^2\zeta = 0$.] You will encounter difficulty in finding a solution for $\mathbf{v}$ that satisfies the necessary boundary conditions at the cylinder's surface $\varpi = a$ and at large radii $\varpi \gg a$. This difficulty is called *Stokes' paradox*, and the resolution to it by including inertial forces at large radii was given by Carl Wilhelm Oseen (see, e.g., Panton, 2005, Sec. 21.10). The result for the drag force per unit length is $\mathbf{F} = -2\pi\eta\mathbf{V}(\alpha^{-1} - 0.87\alpha^{-3} + \ldots)$, where $\alpha = \ln(3.703/\text{Re}_d)$, and $\text{Re}_d = 2aV/\nu$ is the Reynolds number computed from the cylinder's diameter $d = 2a$. The logarithmic dependence on the Reynolds number and thence on the cylinder's diameter is a warning of the subtle mixture of near-cylinder viscous flow and far-distance inertial flow that influences the drag.

14.3.3

14.3.3 Sedimentation Rate

Now we return to the problem that motivated our study of Stokes flow: the rate of sedimentation of soot particles (the rate at which they sink to the ground) after a gigantic volcanic eruption. To analyze this, we must restore gravity to our analysis. We can do so by restoring to the Navier-Stokes equation the uniform pressure gradient and balancing gravitational term that we removed just before Eq. (14.20). Gravity and the buoyancy (Archimedes) force from the uniform pressure gradient exert a net downward force $(4\pi a^3/3)(\rho_s - \rho)g$ on the soot particle, which must balance the upward resistive force (14.34). Here $\rho_s \sim 2{,}000 \text{ kg m}^{-3}$ is the density of soot and $\rho \sim 1 \text{ kg m}^{-3}$ is the far smaller (and here negligible) density of air. Equating these forces, we obtain

$$V = \frac{2\rho_s a^2 g}{9\eta}. \qquad (14.36)$$

The kinematic viscosity of air at sea level is, according to Table 13.2, $\nu \sim 10^{-5} \text{ m}^2 \text{ s}^{-1}$, and the density is $\rho_a \sim 1 \text{ kg m}^{-3}$, so the coefficient of viscosity is $\eta = \rho_a\nu \sim 10^{-5} \text{ kg m}^{-1} \text{ s}^{-1}$. This viscosity is proportional to the square root of temperature and is independent of the density [cf. Eq. (13.72)]; however, the temperature does not vary by more than about 25% up to the stratosphere (Fig. 13.2), so for an approximate calculation, we can use its value at sea level. Substituting the above values into Eq. (14.36), we obtain an equilibrium sedimentation speed

$$V \sim 0.5(a/1\,\mu\text{m})^2 \text{ mm s}^{-1}. \tag{14.37}$$

For self-consistency we should also estimate the Reynolds number:

$$\text{Re} \sim \frac{2\rho_a V a}{\eta} \sim 10^{-4}\left(\frac{a}{1\,\mu\text{m}}\right)^3. \tag{14.38}$$

Our analysis is therefore only likely to be adequate for particles of radius $a \lesssim 10\,\mu\text{m}$.

limitations on validity of analysis

There is also a lower bound to the size of the particle for validity of this analysis: the mean free path of the nitrogen and oxygen molecules must be smaller than the particle. The mean free path is $\sim0.3\,\mu\text{m}$, so the resistive force is reduced when $a \lesssim 1\,\mu\text{m}$.[5]

The sedimentation speed (14.37) is much smaller than wind speeds in the upper atmosphere: $v_{\text{wind}} \sim 30$ m s^{-1}. However, as the stratosphere is reasonably stratified, the net vertical motion due to the winds is quite small,[6] and so we can estimate the settling time by dividing the stratosphere's height ~30 km by the speed [Eq. (14.37)] to obtain

$$t_{\text{settle}} \sim 6 \times 10^7 \left(\frac{a}{1\,\mu\text{m}}\right)^{-2} \text{s} \sim 2\left(\frac{a}{1\,\mu\text{m}}\right)^{-2} \text{years}. \tag{14.39}$$

This calculation is a simple model for more serious and complex analyses of sedimentation after volcanic eruptions, and the resulting mitigation of global warming. Of course, huge volcanic eruptions are rare, so no matter the result of reliable future analyses, we cannot count on volcanos to save humanity from runaway global warming. And the consequences of "geo-engineering" fixes in which particles are deliberately introduced into the atmosphere are correspondingly uncertain.

EXERCISES

Exercise 14.13 *Problem: Viscosity of Earth's Mantle*
Episodic glaciation subjects Earth's crust to loading and unloading by ice. The last major ice age was 10,000 years ago, and the subsequent unloading produces a nontidal contribution to the acceleration of Earth's rotation rate of order $|\mathbf{\Omega}|/|\dot{\mathbf{\Omega}}| \simeq 6 \times 10^{11}$ yr, detectable from observing the positions of distant stars. Corresponding changes in Earth's oblateness produce a decrease in the rate of nodal line regression of the geodetic satellites LAGEOS.

(a) Estimate the speed with which the polar regions (treated as spherical caps of radius $\sim$1,000 km) are rebounding now. Do you think the speed was much greater in the past?

5. A dimensionless number—the ratio of the mean free path to the lengthscale of the flow (in this case the radius of the particle), called the *Knudsen number* or Kn—has been defined to describe this situation. Corrections to Stokes' law for the drag force are needed when Kn $\gtrsim$ 1.
6. Brownian motion also affects the sedimentation rate for very small particles.

(b) Geological evidence suggests that a particular glaciated region of radius about 1,000 km sank in ~3,000 yr during the last ice age. By treating this as a low-Reynolds-number viscous flow, make an estimate of the coefficient of viscosity for the mantle.

Exercise 14.14 *Example: Undulatory Locomotion in Microorganisms*
Many microorganisms propel themselves, at low Reynolds number, using undulatory motion. Examples include the helical motion of *E. coli*'s corkscrew tail (Box 14.3), and undulatory waves in a forest of cilia attached to an organism's surface, or near a bare surface itself. As a 2-dimensional model for locomotion via surface waves, we idealize the organism's undisturbed surface as the plane $y = 0$, and we assume the surface undulates in and out with displacement $\delta y = (-u/kC) \sin[k(x - Ct)]$ and hence velocity $V_y = u \cos[k(x - Ct)]$. Here u is the amplitude, k the wave number, and C the surface speed in the organism's rest frame. Derive the velocity field $\mathbf{v}$ for the fluid at $y > 0$ produced by this wall motion, and from it deduce the velocity of the organism through the fluid. You could proceed as follows.

(a) The velocity field must satisfy the incompressible, low-Reynolds-number equations $\nabla P = \eta \nabla^2 \mathbf{v}$ and $\nabla \cdot \mathbf{v} = 0$ [Eqs. (14.20) and (14.21)]. Explain why $\mathbf{v}$ (which must point in the x and y directions) can be expressed as the curl of a vector potential that points in the z direction: $\mathbf{v} = \nabla \times (\zeta \mathbf{e}_z)$; and show that $\zeta(t, x, y)$ (the *stream function;* see Box 14.4 in Sec. 14.4.1) satisfies the biharmonic equation: $\nabla^2 \nabla^2 \zeta = 0$.

(b) Show that the following ζ satisfies the biharmonic equation and satisfies the required boundary conditions at the organism's surface, $v_x[x, \delta y(x, t), t] = 0$ and $v_y[x, \delta y(x, t), t] = V_y(x, t)$, and the appropriate boundary conditions at $y \to \infty$:

$$\zeta = -\frac{u}{k}(1 + ky)\exp(-ky)\sin[k(x - Ct)]$$

$$+ \frac{u^2}{2C}y\{1 - \exp(-2ky)\cos[2k(x - Ct)]\} + O(u^3). \qquad (14.40)$$

Explain how $\nabla P = \eta \nabla^2 \mathbf{v}$ is then easily satisfied.

(c) Show that the streamlines (tangent to $\mathbf{v}$) are surfaces of constant ζ. Plot these streamlines at $t = 0$ and discuss how they change as t passes. Explain why they are physically reasonable.

(d) Show that at large y the fluid moves with velocity $\mathbf{v} = [u^2/(2C)]\mathbf{e}_x$, and that therefore, in the asymptotic rest frame of the fluid, the organism moves with velocity $-[u^2/(2C)]\mathbf{e}_x$, opposite to $\mathbf{k}$. Is this physically reasonable? Why does the organism's inertia (and hence its mass) not influence this velocity? That the organism's velocity is second order in the velocity of its surface waves illustrates the difficulty of locomotion at low Reynolds number.

(e) Now consider what happens if the undulation is longitudinal with motion lying in the plane of the undulating surface. Sketch the flow pattern, and show that the organism will now move in the direction of **k**.

14.4 High-Reynolds-Number Flow—Laminar Boundary Layers

As we have described, flow near a solid surface creates vorticity, and consequently, the velocity field near the surface cannot be derived from a scalar potential, $\mathbf{v} = \boldsymbol{\nabla}\psi$. However, if the Reynolds number is high, then the vorticity may be localized in a thin boundary layer adjacent to the surface, as in Fig. 14.6. Then the flow may be very nearly of potential form $\mathbf{v} = \boldsymbol{\nabla}\psi$ outside that boundary layer. In this section, we use the equations of hydrodynamics to model the flow in the simplest example of such a boundary layer: that formed when a long, thin plate is placed in a steady, uniform flow $\mathbf{v} = V\mathbf{e}_x$ with its surface parallel to the flow (Fig. 14.11).

boundary layers in high-Re flow

If the plate is not too long, then the flow will be *laminar*, that is, steady and 2-dimensional—a function only of the distances x along the plate's length and y perpendicular to the plate (both being measured from an origin at the plate's front). We assume the flow to be highly subsonic, so it can be regarded as incompressible. As the viscous stress decelerates the fluid close to the plate, it must therefore be deflected away from the plate to avoid accumulating, thereby producing a small y component of velocity along with the larger x component. As the velocity is uniform well away from the plate, the pressure is constant outside the boundary layer. We use this condition to motivate the approximation that P is also constant in the boundary layer. After solving for the flow, we will check the self-consistency of this ansatz (guess). With

laminar flow

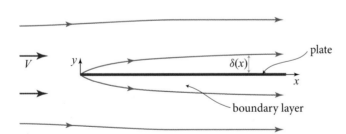

FIGURE 14.11 Laminar boundary layer formed by a long, thin plate in a flow with asymptotic speed V. The length ℓ of the plate must give a Reynolds number $\mathrm{Re}_\ell \equiv V\ell/\nu$ in the range $10 \lesssim \mathrm{Re}_\ell \lesssim 10^6$. If Re_ℓ is much less than 10, the plate will be in or near the regime of low-Reynolds-number flow (Sec. 14.3), and the boundary layer will be so thick everywhere that our analysis will fail. If Re_ℓ is much larger than 10^6, then at sufficiently great distances x down the plate ($\mathrm{Re}_x = Vx/\nu \gtrsim 10^6$), a portion of the boundary layer will become turbulently unstable and its simple laminar structure will be destroyed (see Chap. 15).

P = constant and the flow stationary, only the inertial and viscous terms remain in the Navier-Stokes equation (14.2):

$$(\mathbf{v} \cdot \nabla)\mathbf{v} \simeq \nu\nabla^2\mathbf{v}. \tag{14.41}$$

This equation must be solved in conjunction with $\nabla \cdot \mathbf{v} = 0$ and the boundary conditions $\mathbf{v} \to V\mathbf{e}_x$ as $y \to \infty$ and $\mathbf{v} \to 0$ as $y \to 0$.

The fluid first encounters the no-slip boundary condition at the front of the plate, $x = y = 0$. The flow there abruptly decelerates to vanishing velocity, creating a steep velocity gradient that contains a sharp spike of vorticity. This is the birth of the boundary layer.

Farther downstream, the total flux of vorticity inside the rectangle C of Fig. 14.12, $\int \boldsymbol{\omega} \cdot d\boldsymbol{\Sigma}$, is equal to the circulation $\Gamma_C = \int_C \mathbf{v} \cdot d\mathbf{x}$ around C. The flow velocity is zero on the bottom leg of C, and it is (very nearly) orthogonal to C on the vertical legs, so the only nonzero contribution is from the top leg, which gives $\Gamma_C = V\Delta x$. Therefore, the circulation per unit length (flux of vorticity per unit length $\Gamma_C/\Delta x$) is V everywhere along the plate. This means that there is no new vorticity acquired, and none is lost after the initial spike at the front of the plate.

As the fluid flows down the plate, from $x = 0$ to larger x, the spike of vorticity, created at the plate's leading edge, gradually diffuses outward from the wall into the flow, thickening the boundary layer.

Let us compute the order of magnitude of the boundary layer's thickness $\delta(x)$ as a function of distance x down the plate. Incompressibility, $\nabla \cdot \mathbf{v} = 0$, implies that $v_y \sim v_x\delta/x$. Using this to estimate the relative magnitudes of the various terms in the x component of the force-balance equation (14.41), we see that the dominant inertial term (left-hand side) is $\sim V^2/x$ and the dominant viscous term (right-hand side) is $\sim \nu V/\delta^2$. We therefore obtain the estimate $\delta \sim \sqrt{\nu x/V}$. This motivates us to define the function

$$\delta(x) \equiv \sqrt{\frac{\nu x}{V}} \tag{14.42}$$

for use in our quantitative analysis. Our analysis will reveal that the actual thickness of the boundary layer is several times larger than this $\delta(x)$.

Equation (14.42) shows that the boundary layer has a parabolic shape: $y \sim \delta(x) = \sqrt{\nu x/V}$. To keep the analysis manageable, we confine ourselves to the region, not too close to the front of the plate, where the layer is thin, $\delta \ll x$, and the velocity is nearly parallel to the plate, $v_y \sim (\delta/x)v_x \ll v_x$.

14.4.1

14.4.1 Blasius Velocity Profile Near a Flat Plate: Stream Function and Similarity Solution

To proceed further, we use a technique of widespread applicability in fluid mechanics: we make a *similarity ansatz*, whose validity we elucidate near the end of the calculation.

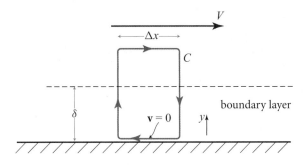

FIGURE 14.12 A rectangle C used in showing that a boundary layer's circulation per unit length $\Gamma_C/\Delta x$ is equal to the flow speed V just above the boundary layer.

We suppose that, once the boundary layer has become thin ($\delta \ll x$), the cross sectional shape of the flow is independent of distance x down the plate (it is "similar" at all x, also called "self-similar"). Stated more precisely, we assume that $v_x(x, y)$ (which has magnitude $\sim V$) and $(x/\delta)v_y$ (which also has magnitude $\sim V$) are functions only of the single transverse, dimensionless variable:

$$\xi = \frac{y}{\delta(x)} = y\sqrt{\frac{V}{\nu x}}. \tag{14.43}$$

Then our task is to compute $\mathbf{v}(\xi)$ subject to the boundary conditions $\mathbf{v} = 0$ at $\xi = 0$, and $\mathbf{v} = V\mathbf{e}_x$ at $\xi \gg 1$. We do so with the aid of a second, very useful calculational device. Recall that any vector field [$\mathbf{v}(\mathbf{x})$ in our case] can be expressed as the sum of the gradient of a scalar potential and the curl of a vector potential: $\mathbf{v} = \boldsymbol{\nabla}\psi + \boldsymbol{\nabla} \times \mathbf{A}$. If our flow were irrotational ($\boldsymbol{\omega} = 0$), we would need only $\boldsymbol{\nabla}\psi$, but it is not; the vorticity in the boundary layer is large. On the other hand, to high accuracy the flow is incompressible, $\theta = \boldsymbol{\nabla} \cdot \mathbf{v} = 0$, which means we need only the vector potential: $\mathbf{v} = \boldsymbol{\nabla} \times \mathbf{A}$. And because the flow is 2-dimensional (depends only on x and y and has $\mathbf{v}$ pointing only in the x and y directions), the vector potential need only have a z component: $\mathbf{A} = A_z\mathbf{e}_z$. We denote its nonvanishing component by $A_z \equiv \zeta(x, y)$ and give it the name *stream function,* since it governs how the laminar flow streams. In terms of the stream function, the relation $\mathbf{v} = \boldsymbol{\nabla} \times A$ takes the simple form

stream function

$$v_x = \frac{\partial\zeta}{\partial y}, \quad v_y = -\frac{\partial\zeta}{\partial x}. \tag{14.44}$$

Equation (14.44) automatically satisfies $\boldsymbol{\nabla} \cdot \mathbf{v} = 0$. Notice that $\mathbf{v} \cdot \boldsymbol{\nabla}\zeta = v_x\partial\zeta/\partial x + v_y\partial\zeta/\partial y = -v_xv_y + v_yv_x = 0$. Thus *the stream function is constant along streamlines.* (As an aside that often will be useful, e.g., in Exs. 14.12 and 14.20, we generalize this stream function in Box 14.4.)

BOX 14.4. STREAM FUNCTION FOR A GENERAL, TWO-DIMENSIONAL, INCOMPRESSIBLE FLOW T2

Consider any orthogonal coordinate system in flat 3-dimensional space, for which the metric coefficients are independent of one of the coordinates, say, x_3:

$$ds^2 = g_{11}(x_1, x_2)\, dx_1^2 + g_{22}(x_1, x_2)\, dx_2^2 + g_{33}(x_1, x_2)\, dx_3^2. \qquad (1)$$

The most common examples are Cartesian coordinates $\{x, y, z\}$ with $g_{11} = g_{22} = g_{33} = 1$; cylindrical coordinates $\{\varpi, z, \phi\}$ with $g_{11} = g_{22} = 1$ and $g_{33} = \varpi^2$; and spherical coordinates $\{r, \theta, \phi\}$ with $g_{11} = 1$, $g_{22} = r^2$, and $g_{33} = r^2 \sin^2 \theta$. Suppose the velocity field is also independent of x_3, so it is effectively 2-dimensional (translation invariant for Cartesian coordinates; axisymmetric for cylindrical or spherical coordinates).

Because the flow is incompressible, $\nabla \cdot \mathbf{v} = 0$, we can write the velocity as the curl of a vector potential: $\mathbf{v} = \nabla \times \mathbf{A}(t, x_1, x_2)$. By imposing the Lorenz gauge on the vector potential (i.e., making it divergence free, as is commonly done in electromagnetism), we can ensure that its only nonvanishing component is $A_3 = \mathbf{A} \cdot \mathbf{e}_3$, where $\mathbf{e}_3$ is the unit vector pointing in the x_3 direction. Now, a special role is played by the vector that generates local translations along the x_3 direction (i.e., that generates the flow's symmetry). If we write a location $\mathcal{P}$ in space as a function of the coordinates $\mathcal{P}(x_1, x_2, x_3)$, then this generator is $\partial \mathcal{P} / \partial x_3 = \sqrt{g_{33}}\, \mathbf{e}_3$. We define the flow's stream function by

$$\boxed{\zeta(t, x_1, x_2) \equiv \mathbf{A} \cdot \partial \mathcal{P} / \partial x_3,} \qquad (2)$$

which implies that the only nonzero component of the vector potential is $A_3 = \zeta / \sqrt{g_{33}}$.

Then it is straightforward to show (Ex. 14.19) the following. (i) The orthonormal components of the velocity field $\mathbf{v} = \nabla \times \mathbf{A}$ are

$$\boxed{v_1 = \frac{1}{\sqrt{g_{22}g_{33}}} \frac{\partial \zeta}{\partial x_2}, \quad v_2 = \frac{-1}{\sqrt{g_{11}g_{33}}} \frac{\partial \zeta}{\partial x_1}.} \qquad (3)$$

These expressions enable one to reduce an analysis of the flow to solving for three scalar functions of (t, x_1, x_2): the stream function ζ, the pressure P, and the density ρ. (ii) The stream function is a constant along flow lines: $\mathbf{v} \cdot \nabla \zeta = 0$. (iii) The stream function is proportional to the flow rate (the amount of fluid volume crossing a surface per unit time). More specifically,

(continued)

BOX 14.4. (continued)

consider a segment of some curve reaching from point $\mathcal{A}$ to point $\mathcal{B}$ in a surface of constant x_3, and expand this curve into a segment of a 2-dimensional surface by translating it through some Δx_3 along the symmetry generator $\partial \mathcal{P} / \partial x_3$. Then the flow rate across this surface is

$$
\mathcal{F} = \int_{\mathcal{A}}^{\mathcal{B}} \mathbf{v} \cdot \left(\Delta x_3 \frac{\partial \mathcal{P}}{\partial x_3} \times d\mathbf{x} \right)
$$

$$
= \int_{\mathcal{A}}^{\mathcal{B}} \mathbf{v} \cdot \left(\sqrt{g_{33}} \Delta x_3 \mathbf{e}_3 \times d\mathbf{x} \right) = [\zeta(\mathcal{B}) - \zeta(\mathcal{A})] \Delta x_3. \tag{4}
$$

Since the stream function varies on the lengthscale δ, to produce a velocity field with magnitude $\sim V$, it must have magnitude $\sim V\delta$. This motivates us to guess that it has the functional form

$$
\zeta = V\delta(x) f(\xi), \tag{14.45}
$$

where $f(\xi)$ is some dimensionless function of order unity. This guess will be good if, when inserted into Eq. (14.44), it produces a self-similar flow [i.e., one with v_x and $(x/\delta)v_y$ depending only on ξ]. Indeed, inserting Eq. (14.45) into Eq. (14.44), we obtain

mathematical form of self-similar boundary layer

$$
v_x = V f', \quad v_y = \frac{\delta(x)}{2x} V(\xi f' - f), \tag{14.46a}
$$

where the prime denotes $d/d\xi$. These equations have the desired self-similar form.

By inserting these self-similar v_x and v_y into the x component of the force-balance equation $\mathbf{v} \cdot \nabla \mathbf{v} = \nu \nabla^2 \mathbf{v}$ [Eq. (14.41)] and neglecting $\partial^2 v_x / \partial x^2$ compared to $\partial^2 v_x / \partial y^2$, we obtain a nonlinear third-order differential equation for $f(\xi)$:

$$
\frac{d^3 f}{d\xi^3} + \frac{f}{2} \frac{d^2 f}{d\xi^2} = 0. \tag{14.46b}
$$

That this equation involves x and y only in the combination $\xi = y\sqrt{V/\nu x}$ confirms that our self-similar ansatz was a good one. Equation (14.46b) must be solved subject to the boundary condition that the velocity vanish at the surface and approach V as $y \to \infty$ ($\xi \to \infty$) [cf. Eqs. (14.46a)]:

$$
f(0) = f'(0) = 0, \quad f'(\infty) = 1. \tag{14.46c}
$$

Not surprisingly, Eq. (14.46b) does not admit an analytic solution. However, it is simple to compute a numerical solution with the boundary conditions (14.46c).

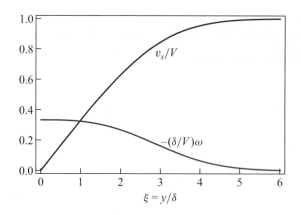

FIGURE 14.13 Laminar boundary layer near a flat plate: the Blasius velocity profile $v_x/V = f'(\xi)$ (blue curve) and vorticity profile $(\delta/V)\omega = -f'''(\xi)$ (red curve) as functions of scaled perpendicular distance $\xi = y/\delta$. Note that the flow speed is 90% of V at a distance of 3δ from the surface, so δ is a good measure of the thickness of the boundary layer.

Blasius profile for self-similar boundary layer

The result for $v_x/V = f'(\xi)$ is shown in Fig. 14.13. This solution, the *Blasius profile*, qualitatively has the form we expected: the velocity v_x rises from 0 to V in a smooth manner as one moves outward from the plate, achieving a sizable fraction of V at a distance several times larger than $\delta(x)$.

This Blasius profile is not only our first example (aside from Ex. 14.12) of the use of a stream function ($v_x = \partial \zeta/\partial y, \quad v_y = -\partial \zeta/\partial x$); it is also our first example of another common procedure in fluid dynamics: *taking account of a natural scaling in the problem to make a self-similar ansatz and thereby transform the partial differential fluid equations into ordinary differential equations*. Solutions of this type are known as *similarity solutions* and also as *self-similar solutions*.

self-similar solutions

The motivation for using similarity solutions is obvious. The nonlinear partial differential equations of fluid dynamics are much harder to solve, even numerically, than are ordinary differential equations. Elementary similarity solutions are especially appropriate for problems where there is no intrinsic characteristic lengthscale or timescale associated with the relevant physical quantities except those explicitly involving the spatial and temporal coordinates. High-Reynolds-number flow past a large plate has a useful similarity solution, whereas flow with $\mathrm{Re}_\ell \sim 1$, where the size of the plate is clearly a significant scale in the problem, does not. We shall encounter more examples of similarity solutions in the following chapters.

Now that we have a solution for the flow, we must examine a key approximation that underlies it: constancy of the pressure P. To do this, we begin with the y component of the force-balance equation (14.41) (a component that we never used explicitly in our analysis). The inertial and viscous terms are both $\mathrm{O}(V^2\delta/x^2)$, so if we reinstate a term $-\nabla P/\rho \sim -\Delta P/\rho\delta$, it can be no larger than $\sim V^2\delta/x^2$. From this we estimate that the pressure difference across the boundary layer is $\Delta P \lesssim \rho V^2 \delta^2 / x^2$. Using this estimate in the x component of force balance (14.41) (the component on which our analysis was based), we verify that the pressure gradient term is smaller than those we kept by a factor $\lesssim \delta^2 / x^2 \ll 1$. For this reason, *when the boundary layer is thin, we can indeed neglect pressure gradients across it when computing its structure from longitudinal force balance.*

14.4.2 Blasius Vorticity Profile

14.4.2

It is illuminating to consider the structure of the Blasius boundary layer in terms of its vorticity. Since the flow is 2-dimensional with velocity $\mathbf{v} = \nabla \times (\zeta \mathbf{e}_z)$, its vorticity is $\boldsymbol{\omega} = \nabla \times \nabla \times (\zeta \mathbf{e}_z) = -\nabla^2 (\zeta \mathbf{e}_z)$, which has as its only nonzero component

$$\omega \equiv \omega_z = -\nabla^2 \zeta = -\frac{V}{\delta} f''(\xi), \tag{14.47}$$

aside from fractional corrections of order δ^2 / x^2. This vorticity is exhibited in Fig. 14.13.

From Eq. (14.46b), we observe that the gradient of vorticity vanishes at the plate. This means that the vorticity is not diffusing out from the plate's surface. Neither is it being convected away from the plate's surface, as the perpendicular velocity vanishes there. This confirms what we already learned from Fig. 14.12: the flux of vorticity per unit length is conserved along the plate, once it has been created as a spike at the plate's leading edge.

If we transform to a frame moving with an intermediate speed $\sim V/2$, and measure time t since passing the leading edge, the vorticity will diffuse a distance $\sim (\nu t)^{1/2} \sim (\nu x / V)^{1/2} = \delta(x)$ away from the surface after time t; see the discussion of vorticity diffusion in Sec. 14.2.5. This behavior exhibits the connection between that diffusion discussion and the similarity solution for the boundary layer in this section.

vorticity diffusion in boundary layer

14.4.3 Viscous Drag Force on a Flat Plate

14.4.3

It is of interest to compute the total drag force exerted on the plate. Let ℓ be the plate's length and $w \gg \ell$ be its width. Noting that the plate has two sides, the drag force produced by the viscous stress acting on the plate's surface is

$$F = 2 \int T_{xy}^{\text{vis}} dx dz = 2 \int (-2\eta \sigma_{xy}) dx dz = 2w \int_0^\ell \rho \nu \left(\frac{\partial v_x}{\partial y} \right)_{y=0} dx. \tag{14.48}$$

Inserting $\partial v_x / \partial y = (V/\delta) f''(0) = V \sqrt{V/(\nu x)} f''(0)$ from Eqs. (14.46a) and performing the integral, we obtain

$$\boxed{F = \frac{1}{2} \rho V^2 \times (2\ell w) \times C_D,} \tag{14.49}$$

where

$$\boxed{C_D = 4 f''(0) \text{Re}_\ell^{-1/2}.} \tag{14.50}$$

Here we have introduced an often-used notation for expressing the drag force of a fluid on a solid body. We have written it as half the incoming fluid's kinetic stress ρV^2 times the surface area of the body $2\ell w$ on which the drag force acts, times a dimensionless drag coefficient C_D. We have expressed the drag coefficient in terms of the Reynolds number

$$\boxed{\mathrm{Re}_\ell = \frac{V\ell}{\nu}} \tag{14.51}$$

formed from the body's relevant dimension, ℓ, and the speed and viscosity of the incoming fluid.

From Fig. 14.13, we estimate that $f''(0) \simeq 0.3$ (an accurate numerical value is 0.332), and so $C_D \simeq 1.328 R_\ell^{-1/2}$. Note that the drag coefficient decreases as the viscosity decreases and the Reynolds number increases. However, as we discuss in Sec. 15.5, this model breaks down for very large Reynolds numbers $\mathrm{Re}_\ell \gtrsim 10^6$, because a portion of the boundary layer becomes turbulent (cf. caption of Fig. 14.11).

14.4.4 Boundary Layer Near a Curved Surface: Separation

Next consider flow past a nonplanar surface (e.g., the aircraft wings of Fig. 14.2a,b). In this case, in general a longitudinal pressure gradient exists along the boundary layer, which cannot be ignored, in contrast to the transverse pressure gradient across the boundary layer. If the pressure decreases along the flow, the flow will accelerate, and so more vorticity will be created at the surface and will diffuse away from the surface. However, if there is an "adverse" pressure gradient causing the flow to decelerate, then negative vorticity must be created at the wall. For a sufficiently adverse gradient, the negative vorticity gets so strong that it cannot diffuse fast enough into and through the boundary layer to maintain a simple boundary-layer-type flow. Instead, the boundary layer *separates* from the surface, as shown in Fig. 14.14, and a backward flow is generated beyond the separation point by the negative vorticity. This phenomenon can occur on an aircraft when the wings' angle of attack (i.e., the inclination of the wings to the horizontal) is too great. An adverse pressure gradient develops on the

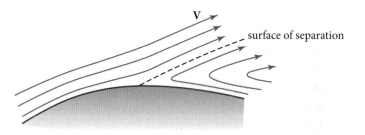

FIGURE 14.14 Separation of a boundary layer in the presence of an adverse pressure gradient.

upper wing surfaces, the flow separates, and the plane stalls. The designers of wings make great efforts to prevent this, as we discuss briefly in Sec. 15.5.2.

Exercise 14.15 *Problem: Reynolds Numbers*
Estimate the Reynolds numbers for the following flows. Make sketches of the flow fields, pointing out any salient features.

(a) A hang glider in flight.

(b) Plankton in the ocean.

(c) A physicist waving her hands.

Exercise 14.16 **Problem: Fluid Dynamical Scaling*
An auto manufacturer wishes to reduce the drag force on a new model by changing its design. She does this by building a one-sixth scale model and putting it into a wind tunnel. How fast must the air travel in the wind tunnel to simulate the flow at 40 mph on the road?

   [Remark: This is our first example of "scaling" relations in fluid dynamics, a powerful concept that we develop and explore in later chapters.]

Exercise 14.17 *Example: Potential Flow around a Cylinder (D'Alembert's Paradox)*
Consider stationary incompressible flow around a cylinder of radius a with sufficiently large Reynolds number that viscosity may be ignored except in a thin boundary layer, which is assumed to extend all the way around the cylinder. The velocity is assumed to have the uniform value **V** at large distances from the cylinder.

(a) Show that the velocity field outside the boundary layer can be derived from a scalar *velocity potential* (introduced in Sec. 13.5.4), $\mathbf{v} = \nabla \psi$, that satisfies Laplace's equation: $\nabla^2 \psi = 0$.

(b) Write down suitable boundary conditions for ψ.

(c) Write the velocity potential in the form

$$\psi = \mathbf{V} \cdot \mathbf{x} + f(\mathbf{x}),$$

and solve for f. Sketch the streamlines and equipotentials.

(d) Use Bernoulli's theorem to compute the pressure distribution over the surface and the net drag force given by this solution. Does your drag force seem reasonable? It did not seem reasonable to d'Alembert in 1752, and it came to be called *d'Alembert's paradox*.

(e) Finally, consider the effect of the pressure distribution on the boundary layer. How do you think this will make the real flow different from the potential solution? How will the drag change?

Exercise 14.18 *Example: Stationary Laminar Flow down a Long Pipe*

Fluid flows down a long cylindrical pipe of length b much larger than radius a, from a reservoir maintained at pressure P_0 (which connects to the pipe at $x = 0$) to a free end at large x, where the pressure is negligible. In this problem, we try to understand the velocity field $v_x(\varpi, x)$ as a function of radius ϖ and distance x down the pipe, for a given discharge (i.e., mass flow per unit time) $\dot{M}$. Assume that the Reynolds number is small enough for the flow to be treated as laminar all the way down the pipe.

(a) Close to the entrance of the pipe (small x), the boundary layer will be thin, and the velocity will be nearly independent of radius. What is the fluid velocity outside the boundary layer in terms of its density and $\dot{M}$?

(b) How far must the fluid travel along the pipe before the vorticity diffuses into the center of the flow and the boundary layer becomes as thick as the radius? An order-of-magnitude calculation is adequate, and you may assume that the pipe is much longer than your estimate.

(c) At a sufficiently great distance down the pipe, the profile will cease evolving with x and settle down into the Poiseuille form derived in Sec. 13.7.6, with the discharge $\dot{M}$ given by the Poiseuille formula. Sketch how the velocity profile changes along the pipe, from the entrance to this final Poiseuille region.

(d) Outline a procedure for computing the discharge in a long pipe of arbitrary cross section.

Exercise 14.19 *Derivation: Stream Function in General* `T2`

Derive results (i), (ii), and (iii) in the last paragraph of Box 14.4. [Hint: The derivation is simplest if one works in a coordinate basis (Sec. 24.3) rather than in the orthonormal bases that we use throughout Parts I–VI of this book.]

Exercise 14.20 *Problem: Stream Function for Stokes Flow around a Sphere* `T2`

Consider low-Reynolds-number flow around a sphere. Derive the velocity field (14.30) using the stream function of Box 14.4. This method is more straightforward but less intuitive than that used in Sec. 14.3.2.

14.5   **14.5 Nearly Rigidly Rotating Flows—Earth's Atmosphere and Oceans**

In Nature one often encounters fluids that rotate nearly rigidly (i.e., fluids with a nearly uniform distribution of vorticity). Earth's oceans and atmosphere are important examples, where the rotation is forced by the underlying rotation of Earth. Such rotating fluids are best analyzed in a rotating reference frame, in which the unperturbed fluid is at rest and the perturbations are influenced by Coriolis forces, resulting in surprising phenomena. We explore some of these phenomena in this section.

As a foundation for this exploration, we transform the Navier-Stokes equation from the inertial frame in which it was derived to a uniformly rotating frame: the mean rest frame of the flows we study.

We begin by observing that the Navier-Stokes equation has the same form as Newton's second law for particle motion:

$$\frac{d\mathbf{v}}{dt} = \mathbf{f}, \tag{14.52}$$

where the force per unit mass is $\mathbf{f} = -\nabla P/\rho - \nabla\Phi + \nu\nabla^2\mathbf{v}$. We transform to a frame rotating with uniform angular velocity $\mathbf{\Omega}$ by adding "fictitious" Coriolis and centrifugal accelerations, given by $-2\mathbf{\Omega}\times\mathbf{v}$ and $-\mathbf{\Omega}\times(\mathbf{\Omega}\times\mathbf{x})$, respectively, and expressing the force $\mathbf{f}$ in rotating coordinates. The fluid velocity transforms as

$$\mathbf{v} \rightarrow \mathbf{v} + \mathbf{\Omega}\times\mathbf{x}. \tag{14.53}$$

It is straightforward to verify that this transformation leaves the expression for the viscous acceleration, $\nu\nabla^2\mathbf{v}$, unchanged. Therefore the expression for the force $\mathbf{f}$ is unchanged, and the Navier-Stokes equation in rotating coordinates becomes

$$\frac{d\mathbf{v}}{dt} = -\frac{\nabla P}{\rho} - \nabla\Phi + \nu\nabla^2\mathbf{v} - 2\mathbf{\Omega}\times\mathbf{v} - \mathbf{\Omega}\times(\mathbf{\Omega}\times\mathbf{x}). \tag{14.54}$$

The centrifugal acceleration $-\mathbf{\Omega}\times(\mathbf{\Omega}\times\mathbf{x})$ can be expressed as the gradient of a centrifugal potential, $\nabla[\frac{1}{2}(\mathbf{\Omega}\times\mathbf{x})^2] = \nabla[\frac{1}{2}(\Omega\varpi)^2]$, where ϖ is distance from the rotation axis. (The location of the rotation axis is actually arbitrary, aside from the requirement that it be parallel to $\mathbf{\Omega}$; see Box 14.5.) For simplicity we confine ourselves to an incompressible fluid, so that ρ is constant. This allows us to define an *effective pressure*

$$P' = P + \rho\left[\Phi - \frac{1}{2}(\mathbf{\Omega}\times\mathbf{x})^2\right] \tag{14.55}$$

that includes the combined effects of the real pressure, gravity, and the centrifugal force. In terms of P' the Navier-Stokes equation in the rotating frame becomes

$$\frac{d\mathbf{v}}{dt} = -\frac{\nabla P'}{\rho} + \nu\nabla^2\mathbf{v} - 2\mathbf{\Omega}\times\mathbf{v}. \tag{14.56a}$$

The quantity P' will be constant if the fluid is at rest in the rotating frame, $\mathbf{v} = 0$, in contrast to the true pressure P, which does have a gradient. Equation (14.56a) is the most useful form for the Navier-Stokes equation in a rotating frame. In keeping with our assumptions that ρ is constant and the flow speeds are very low in comparison with the speed of sound, we augment Eq. (14.56a) by the incompressibility condition

$\nabla \cdot \mathbf{v} = 0$, which is left unchanged by the transformation (14.53) to a rotating reference frame:

$$\boxed{\nabla \cdot \mathbf{v} = 0.} \tag{14.56b}$$

It should be evident from Eq. (14.56a) that two dimensionless numbers characterize rotating fluids. The first is the *Rossby number,*

Rossby number

$$\boxed{\mathrm{Ro} = \frac{V}{\Omega L},} \tag{14.57}$$

where V is a characteristic velocity of the flow relative to the rotating frame, and L is a characteristic length. Ro measures the relative strength of the inertial acceleration and the Coriolis acceleration:

$$\boxed{\mathrm{Ro} \sim \frac{|(\mathbf{v} \cdot \nabla)\mathbf{v}|}{|2\boldsymbol{\Omega} \times \mathbf{v}|} \sim \frac{\text{inertial force}}{\text{Coriolis force}}.} \tag{14.58}$$

The second dimensionless number is the *Ekman number,*

Ekman number

$$\boxed{\mathrm{Ek} = \frac{\nu}{\Omega L^2},} \tag{14.59}$$

which analogously measures the relative strengths of the viscous and Coriolis accelerations:

$$\boxed{\mathrm{Ek} \sim \frac{|\nu \nabla^2 \mathbf{v}|}{|2\boldsymbol{\Omega} \times \mathbf{v}|} \sim \frac{\text{viscous force}}{\text{Coriolis force}}.} \tag{14.60}$$

Notice that $\mathrm{Ro}/\mathrm{Ek} = \mathrm{Re}$ is the Reynolds number.

storms, ocean currents, and tea cups

The three traditional examples of rotating flows are large-scale storms and other weather patterns on rotating Earth, deep currents in Earth's oceans, and water in a stirred tea cup.

For a typical storm, the wind speed might be $V \sim 25$ mph (~ 10 m s^{-1}), and a characteristic lengthscale might be $L \sim 1{,}000$ km. The effective angular velocity at a temperate latitude is (see Box 14.5) $\Omega_\star = \Omega_\oplus \sin 45° \sim 10^{-4}$ rad s^{-1}, where $\Omega_\oplus$ is Earth's rotational angular velocity. As the air's kinematic viscosity is $\nu \sim 10^{-5}$ m^2 s^{-1}, we find that Ro ~ 0.1 and Ek $\sim 10^{-13}$. This tells us immediately that Coriolis forces are important but not totally dominant, compared to inertial forces, in controlling the weather, and that viscous forces are unimportant except in thin boundary layers.

For deep ocean currents, such as the Gulf Stream, V ranges from ~ 0.01 to ~ 1 m s^{-1}, so we use $V \sim 0.1$ m s^{-1}, lengthscales are $L \sim 1{,}000$ km, and for water $\nu \sim 10^{-6}$ m^2 s^{-1}, so Ro $\sim 10^{-3}$ and Ek $\sim 10^{-14}$. Thus, Coriolis accelerations are far more important than inertial forces, and viscous forces are important only in thin boundary layers.

768   Chapter 14. Vorticity

BOX 14.5. ARBITRARINESS OF ROTATION AXIS; Ω FOR ATMOSPHERIC AND OCEANIC FLOWS

ARBITRARINESS OF ROTATION AXIS

Imagine yourself on the rotation axis $\mathbf{x} = 0$ of a rigidly rotating flow. All fluid elements circulate around you with angular velocity $\boldsymbol{\Omega}$. Now move perpendicular to the rotation axis to a new location $\mathbf{x} = \mathbf{a}$, and ride with the flow there. All other fluid elements will still rotate around you with angular velocity $\boldsymbol{\Omega}$! The only way you can tell you have moved (if all fluid elements look identical) is that you will now experience a centrifugal force $\boldsymbol{\Omega} \times (\boldsymbol{\Omega} \times \mathbf{a})$.

This shows up mathematically in the rotating-frame Navier-Stokes equation (14.54). When we set $\mathbf{x} = \mathbf{x}_{new} - \mathbf{a}$, the only term that changes is the centrifugal force; it becomes $-\boldsymbol{\Omega} \times (\boldsymbol{\Omega} \times \mathbf{x}_{new}) + \boldsymbol{\Omega} \times (\boldsymbol{\Omega} \times \mathbf{a})$. If we absorb the new, constant, centrifugal term $\boldsymbol{\Omega} \times (\boldsymbol{\Omega} \times \mathbf{a})$ into the gravitational acceleration $\mathbf{g} = -\nabla\Phi$, then the Navier-Stokes equation is completely unchanged. In this sense, the choice of rotation axis is arbitrary.

ANGULAR VELOCITY Ω FOR LARGE-SCALE FLOWS IN EARTH'S ATMOSPHERE AND OCEANS

For large-scale flows in Earth's atmosphere and oceans (e.g., storms), the rotation of the unperturbed fluid is that due to the rotation of Earth. One might think that this means we should take as the angular velocity $\boldsymbol{\Omega}$ in the Coriolis term of the Navier-Stokes equation (14.56a) Earth's angular velocity $\boldsymbol{\Omega}_\oplus$. Not so. The atmosphere and ocean are so thin vertically that vertical motions cannot achieve small Rossby numbers: Coriolis forces are unimportant for vertical motions. Correspondingly, the only component of Earth's angular velocity $\boldsymbol{\Omega}_\oplus$ that is important for Coriolis forces is that which couples horizontal flows to horizontal flows: the vertical component $\Omega_* = \Omega_\oplus \sin(\text{latitude})$. (A similar situation occurs for a Foucault pendulum.) Thus, in the Coriolis term of the Navier-Stokes equation, we must set $\boldsymbol{\Omega} = \Omega_* \mathbf{e}_z = \Omega_\oplus \sin(\text{latitude})\mathbf{e}_z$, where $\mathbf{e}_z$ is the vertical unit vector. By contrast, in the centrifugal potential $\frac{1}{2}(\boldsymbol{\Omega} \times \mathbf{x})^2$, $\boldsymbol{\Omega}$ remains the full angular velocity of Earth, $\boldsymbol{\Omega}_\oplus$—unless (as is commonly done) we absorb a portion of it into the gravitational potential as when we change rotation axes, in which case we can use $\boldsymbol{\Omega} = \Omega_* \mathbf{e}_z$ in the centrifugal potential.

For water stirred in a tea cup (with parameters typical of many flows in the laboratory), $L \sim 10$ cm, $\Omega \sim V/L \sim 10$ rad s^{-1}, and $\nu \sim 10^{-6}$ m^2 s^{-1} giving Ro ~ 1 and Ek $\sim 10^{-5}$. Coriolis and inertial forces are comparable in this case, and viscous forces again are confined to boundary layers, but the layers are much thicker relative to the bulk flow than in the atmospheric and oceanic cases.

Notice that for all these flows—atmospheric, oceanic, and tea cup—the (effective) rotation axis is vertical: $\boldsymbol{\Omega}$ is vertically directed (cf. Box 14.5). This will be the case for all nearly rigidly rotating flows considered in this chapter.

14.5.2 Geostrophic Flows

14.5.2

geostrophic flow

Stationary flows $\partial \mathbf{v}/\partial t = 0$ in which both the Rossby and Ekman numbers are small (i.e., with Coriolis forces big compared to inertial and viscous forces) are called *geostrophic*, even in the laboratory. Geostrophic flow is confined to the bulk of the fluid, well away from all boundary layers, since viscosity will become important in those layers. For such geostrophic flows, the Navier-Stokes equation (14.56a) reduces to

Navier-Stokes equation for geostrophic flow

$$2\boldsymbol{\Omega} \times \mathbf{v} = -\frac{\boldsymbol{\nabla} P'}{\rho}. \tag{14.61}$$

This equation states that the velocity $\mathbf{v}$ (measured in the rotating frame) is orthogonal to the body force $\boldsymbol{\nabla} P'$, which drives it. Correspondingly, the streamlines are perpendicular to the gradient of the generalized pressure (i.e., they lie on surfaces of constant P').

An example of geostrophic flow is the motion of atmospheric winds around a low pressure region or *depression*. [Since $P' = P + \rho(\Phi - \frac{1}{2}\Omega^2\varpi^2)$, when the actual pressure P goes up or down at some fixed location, P' goes up or down by the same amount, so a depression of P' is a depression of P.] The geostrophic equation (14.61) tells us that such winds must be counterclockwise in the northern hemisphere as seen from a satellite, and clockwise in the southern hemisphere. For a flow with speed $v \sim 10$ m s^{-1} around a $\sim 1{,}000$-km depression, the drop in effective pressure at the depression's center is $\Delta P' = \Delta P \sim 1$ kPa ~ 10 mbar ~ 0.01 atmosphere ~ 0.3 inches of mercury ~ 4 inches of water. Around a high-pressure region, winds circulate in the opposite direction.

It is here that we can see the power of introducing the effective pressure P'. In the case of atmospheric and oceanic flows, the true pressure P changes significantly vertically, and the pressure scale height is generally much shorter than the horizontal lengthscale. However, the effective pressure will be almost constant vertically, any small variation being responsible for minor updrafts and downdrafts, which we generally ignore when describing the wind or current flow pattern. It is the horizontal pressure gradients that are responsible for driving the flow. When pressures are quoted, they must therefore be referred to some reference equipotential surface: $\Phi - \frac{1}{2}(\boldsymbol{\Omega} \times \mathbf{x})^2 = $ const. The convenient one to use is the equipotential associated

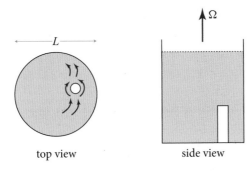

top view   side view

FIGURE 14.15 Taylor column. A solid cylinder (white) is placed in a large container of water, which is then spun up on a turntable to a high enough angular velocity Ω that the Ekman number is small: $\text{Ek} = \nu/(\Omega L^2) \ll 1$. A slow, steady flow relative to the cylinder is then induced. [The flow's velocity $\mathbf{v}$ in the rotating frame must be small enough to keep the Rossby number $\text{Ro} = v/(\Omega L) \ll 1$.] The water in the bottom half of the container flows around the cylinder. The water in the top half does the same as if there were an invisible cylinder present. This is an illustration of the Taylor-Proudman theorem, which states that there can be no vertical gradients in the velocity field. The effect can also be demonstrated with vertical velocity: If the cylinder is slowly made to rise, then the fluid immediately above it will also be pushed upward rather than flow past the cylinder—except at the water's surface, where the geostrophic flow breaks down. The fluid above the cylinder, which behaves as though it were rigidly attached to the cylinder, is called a *Taylor column*.

with the surface of the ocean, usually called "mean sea level." This is the pressure that appears on a meteorological map.

14.5.3 Taylor-Proudman Theorem

There is a simple theorem due to Taylor and Proudman that simplifies the description of 3-dimensional, geostrophic flows. Take the curl of Eq. (14.61) and use $\nabla \cdot \mathbf{v} = 0$; the result is

$$\boxed{(\mathbf{\Omega} \cdot \nabla)\mathbf{v} = 0.}$$ (14.62)

Taylor-Proudman theorem for geostrophic flow

Thus, there can be no vertical gradient (gradient along $\mathbf{\Omega}$) of the velocity under geostrophic conditions. This result provides a good illustration of the stiffness of vortex lines: the vortex lines associated with the rigid rotation $\mathbf{\Omega}$ are frozen in the fluid under geostrophic conditions (where other contributions to the vorticity are small), and they refuse to be bent. The simplest demonstration of this is the Taylor column of Fig. 14.15.

It is easy to see that any vertically constant, divergence-free velocity field $\mathbf{v}(x, y)$ can be a solution to the geostrophic equation (14.61). The generalized pressure P' can be adjusted to make it a solution (see the discussion of pressure adjusting itself to whatever the flow requires, in Ex. 14.7). However, one must keep in mind that to guarantee it is also a true (approximate) solution of the full Navier-Stokes equation (14.56a), its Rossby and Ekman numbers must be $\ll 1$.

14.5.4 Ekman Boundary Layers

As we have seen, Ekman numbers are typically small in the bulk of a rotating fluid. However, as was true in the absence of rotation, the no-slip condition at a solid surface generates a boundary layer that can significantly influence the global velocity field, albeit indirectly.

Ekman boundary layer and its thickness

When the Rossby number is less than one, the structure of a laminar boundary layer is dictated by a balance between viscous and Coriolis forces rather than between viscous and inertial forces. Balancing the relevant terms in Eq. (14.56a), we obtain an estimate of the boundary-layer thickness:

$$\text{thickness} \sim \delta_E \equiv \sqrt{\frac{\nu}{\Omega}}. \tag{14.63}$$

In other words, the thickness of the boundary layer is that which makes the layer's Ekman number unity: $\text{Ek}(\delta_E) = \nu/(\Omega \delta_E^2) = 1$.

Consider such an "Ekman boundary layer" at the bottom or top of a layer of geostrophically flowing fluid. For the same reasons as for ordinary laminar boundary layers (Sec. 14.4), the generalized pressure P' will be nearly independent of height z through the Ekman layer, that is, it will have the value dictated by the flow just outside the layer: $\nabla P' = -2\rho \boldsymbol{\Omega} \times \mathbf{V} = \text{const}$. Here $\mathbf{V}$ is the velocity just outside the layer (the velocity of the bulk flow), which we assume to be constant on scales $\sim \delta_E$. Since $\boldsymbol{\Omega}$ is vertical, $\nabla P'$, like $\mathbf{V}$, will be horizontal (i.e., they both will lie in the x-y plane). To simplify the analysis, we introduce the fluid velocity relative to the bulk flow,

$$\mathbf{w} \equiv \mathbf{v} - \mathbf{V}, \tag{14.64}$$

which goes to zero outside the boundary layer. When rewritten in terms of $\mathbf{w}$, the Navier-Stokes equation (14.56a) [with $\nabla P'/\rho = -2\boldsymbol{\Omega} \times \mathbf{V}$ and with $d\mathbf{v}/dt = \partial \mathbf{v}/\partial t + (\mathbf{v} \cdot \nabla)\mathbf{v} = 0$, because the flow in the thin boundary layer is steady and $\mathbf{v}$ is horizontal and varies only vertically] takes the simple form $d^2\mathbf{w}/dz^2 = (2/\nu)\boldsymbol{\Omega} \times \mathbf{w}$. Choosing Cartesian coordinates with an upward vertical z direction, assuming $\boldsymbol{\Omega} = +\Omega \mathbf{e}_z$ (as is the case for the oceans and atmosphere in the northern hemisphere), and introducing the complex quantity

complex velocity field

$$w = w_x + i w_y \tag{14.65}$$

to describe the horizontal velocity field, we can rewrite $d^2\mathbf{w}/dz^2 = (2/\nu)\boldsymbol{\Omega} \times \mathbf{w}$ as

$$\frac{d^2w}{dz^2} = \frac{2i}{\delta_E^2}w = \left(\frac{1+i}{\delta_E}\right)^2 w. \tag{14.66}$$

Equation (14.66) must be solved subject to $w \to 0$ far from the water's boundary and some appropriate condition at the boundary.

For a first illustration of an Ekman layer, consider the effects of a wind blowing in the $\mathbf{e}_x$ direction above a still ocean, $\mathbf{V} = 0$, and set $z = 0$ at the ocean's surface. The

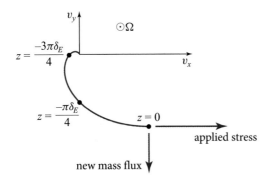

FIGURE 14.16 Ekman spiral (water velocity as a function of depth) at the ocean surface, where the wind exerts a stress.

wind will exert, through a turbulent boundary layer of air, a stress $-T_{xz} > 0$ on the ocean's surface. This stress must be balanced by an equal, viscous stress $\nu\rho\, dw_x/dz$ at the top of the water's boundary layer, $z = 0$. Thus there must be a velocity gradient, $dw_x/dz = -T_{xz}/\nu\rho = |T_{xz}|/\nu\rho$, in the water at $z = 0$. (This replaces the no-slip boundary condition that we have when the boundary is a solid surface.) Imposing this boundary condition along with $w \to 0$ as $z \to -\infty$ (down into the ocean), we find from Eqs. (14.66) and (14.65):

$$v_x = w_x = \left(\frac{|T_{xz}|\delta_E}{\sqrt{2}\,\nu\rho}\right) e^{z/\delta_E}\cos(z/\delta_E - \pi/4),$$

wind-driven Ekman boundary layer at top of a geostrophic flow

$$v_y = w_y = \left(\frac{|T_{xz}|\delta_E}{\sqrt{2}\,\nu\rho}\right) e^{z/\delta_E}\sin(z/\delta_E - \pi/4), \qquad (14.67)$$

for $z \le 0$ (Fig. 14.16). As a function of depth, this velocity field has the form of a spiral—the so-called *Ekman spiral*. When $\boldsymbol{\Omega}$ points toward us (as in Fig. 14.16), the spiral is clockwise and tightens as we move away from the boundary ($z = 0$ in the figure) into the bulk flow.

structure of the boundary layer: Ekman spiral

By integrating the mass flux $\rho\mathbf{w}$ over z, we find for the total mass flowing per unit time per unit length of the ocean's surface

$$\mathbf{F} = \rho \int_{-\infty}^{0} \mathbf{w}\,dz = -\frac{\delta_E^2}{2\nu}|T_{xz}|\mathbf{e}_y; \qquad (14.68)$$

see Fig. 14.16. Thus the wind, blowing in the $\mathbf{e}_x$ direction, causes a net mass flow in the direction of $\mathbf{e}_x \times \boldsymbol{\Omega}/\Omega = -\mathbf{e}_y$. This response (called "Ekman pumping") may seem less paradoxical if one recalls how a gyroscope responds to applied forces.

Ekman pumping

This mechanism is responsible for the creation of *gyres* in the oceans (Ex. 14.21 and Fig. 14.18 below).

gyres

As a second illustration of an Ekman boundary layer, we consider a geostrophic flow with nonzero velocity $\mathbf{V} = V\mathbf{e}_x$ in the bulk of the fluid, and examine this flow's

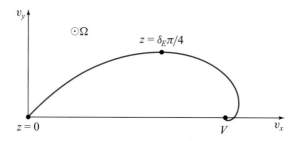

FIGURE 14.17 Ekman spiral (water velocity as a function of height) in the bottom boundary layer, when the bulk flow above it moves geostrophically with speed $V\mathbf{e}_x$.

interaction with a static, solid surface at its bottom. We set $z = 0$ at the bottom, with z increasing upward, in the $\boldsymbol{\Omega}$ direction. The structure of the boundary layer on the bottom is governed by the same differential equation (14.66) as at the wind-blown surface, but with altered boundary conditions. The solution is

Ekman boundary layer at bottom of a geostrophic flow

$$v_x - V = w_x = -V \exp(-z/\delta_E) \cos(z/\delta_E),$$
$$v_y = w_y = +V \exp(-z/\delta_E) \sin(z/\delta_E). \tag{14.69}$$

This solution is shown in Fig. 14.17.

Recall that we have assumed $\boldsymbol{\Omega}$ points in the upward $+z$ direction, which is appropriate for the ocean and atmosphere in the northern hemisphere. If, instead, $\boldsymbol{\Omega}$ points downward (as in the southern hemisphere), then the handedness of the Ekman spiral is reversed.

Ekman boundary layers drive circulation in rotating fluids

Ekman boundary layers are important, because they can circulate rotating fluids faster than viscous diffusion can. Suppose we have a nonrotating container (e.g., a tea cup) of radius L containing a fluid that rotates with angular velocity Ω (e.g., due to stirring; cf. Ex. 14.22). As you will see in your analysis of Ex. 14.22, the Ekman layer at the container's bottom experiences a pressure difference between the wall and the container's center given by $\Delta P \sim \rho L^2 \Omega^2$. This drives a fluid circulation in the Ekman layer, from the wall toward the center, with radial speed $V \sim \Omega L$. The circulating fluid must upwell at the bottom's center from the Ekman layer into the bulk fluid. This produces a poloidal mixing of the fluid on a timescale given by

$$t_E \sim \frac{L^3}{L\delta_E V} \sim \frac{L\delta_E}{\nu}. \tag{14.70}$$

This timescale is shorter than that for simple diffusion of vorticity, $t_\nu \sim L^2/\nu$, by a factor $t_E/t_\nu \sim \sqrt{\text{Ek}}$, which (as we have seen) can be very small. This circulation and mixing are key to the piling up of tea leaves at the bottom center of a stirred tea cup, and to the mixing of the tea or milk into the cup's hot water (Ex. 14.22).

The circulation driven by an Ekman layer is an example of a *secondary flow*—a weakly perturbative bulk flow that is produced by interaction with a boundary. For other examples of secondary flows, see Taylor (1968).

secondary flows

Exercise 14.21 **Example: Winds and Ocean Currents in the North Atlantic*
The north Atlantic Ocean exhibits the pattern of winds and ocean currents shown in Fig. 14.18. Westerly winds blow from west to east at 40° latitude. Trade winds blow from east to west at 20° latitude. In between, around 30° latitude, is the Sargasso Sea: a 1.5-m-high gyre (raised hump of water). The gyre is created by ocean surface currents, extending down to a depth of only about 30 m, that flow northward from the trade-wind region and southward from the westerly wind region (upper inset in Fig. 14.18). A deep ocean current, extending from the surface down to near the bottom, circulates around the Sargasso Sea gyre in a clockwise manner. This current goes under different names in different regions of the ocean: Gulf Stream, West Wind Drift, Canaries Current, and North Equatorial Current. Explain both qualitatively and semiquantitatively (in terms of order of magnitude) how the winds are ultimately responsible for all these features of the ocean. More specifically, do the following.

(a) Explain the surface currents in terms of an Ekman layer at the top of the ocean, with thickness δ_E about 30 m. From this measured δ_E compute the kinematic viscosity ν in the boundary layer. Your result, $\nu \sim 0.03 \text{ m}^2 \text{ s}^{-1}$, is far larger than

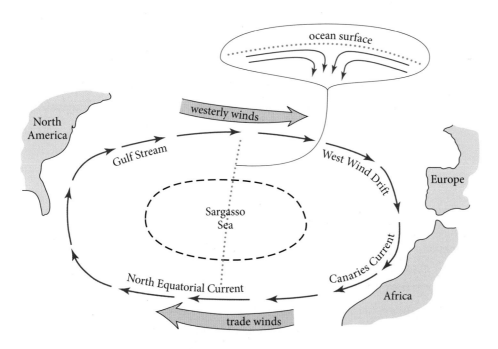

FIGURE 14.18 Winds and ocean currents in the north Atlantic. The upper inset shows the surface currents, along the dotted north-south line, that produce the Sargasso Sea gyre.

the molecular viscosity of water, $\sim 10^{-6}$ m^2 s^{-1} (Table 13.2). The reason is that the boundary layer is turbulent, and its eddies produce this large viscosity (see Sec. 15.4.2).

(b) Explain why the height of the gyre that the surface currents produce in the Sargasso Sea is about 1.5 m.

(c) Explain the deep ocean current (Gulf Stream, etc.) in terms of a geostrophic flow, and estimate the speed of this current. This current, like circulation in a tea cup, is an example of a "secondary flow."

(d) If there were no continents on Earth, but only an ocean of uniform depth, what would be the flow pattern of this deep current—its directions of motion at various locations around Earth, and its speeds? The continents (North America, Europe, and Africa) must be responsible for the deviation of the actual current (Gulf Stream, etc.) from this continent-free flow pattern. How do you think the continents give rise to the altered flow pattern?

Exercise 14.22 **Example: Circulation in a Tea Cup*
Place tea leaves and water in a tea cup, glass, or other larger container. Stir the water until it is rotating uniformly, and then stand back and watch the motion of the water and leaves. Notice that the tea leaves tend to pile up at the cup's center. An Ekman boundary layer on the bottom of the cup is responsible for this phenomenon. In this exercise you explore the origin and consequences of this Ekman layer.

(a) Evaluate the pressure distribution $P(\varpi, z)$ in the bulk flow (outside all boundary layers), assuming that it rotates rigidly. (Here z is height and ϖ is distance from the water's rotation axis.) Perform your evaluation in the water's rotating reference frame. From this $P(\varpi, z)$ deduce the shape of the top surface of the water. Compare your deduced shape with the actual shape in your experiment.

(b) Estimate the thickness of the Ekman layer at the bottom of your container. It is very thin. Show, using the Ekman spiral diagram (Fig. 14.17), that the water in this Ekman layer flows inward toward the container's center, causing the tea leaves to pile up at the center. Estimate the radial speed of this Ekman-layer flow and the mass flux that it carries.

(c) To get a simple physical understanding of this inward flow, examine the radial gradient $\partial P / \partial \varpi$ of the pressure P in the bulk flow just above the Ekman layer. Explain why $\partial P / \partial \varpi$ in the Ekman layer will be the same as in the rigidly rotating flow above it. Then apply force balance in an inertial frame to deduce that the water in the Ekman layer will be accelerated inward toward the center.

(d) Using geostrophic-flow arguments, deduce the fate of the boundary-layer water after it reaches the center of the container's bottom. Where does it go? What is the large-scale circulation pattern that results from the "driving force" of the Ekman layer's mass flux? What is the Rossby number for this large-scale circulation pattern? How and where does water from the bulk, nearly rigidly rotating flow, enter the bottom boundary layer so as to be swept inward toward the center?

(e) Explain how this large-scale circulation pattern can mix much of the water through the boundary layer in the time t_E of Eq. (14.70). What is the value of this t_E for the water in your container? Explain why this, then, must also be the time for the angular velocity of the bulk flow to slow substantially. Compare your computed value of t_E with the observed slow-down time for the water in your container.

Exercise 14.23 **Problem: Water down a Drain in Northern and Southern Hemispheres*

One often hears the claim that water in a bathtub or basin swirls down a drain clockwise in the northern hemisphere and counterclockwise in the southern hemisphere. In fact, on YouTube you are likely to find video demonstrations of this (e.g., by searching on "water down drain at equator"). Show that for Earth-rotation centrifugal forces to produce this effect, it is necessary that the water in the basin initially be moving with a speed smaller than

$$v_{\max} \sim a\Omega_* \sim a\Omega_\oplus \frac{\ell}{R_\oplus} \sim \frac{30 \text{ cm}}{\text{yr}} \left(\frac{a}{1 \text{ m}}\right)\left(\frac{\ell}{1 \text{ km}}\right). \qquad (14.71)$$

Here $\Omega_\oplus$ is Earth's rotational angular velocity, Ω_* is its vertical component (Box 14.5), $R_\oplus$ is Earth's radius, a is the radius of the basin, and ℓ is the distance of the basin from the equator. Even for a basin in Europe or North America, this maximum speed is $\sim 3 \text{ mm min}^{-1}$ for a 1-m-diameter basin—exceedingly difficult to achieve. Therefore, the residual initial motion of the water in any such basin will control the direction in which the water swirls down the drain. There is effectively no difference between northern and southern hemispheres.

Exercise 14.24 **Example: Water down Drain: Experiment*

(a) In a shower or bathtub with the drain somewhere near the center, not the wall, set water rotating so a whirlpool forms over the drain. Perform an experiment to see where the water going down the drain comes from: the surface of the water, its bulk, or its bottom. For example, you could sprinkle baby powder on top of the water, near the whirlpool, and measure how fast the powder is pulled inward and down the drain; put something neutrally buoyant in the bulk and watch its motion; and put sand on the bottom of the shower near the whirlpool and measure how fast the sand is pulled inward and down the drain.

(b) Explain the result of your experiment in part (a). How is it related to the tea cup problem, Ex. 14.22?

(c) Compute the shape of the surface of the water near and in the whirlpool.

14.6 Instabilities of Shear Flows—Billow Clouds and Turbulence in the Stratosphere T2

Kelvin-Helmholtz instability

Here we explore the stability of a variety of shear flows. We begin with the simplest case of two incompressible fluids, one above the other, that move with different uniform speeds, when gravity is negligible. Such a flow has a delta-function spike of vorticity at the interface between the fluids, and, as we shall see, the interface is always unstable against growth of so-called *internal waves*. This is the *Kelvin-Helmholtz instability*. We then explore the ability of gravity to suppress this instability. If the densities of the two fluids are nearly the same, there is no suppression, which is why the Kelvin-Helmholtz instability is seen in a variety of places in Earth's atmosphere and oceans. If the densities are substantially different, then gravity easily suppresses the instability, unless the two flow speeds are very different. Finally, we allow the density and horizontal velocity to change continuously in the vertical direction (e.g., in Earth's stratosphere), and for such a flow we deduce the *Richardson criterion for instability*, which is often satisfied in the stratosphere and leads to turbulence. Along the way we briefly visit several other instabilities that occur in stratified fluids.

14.6.1 Discontinuous Flow: Kelvin-Helmholtz Instability T2

14.6.1

A particularly interesting and simple type of vorticity distribution is one where the vorticity is confined to a thin, plane interface between two immiscible fluids. In other words, one fluid is in uniform motion relative to the other. This type of flow arises quite frequently; for example, when the wind blows over the ocean or when smoke from a smokestack discharges into the atmosphere (a flow that is locally but not globally planar).

We analyze the stability of such flows, initially without gravity (this subsection), then with gravity present (next subsection). Our analysis provides another illustration of the behavior of vorticity and an introduction to techniques that are commonly used to analyze fluid instabilities.

We restrict attention to the simplest version of this flow: an equilibrium with a fluid of density ρ_+ moving horizontally with speed V above a second fluid, which is at rest, with density ρ_-. Let x be a Cartesian coordinate measured along the planar interface in the direction of the flow, and let y be measured perpendicular to it. The equilibrium contains a sheet of vorticity lying in the plane $y = 0$, across which the velocity changes discontinuously. Now, this discontinuity ought to be treated as a boundary layer, with a thickness determined by the viscosity. However, in this problem we analyze disturbances with lengthscales much greater than the thickness of the boundary layer, and so we can ignore it. As a corollary, we can also ignore viscous stresses in the body of the flow. In addition, we specialize to very subsonic speeds, for which the flow can be treated as incompressible; we also ignore the effects of surface tension as well as gravity.

A full description of this flow requires solving the full equations of fluid dynamics, which are quite nonlinear and, as it turns out, for this problem can only be solved numerically. However, we can make progress analytically on an important subproblem. This is the issue of whether this equilibrium flow is stable to small perturbations, and if unstable, the nature of the growing modes. To answer this question, we linearize the fluid equations in the amplitude of the perturbations.

We consider a small vertical perturbation $\delta y = \xi(x, t)$ in the location of the interface (see Fig. 14.19a below). We denote the associated perturbations to the pressure and velocity by δP and $\delta \mathbf{v}$. That is, we write

$$P(\mathbf{x}, t) = P_0 + \delta P(\mathbf{x}, t), \quad \mathbf{v} = V H(y)\mathbf{e}_x + \delta \mathbf{v}(\mathbf{x}, t), \tag{14.72}$$

where P_0 is the constant pressure in the equilibrium flow about which we are perturbing, V is the constant speed of the flow above the interface, and $H(y)$ is the Heaviside step function (1 for $y > 0$ and 0 for $y < 0$). We substitute these $P(\mathbf{x}, t)$ and $\mathbf{v}(\mathbf{x}, t)$ into the governing equations: the incompressibility relation,

$$\nabla \cdot \mathbf{v} = 0, \tag{14.73}$$

and the viscosity-free Navier-Stokes equation (i.e., the Euler equation),

$$\frac{d\mathbf{v}}{dt} = \frac{-\nabla P}{\rho}. \tag{14.74}$$

We then subtract off the equations satisfied by the equilibrium quantities to obtain, for the perturbed variables,

$$\nabla \cdot \delta \mathbf{v} = 0, \tag{14.75}$$

$$\frac{d\delta \mathbf{v}}{dt} = -\frac{\nabla \delta P}{\rho}. \tag{14.76}$$

Combining these two equations, we find, as for Stokes flow (Sec. 14.3.2), that the pressure satisfies Laplace's equation:

$$\nabla^2 \delta P = 0. \tag{14.77}$$

We now follow the procedure used in Sec. 12.4.2 when treating Rayleigh waves on the surface of an elastic medium: we seek an internal wave mode, in which the perturbed quantities vary $\propto \exp[i(kx - \omega t)]f(y)$, with $f(y)$ dying out away from the interface. From Laplace's equation (14.77), we infer an exponential falloff with $|y|$:

$$\delta P = \delta P_0 e^{-k|y|+i(kx-\omega t)}, \tag{14.78}$$

where δP_0 is a constant.

Our next step is to substitute this δP into the perturbed Euler equation (14.76) to obtain

$$\delta v_y = \frac{ik\delta P}{(\omega - kV)\rho_+} \quad \text{for } y > 0, \quad \delta v_y = \frac{-ik\delta P}{\omega \rho_-} \quad \text{for } y < 0. \tag{14.79}$$

We must impose two boundary conditions at the interface between the fluids: continuity of the vertical displacement ξ of the interface (the tangential displacement need not be continuous, since we are examining scales large compared to the boundary layer), and continuity of the pressure P across the interface. [See Eq. (12.44) and associated discussion for the analogous boundary conditions at a discontinuity in an elastic medium.] The vertical interface displacement ξ is related to the velocity perturbation by $d\xi/dt = \delta v_y(y = 0)$, which implies that

$$\xi = \frac{i\delta v_y}{(\omega - kV)} \quad \text{at } y = 0_+ \quad \text{(immediately above the interface)},$$

$$\xi = \frac{i\delta v_y}{\omega} \quad \text{at } y = 0_- \quad \text{(immediately below the interface).} \tag{14.80}$$

Then, by virtue of Eqs. (14.78), (14.80), and (14.79), the continuity of pressure and vertical displacement at $y = 0$ imply that

$$\rho_+(\omega - kV)^2 + \rho_-\omega^2 = 0, \tag{14.81}$$

where ρ_+ and ρ_- are the densities of the fluid above and below the interface, respectively. Solving for frequency ω as a function of horizontal wave number k, we obtain the following dispersion relation for internal wave modes localized at the interface, which are also called *linear Kelvin-Helmholtz modes*:

dispersion relation for linear Kelvin-Helmholtz modes at an interface

$$\boxed{\omega = kV\left(\frac{\rho_+ \pm i(\rho_+\rho_-)^{1/2}}{\rho_+ + \rho_-}\right).} \tag{14.82}$$

This dispersion relation can be used to describe both a sinusoidal perturbation whose amplitude grows in time, and a time-independent perturbation that grows spatially:

TEMPORAL GROWTH

Suppose that we create some small, localized disturbance at time $t = 0$. We can Fourier analyze the disturbance in space, and—as we have linearized the problem—can consider the temporal evolution of each Fourier component separately. What we ought to do is solve the initial value problem carefully, taking account of the initial conditions. However, when there are growing modes, we can usually infer the long-term behavior by ignoring the transients and just considering the growing solutions. In our case, we infer from Eqs. (14.78)–(14.80) and the dispersion relation (14.82) that a mode with spatial frequency k must grow as

temporal growth of mode with wavelength $2\pi/k$

$$\delta P, \ \xi \propto \exp\left[\left(\frac{kV(\rho_+\rho_-)^{1/2}}{(\rho_+ + \rho_-)}\right)t + ik\left(x - V\frac{\rho_+}{\rho_+ + \rho_-}t\right)\right]. \tag{14.83}$$

Thus, this mode grows exponentially with time (Fig. 14.19a). Note that the mode is nondispersive, and if the two densities are equal, it *e*-folds in $1/(2\pi)$ times a period. This means that the fastest modes to grow are those with the shortest periods and hence the shortest wavelengths. (However, the wavelength must not approach the

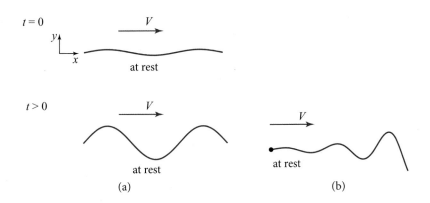

FIGURE 14.19 Kelvin-Helmholtz instability. (a) Temporally growing mode. (b) Spatially growing mode.

thickness of the boundary layer and thereby compromise our assumption that the effects of viscosity are negligible.)

We can understand this growing mode somewhat better by transforming to the center-of-momentum frame, which moves with speed $\rho_+ V/(\rho_+ + \rho_-)$ relative to the frame in which the lower fluid is at rest. In this (primed) frame, the velocity of the upper fluid is $V' = \rho_- V/(\rho_+ + \rho_-)$, and so the perturbations evolve as

$$\delta P, \ \xi \propto \exp[kV'(\rho_+/\rho_-)^{1/2}t] \cos(kx').$$ (14.84)

In this frame the wave is purely growing, whereas in our original frame it oscillated with time as it grew.

SPATIAL GROWTH

An alternative type of mode is one in which a small perturbation is excited temporally at some point where the shear layer begins (Fig. 14.19b). In this case we regard the frequency ω as real and look for the mode with negative imaginary k corresponding to spatial growth. Using Eq. (14.82), we obtain

$$k = \frac{\omega}{V}\left[1 - i\left(\frac{\rho_-}{\rho_+}\right)^{1/2}\right].$$ (14.85)

spatial growth of mode with angular frequency ω

The mode therefore grows exponentially with distance from the point of excitation.

PHYSICAL INTERPRETATION OF THE INSTABILITY

We have performed a normal mode analysis of a particular flow and discovered that there are unstable internal wave modes. However much we calculate the form of the growing modes, though, we cannot be said to understand the instability until we can explain it physically. In the case of this Kelvin-Helmholtz instability, this task is simple.

physical origin of Kelvin-Helmholtz instability

The flow pattern is shown schematically in Fig. 14.20. The upper fluid will have to move faster when passing over a crest in the water wave, because the cross sectional

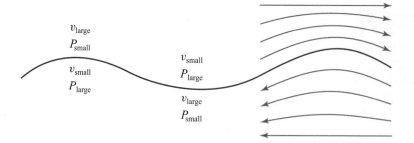

FIGURE 14.20 Physical explanation for the Kelvin-Helmholtz instability.

area of a flow tube diminishes and the flux of fluid must be conserved. By Bernoulli's theorem, the upper pressure will be lower than ambient at this point, and so the crest will rise even higher. Conversely, in the trough of the wave the upper fluid will travel slower, and its pressure will increase. The pressure differential will push the trough downward, making it grow.

Equivalently, we can regard the boundary layer as a plane containing parallel vortex lines that interact with one another, much like magnetic field lines exert pressure on one another. When the vortex lines all lie strictly in a plane, they are in equilibrium, because the repulsive force exerted by one on its neighbor is balanced by an opposite force exerted by the opposite neighbor. However, when this equilibrium is disturbed, the forces become unbalanced and the vortex sheet effectively buckles.

More generally, when there is a large amount of relative kinetic energy available in a high-Reynolds-number flow, there exists the possibility of unstable modes that can tap this energy and convert it into a spectrum of growing modes. These modes can then interact nonlinearly to create fluid turbulence, which ultimately is dissipated as heat. However, the availability of free kinetic energy does not necessarily imply that the flow is unstable; sometimes it is stable. Instability must be demonstrated—often a very difficult task.

14.6.2 Discontinuous Flow with Gravity T2

What happens to this Kelvin-Helmholtz instability if we turn on gravity? To learn the answer, we insert a downward gravitational acceleration **g** into the above analysis. The result (Ex. 14.25) is the following modification of the dispersion relation (14.82):

$$\frac{\omega}{k} = \frac{\rho_+ V}{\rho_+ + \rho_-} \pm \left[\frac{g}{k} \left(\frac{\rho_- - \rho_+}{\rho_+ + \rho_-} \right) - \frac{\rho_+ \rho_-}{(\rho_+ + \rho_-)^2} V^2 \right]^{1/2}. \tag{14.86}$$

gravity suppresses Kelvin-Helmholtz instability

If the lower fluid is sufficiently more dense than the upper fluid (ρ_- sufficiently bigger than ρ_+), then gravity g will change the sign of the quantity inside the square root, making ω/k real, which means the Kelvin-Helmholtz instability is suppressed. In other words, *for the flow to be Kelvin-Helmholtz unstable in the presence of gravity, the two fluids must have nearly the same density:*

FIGURE 14.21 Billow clouds above San Francisco. The cloud structure is generated by the Kelvin-Helmholtz instability. These types of billow clouds may have inspired the swirls in Vincent van Gogh's famous painting "Starry Night." Courtesy of Science Source.

$$\frac{\rho_- - \rho_+}{\rho_+ + \rho_-} < \frac{\rho_+ \rho_-}{(\rho_+ + \rho_-)^2} \frac{kV^2}{g}. \tag{14.87}$$

This is the case for many interfaces in Nature, which is why the Kelvin-Helmholtz instability is often seen. An example is the interface between a water-vapor-laden layer of air under a fast-moving, drier layer. The result is the so-called "billow clouds" shown in Fig. 14.21. Other examples are flow interfaces in the ocean and the edges of dark smoke pouring out of a smoke stack.

As another application of the dispersion relation (14.86), consider the excitation of ocean waves by a laminar-flowing wind. In this case, the "+" fluid is air, and the "−" fluid is water, so the densities are very different: $\rho_+/\rho_- \simeq 0.001$. The instability criterion (14.87) tells us the minimum wind velocity V required to make the waves grow:

$$V_{\min} \simeq \left(\frac{g}{k} \frac{\rho_-}{\rho_+} \right)^{1/2} \simeq 450 \ \text{km h}^{-1} \sqrt{\lambda/10 \ \text{m}}, \tag{14.88}$$

where $\lambda = 2\pi/k$ is the waves' wavelength.

Obviously, this answer is physically wrong. Water waves are easily driven by winds that are far slower than this. Evidently, some other mechanism of interaction between wind and water must drive the waves much more strongly. Observations of wind over water reveal the answer: The winds near the sea's surface are typically quite turbulent (Chap. 15), not laminar. The randomly fluctuating pressures in a turbulent wind are far more effective than the smoothly varying pressure of a laminar wind in driving ocean waves. For two complementary models of this, see Phillips (1957) and Miles (1993).

As another, very simple application of the dispersion relation (14.86), set the speed V of the upper fluid to zero. In this case, the interface is unstable if and only if the upper

Rayleigh-Taylor instability

fluid has higher density than the lower fluid: $\rho_+ > \rho_-$. This is called the *Rayleigh-Taylor instability* for incompressible fluids.

EXERCISES

Exercise 14.25 *Problem: Discontinuous Flow with Gravity* T2

Insert gravity into the analysis of the Kelvin-Helmholtz instability (with the uniform gravitational acceleration **g** pointing perpendicularly to the fluid interface, from the upper "+" fluid to the lower "−" fluid). Thereby derive the dispersion relation (14.86).

14.6.3

14.6.3 Smoothly Stratified Flows: Rayleigh and Richardson Criteria for Instability T2

Sometimes one can diagnose a fluid instability using simple physical arguments rather than detailed mathematical analysis. We conclude this chapter with two examples.

rotating Couette flow

First, we consider *rotating Couette flow*: the azimuthal flow of an incompressible, effectively inviscid fluid, rotating axially between two coaxial cylinders (Fig. 14.22).

We explore the stability of this flow to purely axisymmetric perturbations by using a thought experiment in which we interchange two fluid rings. As there are no azimuthal forces (no forces in the ϕ direction), the interchange will occur at constant angular momentum per unit mass. Suppose that the ring that moves outward in radius ϖ has lower specific angular momentum j than the surroundings into which it has moved; then it will have less centrifugal force per unit mass j^2/ϖ^3 than its surroundings and will thus experience a restoring force that drives it back to its original position. Conversely, if the surroundings' angular momentum per unit mass

Rayleigh criterion for instability of rotating flows

decreases outward, then the displaced ring will continue to expand. We conclude on this basis that *Couette and similar flows are unstable when the angular momentum per unit mass decreases outward*. This is known as the *Rayleigh criterion*. We return to Couette flow in Sec. 15.6.1.

Compact stellar objects (black holes, neutron stars, and white dwarfs) are sometimes surrounded by orbiting *accretion disks* of gas (Exs. 26.17 and 26.18). The gas in these disks orbits with its angular velocity approximately dictated by Kepler's laws. Therefore, the specific angular momentum of the gas increases radially outward, approximately proportional to the square root of the orbital radius. Consequently, accretion disks are stable to this type of instability.

For our second example of a simple physical diagnosis of instability, consider the situation analyzed in Ex. 14.25 (Kelvin-Helmholtz instability with gravity present) but with the density and velocity changing continuously instead of discontinuously, as one moves upward. More specifically, focus on Earth's stratosphere, which extends

stability of a superadiabatic density gradient

from ~10 km height to ~40 km. The density in the stratosphere decreases upward faster than if the stratosphere were isentropic. This means that, when a fluid element moves upward adiabatically, its pressure-induced density reduction is smaller than the density decrease in its surroundings. Since it is now more dense than its surroundings, the downward pull of gravity on the fluid element will exceed the upward buoyant

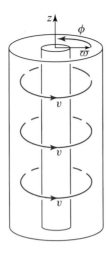

FIGURE 14.22 Rotating Couette flow.

force, so the fluid element will be pulled back down. Therefore, the stratosphere is stably stratified. (We shall return to this phenomenon, in the context of stars, in Sec. 18.5.)

However, it may be possible for the stratosphere to tap the relative kinetic energy in its horizontal winds, so as to mix the air vertically. Consider the pedagogical example of two thin stream tubes of horizontally flowing air in the stratosphere, separated vertically by a distance $\delta\ell$ large compared to their thicknesses. The speed of the upper stream will exceed that of the lower stream by $\delta V = V'\delta\ell$, where $V' = dV/dz$ is the velocity gradient. In the center of velocity frame, the streams each have speed $\delta V/2$, and they differ in density by $\delta\rho = |\rho'\delta\ell|$. To interchange these streams requires doing work per unit mass[7] $\delta W = g[\delta\rho/(2\rho)]\delta\ell$ against gravity (where the factor 2 comes from the fact that there are two unit masses involved in the interchange, one going up and the other coming down). This work can be supplied by the streams' kinetic energy, if the available kinetic energy per unit mass, $\delta E_k = (\delta V/2)^2/2$, exceeds the required work. *A necessary condition for instability* is then that

$$\delta E_k = (\delta V)^2/8 > \delta W = g|\rho'/(2\rho)|\delta\ell^2, \tag{14.89}$$

or

$$\boxed{\mathrm{Ri} = \frac{|\rho'|g}{\rho V'^2} < \frac{1}{4},} \tag{14.90}$$

Richardson criterion for instability of a stratified shear flow

7. Here, for simplicity we idealize the streams' densities as not changing when they move vertically—an idealization that makes gravity be more effective at resisting the instability. In the stratosphere, where the temperature is constant (Fig. 13.2) so the density drops vertically much faster than if it were isentropic, this idealization is pretty good.

where $V' = dV/dz$ is the velocity gradient. This is known as the *Richardson criterion for instability*, and Ri is the *Richardson number*. Under a wide variety of circumstances this criterion turns out to be sufficient for instability.

The density scale height in the stratosphere is $\rho/|\rho'| \sim 10$ km. Therefore the maximum velocity gradient allowed by the Richardson criterion is

$$V' \lesssim 60 \frac{\text{m s}^{-1}}{\text{km}}. \tag{14.91}$$

Larger velocity gradients are rapidly disrupted by instabilities. This instability is responsible for much of the clear air turbulence encountered by airplanes, and it is to a discussion of turbulence that we turn in the next chapter.

Bibliographic Note

Vorticity is so fundamental to fluid mechanics that it and its applications are treated in detail in all fluid-dynamics textbooks. Among those with a physicist's perspective, we particularly like Acheson (1990) and Lautrup (2005) at an elementary level, and Landau and Lifshitz (1959), Lighthill (1986), and Batchelor (2000) at a more advanced level. Tritton (1987) is especially good for physical insight. Panton (2005) is almost encyclopedic and nicely bridges the viewpoints of physicists and engineers. For an engineering emphasis, we like Munson, Young, and Okiishi (2006) and Potter, Wiggert, and Ramadan (2012); see also the more advanced text on viscous flow, White (2006). For the viewpoint of an applied mathematician, see Majda and Bertozzi (2002).

To build up physical intuition, we recommend Tritton (1987) and the movies listed in Box 14.2. For a textbook treatment of rotating flows and weak perturbations of them, we recommend Greenspan (1973). For geophysical applications at an elementary level, we like Cushman-Roisin and Beckers (2011), and for a more mathematical treatment Pedlosky (1987). A more encyclopedic treatment at an elementary level (including some discussion of simulations) can be found in Gill (1982) and Vallis (2006).

15

Turbulence

Big whirls have little whirls, which feed on their velocity.
Little whirls have lesser whirls, and so on to viscosity.

LEWIS RICHARDSON (1922)

15.1 Overview

In Sec. 13.7.6, we derived the Poiseuille formula for the flow of a viscous fluid down a pipe by assuming that the flow is laminar (i.e., that it has a velocity parallel to the pipe wall). We showed how balancing the stress across a cylindrical surface led to a parabolic velocity profile and a rate of flow proportional to the fourth power of the pipe diameter d. We also defined the Reynolds number; for pipe flow it is $\mathrm{Re}_d \equiv \bar{v}d/\nu$, where $\bar{v}$ is the mean speed in the pipe, and ν is the kinematic viscosity. Now, it turns out experimentally that the pipe flow only remains laminar up to a critical Reynolds number that has a value in the range $\sim 10^3$–10^5, depending on the smoothness of the pipe's entrance and roughness of its walls. If the pressure gradient is increased further (and thence the mean speed $\bar{v}$ and Reynolds number Re_d are increased), then the velocity field in the pipe becomes irregular both temporally and spatially, a condition known as *turbulence*.

Turbulence is common in high-Reynolds-number flows. Much of our experience with fluids involves air or water, for which the kinematic viscosities are $\sim 10^{-5}$ and 10^{-6} m^2 s^{-1}, respectively. For a typical everyday flow with a characteristic speed of $v \sim 10$ m s^{-1} and a characteristic length of $d \sim 1$ m, the Reynolds number is huge: $\mathrm{Re} = vd/\nu \sim 10^6 - 10^7$. It is therefore not surprising that we see turbulent flows all around us. Smoke in a smokestack, a cumulus cloud, and the wake of a ship are examples.

In Sec. 15.2 we illustrate the phenomenology of the transition to turbulence as the Reynolds number increases using a particularly simple example: the flow of a fluid past a circular cylinder oriented perpendicular to the flow's incoming velocity. We shall see how the flow pattern is dictated by the Reynolds number, and how the velocity changes from steady creeping flow at low Re to fully developed turbulence at high Re.

What is turbulence? Fluid dynamicists can recognize it, but they have a hard time defining it precisely[1] and an even harder time describing it quantitatively.[2] So typically for a definition they rely on empirical, qualitative descriptions of its physical properties (Sec. 15.3). Closely related to this description is the crucial role of vorticity in driving turbulent energy from large scales to small (Sec. 15.3.1).

At first glance, a quantitative description of turbulence appears straightforward. Decompose the velocity field into Fourier components just as is done for the electromagnetic field when analyzing electromagnetic radiation. Then recognize that the equations of fluid dynamics are nonlinear, so there will be coupling between different modes (akin to wave-wave coupling between optical modes in a nonlinear crystal, discussed in Chap. 10). Analyze that coupling perturbatively. The resulting *weak-turbulence formalism* is sketched in Secs. 15.4.1 and 15.4.2 and in Ex. 15.5.[3]

However, most turbulent flows come under the heading of *fully developed* or *strong turbulence* and cannot be well described by weak-turbulence models. Part of the problem is that the $(\mathbf{v} \cdot \nabla)\mathbf{v}$ term in the Navier-Stokes equation is a strong nonlinearity, not a weak coupling between linear modes. As a consequence, eddies of size ℓ persist for typically no more than one turnover timescale $\sim \ell/v$ before they are broken up, and so they do not behave like weakly coupled normal modes.

In the absence of a decent quantitative theory of strong turbulence, fluid dynamicists sometimes simply push the weak-turbulence formalism into the strong-turbulence regime and use it there to gain qualitative or semiquantitative insights (e.g., Fig. 15.7 below and associated discussion in the text). A simple alternative (which we explore in Sec. 15.4.3 in the context of wakes and jets and in Sec. 15.5 for turbulent boundary layers) is intuitive, qualitative, and semiquantitative approaches to the *physical* description of turbulence. We emphasize the adjective "physical," because our goal is to start to comprehend the underlying physical character of turbulence, going beyond empirical descriptions of its consequences on the one hand and uninstruc-

---

1. The analogy to Justice Potter Stewart's definition of pornography should be resisted.
2. Werner Heisenberg's dissertation was "On the Stability and Turbulence of Fluid Flow." He was disappointed with his progress and was glad to change to a more tractable problem.
3. Another weak-turbulence formalism that is developed along similar lines is the *quasilinear* theory of nonlinear plasma interactions, which we discuss in Chap. 23.

tive mathematical expansions on the other. Much modern physics has this character. An important feature that we meet in Sec. 15.4.3, when we discuss wakes and jets, is *entrainment*. This leads to irregular boundaries of turbulent flows caused by giant eddies and to dramatic time dependence, including *intermittency*.

One triumph of this approach (Sec. 15.4.4) is the Kolmogorov analysis of the shape of the time-averaged turbulent energy spectrum (the turbulent energy per unit wave number as a function of wave number) in a stationary turbulent flow. This spectrum has been verified experimentally under many different conditions. The arguments used to justify it are characteristic of many semiempirical derivations of scaling relations that find confident practical application in the world of engineering.

In the context of turbulent boundary layers, our physical approach will reveal semiquantitatively the structures of such boundary layers (Sec. 15.5.1), and it will explain why turbulent boundary layers generally exert more shear stress on a surface than do laminar boundary layers, but nevertheless usually produce less total drag on airplane wings, baseballs, etc. (Sec. 15.5.2).

Whether or not a flow becomes turbulent can have a major influence on how fast chemical reactions occur in liquids and gases. In Sec. 15.5.3, we briefly discuss how turbulence can arise through instability of a laminar boundary layer.

One can gain additional insight into turbulence by a technique that is often useful when struggling to understand complex physical phenomena: Replace the system being studied by a highly idealized model system that is much simpler than the original one, both conceptually and mathematically, but that retains at least one central feature of the original system. Then analyze the model system completely, with the hope that the quantitative insight so gained will be useful in understanding the original problem. Since the 1970s, new insights into turbulence have come from studying idealized dynamical systems that have very few degrees of freedom but have the same kinds of nonlinearities as produce turbulence in fluids (e.g., Ott, 1982; Ott, 1993). We examine several such low-dimensional dynamical systems and the insights they give in Sec. 15.6.

The most useful of those insights deal with the onset of weak turbulence and the observation that it seems to have much in common with the onset of chaos (irregular and unpredictable dynamical behavior) in a wide variety of other dynamical systems. A great discovery of modern classical physics/mathematics has been that there exist organizational principles that govern the behavior of these seemingly quite different chaotic physical systems.

In parallel with studying this chapter, to build up physical intuition the reader is urged to watch movies and study photographs that deal with turbulence; see Box 15.2.

15.2 The Transition to Turbulence—Flow Past a Cylinder 15.2

We illustrate qualitatively how a flow (and especially its transition to turbulence) depends on its Reynolds number by considering a specific problem: the flow of a uniformly moving fluid past a cylinder oriented transversely to the flow's incoming velocity (Fig. 15.1). We assume that the flow velocity is small compared with the speed of sound, so the effects of compressibility can be ignored. Let the cylinder diameter be d, and choose this as the characteristic length in the problem. Similarly, let the velocity far upstream be V and choose this as the characteristic velocity, so the Reynolds number is[4]

Reynolds number

$$\mathrm{Re}_d = \frac{Vd}{\nu}. \tag{15.1}$$

equations governing stationary, unperturbed flows

Initially, we assume that the flow is stationary (no turbulence) as well as incompressible, and the effects of gravity are negligible. Then the equations governing the flow are incompressibility,

$$\nabla \cdot \mathbf{v} = 0, \tag{15.2a}$$

and the time-independent Navier-Stokes equation (13.69) with $\partial \mathbf{v}/\partial t = 0$:

$$(\mathbf{v} \cdot \nabla)\mathbf{v} = -\frac{\nabla P}{\rho} + \nu\nabla^2\mathbf{v}. \tag{15.2b}$$

These four equations (one for incompressibility, three for the components of Navier-Stokes) can be solved for the pressure and the three components of velocity subject to the velocity vanishing on the surface of the cylinder and becoming uniform far upstream.

From the parameters of the flow (the cylinder's diameter d; the fluid's incoming velocity V, its density ρ, and its kinematic viscosity ν) we can construct only one dimensionless number, the Reynolds number $\mathrm{Re}_d = Vd/\nu$. (If the flow speed were high enough that incompressibility fails, then the sound speed c_s would also be a relevant parameter, and a second dimensionless number could be formed: the Mach number, $M = V/c_s$; Chap. 17.) With Re_d the only dimensionless number, we are

---

4. The subscript d is just a reminder that in this instance we have chosen the diameter as the characteristic lengthscale; for other applications, the length of the cylinder might be more relevant.

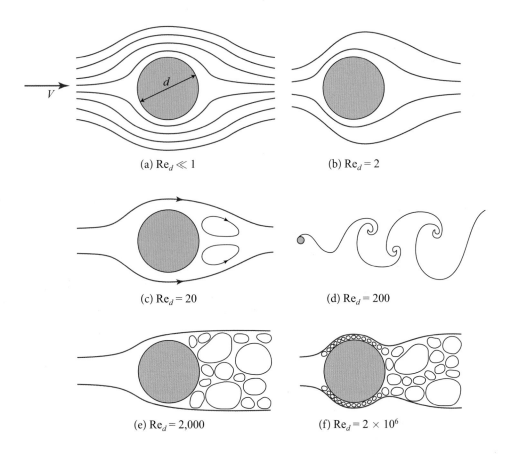

(a) $\mathrm{Re}_d \ll 1$

(b) $\mathrm{Re}_d = 2$

(c) $\mathrm{Re}_d = 20$

(d) $\mathrm{Re}_d = 200$

(e) $\mathrm{Re}_d = 2{,}000$

(f) $\mathrm{Re}_d = 2 \times 10^6$

FIGURE 15.1 Schematic depiction of flow past a cylinder for steadily increasing values of the Reynolds number $\mathrm{Re}_d = Vd/\nu$ as labeled. There are many photographs, drawings, and simulations of this flow on the web, perhaps best found by doing a search on "Kármán vortex street."

guaranteed on dimensional grounds that the solution to the flow equations can be expressed as

$$\mathbf{v}/V = \mathbf{U}(\mathbf{x}/d,\, \mathrm{Re}_d).$$

(15.3)

dimensional analysis gives functional form of flow

Here $\mathbf{U}$ is a dimensionless function of the dimensionless $\mathbf{x}/d$, and it can take wildly different forms, depending on the value of the Reynolds number Re_d (cf. Fig. 15.1, which we discuss below).

The functional form of $\mathbf{v}$ [Eq. (15.3)] has important implications. If we compute the flow for specific values of the upstream velocity V, the cylinder's diameter d, and the kinematic viscosity ν and then double V and d and quadruple ν so that Re_d is unchanged, then the new solution will be *similar* to the original one. It can be produced from the original by rescaling the flow velocity to the new upstream velocity and the distance to the new cylinder diameter. [For this reason, Eq. (15.3) is sometimes called a *scaling relation*.] By contrast, if we had only doubled the kinematic

scaling relation

viscosity, the Reynolds number would have also doubled, and we could be dealing with a qualitatively different flow.

When discussing flow past the cylinder, a useful concept is the *stagnation pressure* in the upstream flow. This is the pressure the fluid would have, according to the Bernoulli principle ($v^2/2 + \int dP/\rho = $ const), if it were brought to rest at the leading edge of the cylinder without significant action of viscosity. Ignoring the effects of compressibility (so ρ is constant), this stagnation pressure is

stagnation pressure

$$\boxed{P_{\text{stag}} = P_0 + \frac{1}{2}\,\rho V^2\,,}$$

(15.4)

where P_0 is the upstream pressure. Suppose that this stagnation pressure were to act over the whole front face of the cylinder, while the pressure P_0 acted on the downstream face. The net force on the cylinder per unit length, F_D, would then be $\frac{1}{2}\rho V^2 d$. This is a first rough estimate for the drag force. It is conventional to define a *drag coefficient* as the ratio of the actual drag force per unit length to this rough estimate:

drag coefficient

$$C_D \equiv \frac{F_D}{\frac{1}{2}\rho V^2 d}.$$

(15.5)

This drag coefficient, being a dimensionless feature of the flow (15.3), can depend **drag coefficient is function** only on the dimensionless Reynolds number Re_d: $C_D = C_D(\text{Re}_d)$; see Fig. 15.2. **of Reynolds number** Similarly for flow past a 3-dimensional body with cross sectional area A perpendicular to the flow and with any other shape, the drag coefficient

$$\boxed{C_D \equiv \frac{F_D}{\frac{1}{2}\rho V^2 A}}$$

(15.6)

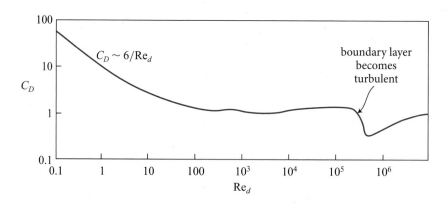

FIGURE 15.2 Drag coefficient C_D for flow past a cylinder as a function of Reynolds number $\text{Re}_d = Vd/\nu$. This graph, adapted from Tritton (1987, Fig. 3.15), is based on experimental measurements.

will be a function only of Re. However, the specific functional form of $C_D(\text{Re})$ will depend on the body's shape and orientation.

Now, turn to the details of the flow around a cylinder as described in Figs. 15.1 and 15.2. At low Reynolds number, $\text{Re}_d \ll 1$, there is creeping flow (Fig. 15.1a) just like that analyzed in detail for a spherical obstacle in Sec. 14.3.2. As you might have surmised by tackling Ex. 14.12, the details of low-Reynolds-number flow past a long object, such as a cylinder, are subtly different from those of flow past a short one, such as a sphere. This is because, for a cylinder, inertial forces become comparable with viscous and pressure forces at distances $\sim d/\text{Re}_d$, where the flow is still significantly perturbed from uniform motion, while for short objects inertial forces become significant only at much larger radii, where the flow is little perturbed by the object's presence. Despite this, the flow streamlines around a cylinder at $\text{Re}_d \ll 1$ (Fig. 15.1a) are similar to those for a sphere (Fig. 14.10) and are approximately symmetric between upstream and downstream. The fluid is decelerated by viscous stresses as it moves past the cylinder along these streamlines, and the pressure is higher on the cylinder's front face than on its back. Both effects contribute to the net drag force acting on the cylinder. The momentum removed from the flow is added to the cylinder. At cylindrical radius $\varpi \ll d/\text{Re}_d$ the viscous stress dominates over the fluid's inertial stress, and the fluid momentum therefore is being transferred largely to the cylinder at a rate per unit area $\sim \rho V^2$. In contrast, for $\varpi \gtrsim d/\text{Re}_d$ the viscous and inertial stresses are comparable and balance each other, and the flow's momentum is not being transferred substantially to the cylinder. This implies that the effective cross sectional width over which the cylinder extracts the fluid's momentum is $\sim d/\text{Re}_d$, and correspondingly, the net drag force per unit length is $F \sim \rho V^2 d/\text{Re}_d$, which implies [cf. Eq. (15.5)] a drag coefficient $\sim 1/\text{Re}_d$ at low Reynolds numbers ($\text{Re}_d \ll 1$). A more careful analysis gives $C_D \sim 6/\text{Re}_d$, as shown experimentally in Fig. 15.2.

As the Reynolds number is increased to ~ 1 (Fig. 15.1b), the effective cross section gets reduced to roughly the cylinder's diameter d; correspondingly, the drag coefficient decreases to $C_D \sim 1$. At this Reynolds number, $\text{Re}_d \sim 1$, the velocity field begins to appear asymmetric from front to back.

With a further increase in Re_d, a laminar boundary layer of thickness $\delta \sim d/\sqrt{\text{Re}_d}$ starts to form. The viscous force per unit length due to this boundary layer is $F \sim \rho V^2 d/\sqrt{\text{Re}_d}$ [cf. Eqs. (14.49)–(14.51) divided by the transverse length w, and with $\ell \sim d$ and $v_o = V$]. It might therefore be thought that the drag would continue to decrease as $C_D \sim 1/\sqrt{\text{Re}_d}$, when Re_d increases substantially above unity, making the boundary layer thin and the external flow start to resemble potential flow. However, this does *not* happen. Instead, at $\text{Re}_d \sim 5$, the flow begins to separate from the back side of the cylinder and is there replaced by two retrograde eddies (Fig. 15.1c). As described in Sec. 14.4.4, this separation occurs because an adverse pressure gradient $(\mathbf{v} \cdot \nabla) P > 0$ develops outside the boundary layer, near the cylinder's downstream face, and causes the separated boundary layer to be replaced by these two counter-rotating eddies. The pressure in these eddies, and thus also on the cylinder's back face,

at $\text{Re}_d \ll 1$: creeping flow and $C_D \sim 6/\text{Re}_d$

at $\text{Re}_d \sim 1$: $C_D \sim 1$ and laminar boundary layer starts to form

at $\text{Re}_d \sim 5$: separation begins

is of order the flow's incoming pressure P_0 and is significantly less than the stagnation pressure $P_{stag} = P_0 + \frac{1}{2}\rho V^2$ at the cylinder's front face, so the drag coefficient stabilizes at $C_D \sim 1$.

at $\text{Re}_d \sim 100$: eddies shed dynamically; form Kármán vortex street

As the Reynolds number increases above $\text{Re}_d \sim 5$, the size of the two eddies increases until, at $\text{Re}_d \sim 100$, the eddies are shed dynamically, and the flow becomes nonstationary. The eddies tend to be shed alternately in time, first one and then the other, producing a beautiful pattern of alternating vortices downstream known as a *Kármán vortex street* (Fig. 15.1d).

at $\text{Re}_d \sim 1,000$: turbulent wake

When $\text{Re}_d \sim 1,000$, the downstream vortices are no longer visible, and the wake behind the cylinder contains a velocity field irregular on all macroscopic scales (Fig. 15.1e). This downstream flow has become turbulent. Finally, at $\text{Re}_d \sim 3 \times 10^5$, the boundary layer itself, which has been laminar up to this point, becomes turbulent (Fig. 15.1f), reducing noticeably the drag coefficient (Fig. 15.2). We explore the cause of this reduction in Sec. 15.5. [The physically relevant Reynolds number for onset of turbulence in the boundary layer is that computed not from the cylinder diameter d, $\text{Re}_d = Vd/\nu$, but rather from the boundary layer thickness $\delta \sim d/\text{Re}_d^{1/2}$:

at $\text{Re}_d \sim 3 \times 10^5$: turbulent boundary layer

$$\text{Re}_\delta = \frac{V\delta}{\nu} \sim \frac{Vd\text{Re}_d^{-1/2}}{\nu} = \sqrt{\text{Re}_d}. \tag{15.7}$$

The onset of boundary-layer turbulence is at $\text{Re}_\delta \sim \sqrt{3 \times 10^5} \sim 500$, about the same as the $\text{Re}_d \sim 1,000$ for the onset of turbulence in the wake.]

turbulence is 3-dimensional

An important feature of this changing flow pattern is that at $\text{Re}_d \ll 1,000$ (Figs. 15.1a–d), before any turbulence sets in, the flow (whether steady or dynamical) is translation symmetric—it is independent of distance z down the cylinder (i.e., it is 2-dimensional). This is true even of the Kármán vortex street. By contrast, the turbulent velocity field at $\text{Re}_d \gtrsim 1,000$ is fully 3-dimensional. At these high Reynolds numbers, small, nontranslation-symmetric perturbations of the translation-symmetric flow grow into vigorous, 3-dimensional turbulence. This is a manifestation of the inability of 2-dimensional flows to exhibit all the chaotic motion associated with 3-dimensional turbulence (Sec. 15.4.4).

critical Reynolds number for onset of turbulence

The most important feature of this family of flows—one that is characteristic of most such families—is that there is a critical Reynolds number for the onset of turbulence. That critical number can range from ~ 30 to $\sim 10^5$, depending on the geometry of the flow and on precisely what length and speed are used to define the Reynolds number.

EXERCISES

Exercise 15.1 **Example: The 2-Dimensional Laminar Wake behind a Cylinder*
In Sec. 15.4.3, we explore the structure of the wake behind the cylinder when the Reynolds number is high enough that the flow is turbulent. For comparison, here we

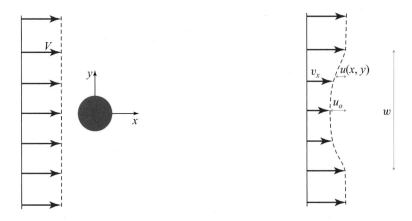

FIGURE 15.3 The 2-dimensional laminar wake behind an infinitely long cylinder. (This figure also describes a turbulent wake, at high Reynolds numbers, if v_x is replaced by the time-averaged $\bar{v}_x$; Sec. 15.4.3.)

compute the wake's structure at lower Reynolds numbers, when the wake is laminar. This computation is instructive: using order-of-magnitude estimates first, followed by detailed calculations, it illustrates the power of momentum conservation. It is our first encounter with the velocity field in a wake.

(a) Begin with an order-of-magnitude estimate. Characterize the wake by its width $w(x)$ at distance x downstream from the cylinder and by the reduction in the flow velocity (the "velocity deficit"), $u_o(x) \equiv V - v_x(x)$, at the center of the wake; see Fig. 15.3. From the diffusion of vorticity show that $w \simeq 2\sqrt{\nu x / V}$.

(b) Explain why momentum conservation requires that the force per unit length on the cylinder, $F_D = C_D \frac{1}{2}\rho V^2 d$ [Eq. (15.5)], equals the transverse integral $\int T_{xx}dy$ of the fluid's kinetic stress ($T_{xx} = \rho v_x v_x$) before the fluid reaches the sphere, minus that integral at distance x after the sphere. Use this requirement to show that the fractional velocity deficit at the center of the wake is $u_o/V \simeq \frac{1}{4}C_D d/w \simeq C_D\sqrt{d\,\mathrm{Re}_d/(64x)}$.

(c) For a more accurate description of the flow, solve the Navier-Stokes equation to obtain the profile of the velocity deficit, $u(x, y) \equiv V - v_x(x, y)$. [Hint: Ignoring the pressure gradient, which is negligible (Why?), the Navier-Stokes equation should reduce to the 1-dimensional diffusion equation, which we have met several times previously in this book.] Your answer should be

$$u = u_o e^{-(2y/w)^2}, \quad w = 4\left(\frac{\nu x}{V}\right)^{1/2}, \quad u_o = V C_D \left(\frac{d\,\mathrm{Re}_d}{16\pi x}\right)^{1/2}, \quad (15.8)$$

where w and u_o are more accurate values of the wake's width and its central velocity deficit.

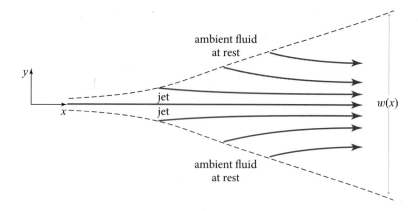

FIGURE 15.4 2-dimensional laminar jet. As the jet widens, it entrains ambient fluid.

Exercise 15.2 *Problem: The 3-Dimensional Laminar Wake behind a Sphere*
Repeat Ex. 15.1 for the 3-dimensional laminar wake behind a sphere.

Exercise 15.3 *Example: Structure of a 2-Dimensional Laminar Jet; Entrainment*
Consider a narrow, 2-dimensional, incompressible (i.e., subsonic) jet emerging from a 2-dimensional nozzle into ambient fluid at rest with the same composition and pressure. (By 2-dimensional we mean that the nozzle and jet are translation symmetric in the third dimension.) Let the Reynolds number be low enough for the flow to be laminar; we study the turbulent regime in Ex. 15.7. We want to understand how rapidly this laminar jet spreads.

(a) Show that the pressure forces far downstream from the nozzle are likely to be much smaller than the viscous forces and can therefore be ignored.

(b) Let the jet's thrust per unit length (i.e., the momentum per unit time per unit length flowing through the nozzle) be $\mathcal{F}$. Introduce Cartesian coordinates x, y, with x parallel to and y perpendicular to the jet (cf. Fig. 15.4). As in Ex. 15.1, use vorticity diffusion (or the Navier-Stokes equation) and momentum conservation to estimate the speed v_x of the jet and its width w as functions of distance x downstream.

(c) Use the scalings from part (b) to modify the self-similarity analysis of the Navier-Stokes equation that we used for the laminar boundary layer in Sec. 14.4, and thereby obtain the following approximate solution for the jet's velocity profile:

$$v_x = \left(\frac{3\mathcal{F}^2}{32\rho^2 \nu x}\right)^{1/3} \mathrm{sech}^2\left[\left(\frac{\mathcal{F}}{48\rho\nu^2 x^2}\right)^{1/3} y\right]. \tag{15.9}$$

(d) Equation (15.9) shows that the jet width w increases downstream as $x^{2/3}$. As the jet widens, it scoops up (entrains) ambient fluid, as depicted in Fig. 15.4. This entrainment actually involves pulling fluid inward in a manner described by the y component of velocity, v_y. Solve the incompressibility equation $\nabla \cdot \mathbf{v} = 0$ to obtain the following expression for v_y:

$$v_y = -\frac{1}{3x} \left(\frac{3\mathcal{F}^2}{32\rho^2 \nu x} \right)^{1/3}$$ (15.10)

$$\times \left\{ \left(\frac{48\rho\nu^2 x^2}{\mathcal{F}} \right)^{1/3} \tanh \left[\left(\frac{\mathcal{F}}{48\rho\nu^2 x^2} \right)^{1/3} y \right] - 2y \; \mathrm{sech}^2 \left[\left(\frac{\mathcal{F}}{48\rho\nu^2 x^2} \right)^{1/3} y \right] \right\}$$

$$\simeq -\left(\frac{1}{6} \frac{\mathcal{F}\nu}{\rho x^2} \right)^{1/3} \mathrm{sign}(y) \quad \text{for} \quad |y| \gg \frac{1}{2} w(x) = \left(\frac{48\rho\nu^2 x^2}{\mathcal{F}} \right)^{1/3}.$$

Thus ambient fluid is pulled inward from both sides to satisfy the jet's entrainment appetite.

Exercise 15.4 *Example: Marine Animals*

One does not have to be a biologist to appreciate the strong evolutionary advantage that natural selection confers on animals that can reduce their drag coefficients. It should be no surprise that the shapes and skins of many animals are highly streamlined. This is particularly true for aquatic animals. Of course, an animal minimizes drag while developing efficient propulsion and lift, which change the flow pattern. Comparisons of the properties of flows past different stationary (e.g., dead or towed) animals are, therefore, of limited value. The species must also organize its internal organs in 3 dimensions, which is an important constraint. The optimization varies from species to species and is suited to the environment. There are some impressive performers in the animal world.

(a) First, idealize our animal as a thin rectangle with thickness t, length ℓ, and width w such that $t \ll \ell \ll w$.[5] Let the animal be aligned with the flow parallel to the ℓ direction. Assuming an area of $A = 2\ell w$, the drag coefficient is $C_D = 1.33\mathrm{Re}_\ell^{-0.5}$ [Eq. (14.50)]. This assumes that the flow is laminar. The corresponding result for a turbulent flow is $C_D \simeq 0.072\mathrm{Re}_\ell^{-0.2}$. Show that the drag can be considerably reduced if the transition to turbulence takes place at high Re_ℓ. Estimate the effective Reynolds number for an approximately flat fish like a flounder of size ~ 0.3 m that can move with a speed ~ 0.3 m s^{-1}, and then compute the drag force. Express your answer as a stopping length.

(b) One impressive performer is the mackerel (a highly streamlined fish), for which the reported drag coefficient is 0.0043 at $\mathrm{Re} \sim 10^5$. Compare this with a thin plate and a sphere. (The drag coefficient for a sphere is $C_D \sim 0.5$, assuming a reference area equal to the total area of the sphere. The drag decreases abruptly when there is a transition to turbulence, just as we found with the cylinder.)

---

5. The aquatic counterpart to the famous Spherical Cow!

(c) Another fine swimmer is the California sea lion, which has a drag coefficient of $C_D = 0.0041$ at Re $\sim 2 \times 10^6$. How does this compare with a plate and a sphere?

Further details can be found in Vogel (1994) and references therein.

15.3 Empirical Description of Turbulence

Empirical studies of turbulent flows have revealed some universal properties that are best comprehended through movies (Box 15.2). Here we simply list the most important of them and comment on them briefly. We revisit most of them in more detail in the remainder of the chapter. Throughout, we restrict ourselves to turbulence with velocities that are very subsonic, and thus the fluid is incompressible.

Turbulence is characterized by:

characteristics of turbulence

- *Disorder, irreproducible in detail but with rich, nonrandom structure.* This disorder is intrinsic to the flow. It appears to arise from a variety of instabilities. No forcing by external agents is required to produce it. If we try to resolve the flow into modes, however, we find that the phases of the modes are not fully random, either spatially or temporally: there are strong correlations.[6] Correspondingly, if we look at a snapshot of a turbulent flow, we frequently observe large, well-defined coherent structures like eddies and jets, which suggests that the flow is more organized than a purely random superposition of modes, just as the light reflected from the surface of a painting differs from that emitted by a blackbody. If we monitor the time variation of some fluid variable, such as one component of the velocity at a given point in the flow, we observe *intermittency*—the irregular starting and ceasing of strong turbulence. Again, this effect is so pronounced that more than a random-mode superposition is at work, reminiscent of the distinction between noise and music (at least some music). [A major consequence that we shall have to face is that strong turbulence is *not* well treated by perturbation theory. As an alternative, semiquantitative techniques of analysis must be devised.]

- *A wide range of interacting scales.* When the fluid velocity and pressure are Fourier analyzed, one finds them varying strongly over many decades of wave number and frequency. We can think of these variations as due to *eddies* with a huge range of sizes. These eddies interact strongly. Typically, large eddies appear to feed their energy to smaller eddies, which in turn feed energy to still smaller eddies, and so forth. Occasionally, amazingly, the flow of energy appears to reverse: small-scale turbulent structures give rise

---

6. As we discuss in Chap. 28, many cosmologists suspected that the primordial fluctuations, out of which galaxies and stars eventually grew, would share this characteristic with turbulence. Insofar as we can measure, this is not the case, and the fluctuations appear to be quite random and uncorrelated.

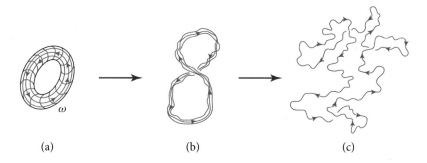

FIGURE 15.5 Schematic illustration of the propagation of turbulence by the stretching of vortex lines. The tube of vortex lines in (a) is stretched and thereby forced into a reduced cross section by the turbulent evolution from (a) to (b) to (c). The reduced cross section means an enhanced vorticity on smaller scales.

to large-scale structures,[7] resulting in intermittency. A region of the flow that appears to have calmed down may suddenly and unexpectedly become excited again.

- *Vorticity, irregularly distributed in three dimensions.* This vorticity varies in magnitude and direction over the same wide range of scales as for the fluid velocity. It appears to play a central role in coupling large scales to small; see Sec. 15.3.1.

- *Large dissipation.* Typically, turbulent energy is fed from large scales to small in just one turnover time of a large eddy. This is extremely fast. The energy cascades down to smaller and smaller lengthscales in shorter and shorter timescales until it reaches eddies so small that their shear, coupled to molecular viscosity, converts the turbulent energy into heat.

- *Efficient mixing and transport.* Most everything that can be transported is efficiently mixed and moved: momentum, heat, salt, chemicals, and so forth.

15.3.1 The Role of Vorticity in Turbulence

Turbulent flows contain tangled vorticity. As we discussed in Sec. 14.2.2, when viscosity is unimportant, vortex lines are frozen into the fluid and can be stretched by the action of neighboring vortex lines. As a bundle of vortex lines is stretched and twisted (Fig. 15.5), the incompressibility of the fluid causes the bundle's cross section to decrease and correspondingly causes the magnitude of its vorticity to increase and the lengthscale on which the vorticity changes to decrease (cf. Sec. 14.2). The continuous lengthening and twisting of fluid elements therefore creates vorticity on progressively smaller lengthscales.

evolution of vorticity in turbulence: toward progressively smaller scales

---

7. The creation of larger structures from smaller ones—sometimes called an *inverse cascade*—is a strong feature of 2-dimensional fluid turbulence.

Note that, when the flow is 2-dimensional (i.e., has translation symmetry), the vortex lines point in the translational direction, so they do not get stretched, and there is thus no inexorable driving of the turbulent energy to smaller and smaller lengthscales. This is one reason why true turbulence does not occur in 2 dimensions, only in 3.

However, something akin to turbulence but with much less richness and small-scale structure, *does* occur in 2 dimensions (e.g., in 2-dimensional simulations of the Kelvin-Helmholtz instability; Box 15.3). But in the real world, once the Kelvin-Helmholtz instability is fully developed, 3-dimensional instabilities grow strong, vortex-line stretching increases, and the flow develops full 3-dimensional turbulence. The same happens in other ostensibly 2-dimensional flows (e.g., the "2-dimensional" wake behind a long circular cylinder and "2-dimensional" jets and boundary layers; Secs. 15.4.3 and 15.5).

15.4

15.4   **15.4 Semiquantitative Analysis of Turbulence**

In this section, we develop a semiquantitative mathematical analysis of turbulence and explore a few applications. This analysis is fairly good for weak turbulence. However, for the much more common strong turbulence, it is at best semiquantitative—but nonetheless widely used for lack of anything simple that is much better.

weak turbulence versus strong turbulence

15.4.1

15.4.1   15.4.1 Weak-Turbulence Formalism

The meaning of weak turbulence can be explained in terms of interacting eddies (a concept we exploit in Sec. 15.4.4 when studying the flow of turbulent energy from large scales to small). One can regard turbulence as weak if the timescale τ_* for a large eddy to feed most of its energy to smaller eddies is long compared to the large eddy's turnover time τ (i.e., its rotation period). The weak-turbulence formalism (model) that we sketch here can be thought of as an expansion in τ/τ_*.

weak turbulence

Unfortunately, for most turbulence seen in fluids, the large eddies' energy-loss time is of order its turnover time, $\tau/\tau_* \sim 1$, which means the eddy loses its identity in roughly one turnover time and the turbulence is strong. In this case, the weak-turbulence formalism that we sketch here is only semiquantitatively accurate.

Our formalism for weak turbulence (with gravity negligible and the flow highly subsonic, so it can be regarded as incompressible) is based on the standard incompressibility equation and the time-dependent Navier-Stokes equation, which we write as

$$\nabla \cdot \mathbf{v} = 0, \tag{15.11a}$$

$$\rho \frac{\partial \mathbf{v}}{\partial t} + \nabla \cdot (\rho \mathbf{v} \otimes \mathbf{v}) = -\nabla P + \rho \nu \nabla^2 \mathbf{v}, \tag{15.11b}$$

assuming for simplicity that ρ is constant. [Equation (15.11b) is equivalent to (15.2b) with $\partial \mathbf{v}/\partial t$ added to account for time dependence and with the inertial force term rewritten using $\nabla \cdot (\rho \mathbf{v} \otimes \mathbf{v}) = \rho (\mathbf{v} \cdot \nabla)\mathbf{v}$, or equivalently in index notation, $(\rho v_i v_j)_{;i} = \rho_{,i} v_i v_j + \rho(v_{i;i} v_j + v_i v_{j;i}) = \rho v_i v_{j;i}$.] Equations (15.11) are four scalar

The Kelvin-Helmholtz instability arises when two fluids with nearly the same density are in uniform motion past each other (Sec. 14.6.1). Their interface (a vortex sheet) develops corrugations that grow (Figs. 14.19 and 14.20). That growth bends the corrugations more and more sharply. Along the corrugated interface the fluids on each side are still sliding past each other, so the instability arises again on this smaller scale—and again and again, somewhat like the cascade of turbulent energy from large scales to small.

In the real world, 3-dimensional instabilities also arise, and the flow becomes fully turbulent. However, much insight is gained into the difference between 2- and 3-dimensional flow by artificially constraining the Kelvin-Helmholtz flow to remain 2-dimensional. This is easily done in numerical simulations. Movies of such simulations abound on the web (e.g., on the Wikipedia web page for the Kelvin-Helmholtz instability; the following picture is a still from that movie).

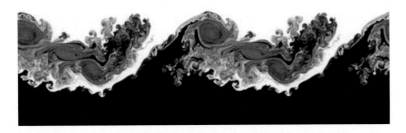

By Bdubb12 (Own work) [Public domain], via Wikimedia Commons.
https://commons.wikimedia.org/wiki/File%3AKHI.gif.

Although the structures in this simulation are complex, they are much less rich than those that appear in fully 3-dimensional turbulence—perhaps largely due to the absence of stretching and twisting of vortex lines when the flow is confined to 2 dimensions. (Not surprisingly, the structures in this simulation resemble some in Jupiter's atmosphere, which also arise from a Kelvin-Helmholtz instability.)

---

equations for four unknowns, $P(\mathbf{x}, t)$ and the three components of $\mathbf{v}(\mathbf{x}, t)$; ρ and ν can be regarded as constants.

To obtain the weak-turbulence versions of these equations, we split the velocity field $\mathbf{v}(\mathbf{x}, t)$ and pressure $P(\mathbf{x}, t)$ into steady parts $\bar{\mathbf{v}}$, $\bar{P}$, plus fluctuating parts, $\delta\mathbf{v}$, δP:

foundations for weak-turbulence theory

$$\mathbf{v} = \bar{\mathbf{v}} + \delta\mathbf{v}, \qquad P = \bar{P} + \delta P. \tag{15.12}$$

We can think of (or, in fact, define) $\bar{\mathbf{v}}$ and $\bar{P}$ as the time averages of $\mathbf{v}$ and P, and define $\delta\mathbf{v}$ and δP as the difference between the exact quantities and the time-averaged quantities.

The time-averaged variables $\bar{\mathbf{v}}$ and $\bar{P}$ are governed by the time-averaged incompressibility and Navier-Stokes equations (15.11). Because the incompressibility equation is linear, its time average,

$$\nabla \cdot \bar{\mathbf{v}} = 0,$$

(15.13a)

entails no coupling of the steady variables to the fluctuating ones. By contrast, the nonlinear inertial term $\nabla \cdot (\rho\mathbf{v} \otimes \mathbf{v})$ in the Navier-Stokes equation gives rise to such a coupling in the (time-independent) time-averaged equation:

$$\rho(\bar{\mathbf{v}} \cdot \nabla)\bar{\mathbf{v}} = -\nabla\bar{P} + \rho\nu\nabla^2\bar{\mathbf{v}} - \nabla \cdot \mathbf{T}_R.$$

(15.13b)

time-averaged Navier-Stokes equation: couples turbulence (via Reynolds stress tensor) to time-averaged flow

Here

$$\mathbf{T}_R \equiv \rho\overline{\delta\mathbf{v} \otimes \delta\mathbf{v}}$$

(15.13c)

is known as the *Reynolds stress tensor*. It serves as a driving term in the time-averaged Navier-Stokes equation (15.13b)—a term by which the fluctuating part of the flow acts back on and so influences the time-averaged flow.

This Reynolds stress $\mathbf{T}_R$ can be regarded as an additional part of the total stress tensor, analogous to the gas pressure computed in kinetic theory,[8] $P = \frac{1}{3}\rho\overline{v^2}$, where v is the molecular speed. $\mathbf{T}_R$ will be dominated by the largest eddies present, and it can be anisotropic, especially when the largest-scale turbulent velocity fluctuations are distorted by interaction with an averaged shear flow [i.e., when $\bar{\sigma}_{ij} = \frac{1}{2}(\bar{v}_{i;j} + \bar{v}_{j;i})$ is large].

Reynolds stress for stationary, homogeneous turbulence

If the turbulence is both stationary and homogeneous (a case we specialize to below when studying the Kolmogorov spectrum), then the Reynolds stress tensor can be written in the form $\mathbf{T}_R = P_R\mathbf{g}$, where P_R is the Reynolds pressure, which is independent of position, and $\mathbf{g}$ is the metric, so $g_{ij} = \delta_{ij}$. In this case, the turbulence exerts no force density on the mean flow, so $\nabla \cdot \mathbf{T}_R = \nabla P_R$ will vanish in the time-averaged Navier-Stokes equation (15.13b). By contrast, near the edge of a turbulent region (e.g., near the edge of a turbulent wake, jet, or boundary layer), the turbulence is inhomogeneous, and thereby (as we shall see in Sec. 15.4.2) exerts an important influence on the time-independent, averaged flow.

correlation functions in turbulence

Notice that the Reynolds stress tensor is the tensorial *correlation function* (also called "autocorrelation function") of the velocity fluctuation field at zero time delay (multiplied by density ρ; cf. Secs. 6.4.1 and 6.5.1). Notice also that it involves the

---

8. Deducible from Eq. (3.37c) or from Eqs. (3.39b) and (3.39c) with mean energy per particle $\bar{E} = \frac{1}{2}m\overline{v^2}$.

temporal cross-correlation function of components of the velocity fluctuation [e.g., $\overline{\delta v_x(\mathbf{x}, t)\delta v_y(\mathbf{x}, t)}$; Sec. 6.5.1]. It is possible to extend this weak-turbulence formalism so it probes the statistical properties of turbulence more deeply, with the aid of correlation functions with finite time delays and correlation functions of velocity components (or other relevant physical quantities) at two different points in space simultaneously; see, for example, Sec. 28.5.3. (It is relatively straightforward experimentally to measure these correlation functions.) As we discuss in greater detail in Sec. 15.4.4 (and also saw for 1- and 2-dimensional random processes in Secs. 6.4.4 and 6.5.2, and for multidimensional, complex random processes in Ex. 9.8), the Fourier transforms of these correlation functions give the spatial and temporal spectral densities of the fluctuating quantities.

Just as the structure of the time-averaged flow is governed by the time-averaged incompressibility and Navier-Stokes equations (15.13) (with the fluctuating variables acting on the time-averaged flow through the Reynolds stress), so also the fluctuating part of the flow is governed by the fluctuating (difference between instantaneous and time-averaged) incompressibility and Navier-Stokes equations. For details, see Ex. 15.5. This exercise is important; it exhibits the weak-turbulence formalism in action and underpins the application to spatial energy flow in a 2-dimensional, turbulent wake in Fig. 15.7 below.

EXERCISES

Exercise 15.5 **Example: Reynolds Stress; Fluctuating Part**
of Navier-Stokes Equation in Weak Turbulence

(a) Derive the time-averaged Navier-Stokes equation (15.13b) from the time-dependent form [Eq. (15.11b)], and thereby infer the definition (15.13c) for the Reynolds stress. Equation (15.13b) shows how the Reynolds stress affects the evolution of the mean velocity. However, it does not tell us how the Reynolds stress evolves.

(b) Explain why an equation for the evolution of the Reynolds stress must involve averages of triple products of the velocity fluctuation. Similarly, the time evolution of the averaged triple products will involve averaged quartic products, and so on (cf. the BBGKY hierarchy of equations in plasma physics, Sec. 22.6.1). How do you think you might "close" this sequence of equations (i.e., terminate it at some low order) and get a fully determined system of equations? [Answer: The simplest way is to use the concept of turbulent viscosity discussed in Sec. 15.4.2.]

(c) Show that the fluctuating part of the Navier-Stokes equation (the difference between the exact Navier-Stokes equation and its time average) takes the following form:

$$\frac{\partial \delta \mathbf{v}}{\partial t} + (\bar{\mathbf{v}} \cdot \boldsymbol{\nabla})\delta\mathbf{v} + (\delta\mathbf{v} \cdot \boldsymbol{\nabla})\bar{\mathbf{v}} + [(\delta\mathbf{v} \cdot \boldsymbol{\nabla})\delta\mathbf{v} - \overline{(\delta\mathbf{v} \cdot \boldsymbol{\nabla})\delta\mathbf{v}}] = \\ -\frac{1}{\rho}\boldsymbol{\nabla}\delta P + \nu\nabla^2(\delta\mathbf{v})$$ (15.14a)

This equation and the fluctuating part of the incompressibility equation

$$\boxed{\nabla \cdot \delta \mathbf{v} = 0}$$

(15.14b)

govern the evolution of the fluctuating variables $\delta \mathbf{v}$ and δP. [The challenge, of course, is to devise ways to solve these equations despite the nonlinearities and the coupling to the mean flow $\bar{\mathbf{v}}$ that show up strongly in Eq. (15.14a).]

(d) By dotting $\delta \mathbf{v}$ into Eq. (15.14a) and then taking its time average, derive the following law for the spatial evolution of the turbulent energy density $\frac{1}{2}\rho\overline{\delta v^2}$:

spatial evolution equation for turbulent energy density

$$\boxed{\bar{\mathbf{v}} \cdot \nabla(\frac{1}{2}\rho\overline{\delta v^2}) + \nabla \cdot \overline{\left(\frac{1}{2}\rho\delta v^2 \delta\mathbf{v} + \delta P \delta\mathbf{v}\right)} = -T_R^{ij}\bar{v}_{i;j} + \nu\rho\overline{\delta\mathbf{v} \cdot (\nabla^2 \delta\mathbf{v})}.}$$

(15.15)

Here $T_R^{ij} = \rho\overline{\delta v_i \delta v_j}$ is the Reynolds stress [Eq. (15.13c)]. Interpret each term in this equation. [The four interpretations will be discussed below, for a 2-dimensional turbulent wake, in Sec. 15.4.3.]

(e) Now derive a similar law for the spatial evolution of the energy density of ordered motion, $\frac{1}{2}\rho\bar{\mathbf{v}}^2$. Show that the energy lost by the ordered motion is compensated for by the energy gained by the turbulent motion.

15.4.2

15.4.2 Turbulent Viscosity

Additional tools that are often introduced in the theory of weak turbulence come from taking the analogy with the kinetic theory of gases one stage further and defining *turbulent transport coefficients,* most importantly a turbulent viscosity that governs the turbulent transport of momentum. These turbulent transport coefficients are derived by simple analogy with the kinetic-theory transport coefficients (Sec. 3.7).

Momentum, heat, and so forth are transported most efficiently by the largest turbulent eddies in the flow; therefore, when estimating the transport coefficients, we replace the particle mean free path by the size ℓ of the largest eddies and the mean particle speed by the magnitude v_ℓ of the fluctuations of velocity in the largest eddies. The result, for momentum transport, is a model turbulent viscosity:

turbulent viscosity determined by largest eddies

$$\boxed{\nu_t \simeq \frac{1}{3}v_\ell\ell}$$

(15.16)

[cf. Eq. (13.72) for molecular viscosity, with $\nu = \eta/\rho$]. The Reynolds stress is then approximated as a turbulent shear stress of the standard form:

$$\boxed{\mathbf{T}_R \simeq -2\rho\nu_t\bar{\boldsymbol{\sigma}}.}$$

(15.17)

Here $\bar{\boldsymbol{\sigma}}$ is the rate of shear tensor (13.67b) evaluated using the mean velocity field $\bar{\mathbf{v}}$. Note that the turbulent kinematic viscosity defined in this manner, ν_t, is a property of the turbulent flow and not an intrinsic property of the fluid; it differs from molecular viscosity in this important respect.

We have previously encountered turbulent viscosity in our study of the physical origin of the Sargasso Sea gyre in the north Atlantic Ocean and the gyre's role in generating the Gulf Stream (Ex. 14.21). The gyre is produced by water flowing in a wind-driven Ekman boundary layer at the ocean's surface. From the measured thickness of that boundary layer, $\delta_E \sim 30$ m, we deduced that the boundary layer's viscosity is $\nu \sim 0.03$ m^2 s^{-1}, which is 30,000 times larger than water's molecular viscosity; it is the turbulent viscosity of Eq. (15.16).

By considerations similar to those above for turbulent viscosity, one can define and estimate a turbulent thermal conductivity for the spatial transport of time-averaged heat (cf. Sec. 3.7.2) and a turbulent diffusion coefficient for the spatial transport of one component of a time-averaged fluid through another, for example, an odor crossing a room (cf. Ex. 3.20).

turbulent thermal conductivity and turbulent diffusion coefficient

The turbulent viscosity ν_t and the other turbulent transport coefficients can be far larger than their kinetic-theory counterparts. Besides the Sargasso Sea gyre, another example is air in a room subjected to typical uneven heating and cooling. The air may circulate with an average largest eddy velocity of $v_\ell \sim 1$ cm s^{-1} and an associated eddy size of $\ell \sim 3$ m. (The values can be estimated by observing the motion of illuminated dust particles.) The turbulent viscosity ν_t and the turbulent diffusion coefficient D_t (Ex. 3.20) associated with these motions are $\nu_t \sim D_t \sim 10^{-2}$ m^2 s^{-1}, some three orders of magnitude larger than the molecular values.

15.4.3 Turbulent Wakes and Jets; Entrainment; the Coanda Effect

15.4.3

As instructive applications of turbulent viscosity and related issues, we now explore in an order-of-magnitude way the structures of turbulent wakes and jets. The more complicated extension to a turbulent boundary layer will be explored in Sec. 15.5. In the text of this section we focus on the 2-dimensional wake behind a cylinder. In Exs. 15.6, 15.7, and 15.8, we study the 3-dimensional wake behind a sphere and 2- and 3-dimensional jets.

TWO-DIMENSIONAL TURBULENT WAKE: ORDER-OF-MAGNITUDE
COMPUTATION OF WIDTH AND VELOCITY DEFICIT; ENTRAINMENT

For the turbulent wake behind a cylinder at high Reynolds number (see Fig. 15.3), we begin by deducing the turbulent viscosity $\nu_t \sim \frac{1}{3} v_\ell \ell$. It is reasonable to expect (and observations confirm) that the largest eddies in a turbulent wake, at distance x past the cylinder, extend transversely across nearly the entire width $w(x)$ of the wake, so their size is $\ell \sim w(x)$. What is the largest eddies' circulation speed v_ℓ? Because these eddies' energies are fed into smaller eddies in (roughly) one eddy turnover time, these eddies must be continually regenerated by interaction between the wake and the uniform flow at its transverse boundaries. This means that the wake's circulation speed v_ℓ cannot be sensitive to the velocity difference V between the incoming flow and the cylinder far upstream; the wake has long since lost memory of that difference. The only characteristic speed that influences the wake is the difference between its own mean downstream speed $\bar{v}_x$ and the speed V of the uniform flow at its boundaries. It seems physically reasonable that the interaction between these two flows will drive the eddy to circulate with that difference speed, the deficit u_o depicted in Fig. 15.3, which observations show to be true. Thus we have $v_\ell \sim u_o$. This means that the turbulent viscosity is $v_t \sim \frac{1}{3} v_\ell \ell \sim \frac{1}{3} u_o(x) w(x)$.

Knowing the viscosity, we can compute our two unknowns, the wake's velocity deficit $u_o(x)$ and its width $w(x)$, from vorticity diffusion and momentum conservation, as we did for a laminar wake in Ex. 15.1a,b.

First, consider momentum conservation. Here, as in Ex. 15.1b, the drag force per unit length on the cylinder, $F_D = C_D \frac{1}{2} \rho V^2 d$, must be equal to the difference between the momentum per unit length ($\int \rho V^2 dy$) in the water impinging on the cylinder and that ($\int \rho v_x^2 dy$) at any chosen distance x behind the cylinder. That difference is easily seen, from Fig. 15.3, to be $\sim \rho V u_o w$. Equating this to F_D, we obtain the product $u_o w$ and thence $v_t \sim \frac{1}{3} u_o w \sim \frac{1}{6} C_D V d$.

Thus, remarkably, the turbulent viscosity is independent of distance x downstream from the cylinder. This means that the wake's vorticity (which is contained primarily in its large eddies) will diffuse transversely in just the manner we have encountered several times before [Sec. 14.2.5, Eq. (14.42), and Ex. 15.1a], causing its width to grow as $w \sim \sqrt{v_t x / V}$. Inserting $v_t \sim \frac{1}{6} C_D V d \sim \frac{1}{3} u_o w$, we obtain

$$ w \sim \left(\frac{C_D d}{6} x \right)^{1/2}, \qquad u_o \sim V \left(\frac{3 C_D d}{2x} \right)^{1/2} \tag{15.18} $$

for the width and velocity deficit in the turbulent wake.

In this analysis, our appeal to vorticity diffusion obscures the physical mechanism by which the turbulent wake widens downstream. That mechanism is entrainment— the capture of fluid from outside the wake into the wake (a phenomenon we met for a laminar jet in Ex. 15.3d). The inertia of the largest eddies enables them to sweep outward into the surrounding flow with a transverse speed that is a sizable fraction of their turnover speed, say, $\frac{1}{6} v_\ell$ (the 1/6 ensures agreement with the diffusion argument). This means that, moving with the flow at speed V, the eddy widens by entrainment at a rate $dw/dt = V dw/dx \sim \frac{1}{6} v_\ell \sim \frac{1}{6} u_o$. Inserting $u_o \sim \frac{1}{2} V C_D d / w$ from momentum conservation and solving the resulting differential equation, we obtain the same wake width $w \sim \sqrt{C_D d\, x/6}$ as we got from our diffusion argument.

DISTRIBUTION OF VORTICITY IN THE WAKE; IRREGULARITY OF THE WAKE'S EDGE

The wake's fluid can be cleanly distinguished from the exterior, ambient fluid by vorticity: the ambient flow velocity $\mathbf{v}(\mathbf{x}, t)$ is vorticity free; the wake has nonzero vorticity. If we look at the actual flow velocity and refrain from averaging, the only way a fluid element in the wake can acquire vorticity is by molecular diffusion. Molecular diffusion is so slow that the boundary between a region with vorticity and one without

FIGURE 15.6 Contours of constant magnitude of vorticity $|\boldsymbol{\omega}|$ in the 2-dimensional turbulent wake between and behind a pair of cylinders. The outer edge of the blue region is the edge of the wake. From a numerical simulation by the research group of Sanjiva K. Lele at Stanford University, http://flowgallery.stanford.edu/research.html.

(the boundary between the wake fluid and the ambient fluid) is very sharp. Sharp, yes, but straight, no! The wake's eddies, little as well as big, drive the boundary into a very convoluted shape (Fig. 15.6). Correspondingly, the order-of-magnitude wake width $w \sim \sqrt{C_D d\, x/6}$ that we have derived is only the width of a thoroughly averaged flow and is not at all the instantaneous, local width of the wake.

Intermittency is one consequence of the wake's highly irregular edge. A fixed location behind the cylinder that is not too close to the wake's center will sometimes be outside the wake and sometimes inside it. This will show up in one's measurement of any flow variable [e.g., $v_y(t)$, $\omega_x(t)$, or pressure P] at the fixed location. When outside the wake, the measured quantities will be fairly constant; when inside, they will change rapidly and stochastically. The quiet epochs combined with interspersed stochastic epochs are a form of intermittency.

intermittency

AVERAGED ENERGY FLOW IN THE WAKE

The weak-turbulence formalism of Sec. 15.4.1 and Ex. 15.5 can be used to explore the generation and flow of turbulent energy in the 2-dimensional wake behind a cylinder. This formalism, even when extended, is not good enough to make definitive predictions, but it *can* be used to deduce the energy flow from measurements of the mean (time-averaged) flow velocity $\bar{\mathbf{v}}$, the turbulent velocity $\delta\mathbf{v}$, and the turbulent pressure δP. A classic example of this was carried out long ago by Townsend (1949) and is summarized in Fig. 15.7.

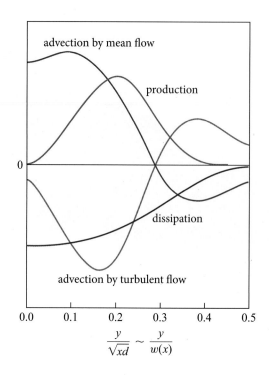

FIGURE 15.7 The four terms in the rate of change of the time-averaged turbulent energy density [Eq. (15.15)] for the 2-dimensional turbulent wake behind a cylinder. The horizontal axis measures the distance y across the wake in units of the wake's mean width $w(x)$. The vertical axis shows the numerical value of each term. For a discussion of the four terms, see the four bulleted points in the text. Energy conservation and stationarity of the averaged flow guarantee that the sum of the four terms vanishes at each y [Eq. (15.15)]. Adapted from Tritton (1987, Fig. 21.8), which is based on measurements and analysis by Townsend (1949) at $\mathrm{Re}_d = 1{,}360$ and $x/d > 500$.

averaged energy flow in a wake: production, advection, and dissipation

The time-averaged turbulent energy density changes due to four processes that are graphed in that figure as a function of distance y across the wake:

- *Production.* Energy in the organized bulk flow (mean flow) $\bar{\mathbf{v}}$ is converted into turbulent energy by the interaction of the mean flow's shear with the turbulence's Reynolds stress, at a rate per unit volume of $T_R^{ij}\bar{v}_{i;j}$. This production vanishes at the wake's center ($y = 0$), because the shear of the mean flow vanishes there; it also vanishes at the edge of the wake, because both the mean-flow shear and the Reynolds stress go to zero there.

- *Advection by mean flow.* Once produced, the turbulent energy is advected across the wake by the mean flow. This causes an increase in turbulent energy density in the center of the wake and a decrease in the wake's outer parts at the rate $\boldsymbol{\nabla} \cdot (\frac{1}{2}\rho\overline{\delta v^2}\bar{\mathbf{v}}) = (\bar{\mathbf{v}} \cdot \boldsymbol{\nabla})(\frac{1}{2}\rho\overline{\delta v^2})$.

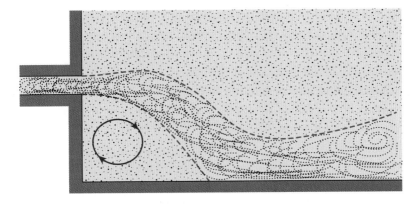

FIGURE 15.8 The Coanda effect. A turbulent jet emerging from an orifice in the left wall is attracted to the solid bottom wall.

- *Advection by turbulent flow.* The turbulent energy also is advected by the turbulent part of the flow, causing a decrease of turbulent energy density in the central regions of the wake and an increase in the outer regions at the rate $\nabla \cdot (\frac{1}{2}\rho \overline{\delta v^2 \delta \mathbf{v}} + \overline{\delta P \delta \mathbf{v}})$.

- *Dissipation.* The turbulent energy is converted to heat by molecular viscosity at a rate per unit volume given by $-\nu\rho \overline{\delta \mathbf{v} \cdot (\nabla^2 \delta \mathbf{v})}$. This dissipation is largest at the wake's center and falls off gradually toward the wake's averaged edge.

Energy conservation plus stationarity of the averaged flow guarantees that the sum of these four terms vanishes at all locations in the wake (all y of Fig. 15.7). This is the physical content of Eq. (15.15) and is confirmed in Fig. 15.7 by the experimental data.

ENTRAINMENT AND COANDA EFFECT

Notice how much wider the (averaged) turbulent wake is than the corresponding laminar wake of Ex. 15.1. The ratio of their widths [Eqs. (15.18) and (15.8)] is $w_t/w_l \sim \sqrt{C_D V d/\nu}$. For $C_D \sim 1$ in the turbulent wake, $V \sim 1$ m s^{-1}, $d \sim 1$ m, and water's kinematic viscosity $\nu \sim 10^{-6}$ m^2 s^{-1}, the ratio is $w_t/w_l \sim 10^3$, independent of distance x downstream. In this sense, entrainment in the turbulent wake is a thousand times stronger than entrainment in the laminar wake. Turbulent wakes and jets have voracious appetites for ambient fluid!

strong entrainment in turbulent jets and wakes

Entrainment is central to the *Coanda effect,* depicted in Fig. 15.8. Consider a turbulent flow (e.g., the jet of Fig. 15.8) that is widening by entrainment of surrounding fluid. The jet normally widens downstream by pulling surrounding fluid into itself, and the inflow toward the jet extends far beyond the jet's boundaries (see, e.g., Ex. 15.3d). However, when a solid wall is nearby, so there is no source for inflowing ambient fluid, the jet's entrainment causes a drop in the pressure of the ambient fluid near the wall. The resulting pressure gradient pushes the jet toward the wall as depicted in Fig. 15.8.

Similarly, if a turbulent flow is already close to a wall and the wall begins to curve away from the flow, the flow develops a pressure gradient that tends to keep the turbulent region attached to the wall. In other words, turbulent flows are attracted to solid surfaces and tend to stick to them. This is the Coanda effect.

The Coanda effect also occurs for laminar flows, but because entrainment is typically orders of magnitude weaker in laminar flows than in turbulent ones, the effect is also orders of magnitude weaker.

The Coanda effect is important in aeronautics; for example, it is exploited to prevent the separation of the boundary layer from the upper surface of a wing, thereby improving the wing's lift and reducing its drag, as we discuss in Sec. 15.5.2.

EXERCISES

Exercise 15.6 *Problem: Turbulent Wake behind a Sphere*
Compute the width $w(x)$ and velocity deficit $u_o(x)$ for the 3-dimensional turbulent wake behind a sphere.

Exercise 15.7 *Problem: Turbulent Jets in 2 and 3 Dimensions*
Consider a 2-dimensional turbulent jet emerging into an ambient fluid at rest, and contrast it to the laminar jet analyzed in Ex. 15.3.

(a) Find how the mean jet velocity and the jet width scale with distance downstream from the nozzle.

(b) Repeat the exercise for a 3-dimensional jet.

Exercise 15.8 *Problem: Entrainment and Coanda Effect in a 3-Dimensional Jet*
(a) Evaluate the scaling of the rate of mass flow (discharge) $\dot{M}(x)$ along the 3-dimensional turbulent jet of the previous exercise. Show that $\dot{M}$ increases with distance from the nozzle, so that mass must be entrained in the flow and become turbulent.

(b) Compare the entrainment rate for a turbulent jet with that for a laminar jet (Ex. 15.3). Do you expect the Coanda effect to be stronger for a turbulent or a laminar jet?

15.4.4 Kolmogorov Spectrum for Fully Developed, Homogeneous, Isotropic Turbulence

When a fluid exhibits turbulence over a large volume that is well removed from any solid bodies, there will be no preferred directions and no substantial gradients in the statistically averaged properties of the turbulent velocity field. This suggests that the statistical properties of the turbulence will be stationary and isotropic. We derive a semiquantitative description of some of these statistical properties, proceeding in two steps: first we analyze the turbulence's velocity field and then the turbulent distribution of quantities that are transported by the fluid.

TURBULENT VELOCITY FIELD

Our analysis will be based on the following simple physical model. We idealize the turbulent velocity field as made of a set of large eddies, each of which contains a set of smaller eddies, and so on. We suppose that each eddy splits into eddies roughly half its size after a few turnover times. This process can be described mathematically as nonlinear or triple velocity correlation terms [terms like the second one in Eq. (15.15)] producing, in the law of energy conservation, an energy transfer (a "cascade" of energy) from larger eddies to smaller ones. Now, for large enough eddies, we can ignore the effects of molecular viscosity in the flow. However, for sufficiently small eddies, viscous dissipation will convert the eddy bulk kinetic energy into heat. This simple model enables us to derive a remarkably successful formula (the *Kolmogorov spectrum*) for the distribution of turbulent energy over eddy size.

physical model underlying Kolmogorov spectrum for turbulence: cascading eddies

We must first introduce and define the turbulent energy per unit wave number and per unit mass, $u_k(k)$. For this purpose, we focus on a volume $\mathcal{V}$ much larger than the largest eddies. At some moment of time t, we compute the spatial Fourier transform of the fluctuating part of the velocity field $\delta\mathbf{v}(\mathbf{x})$, confined to this volume [with $\delta\mathbf{v}(\mathbf{x})$ set to zero outside $\mathcal{V}$], and also write down the inverse Fourier transform:

$$\delta\tilde{\mathbf{v}}(\mathbf{k}) = \int_{\mathcal{V}} d^3x\, \delta\mathbf{v}(\mathbf{x})e^{-i\mathbf{k}\cdot\mathbf{x}}, \qquad \delta\mathbf{v} = \int \frac{d^3k}{(2\pi)^3}\delta\tilde{v}e^{i\mathbf{k}\cdot\mathbf{x}} \quad \text{inside } \mathcal{V}. \tag{15.19}$$

The total energy per unit mass u in the turbulence, averaged over the box $\mathcal{V}$, is then

$$u = \int \frac{d^3x}{\mathcal{V}}\frac{1}{2}\overline{|\delta\mathbf{v}|^2} = \int \frac{d^3k}{(2\pi)^3}\frac{\overline{|\delta\tilde{\mathbf{v}}|^2}}{2\mathcal{V}} \equiv \int_0^\infty dk\, u_k(k), \tag{15.20}$$

where we have used Parseval's theorem in the second equality, we have used $d^3k = 4\pi k^2 dk$, and we have defined

$$u_k(k) \equiv \frac{\overline{|\delta\tilde{\mathbf{v}}|^2}k^2}{4\pi^2 \mathcal{V}}. \tag{15.21}$$

spectral energy per unit mass in turbulence

Here the bars denote a time average, k is the magnitude of the wave vector: $k \equiv |\mathbf{k}|$ (i.e., it is the wave number or equivalently 2π divided by the wavelength), and $u_k(k)$ is called the *spectral energy per unit mass* of the turbulent velocity field $\delta\mathbf{v}$. In the third equality in Eq. (15.20), we have assumed that the turbulence is isotropic, so the integrand depends only on wave number k and not on the direction of $\mathbf{k}$. Correspondingly, we have defined $u_k(k)$ as the energy per unit wave number rather than an energy per unit volume of $\mathbf{k}$-space. This means that $u_k(k)dk$ is the average kinetic energy per unit mass associated with modes that have k lying in the interval dk; we treat k as positive.

In Chap. 6, we introduced the concepts of a random process and its spectral density. The Cartesian components of the fluctuating velocity δv_x, δv_y, and δv_z obviously are random processes that depend on vectorial location in space $\mathbf{x}$ rather than on time as in Chap. 6. It is straightforward to show that their *double-sided* spectral densities are related to $u_k(k)$ by

spectral densities of velocity field

$$\boxed{S_{v_x}(\mathbf{k}) = S_{v_y}(\mathbf{k}) = S_{v_z}(\mathbf{k}) = \frac{(2\pi)^2}{3k^2} \times u_k(k).}$$

(15.22)

If we fold negative k_x into positive, and similarly for k_y and k_z, so as to get the kind of single-sided spectral density (Sec. 6.4.2) that we used in Chap. 6, then these spectral densities should be multiplied by $2^3 = 8$.

We now use our physical model of turbulence to derive an expression for $u_k(k)$. Denote by $k_{min} = 2\pi/\ell$ the wave number of the largest eddies, and by k_{max} that of the smallest ones (those in which viscosity dissipates the cascading, turbulent energy). Our derivation will be valid (and the result valid) only when $k_{max}/k_{min} \gg 1$: only when there is a long sequence of eddies from the largest to half the largest to a quarter the largest and so forth down to the smallest.

eddy turnover speed and time

As a tool in computing $u_k(k)$, we introduce the root-mean-square turbulent turnover speed of the eddies with wave number k: $v(k) \equiv v$. Ignoring factors of order unity, we treat the size of these eddies as k^{-1}. Then their turnover time is $\tau(k) \sim k^{-1}/v(k) = 1/[kv(k)]$. Our model presumes that in this same time τ (to within a factor of order unity), each eddy of size k^{-1} splits into eddies of half this size (i.e., the turbulent energy cascades from k to $2k$). In other words, our model presumes the turbulence is strong. Since the energy cascade is presumed stationary (i.e., no energy is accumulating at any wave number), the energy per unit mass that cascades in a unit time from k to $2k$ must be independent of k. Denote by q that k-independent, cascading energy per unit mass per unit time. Since the energy per unit mass in the eddies of size k^{-1} is v^2 (aside from a factor 2, which we neglect), and the cascade time is $\tau \sim 1/(kv)$, then $q \sim v^2/\tau \sim v^3 k$. This tells us that the rms turbulent velocity is

$$v(k) \sim (q/k)^{1/3}.$$

(15.23)

Our model lumps together all eddies with wave numbers in a range $\Delta k \sim k$ around k and treats them all as having wave number k. The total energy per unit mass in these eddies is $u_k(k)\Delta k \sim ku_k(k)$ when expressed in terms of the sophisticated quantity $u_k(k)$; it is $\sim v(k)^2$ when expressed in terms of our simple model. Thus our model predicts that $u_k(k) \sim v(k)^2/k$, which by Eqs. (15.23) and (15.22) implies

Kolmogorov spectrum for stationary, isotropic, incompressible turbulence

$$\boxed{u_k(k) \sim q^{2/3}k^{-5/3}, \quad S_{v_j}(\mathbf{k}) \sim q^{2/3}k^{-11/3} \quad \text{for } k_{min} \ll k \ll k_{max};}$$

(15.24)

see Fig. 15.9. This is the *Kolmogorov spectrum* for the spectral energy density of stationary, isotropic, incompressible turbulence. It is valid only in the range $k_{min} \ll k \ll k_{max}$, because only in this range are the turbulent eddies continuously receiving energy from larger lengthscales and passing it on to smaller scales. At the ends of the range, the spectrum has to be modified in the manner illustrated qualitatively in Fig. 15.9.

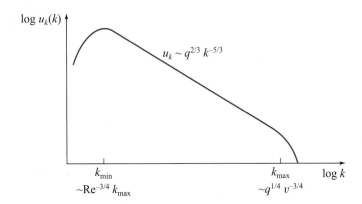

FIGURE 15.9 The Kolmogorov spectral energy per unit mass for stationary, homogeneous turbulence.

At the smallest lengthscales present, k_{max}^{-1}, the molecular viscous forces become competitive with inertial forces in the Navier-Stokes equation; these viscous forces convert the cascading energy into heat. Since the ratio of inertial forces to viscous forces is the Reynolds number, the smallest eddies have a Reynolds number of order unity: $Re_{k_{max}} = v(k_{max})k_{max}^{-1}/\nu \sim 1$. Inserting Eq. (15.23) for $v(k)$, we obtain

$$k_{max} \sim q^{1/4}\nu^{-3/4}. \tag{15.25}$$

The largest eddies have sizes $\ell \sim k_{min}^{-1}$ and turnover speeds $v_\ell = v(k_{min}) \sim (q/k_{min})^{1/3}$. By combining these relations with Eq. (15.25), we see that the ratio of the largest wave numbers present in the turbulence to the smallest is

$$\boxed{\frac{k_{max}}{k_{min}} \sim \left(\frac{v_\ell \ell}{\nu}\right)^{3/4} = Re_\ell^{3/4}.} \tag{15.26}$$

range of wave numbers in turbulence

Here Re_ℓ is the Reynolds number for the flow's largest eddies.

Let us now take stock of our results. If we know the scale ℓ of the largest eddies and their rms turnover speeds v_ℓ (and, of course, the viscosity of the fluid), then we can compute their Reynolds number Re_ℓ; from that, Eq. (15.26), and $k_{min} \sim \ell^{-1}$, we can compute the flow's maximum and minimum wave numbers; and from $q \sim v_\ell^3/\ell$ and Eq. (15.24), we can compute the spectral energy per unit mass in the turbulence.

We can also compute the total time required for energy to cascade from the largest eddies to the smallest. Since $\tau(k) \sim 1/(kv) \sim 1/(q^{1/3}k^{2/3})$, each successive set of eddies feeds its energy downward in a time $2^{-2/3}$ shorter than the preceding set did. As a result, it takes roughly the same amount of time for energy to pass from the second-largest eddies (size $\ell/2$) to the very smallest (size k_{max}^{-1}) as it takes for the second-largest to extract the energy from the very largest. The total cascade occurs in a time of several ℓ/v_ℓ (during which time, of course, the mean flow has fed new energy into the largest eddies and they are sending it downward).

These results are accurate only to within factors of order unity—with one major exception. The $-5/3$ power law in the Kolmogorov spectrum is very accurate. That this ought to be so one can verify in two equivalent ways: (i) Repeat the above derivation inserting arbitrary factors of order unity at every step. These factors will influence the final multiplicative factor in the Kolmogorov spectrum but will not influence the $-5/3$ power. (ii) Use dimensional analysis. Specifically, notice that the only dimensional entities that can influence the spectrum in the region $k_{min} \ll k \ll k_{max}$ are the energy cascade rate q and the wave number k. Then notice that the only quantity with the dimensions of $u_k(k)$ (energy per unit mass per unit wave number) that can be constructed from q and k is $q^{2/3}k^{-5/3}$. Thus, aside from a multiplicative factor of order unity, this must be the form of $u_k(k)$.

phenomena missed by
Kolmogorov spectrum

intermittency

Let us now review and critique the assumptions that went into our derivation of the Kolmogorov spectrum. First, we assumed that the turbulence is stationary and homogeneous. Real turbulence is neither of these, since it exhibits intermittency (Sec. 15.3), and smaller eddies tend to occupy less volume overall than do larger eddies and so cannot be uniformly distributed in space. Second, we assumed that the energy source is large-lengthscale motion and that the energy transport is local in k-space from the large lengthscales to steadily smaller ones. In the language of a Fourier decomposition into normal modes, we assumed that nonlinear coupling between modes with wave number k causes modes with wave number of order $2k$ to grow but does not significantly enhance modes with wave number $100k$ or $0.01k$. Again this behavior is not completely in accord with observations, which reveal the development of *coherent structures*—large-scale regions with correlated vorticity

coherent structures, flow
of energy from small scales
to large, and entrainment

in the flow. These structures are evidence for a reversed flow of energy in k-space from small scales to large ones, and they play a major role in another feature of real turbulence, entrainment—the spreading of an organized motion (e.g., a jet) into the surrounding fluid (Sec. 15.4.3).

Despite these issues with its derivation, the Kolmogorov law is surprisingly accurate and useful. It has been verified in many laboratory flows, and it describes many naturally occurring instances of turbulence. For example, the twinkling of starlight is caused by refractive index fluctuations in Earth's atmosphere, whose power spectrum we can determine optically. The underlying turbulence spectrum turns out to be of Kolmogorov form; see Box 9.2.

KOLMOGOROV SPECTRUM FOR QUANTITIES TRANSPORTED BY THE FLUID

In applications of the Kolmogorov spectrum, including twinkling starlight, one often must deal with such quantities as the index of refraction n that, in the turbulent cascade, are transported passively with the fluid, so that $dn/dt = 0$.[9] As the cascade

---

9. In Earth's atmospheric turbulence, n is primarily a function of the concentration of water molecules and the air density, which is controlled by temperature via thermal expansion and contraction. Both water concentration and temperature are carried along by the fluid in the turbulent cascade, whence so is n.

strings out and distorts fluid elements, it will also stretch and distort any inhomogeneities of n, driving them toward smaller lengthscales (larger k).

What is the k-dependence of the resulting spectral density $S_n(k)$? The quickest route to an answer is dimensional analysis. This S_n can be influenced only by the quantities that characterize the velocity field (its wave number k and its turbulent-energy cascade rate q) plus a third quantity: the rate $q_n = d\sigma_n^2/dt$ at which the variance $\sigma_n^2 = \int S_n \, d^3k/(2\pi)^3$ of n would die out if the forces driving the turbulence were turned off. (This q_n is the analog, for n, of q for velocity.) The only combination of k, q, and q_n that has the same dimensions as S_n is (Ex. 15.9)

$$\boxed{S_n \sim q_n q^{-1/3} k^{-11/3}.}$$

(15.27)

Kolmogorov spectrum for quantities transported with the fluid in its turbulent cascade

This is the same $-11/3$ power law as for S_{v_j} [Eq. (15.24)], and it will hold true also for any other quantity Θ that is transported with the fluid ($d\Theta/dt = 0$) in the turbulent cascade.

As for the velocity's inhomogeneities, so also for the inhomogeneities of n, there is some small lengthscale, $1/k_{n\,\text{max}}$, at which diffusion wipes them out, terminating the cascade. Because n is a function of temperature and water concentration, the relevant diffusion is a combination of thermal diffusion and water-molecule diffusion, which will proceed at a (perhaps modestly) different rate from the viscous diffusion for velocity inhomogeneities. Therefore, the maximum k in the cascade may be different for the index of refraction (and for any other quantity transported by the fluid) than for the velocity.

In applications, one often needs a spatial description of the turbulent spectrum rather than a wave-number description. This is normally given by the Fourier transform of the spectral density, which is the correlation function $C_n(\boldsymbol{\xi}) = \langle \delta n(x) \delta n(x + \boldsymbol{\xi}) \rangle$ (where δn is the perturbation away from the mean and $\langle \cdot \rangle$ is the ensemble average or average over space); see Sec. 6.4. For turbulence, the correlation function has the unfortunate property that $C_n(0) = \langle \delta n^2 \rangle \equiv \sigma_n^2$ is the variance, which is dominated by the largest lengthscales, where the Kolmogorov power-law spectrum breaks down. To avoid this problem it is convenient, when working with turbulence, to use in place of the correlation function the so-called *structure function*:

$$D_n(\boldsymbol{\xi}) = \langle [\delta n(\mathbf{x} + \boldsymbol{\xi}) - \delta n(\mathbf{x})]^2 \rangle = 2[\sigma_n^2 - C_n(\boldsymbol{\xi})].$$

(15.28)

structure function for spatial structure of turbulence

By Fourier transforming the spectral density (15.27), or more quickly by dimensional analysis, one can infer the Kolmogorov spectrum for this structure function:

$$D_n(\boldsymbol{\xi}) \sim (q_n/q^{1/3})\xi^{2/3} \quad \text{for} \quad k_{\text{max}}^{-1} \lesssim \xi \lesssim k_{\text{min}}^{-1}.$$

(15.29)

In Box 9.2, we use this version of the spectrum to analyze the twinkling of starlight due to atmospheric turbulence.

Exercise 15.9 *Derivation and Practice: Kolmogorov Spectrum for Quantities Transported by the Fluid*

(a) Fill in the details of the dimensional-analysis derivation of the Kolmogorov spectrum (15.27) for a quantity such as $\mathfrak{n}$ that is transported by the fluid and thus satisfies $d\mathfrak{n}/dt = 0$. In particular, convince yourself (or refute!) that the only quantities $S_{\mathfrak{n}}$ can depend on are k, q, and $q_{\mathfrak{n}}$, identify their dimensions and the dimension of $S_{\mathfrak{n}}$, and use those dimensions to deduce Eq. (15.27).

(b) Derive the spatial version (15.29) of the Kolmogorov spectrum for a transported quantity by two methods: (i) Fourier transforming the wave-number version (15.27), and (ii) dimensional analysis.

Exercise 15.10 *Example: Excitation of Earth's Normal Modes by Atmospheric Turbulence*[10]

Earth has normal modes of oscillation, many of which are in the milli-Hertz frequency range. Large earthquakes occasionally excite these modes strongly, but the quakes are usually widely spaced in time compared to the ringdown time of a particular mode (typically a few days). There is evidence of a background level of continuous excitation of these modes, with an rms ground acceleration per mode of $\sim 10^{-10}$ cm s^{-2} at seismically "quiet" times. The excitation mechanism is suspected to be stochastic forcing by the pressure fluctuations associated with atmospheric turbulence. This exercise deals with some aspects of this hypothesis.

(a) Estimate the rms pressure fluctuations $P(f)$ at frequency f, in a bandwidth equal to frequency $\Delta f = f$, produced on Earth's surface by atmospheric turbulence, assuming a Kolmogorov spectrum for the turbulent velocities and energy. Make your estimate using two methods: (i) using dimensional analysis (what quantity can you construct from the energy cascade rate q, atmospheric density ρ, and frequency f that has dimensions of pressure?) and (ii) using the kinds of arguments about eddy sizes and speeds developed in Sec. 15.4.4.

(b) Your answer using method (i) in part (a) should scale with frequency as $P(f) \propto 1/f$. In actuality, the measured pressure spectra have a scaling law more nearly like $P(f) \propto 1/f^{2/3}$, not $P(f) \propto 1/f$ (e.g., Tanimoto and Um, 1999, Fig. 2a). Explain this discrepancy [i.e., what is wrong with the argument in method (i) and how can you correct it to give $P(f) \propto 1/f^{2/3}$?].

(c) The low-frequency cutoff for the $P(f) \propto 1/f^{2/3}$ pressure spectrum is about 0.5 mHz, and at 1 mHz, $P(f)$ has the value $P(f = 1\,\text{mHz}) \sim 0.3$ Pa, which is

10. Problem devised by David Stevenson, based in part on Tanimoto and Um (1999), who, however, used the pressure spectrum deduced in method (i) of part (a) rather than the more nearly correct spectrum of part (b). The difference in spectra does not much affect their conclusions.

about 3×10^{-6} of atmospheric pressure. Assuming that 0.5 mHz corresponds to the largest eddies, which have a lengthscale of a few kilometers (a little less than the scale height of the atmosphere), derive an estimate for the eddies' turbulent viscosity ν_t in the lower atmosphere. By how many orders of magnitude does this exceed the molecular viscosity? What fraction of the Sun's energy input to Earth ($\sim 10^6$ erg cm^{-2} s^{-1}) goes into maintaining this turbulence (assumed to be distributed over the lowermost 10 km of the atmosphere)?

(d) At $f = 1$ mHz, what is the characteristic spatial scale (wavelength) of the relevant normal modes of Earth? [Hint: The relevant modes have few or no nodes in the radial direction. All you need to answer this question is a typical wave speed for seismic shear waves, which you can take to be 5 km s^{-1}.] What is the characteristic spatial scale (eddy size) of the atmospheric pressure fluctuations at this same frequency, assuming isotropic turbulence? Suggest a plausible estimate for the rms amplitude of the pressure fluctuation averaged over a surface area equal to one square wavelength of Earth's normal modes. (You must keep in mind the random spatially and temporally fluctuating character of the turbulence.)

(e) Challenge: Using your answer from part (d) and a characteristic shear and bulk modulus for Earth's deformations of $K \sim \mu \sim 10^{12}$ dyne cm^{-2}, comment on how the observed rms normal-mode acceleration (10^{-10} cm s^{-2}) compares with that expected from stochastic forcing due to atmospheric turbulence. You may need to review Chaps. 11 and 12, and think about the relationship between surface force and surface deformation. [Note: Several issues emerge when doing this assessment accurately that have not been dealt with in this exercise (e.g., number of modes in a given frequency range), so don't expect to be able to get an answer more accurate than an order of magnitude.]

15.5 Turbulent Boundary Layers

Great interest surrounds the projection of spheres of cork, rubber, leather, and string by various parts of the human anatomy, with and without the mechanical advantage of levers of willow, ceramic, and the finest Kentucky ash. As is well known, changing the surface texture, orientation, and spin of a sphere in various sports can influence its trajectory markedly. Much study has been made of ways to do this both legally and illegally. Some procedures used by professional athletes are pure superstition, but many others find physical explanations that are good examples of the behavior of boundary layers. Many sports involve the motion of balls at Reynolds numbers where the boundary layers can transition between laminar and turbulent, which allows opportunities for controlling the flow. With the goal of studying this transition, let us now consider the structure of a turbulent boundary layer—first along a straight wall and later along a ball's surface.

15.5.1 Profile of a Turbulent Boundary Layer

In Sec. 14.4.1, we derived the Blasius profile for a laminar boundary layer and showed that its thickness a distance x downstream from the start of the boundary layer was roughly $3\delta = 3(\nu x/V)^{1/2}$, where V is the free-stream speed (cf. Fig. 14.13). As we have described, when the Reynolds number is large enough—$\mathrm{Re}_d = Vd/\nu \sim 3 \times 10^5$ or $\mathrm{Re}_\delta \sim \sqrt{\mathrm{Re}_d} \sim 500$ in the case of flow past a cylinder (Figs. 15.1 and 15.2)—the boundary layer becomes turbulent.

turbulent boundary layer's laminar sublayer and turbulent zone

A turbulent boundary layer consists of a thin laminar sublayer of thickness δ_{ls} close to the wall and a much thicker turbulent zone of thickness δ_t (Fig. 15.10).

In the following paragraphs we use the turbulence concepts developed in the previous sections to compute the order of magnitude of the structures of the laminar sublayer and the turbulent zone, and the manner in which those structures evolve along the boundary. We let the wall have any shape, so long as its radius of curvature is large compared to the boundary layer's thickness, so it looks locally planar. We also use locally Cartesian coordinates with distance y measured perpendicular to the boundary and distance x along it, in the direction of the near-wall flow.

One key to the structure of the boundary layer is that, in the x component of the time-averaged Navier-Stokes equation, the stress-divergence term $T_{xy,y}$ has the *potential* to be so large (because of the boundary layer's small thickness) that no other term can compensate for it. This is true in the turbulent zone, where T_{xy} is the huge Reynolds stress; it is also true in the laminar sublayer, where T_{xy} is the huge viscous stress produced by a large shear that results from the thinness of the layer. (One can check at the end of the following analysis that, for the computed boundary-layer structure, other terms in the x component of the Navier-Stokes equation are indeed so small that they could not compensate a significantly nonzero $T_{xy,y}$.) This potential dominance by $T_{xy,y}$ implies that the flow must adjust itself to make $T_{xy,y}$ nearly zero, so T_{xy} must be (very nearly) independent of distance y from the boundary.

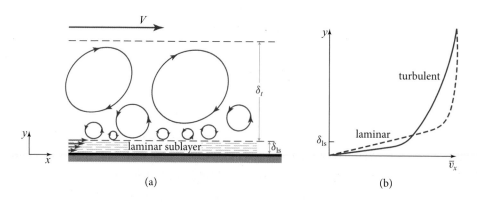

(a)   (b)

FIGURE 15.10 (a) Physical structure of a turbulent boundary layer. (b) Mean flow speed $\bar{v}_x$ as a function of distance from the wall for the turbulent boundary layer [solid curve, Eqs. (15.30)] and for a laminar boundary layer [dashed curve; the Blasius profile, Eqs. (14.46)].

In the turbulent zone, T_{xy} is the Reynolds stress, ρv_ℓ^2, where v_ℓ is the turbulent velocity of the largest eddies at a distance y from the wall; therefore, constancy of T_{xy} implies constancy of v_ℓ. The largest eddies at y have a size ℓ of order the distance y from the wall; correspondingly, the turbulent viscosity is $v_t \sim v_\ell y/3$. Equating the expression ρv_ℓ^2 for the Reynolds stress to the alternative expression $2\rho v_t \frac{1}{2}\bar{v}_{,y}$ (where $\bar{v}$ is the mean-flow speed at y, and $\frac{1}{2}\bar{v}_{,y}$ is the shear), and using $v_t \sim v_\ell y/3$ for the turbulent viscosity, we discover that in the turbulent zone the mean flow speed varies logarithmically with distance from the wall: $\bar{v} \sim v_\ell \ln y + \text{constant}$. Since the turbulence is created at the inner edge of the turbulent zone, $y \sim \delta_{ls}$ (Fig. 15.10) by interaction of the mean flow with the laminar sublayer, the largest turbulent eddies there must have their turnover speeds v_ℓ equal to the mean-flow speed there: $\bar{v} \sim v_\ell$ at $y \sim \delta_{ls}$. This tells us the normalization of the logarithmically varying mean-flow speed:

$$\boxed{\bar{v} \sim v_\ell[1 + \ln(y/\delta_{ls})] \quad \text{at } y \gtrsim \delta_{ls}.}$$
(15.30a)

profile of mean flow speed in turbulent zone

Turn next to the structure of the laminar sublayer. There the constant shear stress is viscous, $T_{xy} = \rho v \bar{v}_{,y}$. Stress balance at the interface between the laminar sublayer and the turbulent zone requires that this viscous stress be equal to the turbulent zone's ρv_ℓ^2. This equality implies a linear profile for the mean-flow speed in the laminar sublayer, $\bar{v} = (v_\ell^2/v)y$. The thickness of the sublayer is then fixed by continuity of $\bar{v}$ at its outer edge: $(v_\ell^2/v)\delta_{ls} = v_\ell$. Combining these last two relations, we obtain the following profile and laminar-sublayer thickness:

$$\boxed{\bar{v} \sim v_\ell \left(\frac{y}{\delta_{ls}}\right) \quad \text{at } y \lesssim \delta_{ls} \sim v/v_\ell.}$$
(15.30b)

mean flow velocity in laminar sublayer

Having deduced the internal structure of the boundary layer, we turn to the issue of what determines the y-independent turbulent velocity v_ℓ of the largest eddies. This v_ℓ is fixed by matching the turbulent zone to the free-streaming region outside it. The free-stream velocity V (which may vary slowly with x due to curvature of the wall) must be equal to the mean flow velocity $\bar{v}$ [Eq. (15.30a)] at the outer edge of the turbulent zone. The logarithmic term dominates, so $V = v_\ell \ln(\delta_t/\delta_{ls})$. Introducing an overall Reynolds number for the boundary layer,

$$\boxed{\text{Re}_\delta \equiv V\delta_t/v,}$$
(15.31)

and noting that turbulence requires a huge value ($\gtrsim 1{,}000$) of this Re_δ, we can reexpress V as $V \sim v_\ell \ln \text{Re}_\delta$. This should actually be regarded as an equation for the turbulent velocity of the largest-scale eddies in terms of the free-stream velocity:

$$\boxed{v_\ell \sim \frac{V}{\ln \text{Re}_\delta}.}$$
(15.32)

turbulent velocity of largest-scale eddies in turbulent zone

If the free-stream velocity $V(x)$ and the thickness $\delta_t + \delta_{ls} \simeq \delta_t$ of the entire boundary layer are given, then Eq. (15.31) determines the boundary layer's Reynolds number, Eq. (15.32) then determines the turbulent velocity, and Eqs. (15.30) determine the layer's internal structure.

Finally, we turn to the issue of how the boundary layer thickness δ_t evolves with distance x down the wall (and, correspondingly, how the rest of the boundary layer's structure, which is fixed by δ_t, evolves). The key to the evolution of δ_t is entrainment, which we met in our discussion of turbulent wakes and jets (Sec. 15.4.3). At the turbulent zone's outer edge, the largest turbulent eddies move with speed $\sim v_\ell$ into the free-streaming fluid, entraining that fluid into themselves. Correspondingly, the thickness grows at a rate

thickness of turbulent
boundary layer

$$
\boxed{\frac{d\delta_t}{dx} \sim \frac{v_\ell}{V} \sim \frac{1}{\ln \mathrm{Re}_\delta}.}
\tag{15.33}
$$

Since $\ln \mathrm{Re}_\delta$ depends only extremely weakly on δ_t, the turbulent boundary layer expands essentially linearly with distance x, by contrast with a laminar boundary layer's $\delta \propto x^{1/2}$.

15.5.2 Coanda Effect and Separation in a Turbulent Boundary Layer

15.5.2

One can easily verify that not only does the turbulent boundary layer expand more rapidly than the corresponding laminar boundary layer would (if it were stable) but also that the turbulent layer is thicker at all locations down the wall. Physically, this distinction can be traced, in part, to the fact that the turbulent boundary layer involves a 3-dimensional velocity field, whereas the corresponding laminar layer would involve only a 2-dimensional field. The enhanced thickness and expansion contribute to the Coanda effect for a turbulent boundary layer—the layer's ability to stick to the wall under adverse conditions (Sec. 15.4.3).

However, there is a price to be paid for this benefit. Since the velocity gradient is increased near the surface, the actual surface shear stress exerted by the turbulent layer, through its laminar sublayer, is significantly larger than in the corresponding laminar boundary layer. As a result, if the entire boundary layer were to remain laminar, the portion that would adhere to the surface would produce less viscous drag than the near-wall laminar sublayer of the turbulent boundary layer. Correspondingly, in a long, straight pipe, the drag on the pipe wall goes up when the boundary layer becomes turbulent.

influence of turbulence on
separation of boundary
layer and thence on drag

However, for flow around a cylinder or other confined body, the drag goes down! (Cf. Fig. 15.2.) The reason is that in the separated, laminar boundary layer the dominant source of drag is not viscosity but rather a pressure differential between the front face of the cylinder (where the layer adheres) and the back face (where the reverse eddies circulate). The pressure is much lower in the back-face eddies than in the front-face boundary layer, and that pressure differential gives rise to a significant drag, which is reduced when the boundary layer goes turbulent and adheres to the back

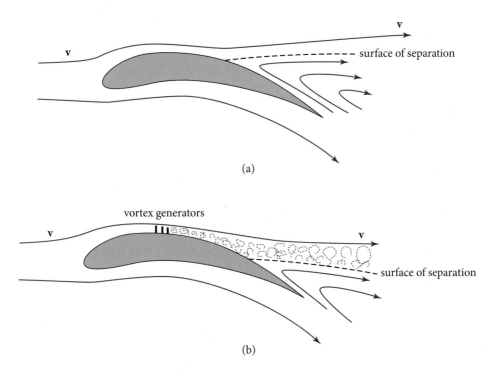

FIGURE 15.11 (a) A laminar boundary layer separating from an airplane wing due to an adverse pressure gradient in the wing's back side. (b) Vortex generators attached to the wing's top face generate turbulence. The turbulent boundary layer sticks to the wing more effectively than does the laminar boundary layer (the Coanda effect). Separation from the wing is delayed, and the wing's lift is increased and drag is decreased.

face. Therefore, if one's goal is to reduce the overall drag and the laminar flow is prone to separation, a nonseparating (or delayed-separation) turbulent layer is preferred to the laminar layer. Similarly (and for essentially the same reason), for an airplane wing, if one's goal is to maintain a large lift, then a nonseparating (or delayed-separation) turbulent layer is superior to a separating, laminar one.[11]

For this reason, steps are often taken in engineering flows to ensure that boundary layers become and remain turbulent. A crude but effective example is provided by the vortex generators that are installed on the upper surfaces of some airplane wings (Fig. 15.11). These structures are small obstacles on the wing that penetrate through a laminar boundary layer into the free flow. By changing the pressure distribution, they force air into the boundary layer and initiate 3-dimensional vortical motion in the boundary layer, forcing it to become partially turbulent. This turbulence improves the

vortex generators on airplane wings

11. Another example of separation occurs in "lee waves," which can form when wind blows over a mountain range. These consist of standing-wave eddies in the separated boundary layer, somewhat analogous to the Kármán vortex street of Fig. 15.1d. Lee waves are sometimes used by glider pilots to regain altitude.

wing's lift, it allows the airplane to climb more steeply without stalling from boundary-layer separation, and it helps reduce aerodynamical drag.

15.5.3 Instability of a Laminar Boundary Layer

15.5.3

Much work has been done on the linear stability of laminar boundary layers. The principles of such stability analyses should now be familiar, although the technical details are formidable. In the simplest case, an equilibrium flow is identified (e.g., the Blasius profile), and the equations governing the time evolution of small perturbations are written down. The individual Fourier components are assumed to vary in space and time as $\exp i(\mathbf{k} \cdot \mathbf{x} - \omega t)$, and we seek modes that have zero velocity perturbation on the solid surface past which the fluid flows and that decay to zero in the free stream. We ask whether there are unstable modes (i.e., modes with real $\mathbf{k}$ for which the imaginary part of ω is positive, so they grow exponentially in time). Such exponential growth drives the perturbation to become nonlinear, which then typically triggers turbulence.

The results can generally be expressed in the form of a diagram like that in Fig. 15.12. As shown in that figure, there is generally a critical Reynolds number $\mathrm{Re}_{\mathrm{crit}} \sim 500$ at which one mode becomes linearly unstable. At higher values of the Reynolds number, modes with a range of k-vectors are linearly unstable. One interesting result of these calculations is that in the absence of viscous forces (i.e., in the limit $\mathrm{Re}_{\delta} \to \infty$), the boundary layer is unstable if and only if there is a point of inflection in the velocity profile (a point where $d^2 v_x/dy^2$ changes sign; cf. Fig. 15.12 and Ex. 15.11).

dynamically stable and unstable modes in laminar boundary layer

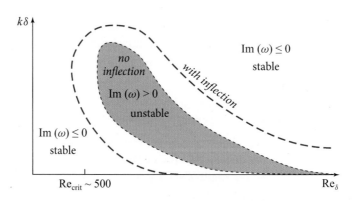

FIGURE 15.12 Values of wave number k for stable and unstable wave modes in a laminar boundary layer with thickness δ, as a function of the boundary layer's Reynolds number $\mathrm{Re}_{\delta} = V\delta/\nu$. If the unperturbed velocity distribution $v_x(y)$ has no inflection point (i.e., if $d^2 v_x/dy^2 < 0$ everywhere, as is the case for the Blasius profile; Fig. 14.13), then the unstable modes are confined to the shaded region. If there is an inflection point (so $d^2 v_x/dy^2$ is positive near the wall but becomes negative farther from the wall, as is the case near a surface of separation; Fig. 14.14), then the unstable region is larger (dashed boundary) and does not asymptote to $k = 0$ as $\mathrm{Re}_{\delta} \to \infty$.

Although, in the absence of an inflection, an inviscid flow $v_x(y)$ is stable, for some such profiles even the slightest viscosity can trigger instability. Physically, this is because viscosity can tap the relative kinetic energies of adjacent flow lines. Viscous-triggered instabilities of this sort are sometimes called *secular instabilities* by contrast with the *dynamical instabilities* that arise in the absence of viscosity. Secular instabilities are quite common in fluid mechanics.

secular instabilities contrasted with dynamical instabilities

EXERCISES

Exercise 15.11 *Problem: Tollmien-Schlichting Waves*
Consider an inviscid ($v = 0$), incompressible flow near a plane wall where a laminar boundary layer is established. Introduce coordinates x parallel to the wall and y perpendicular to it. Let the components of the equilibrium velocity be $v_x(y)$.

(a) Show that a weak propagating-wave perturbation in the velocity, $\delta v_y \propto \exp ik(x - Ct)$, with k real and frequency Ck possibly complex, satisfies the differential equation

$$\frac{\partial^2 \delta v_y}{\partial y^2} = \left[\frac{1}{(v_x - C)} \frac{d^2 v_x}{dy^2} + k^2 \right] \delta v_y. \qquad (15.34)$$

These are called *Tollmien-Schlichting waves.*

(b) Hence argue that a sufficient condition for unstable wave modes [$\text{Im}(C) > 0$] is that the velocity field possess a point of inflection (i.e., a point where $d^2 v_x/dy^2$ changes sign; cf. Fig. 15.12). The boundary layer can also be unstable in the absence of a point of inflection, but viscosity must then trigger the instability.

15.5.4 Flight of a Ball

15.5.4

physics of sports balls

Having developed some insights into boundary layers and their stability, we now apply those insights to the balls used in various sports.

The simplest application is to the dimples on a golf ball (Fig. 15.13a). The dimples provide finite-amplitude disturbances in the flow that can trigger growing wave modes and then turbulence in the boundary layer. The adherence of the boundary layer to the ball is improved, and separation occurs farther behind the ball, leading to a lower drag coefficient and a greater range of flight; see Figs. 15.2 and 15.13a.

golf ball: turbulent boundary layer reduces drag

A variant on this mechanism is found in the game of cricket, which is played with a ball whose surface is polished leather with a single equatorial seam of rough stitching. When the ball is "bowled" in a nonspinning way with the seam inclined to the direction of motion, a laminar boundary layer exists on the smooth side and a turbulent one on the side with the rough seam (Fig. 15.13b). These two boundary layers separate at different points behind the flow, leading to a net deflection of the air. The ball therefore swerves toward the side with the leading seam. (The effect is strongest when the ball is new and still shiny and on days when the humidity is high, so the thread in the seam swells and is more efficient at making turbulence.)

cricket ball: two boundary layers, turbulent and laminar, produce deflection

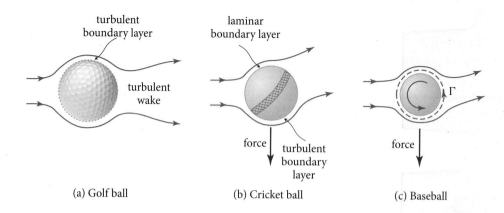

turbulent
boundary layer

laminar
boundary layer

turbulent
wake

force

turbulent
boundary
layer

force

Γ

(a) Golf ball

(b) Cricket ball

(c) Baseball

FIGURE 15.13 Boundary layers around (a) golf ball, (b) cricket ball, and (c) baseball as they move leftward relative to the air (i.e., as the air flows rightward as seen in their rest frames).

This mechanism is different from that used to throw a slider or curveball in baseball, in which the pitcher causes the ball to spin about an axis roughly perpendicular to the direction of motion. In the slider the axis is vertical; for a curveball it is inclined at about 45° to the vertical. The spin of the ball creates circulation (in a nonrotating, inertial frame) like that around an airfoil. The pressure forces associated with this circulation produce a net sideways force in the direction of the baseball's rotational velocity on its leading hemisphere (i.e., as seen by the hitter; Fig. 15.13c). The physical origin of this effect is actually quite complex and is only properly described with reference to experimental data. The major effect is that separation is delayed on the side of the ball where the rotational velocity is in the same direction as the airflow and happens sooner on the opposite side (Fig. 15.13c), leading to a pressure differential. The reader may be curious as to how this circulation can be established in view of Kelvin's theorem, Eq. (14.16), which states that if we use a closed curve $\partial\mathcal{S}$, Eq. (14.12a), that is so far from the ball and its wake that viscous forces cannot cause the vorticity to diffuse to it, then the circulation must be zero. What actually happens is similar to an airplane wing (Fig. 14.2b). When the flow is initiated, starting vortices are shed by the baseball and are then convected downstream, leaving behind the net circulation Γ that passes through the ball (Fig. 15.14). This effect is very much larger in 2 dimensions with a rotating cylinder than in 3 dimensions, because the magnitude of the shed vorticity is much larger. It goes by the name of *Magnus effect* in 2 dimensions and *Robins effect* in 3, and it underlies Kutta-Joukowski's theorem for the lift on an airplane wing (Ex. 14.8).

In table tennis, a drive is often hit with *topspin,* so that the ball rotates about a horizontal axis perpendicular to the direction of motion. In this case, the net force is downward, and the ball falls faster toward the table, the effect being largest after it has somewhat decelerated. This allows a ball to be hit hard over the net and bounce before passing the end of the table, increasing the margin for errors in the direction of the hit.

baseball and table tennis: circulation produces deflection

Magnus and Robins effects

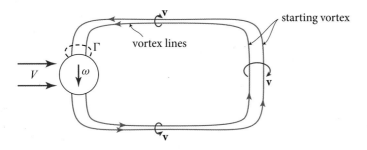

FIGURE 15.14 Vortex lines passing through a spinning ball. The starting vortex is created and shed when the ball is thrown and is carried downstream by the flow as seen in the ball's frame of reference. The vortex lines connecting this starting vortex to the ball lengthen as the flow continues.

Those wishing to improve their curveballs or cure a bad slice are referred to the monographs by Adair (1990), Armenti (1992), and Lighthill (1986).

Exercise 15.12 *Problem: Effect of Drag*
A well-hit golf ball travels about 300 yards. A fast bowler or fastball pitcher throws a cricket ball or baseball at more than 90 mph (miles per hour). A table-tennis player can hit a forehand return at about 30 mph. The masses and diameters of each of these four types of balls are $m_g \sim 46$ g, $d_g \sim 43$ mm; $m_c \sim 160$ g, $d_c \sim 70$ mm; $m_b \sim 140$ g, $d_b \sim 75$ mm; and $m_{tt} \sim 2.5$ g, $d_{tt} \sim 38$ mm.

(a) For golf, cricket (or baseball), and table tennis, estimate the Reynolds number of the flow and infer the drag coefficient C_D. (The variation of C_D with Re_d can be assumed to be similar to that in flow past a cylinder; Fig. 15.2.)

(b) Hence estimate the importance of aerodynamic drag in determining the range of a ball in each of these three cases.

15.6 The Route to Turbulence—Onset of Chaos

15.6.1 Rotating Couette Flow

Let us examine qualitatively how a viscous flow becomes turbulent. A good example is rotating Couette flow between two long, concentric, differentially rotating cylinders, as introduced in Sec. 14.6.3 and depicted in Fig. 15.15a. This flow was studied in a classic set of experiments by Gollub, Swinney, and collaborators (Fenstermacher, Swinney, and Gollub, 1979; Gorman and Swinney, 1982, and references therein). In these experiments, the inner cylinder rotates and the outer one does not, and the fluid is a liquid whose viscosity is gradually decreased by heating it, so the Reynolds number gradually increases. At low Reynolds numbers, the equilibrium flow is stable,

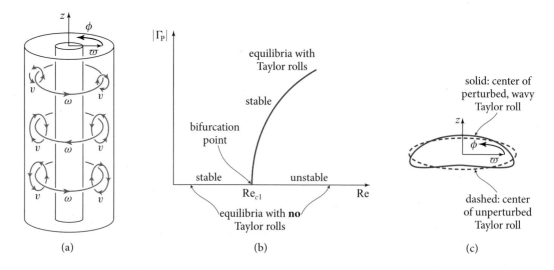

FIGURE 15.15 Bifurcation of equilibria in rotating Couette flow. (a) Equilibrium flow with Taylor rolls. (b) Bifurcation diagram in which the amplitude of the poloidal circulation $|\Gamma_P|$ in a Taylor roll is plotted against the Reynolds number. At low Reynolds numbers (Re $<$ Re$_{c1}$), the only equilibrium flow configuration is smooth, azimuthal flow. At larger Reynolds numbers (Re$_{c1}$ $<$ Re $<$ Re$_{c2}$), there are two equilibria, one with Taylor rolls and stable; the other with smooth, azimuthal flow, which is unstable. (c) Shape of a Taylor roll at Re$_{c1}$ $<$ Re $<$ Re$_{c2}$ (dashed ellipse) and at higher values, Re$_{c2}$ $<$ Re $<$ Re$_{c3}$ (solid, wavy curve).

stationary, and azimuthal (strictly in the ϕ direction; Fig. 15.15a). At very high Reynolds numbers, the flow is unstable, according to the Rayleigh criterion (angular momentum per unit mass decreases outward; Sec. 14.6.3). Therefore, as the Reynolds number is gradually increased, at some critical value Re$_{c1}$, the flow becomes unstable to the growth of small perturbations. These perturbations drive a transition to a new stationary equilibrium whose form is what one might expect from the Rayleigh-criterion thought experiment (Sec. 14.6.3): it involves poloidal circulation (quasi-circular motions in the r and z directions, called *Taylor rolls*; see Fig. 15.15a).

Thus an equilibrium with a high degree of symmetry has become unstable, and a new, lower-symmetry, stable equilibrium has been established; see Fig. 15.15b. Translational invariance along the cylinder axis has been lost from the flow, even though the boundary conditions remain translationally symmetric. This transition is another example of the bifurcation of equilibria that we discussed when treating the buckling of beams and playing cards (Secs. 11.6 and 12.3.5).

As the Reynolds number is increased further, this process repeats. At a second critical Reynolds number Re$_{c2}$, a second bifurcation of equilibria occurs in which the azimuthally smooth Taylor rolls become unstable and are replaced by new, azimuthally wavy Taylor rolls; see Fig. 15.15c. Again, an equilibrium with higher symmetry (rotation invariance) has been replaced at a bifurcation point by one of lower symmetry (no rotation invariance). A fundamental frequency f_1 shows up in the fluid's velocity $\mathbf{v}(\mathbf{x}, t)$ as the wavy Taylor rolls circulate around the central cylinder.

Taylor rolls

bifurcation to a lower symmetry state

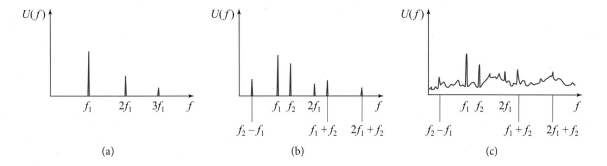

FIGURE 15.16 The energy spectrum of velocity fluctuations in rotating Couette flow (schematic). (a) Spectrum for a moderate Reynolds number, $Re_{c2} < Re < Re_{c3}$, at which the stable equilibrium flow is that with the wavy Taylor rolls of Fig. 15.15c. (b) Spectrum for a higher Reynolds number, $Re_{c3} < Re < Re_{c4}$, at which the stable flow has wavy Taylor rolls with two incommensurate fundamental frequencies present. (c) Spectrum for a still higher Reynolds number, $Re > Re_{c4}$, at which turbulence has set in.

Since the waves are nonlinearly large, harmonics of this fundamental are also seen when the velocity field is Fourier decomposed (cf. Fig. 15.16a). When the Reynolds number is increased still further to some third critical value Re_{c3}, there is yet another bifurcation. The Taylor rolls now develop a second set of waves, superimposed on the first, with a corresponding new fundamental frequency f_2 that is incommensurate with f_1. In the energy spectrum one now sees various harmonics of f_1 and of f_2, as well as sums and differences of these two fundamentals (Fig. 15.16b).

wavy Taylor rolls: two frequencies

It is exceedingly difficult to construct an experimental apparatus that is clean enough—and sufficiently free from the effects of finite lengths of the cylinders—to reveal what happens next as the Reynolds number increases. However, despite the absence of clean experiments, it seemed obvious before the 1970s what would happen. The sequence of bifurcations would continue, with ever-decreasing intervals of Reynolds number ΔRe between them, eventually producing such a complex maze of frequencies, harmonics, sums, and differences, as to be interpreted as turbulence. Indeed, one finds the onset of turbulence described in just this manner in the classic fluid mechanics textbook of Landau and Lifshitz (1959), based on earlier research by Landau (1944).

The 1970s and 1980s brought a major breakthrough in our understanding of the onset of turbulence. This breakthrough came from studies of model dynamical systems with only a few degrees of freedom, in which nonlinear effects play similar roles to the nonlinearities of the Navier-Stokes equation. These studies revealed only a handful of routes to irregular or unpredictable behavior known as chaos, and none were of the Landau type. However, one of these routes starts out in the same manner as does rotating Couette flow: As a control parameter (the Reynolds number in the case of Couette flow) is gradually increased, first oscillations with one fundamental frequency f_1 and its harmonics turn on; then a second frequency f_2 (incommensurate with the first) and its harmonics turn on, along with sums and differences of f_1

route to turbulence in rotating Couette flow: one frequency, two frequencies, then turbulence (chaos)

and f_2; and then, suddenly, chaos sets in. Moreover, the chaos is clearly not being produced by a complicated superposition of new frequencies; it is fundamentally different from that.

Remarkably, the experiments of Gollub, Swinney, and colleagues gave convincing evidence that the onset of turbulence in rotating Couette flow takes precisely this route (Fig. 15.16c).

15.6.2

15.6.2 Feigenbaum Sequence, Poincaré Maps, and the Period-Doubling Route to Turbulence in Convection

The simplest of systems in which one can study the several possible routes to chaos are 1-dimensional mathematical maps. A lovely example is the *Feigenbaum sequence,* explored by Mitchell Feigenbaum (1978).

The Feigenbaum sequence is a sequence $\{x_1, x_2, x_3, \ldots\}$ of values of a real variable x, given by the rule (sometimes called the *logistic equation*)[12]

logistic equation and its
Feigenbaum sequence

$$x_{n+1} = 4ax_n(1 - x_n). \tag{15.35}$$

Here a is a fixed control parameter. It is easy to compute Feigenbaum sequences $\{x_1, x_2, x_3, \ldots\}$ for different values of a on a personal computer (Ex. 15.13). What is found is that there are critical parameters $a_1, a_2, \ldots$ at which the character of the sequence changes sharply. For $a < a_1$, the sequence asymptotes to a stable fixed point. For $a_1 < a < a_2$, the sequence asymptotes to stable, periodic oscillations between two fixed points. If we increase the parameter further, so that $a_2 < a < a_3$, the sequence becomes a periodic oscillation between four fixed points. The period of the oscillation has doubled. This *period doubling* (not to be confused with frequency doubling) happens again: When $a_3 < a < a_4$, x asymptotes to regular motion between eight fixed points. Period doubling increases with shorter and shorter intervals of a until at some value a_∞, the period becomes infinite, and the sequence does not repeat. Chaos has set in.

period doubling route to
chaos for Feigenbaum
sequence

This period doubling is a second route to chaos, very different in character from the one-frequency, two-frequencies, chaos route that appears to occur in rotating Couette flow. Remarkably, fluid dynamical turbulence can set in by this second route as well as by the first. It does so in certain very clean experiments on convection. We return to this phenomenon at the end of this section and then again in Sec. 18.4.

How can so starkly simple and discrete a model as a 1-dimensional map bear any relationship to the continuous solutions of the fluid dynamical differential equations? The answer is quite remarkable.

---

12. This equation first appeared in discussions of population biology (Verhulst, 1838). If we consider x_n as being proportional to the number of animals in a species (traditionally rabbits), the number in the next season should be proportional to the number of animals already present and to the availability of resources, which will decrease as x_n approaches some maximum value, in this case unity. Hence the terms x_n and $1 - x_n$ in Eq. (15.35).

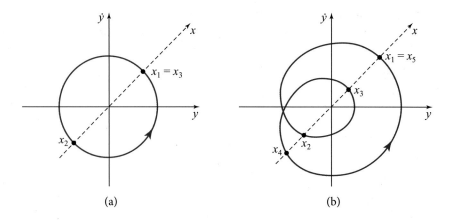

FIGURE 15.17 (a) Representation of a single periodic oscillation as motion in phase space. (b) Motion in phase space after period doubling. The behavior of the system may also be described by using the coordinate x of the Poincaré map.

Consider a steady flow in which one parameter a (e.g., the Reynolds number) can be adjusted. Now, as we change a and approach turbulence, the flow may develop a periodic oscillation with a single frequency f_1. We could measure this by inserting some probe at a fixed point in the flow to measure a fluid variable y (e.g., one component of the velocity). We can detect the periodicity either by inspecting the readout $y(t)$ or its Fourier transform $\tilde{y}$. However, there is another way, which may be familiar from classical mechanics. This is to regard $\{y, \dot{y}\}$ as the two coordinates of a 2-dimensional phase space. (Of course, instead one could measure many variables and their time derivatives, resulting in an arbitrarily large phase space, but let us keep matters as simple as possible.) For a single periodic oscillation, the system will follow a closed path in this phase space (Fig. 15.17a). As we increase a further, a period doubling may occur, and the trajectory in phase space may look like Fig. 15.17b. As we are primarily interested in the development of the oscillations, we need only keep one number for every fundamental period $P_1 = 1/f_1$. Let us do this by taking a section through phase space and introducing a coordinate x on this section, as shown in Fig. 15.17. The nth time the trajectory crosses this section, its crossing point is x_n, and the mapping from x_n to x_{n+1} can be taken as a representative characterization of the flow. When only the frequency f_1 is present, the map reads $x_{n+2} = x_n$ (Fig. 15.17a). When f_1 and $f_2 = \frac{1}{2}f_1$ are present, the map reads $x_{n+4} = x_n$ (Fig. 15.17b). (These specific maps are overly simple compared to what one may encounter in a real flow, but they illustrate the idea.)

phase-space representation of dynamics

To reiterate, instead of describing the flow by the full solution $\mathbf{v}(\mathbf{x}, t)$ to the Navier-Stokes equation and the flow's boundary conditions, we can construct the simple map $x_n \to x_{n+1}$ to characterize the flow. This procedure is known as a *Poincaré map*. The mountains have labored and brought forth a mouse! However, this mouse turns out to be all that we need. For some convection experiments, the same period-

reducing the dynamics of a fluid's route to turbulence to a 1-dimensional map: the Poincaré map

doubling behavior and approach to chaos are present in these maps as in the 2-dimensional phase-space diagram and in the full solution to the fluid dynamical equations. Furthermore, when observed in the Poincaré maps, the transition looks qualitatively the same as in the Feigenbaum sequence. It is remarkable that for a system with so many degrees of freedom, chaotic behavior can be observed by suppressing almost all of them.

If, in the period-doubling route to chaos, we compute the limiting ratio of successive critical parameters,

Feigenbaum number for period doubling route to chaos, and its universality

$$\mathcal{F} = \lim_{j \to \infty} \frac{a_j - a_{j-1}}{a_{j+1} - a_j}, \tag{15.36}$$

we find that it has the value 4.66920160910 .... This *Feigenbaum number* seems to be a universal constant characteristic of most period-doubling routes to chaos, independent of the particular map that was used. For example, if we had used

$$x_{n+1} = a \sin \pi x_n \tag{15.37}$$

we would have gotten the same constant.

period doubling route to turbulence in Libchaber convection experiments

The most famous illustration of the period-doubling route to chaos is a set of experiments by Libchaber and colleagues on convection in liquid helium, water, and mercury, culminating with the mercury experiment by Libchaber, Laroche, and Fauve (1982). In each experiment (depicted in Fig. 18.1), the temperature at a point was monitored with time as the temperature difference ΔT between the fluid's bottom and top surfaces was slowly increased. In each experiment, at some critical temperature difference ΔT_1 the temperature began to oscillate with a single period; then at some ΔT_2 that oscillation was joined by another at twice the period; at ΔT_3 another period doubling occurred; at ΔT_4 another; and at ΔT_5 yet another. The frequency doubling could not be followed beyond this because the signal was too weak, but shortly thereafter the convection became turbulent. In each experiment it was

universality in the route to chaos

possible to estimate the Feigenbaum ratio (15.36), with a_j being the jth critical ΔT. For the highest-accuracy (mercury) experiment, the experimental result agreed with the Feigenbaum number 4.669 ... to within the experimental accuracy (about 6%); the helium and water experiments also agreed with Feigenbaum to within their experimental accuracies. This result was remarkable. *There truly is a deep universality in the route to chaos, a universality that extends even to fluid-dynamical convection!* This work won Libchaber and Feigenbaum together the prestigious 1986 Wolf Prize.

EXERCISES

Exercise 15.13 *Problem: Feigenbaum Number for the Logistic Equation*
Use a computer to calculate the first five critical parameters a_j for the sequence of numbers generated by the logistic equation (15.35). Hence verify that the ratio of successive differences tends toward the Feigenbaum number $\mathcal{F}$ quoted in Eq. (15.36). (Hint: To find suitable starting values x_1 and starting parameter a, you might find it

helpful to construct a graph.) For insights into the universality of the Feigenbaum number, based in part on the renormalization group (Sec. 5.8.3), see Sethna (2006, Ex. 12.9).

15.6.3 Other Routes to Turbulent Convection

Some other routes to turbulence have been seen in convection experiments, in addition to the Feigenbaum/Libchaber period-doubling route. Particularly impressive were convection experiments by Gollub and Benson (1980), which showed—depending on the experimental parameters and the nature of the initial convective flow—four different routes:

four different routes to turbulence in convection experiments

1. The period-doubling route.

2. A variant of the one-frequency, two-frequencies, chaos route. In this variant, a bit after (at higher ΔT) the second frequency appears, incommensurate with the first, the two frequencies become commensurate and their modes become phase locked (one entrains the other); and then a bit later (higher ΔT) the phase lock is broken and simultaneously turbulence sets in.

3. A one-frequency, two-frequencies, three-frequencies, chaos route.

4. An intermittency route, in which, as ΔT is increased, the fluid oscillates between a state with two or three incommensurate modes and a state with turbulence.

These four routes to chaos are all seen in simple mathematical maps or low-dimensional dynamical systems; for example, the intermittency route is seen in the Lorenz equations of Ex. 15.16.

Note that the convective turbulence that is triggered by each of these routes is *weak*; the control parameter ΔT must be increased further to drive the fluid into a fully developed, strong-turbulence state.

weakness and confinement of the turbulence that follows these routes

Also important is the fact that these experimental successes, which compare the onset of turbulence with the behaviors of simple mathematical maps or low-dimensional dynamical systems, all entail fluids that are confined by boundaries and have a modest aspect ratio (ratio of the largest to the smallest dimension), 20:1 for rotating Couette flow and 5:1 for convection. For confined fluids with much larger aspect ratios, and for unconfined fluids, there has been no such success. It may be that the successes are related to the small number of relevant modes in a system with modest aspect ratio.

These successes occurred in the 1970s and 1980s. Much subsequent research focuses on the mechanism by which the final transition to turbulence occurs [e.g., the role of the Navier-Stokes nonlinear term $(\mathbf{v} \cdot \mathbf{\nabla})\mathbf{v}$]. This research involves a mixture of analytical work and numerical simulations, plus some experiment (see, e.g., Grossman, 2000).

15.6.4 Extreme Sensitivity to Initial Conditions

The evolution of a dynamical system becomes essentially incalculable after the onset of chaos. This is because, as can be shown mathematically, the state of the system (as measured by the value of a map, or in a fluid by the values of a set of fluid variables) at some future time becomes highly sensitive to the assumed initial state. Paths (in the map or in phase space) that start arbitrarily close together diverge from each other exponentially rapidly with time.

It is important to distinguish this unpredictability of classical chaos from unpredictability in the evolution of a quantum mechanical system. A classical system evolves under precisely deterministic differential equations. Given a full characterization of the system at any time t, the system is fully specified at a later time $t + \Delta t$, for any Δt. However, what characterizes chaos is that the evolution of two identical systems in neighboring initial states will eventually evolve so that they follow totally different histories. The time it takes for this to happen is called the *Lyapunov time*. The practical significance of this essentially mathematical feature is that if, as will always be the case, we can only specify the initial state up to a given accuracy (due to practical considerations, not issues of principle), then the true initial state could be any one of those lying in some region, so we have no way of predicting what the state will be after a few Lyapunov times have elapsed.

Quantum mechanical indeterminacy is different. If we can prepare a system in a given state described by a wave function, then the wave function's evolution will be governed fully deterministically by the time-dependent Schrödinger equation. However, if we choose to make a measurement of an observable, many quite distinct outcomes are immediately possible, and (for a high-precision measurement) the system will be left in an eigenstate corresponding to the actual measured outcome. The quantum mechanical description of classical chaos is the subject of *quantum chaos* (e.g., Gutzwiller, 1990).

The realization that many classical systems have an intrinsic unpredictability despite being deterministic from instant to instant has been widely publicized in popularizations of research on chaos. However, the concept is not particularly new. It was well understood, for example, by Poincaré around 1900, and watching the weather report on the nightly news bears witness to its dissemination into popular culture! What *is* new and intriguing is the manner in which a system transitions from a deterministic to an unpredictable evolution.

Chaotic behavior is well documented in a variety of physical dynamical systems: electrical circuits, nonlinear pendula, dripping faucets, planetary dynamics, and so on. The extent to which the principles that have been devised to describe chaos in these systems can also be applied to general fluid turbulence remains a matter for debate. There is no question that similarities exist, and (as we have seen) quantitative success has been achieved by applying chaos results to particular forms of turbulent convection. However, most forms of turbulence are not so easily described, and

Lyapunov time: characterizes sensitivity of a classical system to initial conditions

quantum indeterminacy

examples of chaotic behavior

there is still a huge gap between the intriguing mathematics of chaotic dynamics and practical applications to natural and technological flows.

Exercise 15.14 *Example: Lyapunov Exponent*

Consider the logistic equation (15.35) for the special case $a = 1$, which is large enough to ensure that chaos has set in.

(a) Make the substitution $x_n = \sin^2 \pi \theta_n$, and show that the logistic equation can be expressed in the form $\theta_{n+1} = 2\theta_n \pmod{1}$; that is, θ_{n+1} equals the fractional part of $2\theta_n$.

(b) Write θ_n as a "binimal" (binary decimal). For example, $11/16 = 1/2 + 0/4 + 1/8 + 1/16$ has the binary decimal form 0.1011. Explain what happens to this number in each successive iteration.

(c) Now suppose that an error is made in the ith digit of the starting binimal. When will it cause a major error in the predicted value of x_n?

(d) If the error after n iterations is written ϵ_n, show that the Lyapunov exponent p defined by

$$p = \lim_{n \to \infty} \frac{1}{n} \ln \left| \frac{\epsilon_n}{\epsilon_0} \right| \tag{15.38}$$

is ln 2 (so $\epsilon_n \simeq 2^n \epsilon_0$ for large enough n). Lyapunov exponents play an important role in the theory of dynamical systems.

Exercise 15.15 *Example: Strange Attractors*

Another interesting 1-dimensional map is provided by the recursion relation

$$x_{n+1} = a \left(1 - 2 \left| x_n - \frac{1}{2} \right| \right). \tag{15.39}$$

(a) Consider the asymptotic behavior of the variable x_n for different values of the parameter a, with both x_n and a being confined to the interval $[0, 1]$. In particular, find that for $0 < a < a_{\text{crit}}$ (for some a_{crit}), the sequence x_n converges to a stable fixed point, but for $a_{\text{crit}} < a < 1$, the sequence wanders chaotically through some interval $[x_{\min}, x_{\max}]$.

(b) Using a computer, calculate the value of a_{crit} and the interval $[x_{\min}, x_{\max}]$ for $a = 0.8$.

(c) The interval $[x_{\min}, x_{\max}]$ is an example of a *strange attractor*. It has the property that if we consider sequences with arbitrarily close starting values, their values of x_n in this range will eventually diverge. Show that the attractor is strange by computing the sequences with $a = 0.8$ and starting values $x_1 = 0.5, 0.51, 0.501,$ and 0.5001. Determine the number of iterations n_ϵ required to produce significant divergence as a function of $\epsilon = x_1 - 0.5$. It is claimed that $n_\epsilon \sim -\ln_2(\epsilon)$. Can you verify this? Note that the onset of chaos at $a = a_{\text{crit}}$ is quite sudden in this case, unlike the behavior exhibited by the Feigenbaum sequence. See Ruelle (1989) for more on strange attractors.

Exercise 15.16 *Problem: Lorenz Equations*

One of the first discoveries of chaos in a mathematical model was by Lorenz (1963), who made a simple model of atmospheric convection. In this model, the temperature and velocity field are characterized by three variables, x, y, and z, which satisfy the coupled, nonlinear differential equations

$$
\begin{aligned}
dx/dt &= 10(y - x), \\
dy/dt &= -xz + 28x - y, \\
dz/dt &= xy - 8z/3.
\end{aligned}
\tag{15.40}
$$

(The precise definitions of x, y, and z need not concern us here.) Integrate these equations numerically to show that x, y, and z follow nonrepeating orbits in the 3-dimensional phase space that they span, and quickly asymptote to a 2-dimensional strange attractor. (It may be helpful to plot out the trajectories of pairs of the dependent variables.)

These Lorenz equations are often studied with the numbers 10, 28, 8/3 replaced by parameters σ, ρ, and β. As these parameters are varied, the behavior of the system changes.

Bibliographic Note

For physical insight into turbulence, we strongly recommend the movies cited in Box 15.2 and the photographs in Van Dyke (1982) and Tritton (1987).

Many fluid mechanics textbooks have good treatments of turbulence. We particularly like White (2008), and also recommend Lautrup (2005) and Panton (2005). Some treatises on turbulence go into the subject much more deeply (though the deeper treatments often have somewhat limited applicability); among these, we particularly like Tennekes and Lumley (1972) and also recommend the more up-to-date Davidson (2005) and Pope (2000). Standard fluid mechanics textbooks for engineers focus particularly on turbulence in pipe flow and in boundary layers (see, e.g., Munson, Young, and Okiishi, 2006; White, 2006; Potter, Wiggert, and Ramadan, 2012).

For the influence of boundary layers and turbulence on the flight of balls of various sorts, see Lighthill (1986), Adair (1990), and Armenti (1992).

For the onset of turbulence in unstable laminar flows, we particularly like Sagdeev, Usikov, and Zaslovsky (1988, Chap. 11) and Acheson (1990, Chap. 9). For the route to chaos in low-dimensional dynamical systems with explicit connections to the onset of turbulence, see Ruelle (1989); for chaos theory with little or no discussion of turbulence, Baker and Gollub (1990), Alligood, Sauer, and Yorke (1996), and Strogatz (2008).

16

Waves

An ocean traveller has even more vividly the impression
that the ocean is made of waves than that it is made of water.

ARTHUR EDDINGTON (1927)

16.1 Overview

In the preceding chapters, we have derived the basic equations of fluid dynamics and developed a variety of techniques to describe stationary flows. We have also demonstrated how, even if there exists a rigorous, stationary solution of these equations for a time-steady flow, instabilities may develop, in which the amplitude of an oscillatory disturbance grows with time. These unstable modes of an unstable flow can usually be thought of as waves that interact strongly with the flow and extract energy from it. Of course, wave modes can also be stable and can be studied as independent, individual modes.

Fluid dynamical waves come in a wide variety of forms. They can be driven by a combination of gravitational, pressure, rotational, and surface-tension stresses and also by mechanical disturbances, such as water rushing past a boat or air passing through a larynx. In this chapter, we describe a few examples of wave modes in fluids, chosen to illustrate general wave properties.

The most familiar types of wave are probably *gravity waves* on the surface of a large body of water (Sec. 16.2), such as ocean waves and waves on lakes and rivers. We consider them in the linear approximation and find that they are dispersive in general, though they become nondispersive in the long-wavelength (shallow-water) limit (i.e., when they are influenced by the water's bottom). We also examine the effects of surface tension on gravity waves, which converts them into *capillary waves*, and in this connection we develop a mathematical description of surface tension (see Box 16.4). Boundary conditions can give rise to a discrete spectrum of normal modes, which we illustrate by *helioseismology*: the study of coherent-wave modes excited in the Sun by convective overturning motions.

By contrast with the elastodynamic waves of Chap. 12, waves in fluids often develop amplitudes large enough that nonlinear effects become important (Sec. 16.3). The nonlinearities can cause the front of a wave to steepen and then break—a phenomenon we have all seen at the sea shore. It turns out that, at least under some restrictive conditions, nonlinear waves have very surprising properties. There exist *soliton* or *solitary-wave* modes, in which the front-steepening due to nonlinearity is stably held in check by dispersion, so particular wave profiles are quite robust and propagate for long intervals of time without breaking or dispersing. We demonstrate this behavior by studying flow in a shallow channel. We also explore the remarkable behaviors of such solitons when they pass through each other.

In a nearly rigidly rotating fluid, there are remarkable waves in which the restoring force is the Coriolis effect; they have the unusual property that their group and phase velocities are oppositely directed. These so-called *Rossby waves,* studied in Sec. 16.4, are important in both the oceans and the atmosphere.

The simplest fluid waves of all are small-amplitude *sound waves*—a paradigm for scalar waves. They are nondispersive, just like electromagnetic waves, and are therefore sometimes useful for human communication. We shall study sound waves in Sec. 16.5 and use them to explore (i) the radiation reaction force that acts back on a wave-emitting object (a fundamental physics issue) and (ii) matched asymptotic expansions (a mathematical physics technique). We also describe how sound waves can be produced by fluid flows. This process will be illustrated with the problem of sound generation by high-speed turbulent flows—a problem that provides a good starting point for the topic of the following chapter, compressible flows.

Other examples of fluid waves are treated elsewhere in Part V: Kelvin-Helmholtz waves at the interface between two fluids that move relative to each other (Sec. 14.6.1); Tollmien-Schlichting waves in a laminar boundary layer (Ex. 15.11); lee waves in a separated boundary layer (footnote in Sec. 15.5.2); wavy Taylor rolls in rotating Couette flow (Sec. 15.6.1); shock waves (Sec. 17.5); Sedov-Taylor blast waves (Sec. 17.6); hydraulic jumps (Ex. 17.10); internal waves, which propagate when fluid is stratified in a gravitational field (Ex. 18.8); and various types of magnetohydrodynamic waves (Chap. 19).

As in Chaps. 14 and 15, readers are urged to watch movies in parallel with reading this chapter; see Box 16.2.

16.2 Gravity Waves on and beneath the Surface of a Fluid

Gravity waves[1] are waves on and beneath the surface of a fluid, for which the restoring force is the downward pull of gravity. Familiar examples are ocean waves and the waves produced on the surface of a pond when a pebble is thrown in. Less familiar examples are "g modes" of vibration inside the Sun, discussed at the end of this section.

Consider a small-amplitude wave propagating along the surface of a flat-bottomed lake with depth h_o, as shown in Fig. 16.1. As the water's displacement is small, we can describe the wave as a linear perturbation about equilibrium. The equilibrium water is at rest (i.e., it has velocity $\mathbf{v} = 0$). The water's perturbed motion is essentially inviscid and incompressible, so $\nabla \cdot \mathbf{v} = 0$. A simple application of the equation of vorticity transport, Eq. (14.6), assures us that, since the water is static and thus irrotational before and after the wave passes, it must also be irrotational in the wave. Therefore, we can describe the wave inside the water by a velocity potential ψ whose gradient is the velocity field:

$$\mathbf{v} = \nabla \psi. \tag{16.1}$$

Incompressibility, $\nabla \cdot \mathbf{v} = 0$, applied to this equation, implies that ψ satisfies Laplace's equation:

$$\nabla^2 \psi = 0. \tag{16.2}$$

We introduce horizontal coordinates x, y and a vertical coordinate z measured upward from the lake's equilibrium surface (Fig. 16.1). For simplicity we confine

16.2

gravity waves

velocity potential

Laplace's equation for irrotational, incompressible flow

---

1. Not to be confused with gravitational waves, which are waves in the relativistic gravitational field (spacetime curvature) that propagate at the speed of light, and which we meet in Chap. 27.

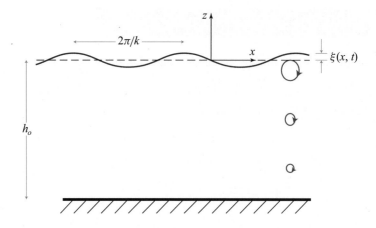

FIGURE 16.1 Gravity waves propagating horizontally across a lake with constant depth h_o.

attention to a sinusoidal wave propagating in the x direction with angular frequency ω and wave number k. Then ψ and all other perturbed quantities have the form $f(z) \exp[i(kx - \omega t)]$ for some function $f(z)$. More general disturbances can be expressed as a superposition of many of these elementary wave modes propagating in various horizontal directions (and in the limit, as a Fourier integral over modes). All the properties of such superpositions follow straightforwardly from those of our elementary plane-wave mode (see Secs. 7.2.2 and 7.3), so we continue to focus on it.

We can use Laplace's equation (16.2) to solve for the vertical variation $f(z)$ of the velocity potential. As the horizontal variation at a particular time is $\propto \exp(ikx)$, direct substitution into Eq. (16.2) gives two possible vertical variations: $\psi \propto \exp(\pm kz)$. The precise linear combination of these two forms is dictated by the boundary conditions. The boundary condition we need is that the vertical component of velocity $v_z = \partial \psi / \partial z$ vanish at the bottom of the lake ($z = -h_o$). The only combination that can vanish is a sinh function. Its integral, the velocity potential, therefore involves a cosh function:

boundary condition at bottom of lake

velocity potential for gravity waves

$$\psi = \psi_0 \cosh[k(z + h_o)] \exp[i(kx - \omega t)]. \tag{16.3}$$

An alert reader might note at this point that, for this ψ, the horizontal component of velocity $v_x = \psi_{,x} = ik\psi$ does not vanish at the lake bottom, in violation of the no-slip boundary condition. In fact, as we discussed in Sec. 14.4, a thin, viscous boundary layer along the bottom of the lake will join our potential-flow solution (16.3) to nonslip fluid at the bottom. We ignore the boundary layer under the (justifiable) assumption that for our oscillating waves, it is too thin to affect much of the flow.

Returning to the potential flow, we must also impose a boundary condition at the surface. This can be obtained from Bernoulli's law. The version of Bernoulli's law that we need is that for an irrotational, isentropic, time-varying flow:

$$\mathbf{v}^2/2 + h + \Phi + \partial \psi / \partial t = \text{const everywhere in the flow} \tag{16.4}$$

[Eqs. (13.51), (13.54)]. We apply this law at the surface of the perturbed water. Let us examine each term:

1. The term $\mathbf{v}^2/2$ is quadratic in a perturbed quantity and therefore can be dropped.

2. The enthalpy $h = u + P/\rho$ (cf. Box 13.2) is a constant, since u and ρ are constants throughout the fluid and P is constant on the surface (equal to the atmospheric pressure).[2]

3. The gravitational potential at the fluid surface is $\Phi = g\xi$, where $\xi(x, t)$ is the surface's vertical displacement from equilibrium, and we ignore an additive constant.

4. The constant on the right-hand side, which could depend on time $[C(t)]$, can be absorbed in the velocity potential term $\partial\psi/\partial t$ without changing the physical observable $\mathbf{v} = \nabla\psi$.

Bernoulli's law applied at the surface therefore simplifies to give

$$g\xi + \frac{\partial\psi}{\partial t} = 0. \tag{16.5}$$

The vertical component of the surface velocity in the linear approximation is just $v_z(z = 0, t) = \partial\xi/\partial t$. Expressing v_z in terms of the velocity potential, we then obtain

$$\frac{\partial\xi}{\partial t} = v_z = \frac{\partial\psi}{\partial z}. \tag{16.6}$$

Combining this expression with the time derivative of Eq. (16.5), we obtain an equation for the vertical gradient of ψ in terms of its time derivative:

$$g\frac{\partial\psi}{\partial z} = -\frac{\partial^2\psi}{\partial t^2}. \tag{16.7}$$

Finally, substituting Eq. (16.3) into Eq. (16.7) and setting $z = 0$ [because we derived Eq. (16.7) only at the water's surface], we obtain the dispersion relation[3] for linearized gravity waves:

$$\boxed{\omega^2 = gk \; \tanh(kh_o).} \tag{16.8}$$

How do the individual elements of fluid move in a gravity wave? We can answer this question (Ex. 16.1) by computing the vertical and horizontal components of the velocity $v_x = \psi_{,x}$, and $v_z = \psi_{,z}$, with ψ given by Eq. (16.3). We find that the fluid elements undergo forward-rotating elliptical motion, as depicted in Fig. 16.1, similar

---

2. Actually, a slight variation of the surface pressure is caused by the varying weight of the air above the surface, but as the density of air is typically $\sim 10^{-3}$ that of water, this correction is very small.

3. For a discussion of the dispersion relation, phase velocity, and group velocity for waves, see Sec. 7.2.

to that for Rayleigh waves on the surface of a solid (Sec. 12.4.2). However, in gravity waves, the sense of rotation is the same (forward) at all depths, in contrast to reversals with depth found in Rayleigh waves [cf. the discussion following Eq. (12.62)].

We now consider two limiting cases: deep water and shallow water.

16.2.1 Deep-Water Waves and Their Excitation and Damping

When the water is deep compared to the wavelength of the waves, $kh_o \gg 1$, the dispersion relation (16.8) becomes

$$\boxed{\omega = \sqrt{gk}.} \tag{16.9}$$

Thus deep-water waves are dispersive (see Sec. 7.2); their group velocity $V_g \equiv d\omega/dk = \frac{1}{2}\sqrt{g/k}$ is half their phase velocity, $V_{\mathrm{ph}} \equiv \omega/k = \sqrt{g/k}$. [Note that we could have deduced the deep-water dispersion relation (16.9), up to a dimensionless multiplicative constant, by dimensional arguments: The only frequency that can be constructed from the relevant variables g, k, and ρ is $\sqrt{gk}$.]

The quintessential example of deep-water waves is waves on an ocean or a lake before they near the shore, so the water's depth is much greater than their wavelength. Such waves, of course, are excited by wind. We have discussed this excitation in Sec. 14.6.2. There we found that the Kelvin-Helmholtz instability (where air flowing in a laminar fashion over water would raise waves) is strongly suppressed by gravity, so this mechanism does not work. Instead, the excitation is by a turbulent wind's randomly fluctuating pressures pounding on the water's surface. Once the waves have been generated, they can propagate great distances before viscous dissipation damps them; see Ex. 16.2.

16.2.2 Shallow-Water Waves

For shallow-water waves ($kh_o \ll 1$), the dispersion relation (16.8) becomes

$$\boxed{\omega = \sqrt{gh_o}\, k.} \tag{16.10}$$

Thus these waves are nondispersive; their phase and group velocities are equal: $V_{\mathrm{ph}} = V_g = \sqrt{gh_o}$.

Later, when studying solitons, we shall need two special properties of shallow-water waves. First, when the depth of the water is small compared with the wavelength, but not very small, the waves will be slightly dispersive. We can obtain a correction to Eq. (16.10) by expanding the tanh function of Eq. (16.8) as $\tanh x = x - x^3/3 + \ldots$. The dispersion relation then becomes

$$\omega = \sqrt{gh_o}\left(1 - \frac{1}{6}k^2 h_o^2\right) k. \tag{16.11}$$

Second, by computing $\mathbf{v} = \boldsymbol{\nabla}\psi$ from Eq. (16.3), we find that in the shallow-water limit the water's horizontal motions are much larger than its vertical motions and are essentially independent of depth. The reason, physically, is that the fluid acceleration is produced almost entirely by a horizontal pressure gradient (caused by spatially variable water depth) that is independent of height; see Ex. 16.1.

Often shallow-water waves have heights ξ that are comparable to the water's undisturbed depth h_o, and h_o changes substantially from one region of the flow to another. A familiar example is an ocean wave nearing a beach. In such cases, the wave equation is modified by nonlinear and height-dependent effects. In Box 16.3 we derive the equations that govern such waves, and in Ex. 16.3 and Sec. 16.3 we explore properties of these waves.

BOX 16.3. NONLINEAR SHALLOW-WATER WAVES WITH VARIABLE DEPTH

Consider a nonlinear shallow-water wave propagating on a body of water with variable depth. Let $h_o(x, y)$ be the depth of the undisturbed water at location (x, y), and let $\xi(x, y, t)$ be the height of the wave, so the depth of the water in the presence of the wave is $h = h_o + \xi$. As in the linear-wave case, the transverse fluid velocity (v_x, v_y) inside the water is nearly independent of height z, so the wave is characterized by three functions $\xi(x, y, t)$, $v_x(x, y, t)$, and $v_y(x, y, t)$. These functions are governed by the law of mass conservation and the inviscid Navier-Stokes equation (i.e., the Euler equation).

The mass per unit area is $\rho h = \rho(h_o + \xi)$, and the corresponding mass flux (mass crossing a unit length per unit time) is $\rho h \mathbf{v} = \rho(h_o + \xi)\mathbf{v}$, where $\mathbf{v}$ is the 2-dimensional, horizontal vectorial velocity $\mathbf{v} = v_x \mathbf{e}_x + v_y \mathbf{v}_y$. Mass conservation, then, requires that $\partial[\rho(h_o + \xi)]/\partial t + {}^{(2)}\boldsymbol{\nabla} \cdot [\rho(h_o + \xi)\mathbf{v}] = 0$, where ${}^{(2)}\boldsymbol{\nabla}$ is the 2-dimensional gradient operator that acts solely in the horizontal $(x\text{-}y)$ plane. Since ρ is constant, and h_o is time independent, this expression becomes

$$\partial \xi/\partial t + {}^{(2)}\boldsymbol{\nabla} \cdot [(h_o + \xi)\mathbf{v}] = 0. \qquad \text{(1a)}$$

The Euler equation for $\mathbf{v}$ at an arbitrary height z in the water is $\partial \mathbf{v}/\partial t + (\mathbf{v} \cdot {}^{(2)}\boldsymbol{\nabla})\mathbf{v} = -(1/\rho){}^{(2)}\boldsymbol{\nabla}P$, and hydrostatic equilibrium gives the pressure as the weight per unit area of the overlying water: $P = g(\xi - z)\rho$ (where height z is measured from the water's undisturbed surface). Combining these equations, we obtain

$$\partial \mathbf{v}/\partial t + (\mathbf{v} \cdot {}^{(2)}\boldsymbol{\nabla})\mathbf{v} + g\,{}^{(2)}\boldsymbol{\nabla}\xi = 0. \qquad \text{(1b)}$$

Equations (1) are used, for example, in theoretical analyses of tsunamis (Ex. 16.3).

Exercise 16.1 *Example: Fluid Motions in Gravity Waves*

(a) Show that in a gravity wave in water of arbitrary depth (deep, shallow, or in between), each fluid element undergoes forward-rolling elliptical motion as shown in Fig. 16.1. (Assume that the amplitude of the water's displacement is small compared to a wavelength.)

(b) Calculate the longitudinal diameter of the motion's ellipse, and the ratio of vertical to longitudinal diameters, as functions of depth.

(c) Show that for a deep-water wave, $kh_o \gg 1$, the ellipses are all circles with diameters that die out exponentially with depth.

(d) We normally think of a circular motion of fluid as entailing vorticity, but a gravity wave in water has vanishing vorticity. How can this vanishing vorticity be compatible with the circular motion of fluid elements?

(e) Show that for a shallow-water wave, $kh_o \ll 1$, the motion is (nearly) horizontal and is independent of height z.

(f) Compute the fluid's pressure perturbation $\delta P(x, z, t)$ inside the fluid for arbitrary depth. Show that, for a shallow-water wave, the pressure is determined by the need to balance the weight of the overlying fluid, but for greater depth, vertical fluid accelerations alter this condition of weight balance.

Exercise 16.2 *Problem: Viscous Damping of Deep-Water Waves*

(a) Show that viscosity damps a monochromatic deep-water wave with an amplitude e-folding time $\tau_* = (2\nu k^2)^{-1}$, where k is the wave number, and ν is the kinematic viscosity. [Hint: Compute the energy E in the wave and the rate of loss of energy to viscous heating $\dot{E}$, and argue that $\tau_* = -2E/\dot{E}$. Recall the discussions of energy in Sec. 13.5.5 and viscous heating in Sec. 13.7.4.]

(b) As an example, consider the ocean waves that one sees at an ocean beach when the surf is "up" (large-amplitude waves). These are usually generated by turbulent winds in storms in the distant ocean, 1,000 km or farther from shore. The shortest wavelengths present should be those for which the damping length $C\tau_*$ is about 1,000 km; shorter wavelengths than that will have been dissipated before reaching shore. Using the turbulent viscosity $\nu \sim 0.03 \text{ m}^2 \text{ s}^{-1}$ that we deduced from the observed thicknesses of wind-driven Ekman layers in the ocean (Ex. 14.21), compute this shortest wavelength for the large-amplitude waves, and compare with the wavelengths you have observed at an ocean beach. You should get pretty good agreement, thanks to a weak sensitivity of the wavelength to the rather uncertain turbulent viscosity.

(c) Make similar comparisons of theory and observation for (i) the choppy, short-wavelength ocean waves that one sees when (and only when) a local wind is blowing, and (ii) waves on a small lake.

Exercise 16.3 *Example: Shallow-Water Waves with Variable Depth; Tsunamis*[4]
Consider small-amplitude (linear) shallow-water waves in which the height of the
bottom boundary varies, so the unperturbed water's depth is variable: $h_o = h_o(x, y)$.

(a) Using the theory of nonlinear shallow-water waves with variable depth (Box 16.3),
show that the wave equation for the perturbation $\xi(x, y, t)$ of the water's height
takes the form

$$\frac{\partial^2 \xi}{\partial t^2} - {}^{(2)}\nabla \cdot (gh_o {}^{(2)}\nabla \xi) = 0. \tag{16.12}$$

Here $^{(2)}\nabla$ is the 2-dimensional gradient operator that acts in the horizontal (i.e.,
x-y) plane. Note that gh_o is the square of the wave's propagation speed C^2 (phase
speed and group speed), so this equation takes the same form as Eq. (7.17) from
the geometric-optics approximation in Sec. 7.3.1, with $W = 1$.

(b) Describe what happens to the direction of propagation of a wave as the depth h_o
of the water varies (either as a set of discrete jumps in h_o or as a slowly varying
h_o). As a specific example, how must the propagation direction change as waves
approach a beach (but when they are sufficiently far out from the beach that
nonlinearities have not yet caused them to begin to break)? Compare with your
own observations at a beach.

(c) Tsunamis are gravity waves with enormous wavelengths ($\sim$100 km or so) that
propagate on the deep ocean. Since the ocean depth is typically $h_o \sim 4$ km,
tsunamis are governed by the shallow-water wave equation (16.12). What would
you have to do to the ocean floor to create a lens that would focus a tsunami, generated by an earthquake near Japan, so that it destroys Los Angeles? (For simulations
of tsunami propagation, see, e.g., http://bullard.esc.cam.ac.uk/~taylor/Tsunami
.html.)

(d) The height of a tsunami, when it is in the ocean with depth $h_o \sim 4$ km, is only
$\sim$1 m or less. Use the geometric-optics approximation (Sec. 7.3) to show that
the tsunami's wavelength decreases as $\lambda \propto \sqrt{h_o}$ and its amplitude increases as
$\max(\xi) \propto 1/h_o^{1/4}$ as the tsunami nears land and the water's depth h_o decreases.

How high [$\max(\xi)$] does the tsunami get when nonlinearities become strongly
important? (Assume a height of 1 m in the deep ocean.) How does this compare
with the heights of historically disastrous tsunamis when they hit land? From
your answer you should conclude that the nonlinearities must play a major role
in raising the height. Equations (16.11) in Box 16.3 are used by geophysicists to
analyze this nonlinear growth of the tsunami height. If the wave breaks, then
these equations fail, and ideas developed (in rudimentary form) in Ex. 17.10
must be used.

---

4. Exercise courtesy of David Stevenson.

16.2.3 Capillary Waves and Surface Tension

surface tension

When the wavelength is short (so k is large), we must include the effects of *surface tension* on the surface boundary condition. Surface tension can be treated as an isotropic force per unit length, γ, that lies in the surface and is unaffected by changes in the shape or size of the surface; see Box 16.4. In the case of a gravity wave traveling in the x direction, this tension produces on the fluid's surface a net downward force per unit area $-\gamma d^2\xi/dx^2 = \gamma k^2\xi$, where k is the horizontal wave number. [This downward force is like that on a violin string; cf. Eq. (12.29) and associated discussion.] This additional force must be included as an augmentation of $\rho g\xi$. Correspondingly, the effect of surface tension on a mode with wave number k is simply to change the true acceleration of gravity to an effective acceleration of gravity:

$$g \to g + \frac{\gamma k^2}{\rho}. \tag{16.13}$$

The remainder of the derivation of the dispersion relation for deep-water gravity waves carries over unchanged, and the dispersion relation becomes

---

BOX 16.4.   SURFACE TENSION

In a water molecule, the two hydrogen atoms stick out from the larger oxygen atom somewhat like Mickey Mouse's ears, with an H-O-H angle of 105°. This asymmetry of the molecule gives rise to a large electric dipole moment. In the interior of a body of water the dipole moments are oriented rather randomly, but near the water's surface they tend to be parallel to the surface and bond with one another to create surface tension—a macroscopically isotropic, 2-dimensional tension force (force per unit length) γ that is confined to the water's surface.

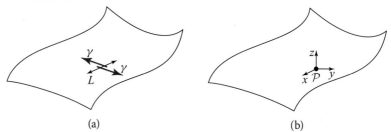

(a)   (b)

More specifically, consider a line L of unit length in the water's surface [drawing (a)]. The surface water on one side of L exerts a tension (pulling) force on the surface water on the other side. The magnitude of this force is γ, and it is orthogonal to the line L regardless of L's orientation. This behavior is analogous to an isotropic pressure P in 3 dimensions, which acts orthogonally across any unit area.

(continued)

BOX 16.4. (continued)

Choose a point $\mathcal{P}$ in the water's surface, and introduce local Cartesian coordinates there with x and y lying in the surface and z orthogonal to it [drawing (b)]. In this coordinate system, the 2-dimensional stress tensor associated with surface tension has components $^{(2)}T_{xx} = {}^{(2)}T_{yy} = -\gamma$, analogous to the 3-dimensional stress tensor for an isotropic pressure: $T_{xx} = T_{yy} = T_{zz} = P$. We can also use a 3-dimensional stress tensor to describe the surface tension: $T_{xx} = T_{yy} = -\gamma\delta(z)$; all other $T_{jk} = 0$. If we integrate this 3-dimensional stress tensor through the water's surface, we obtain the 2-dimensional stress tensor: $\int T_{jk}dz = {}^{(2)}T_{jk}$ (i.e., $\int T_{xx}dz = \int T_{yy}dz = -\gamma$). The 2-dimensional metric of the surface is $^{(2)}\mathbf{g} = \mathbf{g} - \mathbf{e}_z \otimes \mathbf{e}_z$; in terms of this 2-dimensional metric, the surface tension's 3-dimensional stress tensor is $\mathbf{T} = -\gamma\delta(z)^{(2)}\mathbf{g}$.

Water is not the only fluid that exhibits surface tension; all fluids do so, at the interfaces between themselves and other substances. For a thin film (e.g., a soap bubble), there are two interfaces (the top and bottom faces of the film), so if we ignore the film's thickness, its stress tensor is twice as large as for a single surface, $\mathbf{T} = -2\gamma\delta(z)^{(2)}\mathbf{g}$.

The hotter the fluid, the more randomly its surface molecules will be oriented (and hence the smaller the fluid's surface tension γ will be). For water, γ varies from 75.6 dyne/cm at $T = 0\,°C$, to 72.0 dyne/cm at $T = 25\,°C$, to 58.9 dyne/cm at $T = 100\,°C$.

In Exs. 16.4–16.6, we explore some applications of surface tension. In Sec. 16.2.3 and Exs. 16.7 and 16.8, we consider the influence of surface tension on water waves. In Ex. 5.14, we study the statistical thermodynamics of surface tension and its role in the nucleation of water droplets in clouds and fog.

$$\omega^2 = gk + \frac{\gamma k^3}{\rho} \tag{16.14}$$

dispersion relation for gravity waves in deep water with surface tension

[cf. Eqs. (16.9) and (16.13)]. When the second term dominates, the waves are sometimes called *capillary waves*. In Exs. 16.7 and 16.8 we explore some aspects of capillary waves. In Exs. 16.4–16.6 we explore some other aspects of surface tension.

capillary waves

EXERCISES

Exercise 16.4 *Problem: Maximum Size of a Water Droplet*
What is the maximum size of water droplets that can form by water very slowly dripping out of a syringe? Out of a water faucet (whose opening is far larger than that of a syringe)?

Exercise 16.5 *Problem: Force Balance for an Interface between Two Fluids*
Consider a point $\mathcal{P}$ in the curved interface between two fluids. Introduce Cartesian coordinates at $\mathcal{P}$ with x and y parallel to the interface and z orthogonal [as in diagram (b) in Box 16.4], and orient the x- and y-axes along the directions of the interface's principal curvatures, so the local equation for the interface is

$$z = \frac{x^2}{2R_1} + \frac{y^2}{2R_2}.$$ (16.15)

Here R_1 and R_2 are the surface's principal radii of curvature at $\mathcal{P}$; note that each of them can be positive or negative, depending on whether the surface bends up or down along their directions. Show that, in equilibrium, stress balance, $\nabla \cdot \mathbf{T} = 0$, for the surface implies that the pressure difference across the surface is

$$\boxed{\Delta P = \gamma \left(\frac{1}{R_1} + \frac{1}{R_2} \right),}$$ (16.16)

where γ is the surface tension.

Exercise 16.6 *Challenge: Minimum Area of a Soap Film*
For a soap film that is attached to a bent wire (e.g., to the circular wire that a child uses to blow a bubble), the air pressure on the film's two sides is the same. Therefore, Eq. (16.16) (with γ replaced by 2γ, since the film has two faces) tells us that at every point in the film, its two principal radii of curvature must be equal and opposite: $R_1 = -R_2$. It is an interesting exercise in differential geometry to show that this requirement means that the soap film's surface area is an extremum with respect to variations of the film's shape, holding its boundary on the wire fixed. If you know enough differential geometry, prove this extremal-area property of soap films, and then show that for the film's shape to be stable, its extremal area must actually be a minimum.

Exercise 16.7 *Problem: Capillary Waves*
Consider deep-water gravity waves of short enough wavelength that surface tension must be included, so the dispersion relation is Eq. (16.14). Show that there is a minimum value of the group velocity, and find its value together with the wavelength of the associated wave. Evaluate these for water ($\gamma \sim 0.07$ N m$^{-1} = 70$ dyne/cm). Try performing a crude experiment to verify this phenomenon.

Exercise 16.8 *Example: Boat Waves*
A toy boat moves with uniform velocity $\mathbf{u}$ across a deep pond (Fig. 16.2). Consider the wave pattern (time-independent in the boat's frame) produced on the water's surface at distances large compared to the boat's size. Both gravity waves and surface-tension (*capillary*) waves are excited. Show that capillary waves are found both ahead of and behind the boat, whereas gravity waves occur solely inside a trailing wedge. More specifically, do the following.

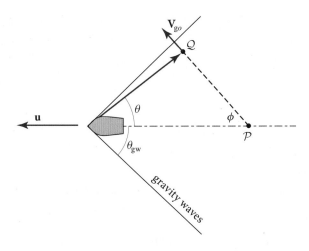

FIGURE 16.2 Capillary and gravity waves excited by a small boat (Ex. 16.8).

(a) In the rest frame of the water, the waves' dispersion relation is Eq. (16.14). Change notation so that ω is the waves' angular velocity as seen in the boat's frame, and ω_o in the water's frame, so the dispersion relation becomes $\omega_o^2 = gk + (\gamma/\rho)k^3$. Use the Doppler shift (i.e., the transformation between frames) to derive the boat-frame dispersion relation $\omega(k)$.

(b) The boat radiates a spectrum of waves in all directions. However, only those with vanishing frequency in the boat's frame, $\omega = 0$, contribute to the time-independent (stationary) pattern. As seen in the water's frame and analyzed in the geometric-optics approximation of Chap. 7, these waves are generated by the boat (at points along its horizontal dash-dot trajectory in Fig. 16.2) and travel outward with the group velocity $\mathbf{V}_{go}$. Regard Fig. 16.2 as a snapshot of the boat and water at a particular moment of time. Consider a wave that was generated at an earlier time, when the boat was at location $\mathcal{P}$, and that traveled outward from there with speed V_{go} at an angle ϕ to the boat's direction of motion. (You may restrict yourself to $0 \leq \phi \leq \pi/2$.) Identify the point $\mathcal{Q}$ that this wave has reached, at the time of the snapshot, by the angle θ shown in the figure. Show that θ is given by

$$\tan\theta = \frac{V_{go}(k)\sin\phi}{u - V_{go}(k)\cos\phi}, \tag{16.17a}$$

where k is determined by the dispersion relation $\omega_0(k)$ together with the vanishing ω condition:

$$\omega_0(k, \phi) = uk\cos\phi. \tag{16.17b}$$

(c) Specialize to capillary waves [$k \gg \sqrt{g\rho/\gamma}$]. Show that

$$\tan\theta = \frac{3\tan\phi}{2\tan^2\phi - 1}. \tag{16.18}$$

Demonstrate that the capillary-wave pattern is present for all values of θ (including in front of the boat, $\pi/2 < \theta < \pi$, and behind it, $0 \le \theta \le \pi/2$).

(d) Next, specialize to gravity waves, and show that

$$\tan \theta = \frac{\tan \phi}{2 \tan^2 \phi + 1}. \tag{16.19}$$

Demonstrate that the gravity-wave pattern is confined to a trailing wedge with angles $\theta < \theta_{\text{gw}} = \sin^{-1}(1/3) = 19.47°$ (cf. Fig. 16.2). You might try to reproduce these results experimentally.

16.2.4 Helioseismology

The Sun provides an excellent example of the excitation of small-amplitude waves in a fluid body. In the 1960s, Robert Leighton and colleagues at Caltech discovered that the surface of the Sun oscillates vertically with a period of roughly 5 min and a speed of ~ 1 km s^{-1}. This motion was thought to be an incoherent surface phenomenon until it was shown that the observed variation was, in fact, the superposition of thousands of highly coherent wave modes excited in the Sun's interior—normal modes of the Sun. Present-day techniques allow surface velocity amplitudes as small as 2 mm s^{-1} to be measured, and phase coherence for intervals as long as a year has been observed. Studying the frequency spectrum and its variation provides a unique probe of the Sun's interior structure, just as the measurement of conventional seismic waves (Sec. 12.4) probes Earth's interior.

The description of the Sun's normal modes requires some modification of our treatment of gravity waves. We eschew the details and just outline the principles—which are rather similar to those for normal modes of a homogeneous elastic sphere (Sec. 12.4.4 and Ex. 12.12). First, the Sun is (very nearly) spherical. We therefore work in spherical polar coordinates rather than Cartesian coordinates. Second, the Sun is made of hot gas, and it is no longer a good approximation to assume that the fluid is incompressible. We must therefore replace the equation $\nabla \cdot \mathbf{v} = 0$ with the full equation of continuity (mass conservation) together with the equation of energy conservation, which governs the relationship between the perturbations of density and pressure. Third, the Sun is not uniform. The pressure and density in the unperturbed gas vary with radius in a known manner and must be included. Fourth, the Sun has a finite surface area. Instead of assuming a continuous spectrum of waves, we must now anticipate that the boundary conditions will lead to a discrete spectrum of normal modes. Allowing for these complications, it is possible to derive a differential equation for the perturbations to replace Eq. (16.7). It turns out that a convenient dependent variable (replacing the velocity potential ψ) is the pressure perturbation. The boundary conditions are that the displacement vanish at the center of the Sun and the pressure perturbation vanish at the surface.

At this point the problem is reminiscent of the famous solution for the eigenfunctions of the Schrödinger equation for a hydrogen atom in terms of associated Laguerre

properties of the Sun that influence its helioseismic modes

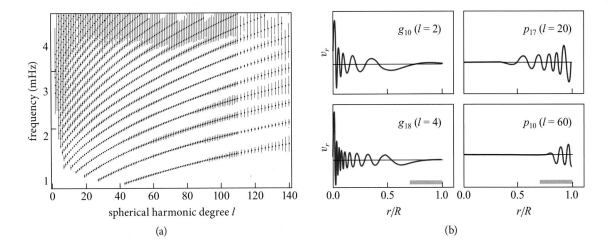

FIGURE 16.3 (a) Measured frequency spectrum for solar p modes with different values of the quantum numbers n and l. The error bars are magnified by a factor of 1,000. The lowest ridge is $n = 0$, the next is $n = 1, \ldots$. (b) Sample eigenfunctions for g and p modes labeled by n (subscripts) and l (in parentheses). The ordinate is the radial velocity, and the abscissa is fractional radial distance from the Sun's center to its surface. The solar convection zone is the shaded region at the bottom. Adapted from Libbrecht and Woodard (1991).

polynomials. The wave frequencies of the Sun's normal modes are given by the eigenvalues of the differential equation. The corresponding eigenfunctions can be classified using three quantum numbers, n, l, m, where n counts the number of radial nodes in the eigenfunction, and the angular variation of the pressure perturbation is proportional to the spherical harmonic $Y_l^m(\theta, \phi)$. If the Sun were precisely spherical, the modes with the same n and l but different m would be degenerate, just as is the case for an atom when there is no preferred direction in space. However, the Sun rotates with a latitude-dependent period in the range $\sim$25–30 days, which breaks the degeneracy, just as an applied magnetic field in an atom breaks the degeneracy of the atom's states (the Zeeman effect). From the observed splitting of the solar-mode spectrum, it is possible to learn about the distribution of rotational angular momentum inside the Sun.

When this problem is solved in detail, it turns out that there are two general classes of modes. One class is similar to gravity waves, in the sense that the forces that drive the gas's motions are produced primarily by gravity (either directly, or indirectly via the weight of overlying material producing pressure that pushes on the gas). These are called *g modes*. In the second class (known as *p modes*),[5] the pressure forces arise mainly from the compression of the fluid just like in sound waves (which we study in Sec. 16.5). It turns out that the *g* modes have large amplitudes in the middle of the Sun, whereas the *p* modes are dominant in the outer layers (Fig. 16.3b). The reasons for this are relatively easy to understand and introduce ideas to which we shall return.

g modes of the Sun or other gravitating fluid spheres

p modes of the Sun or other gravitating fluid spheres

---

5. There are also formally distinguishable f-modes, which for our purposes are just a subset of the p-modes.

The Sun is a hot body, much hotter at its center ($T \sim 1.5 \times 10^7$ K) than on its surface ($T \sim 6{,}000$ K). The sound speed C is therefore much greater in its interior, and so p modes of a given frequency ω can carry their energy flux $\sim \rho \xi^2 \omega^2 C$ (Sec. 16.5) with much smaller amplitudes ξ than near the surface. Therefore the p-mode amplitudes are much smaller in the center of the Sun than near its surface.

The g modes are controlled by different physics and thus behave differently. The outer $\sim 30\%$ (by radius) of the Sun is *convective* (Chap. 18), because the diffusion of heat is inadequate to carry the huge amount of nuclear power being generated in the solar core. The convection produces an equilibrium variation of pressure and density with radius that are just such as to keep the Sun almost neutrally stable, so that regions that are slightly hotter (cooler) than their surroundings will rise (sink) in the solar gravitational field. Therefore there cannot be much of a mechanical restoring force that would cause these regions to oscillate about their average positions, and so the g modes (which are influenced almost solely by gravity) have little restoring force and thus are evanescent in the convection zone; hence their amplitudes decay quickly with increasing radius there.

We should therefore expect only p modes to be seen in the surface motions, which is indeed the case. Furthermore, we should not expect the properties of these modes to be very sensitive to the physical conditions in the core. A more detailed analysis bears this out.

16.3 Nonlinear Shallow-Water Waves and Solitons

In recent decades, *solitons* or solitary waves have been studied intensively in many different areas of physics. However, fluid dynamicists became familiar with them in the nineteenth century. In an oft-quoted passage, John Scott-Russell (1844) described how he was riding along a narrow canal and watched a boat stop abruptly. This deceleration launched a single smooth pulse of water which he followed on horseback for 1 or 2 miles, observing it "rolling on a rate of some eight or nine miles an hour, preserving its original figure some thirty feet long and a foot to a foot and a half in height." This was a soliton—a 1-dimensional, nonlinear wave with fixed profile traveling with constant speed. Solitons can be observed fairly readily when gravity waves are produced in shallow, narrow channels. We use the particular example of a shallow, nonlinear gravity wave to illustrate solitons in general.

16.3.1 Korteweg–de Vries (KdV) Equation

The key to a soliton's behavior is a robust balance between the effects of dispersion and those of nonlinearity. When one grafts these two effects onto the wave equation for shallow-water waves, then to leading order in the strengths of the dispersion and nonlinearity one gets the *Korteweg–de Vries* (KdV) equation for solitons. Since a completely rigorous derivation of the KdV equation is quite lengthy, we content ourselves with a somewhat heuristic derivation that is based on this grafting process and is designed to emphasize the equation's physical content.

We choose as the dependent variable in our wave equation the height ξ of the water's surface above its quiescent position, and we confine ourselves to a plane wave that propagates in the horizontal x direction, so $\xi = \xi(x, t)$.

In the limit of very weak waves, $\xi(x, t)$ is governed by the shallow-water dispersion relation, $\omega = \sqrt{gh_o}\, k$, where h_o is the depth of the quiescent water. This dispersion relation implies that $\xi(x, t)$ must satisfy the following elementary wave equation [cf. Eq. (16.12)]:

$$0 = \frac{\partial^2 \xi}{\partial t^2} - gh_o \frac{\partial^2 \xi}{\partial x^2} = \left(\frac{\partial}{\partial t} - \sqrt{gh_o} \frac{\partial}{\partial x} \right) \left(\frac{\partial}{\partial t} + \sqrt{gh_o} \frac{\partial}{\partial x} \right) \xi. \qquad (16.20)$$

In the second expression, we have factored the wave operator into two pieces, one that governs waves propagating rightward, and the other for those moving leftward. To simplify our derivation and the final wave equation, we confine ourselves to rightward-propagating waves; correspondingly, we can simply remove the left-propagation operator, obtaining

$$\frac{\partial \xi}{\partial t} + \sqrt{gh_o} \frac{\partial \xi}{\partial x} = 0. \qquad (16.21)$$

(Leftward-propagating waves are described by this same equation with a change of sign on one of the terms.)

We now graft the effects of dispersion onto this rightward-wave equation. The dispersion relation, including the effects of dispersion to leading order, is $\omega = \sqrt{gh_o}\, k(1 - \frac{1}{6}k^2 h_o^2)$ [Eq. (16.11)]. Now, this dispersion relation ought to be derivable by assuming a variation $\xi \propto \exp[i(kx - \omega t)]$ and substituting into a generalization of Eq. (16.21) with corrections that take account of the finite depth of the channel. We take a short cut and reverse this process to obtain the generalization of Eq. (16.21) from the dispersion relation. The result is

$$\frac{\partial \xi}{\partial t} + \sqrt{gh_o} \frac{\partial \xi}{\partial x} = -\frac{1}{6}\sqrt{gh_o}\, h_o^2 \frac{\partial^3 \xi}{\partial x^3}, \qquad (16.22)$$

as a simple calculation confirms. This is the linearized KdV equation. It incorporates weak dispersion associated with the finite depth of the channel but is still a linear equation, only useful for small-amplitude waves.

Now let us set aside the dispersive correction and tackle the nonlinearity using the equations derived in Box 16.3. Denoting the depth of the disturbed water by $h = h_o + \xi$, the nonlinear law of mass conservation [Eq. (1a) of Box 16.3] becomes

$$\frac{\partial h}{\partial t} + \frac{\partial(hv)}{\partial x} = 0, \qquad (16.23a)$$

and the Euler equation [Eq. (1b) of Box 16.3] becomes

$$\frac{\partial v}{\partial t} + v\frac{\partial v}{\partial x} + g\frac{\partial h}{\partial x} = 0. \qquad (16.23b)$$

Here we have specialized the equations in Box 16.3 to a 1-dimensional wave in the channel and to a constant depth h_o of the channel's undisturbed water. Equations (16.23a) and (16.23b) can be combined to obtain

$$\frac{\partial \left(v - 2\sqrt{gh}\right)}{\partial t} + \left(v - \sqrt{gh}\right)\frac{\partial \left(v - 2\sqrt{gh}\right)}{\partial x} = 0. \tag{16.23c}$$

This equation shows that the quantity $v - 2\sqrt{gh}$ is constant along characteristics that propagate with speed $v - \sqrt{gh}$. (This constant quantity is a special case of a *Riemann invariant*, a concept that we study in Sec. 17.4.1.) When (as we require below) the nonlinearities are modest, so h does not differ greatly from h_o and the water speed v is small, these characteristics propagate leftward, which implies that for rightward-propagating waves they begin at early times in undisturbed fluid, where $v = 0$ and $h = h_o$. Therefore, the constant value of $v - 2\sqrt{gh}$ is $-2\sqrt{gh_o}$, and correspondingly in regions of disturbed fluid we have

$$v = 2\left(\sqrt{gh} - \sqrt{gh_o}\right). \tag{16.24}$$

Substituting this into Eq. (16.23a), we obtain

$$\frac{\partial h}{\partial t} + \left(3\sqrt{gh} - 2\sqrt{gh_o}\right)\frac{\partial h}{\partial x} = 0. \tag{16.25}$$

We next substitute $\xi = h - h_o$ and expand to second order in ξ to obtain the final form of our wave equation with nonlinearities but no dispersion:

$$\frac{\partial \xi}{\partial t} + \sqrt{gh_o}\frac{\partial \xi}{\partial x} = -\frac{3\xi}{2}\sqrt{\frac{g}{h_o}}\frac{\partial \xi}{\partial x}, \tag{16.26}$$

where the term on the right-hand side is the nonlinear correction.

We now have separate dispersive corrections (16.22) and nonlinear corrections (16.26) to the rightward-wave equation (16.21). Combining the two corrections into a single equation, we obtain

$$\frac{\partial \xi}{\partial t} + \sqrt{gh_o}\left[\left(1 + \frac{3\xi}{2h_o}\right)\frac{\partial \xi}{\partial x} + \frac{h_o^2}{6}\frac{\partial^3 \xi}{\partial x^3}\right] = 0. \tag{16.27}$$

Finally, we substitute

rightward moving spatial coordinate

$$\boxed{\chi \equiv x - \sqrt{gh_o}\, t} \tag{16.28}$$

to transform to a frame moving rightward with the speed of small-amplitude gravity waves. The result is the full *Korteweg–de Vries* or KdV equation:

Korteweg–de Vries (KdV) equation

$$\boxed{\frac{\partial \xi}{\partial t} + \frac{3}{2}\sqrt{\frac{g}{h_o}}\left(\xi\frac{\partial \xi}{\partial \chi} + \frac{1}{9}h_o^3\frac{\partial^3 \xi}{\partial \chi^3}\right) = 0.} \tag{16.29}$$

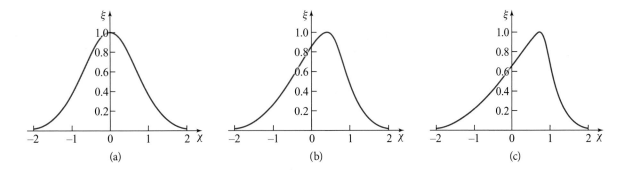

FIGURE 16.4 Steepening of a Gaussian wave profile by the nonlinear term in the KdV equation. The increase of wave speed with amplitude causes the leading part of the profile to steepen with time (going left to right) and the trailing part to flatten. In the full KdV equation (16.29), this effect can be balanced by the effect of dispersion, which causes the high-frequency Fourier components in the wave to travel slightly slower than the low-frequency ones. This allows stable solitons to form.

16.3.2 Physical Effects in the KdV Equation

Before exploring solutions to the KdV equation (16.29), let us consider the physical effects of its nonlinear and dispersive terms. The second (nonlinear) term $\frac{3}{2}\sqrt{g/h_o}\,\xi\partial\xi/\partial\chi$ derives from the nonlinearity in the $(\mathbf{v}\cdot\nabla)\mathbf{v}$ term of the Euler equation. The effect of this nonlinearity is to steepen the leading edge of a wave profile and flatten the trailing edge (Fig. 16.4). Another way to understand the effect of this term is to regard it as a nonlinear coupling of linear waves. Since it is nonlinear in the wave amplitude, it can couple waves with different wave numbers k. For example, if we have a purely sinusoidal wave $\propto \exp(ikx)$, then this nonlinearity leads to the growth of a first harmonic $\propto \exp(2ikx)$. Similarly, when two linear waves with spatial frequencies k and k' are superposed, this term describes the production of new waves at the sum and difference of spatial frequencies. We have already met such wave-wave coupling in our study of nonlinear optics (Chap. 10), and in the route to turbulence for rotating Couette flow (Fig. 15.16). We meet it again in nonlinear plasma physics (Chap. 23).

The third term in Eq. (16.29), $\frac{1}{6}\sqrt{g/h_o}\,h_o^3\,\partial^3\xi/\partial\chi^3$, is linear and is responsible for a weak dispersion of the wave. The higher-frequency Fourier components travel with slower phase velocities than do the lower-frequency components. This has two effects. One is an overall spreading of a wave in a manner qualitatively familiar from elementary quantum mechanics (cf. Ex. 7.2). For example, in a Gaussian wave packet with width Δx, the range of wave numbers k contributing significantly to the profile is $\Delta k \sim 1/\Delta x$. The spread in the group velocity is then $\Delta v_g \sim \Delta k\,\partial^2\omega/\partial k^2 \sim (gh_o)^{1/2}h_o^2\,k\Delta k$ [cf. Eq. (16.11)]. The wave packet will then double in size in a time

$$t_{\text{spread}} \sim \frac{\Delta x}{\Delta v_g} \sim \left(\frac{\Delta x}{h_o}\right)^2 \frac{1}{k\sqrt{gh_o}}. \tag{16.30}$$

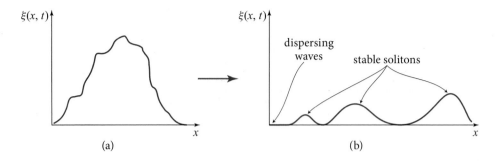

FIGURE 16.5 Production of stable solitons out of an irregular initial wave profile.

The second effect is that since the high-frequency components travel somewhat slower than the low-frequency ones, the profile tends to become asymmetric, with the leading edge less steep than the trailing edge.

Given the opposite effects of these two corrections (nonlinearity makes the wave's leading edge steeper; dispersion reduces its steepness), it should not be too surprising in hindsight that it is possible to find solutions to the KdV equation in which nonlinearity balances dispersion, so there is no change of shape as the wave propagates and no spreading. What is quite surprising, though, is that these solutions, called *solitons,* are very robust and arise naturally out of random initial data. That is to say, if we solve an initial value problem numerically starting with several peaks of random shape and size, then although much of the wave will spread and disappear due to dispersion, we will typically be left with several smooth soliton solutions, as in Fig. 16.5.

solitons in which nonlinearity balances dispersion

16.3.3 Single-Soliton Solution

16.3.3

We can discard some unnecessary algebraic luggage in the KdV equation (16.29) by transforming both independent variables using the substitutions

$$\zeta = \frac{\xi}{h_o}, \quad \eta = \frac{3\chi}{h_o} = \frac{3(x - \sqrt{gh_o}\, t)}{h_o}, \quad \tau = \frac{9}{2}\sqrt{\frac{g}{h_o}}\, t. \tag{16.31}$$

The KdV equation then becomes

simplified form of KdV equation

$$\frac{\partial \zeta}{\partial \tau} + \zeta \frac{\partial \zeta}{\partial \eta} + \frac{\partial^3 \zeta}{\partial \eta^3} = 0. \tag{16.32}$$

There are well-understood mathematical techniques (see, e.g., Whitham, 1974) for solving equations like the KdV equation. However, here we just quote solutions and explore their properties. The simplest solution to the dimensionless KdV equation (16.32) is

single-soliton solution of KdV equation

$$\zeta = \zeta_0 \ \text{sech}^2 \left[\left(\frac{\zeta_0}{12} \right)^{1/2} \left(\eta - \frac{1}{3}\zeta_0 \tau \right) \right]. \tag{16.33}$$

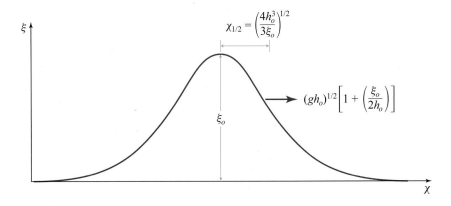

FIGURE 16.6 Profile of the single-soliton solution [Eqs. (16.33) and (16.31)] of the KdV equation. The half-width $\chi_{1/2}$ is inversely proportional to the square root of the peak height ξ_o.

This solution, depicted in Fig. 16.6, describes a one-parameter family of stable solitons. For each such soliton (each ζ_0), the soliton maintains its shape while propagating at speed $d\eta/d\tau = \zeta_0/3$ relative to a weak wave. By transforming to the rest frame of the unperturbed water using Eqs. (16.28) and (16.31), we find for the soliton's speed there:

$$\frac{dx}{dt} = \sqrt{gh_o}\left(1 + \frac{\xi_o}{2h_o}\right).$$

(16.34) **speed of soliton**

The first term is the propagation speed of a weak (linear) wave. The second term is the nonlinear correction, proportional to the wave amplitude $\xi_o = h_o\zeta_o$. A "half-width" of the wave may be defined by setting the argument of the hyperbolic secant to unity. It is $\eta_{1/2} = (12/\zeta_o)^{1/2}$, corresponding to

$$x_{1/2} = \chi_{1/2} = \left(\frac{4h_o^3}{3\xi_o}\right)^{1/2}.$$

(16.35) **half-width of soliton**

The larger the wave amplitude, the narrower its width will be, and the faster it will propagate (cf. Fig. 16.6).

Let us return to Scott-Russell's soliton (start of Sec. 16.3). Converting to SI units, the observed speed was about 4 m s^{-1}, giving an estimate of the depth of the canal of $h_o \sim 1.6$ m. Using the observed half-width $x_{1/2} \sim 5$ m, we obtain a peak height $\xi_o \sim 0.22$ m, somewhat smaller than quoted but within the errors allowing for the uncertainty in the definition of the width.

16.3.4 Two-Soliton Solution

One of the most fascinating properties of solitons is the way that two or more waves interact. The expectation, derived from physics experience with weakly coupled normal modes, might be that, if we have two well-separated solitons propagating in the

16.3.4

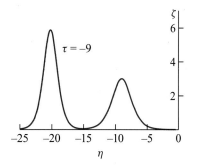 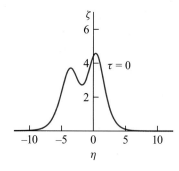 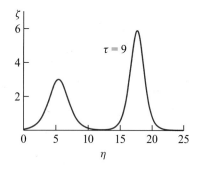

FIGURE 16.7 Two-soliton solution to the dimensionless KdV equation (16.32). This solution describes two waves well separated at $\tau \to -\infty$ that coalesce and then separate, producing the original two waves in reverse order at $\tau \to +\infty$. The notation is that of Eq. (16.36); the values of the parameters in that equation are $\eta_1 = \eta_2 = 0$ (so the solitons will merge at position $\eta = 0$), $\alpha_1 = 1$, and $\alpha_2 = 1.4$.

same direction with the larger wave chasing the smaller one, then the larger will eventually catch up with the smaller, and nonlinear interactions between the two waves will essentially destroy both, leaving behind a single, irregular pulse that will spread and decay after the interaction. However, this is not what happens. Instead, the two waves pass through each other unscathed and unchanged, except that they emerge from the interaction a bit sooner than they would have had they moved with their original speeds during the interaction. See Fig. 16.7. We shall not pause to explain why the two waves survive unscathed, except to remark that there are topological invariants in the solution that must be preserved. However, we can exhibit one such two-soliton solution analytically:

two-soliton solution of KdV equation

$$\zeta = \frac{\partial^2}{\partial \eta^2}[12 \ln F(\eta, \tau)],$$

$$\text{where } F = 1 + f_1 + f_2 + \left(\frac{\alpha_2 - \alpha_1}{\alpha_2 + \alpha_1}\right)^2 f_1 f_2,$$

$$\text{and } f_i = \exp[-\alpha_i(\eta - \eta_i) + \alpha_i^3 \tau], \quad i = 1, 2; \tag{16.36}$$

here α_i and η_i are constants. This solution is depicted in Fig. 16.7.

16.3.5 Solitons in Contemporary Physics

Solitons were rediscovered in the 1960s when they were found in numerical simulations of plasma waves. Their topological properties were soon understood, and general methods to generate solutions were derived. Solitons have been isolated in such different subjects as the propagation of magnetic flux in a Josephson junction, elastic waves in anharmonic crystals, quantum field theory (as *instantons*), and classical general relativity (as solitary, nonlinear gravitational waves). Most classical solitons

some equations with soliton solutions

are solutions to one of a relatively small number of nonlinear partial differential equations, including the KdV equation, the *nonlinear Schrödinger equation* (which governs solitons in optical fibers; Sec. 10.8.3), *Burgers equation,* and the *sine-Gordon* equation. Unfortunately, it has proved difficult to generalize these equations and their soliton solutions to 2 and 3 spatial dimensions.

Just like research into chaos (Sec. 15.6), studies of solitons have taught physicists that nonlinearity need not lead to maximal disorder in physical systems, but instead can create surprisingly stable, ordered structures.

Exercise 16.9 *Example: Breaking of a Dam*
Consider the flow of water along a horizontal channel of constant width after a dam breaks. Sometime after the initial transients have died away[6] the flow may be described by the nonlinear, unidirectional, shallow-water wave equations (16.23a) and (16.23b):

$$\frac{\partial h}{\partial t} + \frac{\partial (hv)}{\partial x} = 0, \quad \frac{\partial v}{\partial t} + v\frac{\partial v}{\partial x} + g\frac{\partial h}{\partial x} = 0. \tag{16.37}$$

Here h is the height of the flow, v is the horizontal speed of the flow, and x is distance along the channel measured from the location of the dam. Solve for the flow, assuming that initially (at $t = 0$) $h = h_o$ for $x < 0$ and $h = 0$ for $x > 0$ (no water). Your solution should have the form shown in Fig. 16.8. What is the speed of the front of the water? [Hint: From the parameters of the problem we can construct only one velocity, $\sqrt{gh_o}$, and no length except h_o. It therefore is a reasonable guess that the solution has the self-similar form $h = h_o \, \tilde{h}(\xi)$, $v = \sqrt{gh_o} \, \tilde{v}(\xi)$, where $\tilde{h}$ and $\tilde{v}$ are dimensionless functions of the similarity variable

$$\xi = \frac{x/t}{\sqrt{gh_o}}. \tag{16.38}$$

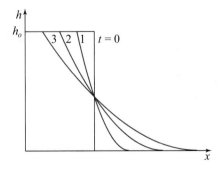

FIGURE 16.8 The water's height $h(x, t)$ after a dam breaks.

---

6. In the idealized case that the dam is removed instantaneously, there are no transients, and Eqs. (16.37) describe the flow from the outset.

Using this ansatz, convert the partial differential equations (16.37) into a pair of ordinary differential equations that can be solved so as to satisfy the initial conditions.]

Exercise 16.10 *Derivation: Single-Soliton Solution*
Verify that expression (16.33) does indeed satisfy the dimensionless KdV equation (16.32).

Exercise 16.11 *Derivation: Two-Soliton Solution*
(a) Verify, using symbolic-manipulation computer software (e.g., Maple, Matlab, or Mathematica) that the two-soliton expression (16.36) satisfies the dimensionless KdV equation. (Warning: Considerable algebraic travail is required to verify this by hand directly.)

(b) Verify analytically that the two-soliton solution (16.36) has the properties claimed in the text. First consider the solution at early times in the spatial region where $f_1 \sim 1$, $f_2 \ll 1$. Show that the solution is approximately that of the single soliton described by Eq. (16.33). Demonstrate that the amplitude is $\zeta_{01} = 3\alpha_1^2$, and find the location of its peak. Repeat the exercise for the second wave and for late times.

(c) Use a computer to follow numerically the evolution of this two-soliton solution as time η passes (thereby filling in timesteps between those shown in Fig. 16.7).

16.4 Rossby Waves in a Rotating Fluid

16.4

Coriolis force as restoring force for Rossby waves

In a nearly rigidly rotating fluid with the rotational angular velocity $\boldsymbol{\Omega}$ parallel or antiparallel to the acceleration of gravity $\mathbf{g} = -g\mathbf{e}_z$, the Coriolis effect observed in the co-rotating reference frame (Sec. 14.5) provides the restoring force for an unusual type of wave motion called *Rossby waves*. These waves are seen in Earth's oceans and atmosphere [with $\boldsymbol{\Omega} =$ (Earth's rotational angular velocity) sin(latitude)$\mathbf{e}_z$; see Box 14.5].

For a simple example, we consider the sea above a sloping seabed; Fig. 16.9. We assume the unperturbed fluid has vanishing velocity $\mathbf{v} = 0$ in Earth's rotating frame, and we study weak waves in the sea with oscillating velocity $\mathbf{v}$. (Since the fluid is at rest in the equilibrium state about which we are perturbing, we write the perturbed velocity as $\mathbf{v}$ rather than $\delta\mathbf{v}$.) We assume that the wavelengths are long enough that viscosity and surface tension are negligible. In this case we also restrict attention to small-amplitude waves, so that nonlinear terms can be dropped from the dynamical equations. The perturbed Navier-Stokes equation (14.56a) then becomes (after linearization)

$$\frac{\partial \mathbf{v}}{\partial t} + 2\boldsymbol{\Omega} \times \mathbf{v} = \frac{-\nabla \delta P'}{\rho}. \tag{16.39}$$

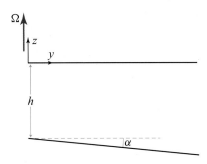

FIGURE 16.9 Geometry of the sea for Rossby waves.

Here, as in Sec. 14.5, $\delta P'$ is the perturbation in the effective pressure [which includes gravitational and centrifugal effects: $P' = P + \rho\Phi - \frac{1}{2}\rho(\boldsymbol{\Omega} \times \mathbf{x})^2$]. Taking the curl of Eq. (16.39), we obtain for the time derivative of the waves' vorticity:

$$\frac{\partial\boldsymbol{\omega}}{\partial t} = 2(\boldsymbol{\Omega} \cdot \nabla)\mathbf{v}. \tag{16.40}$$

We seek a wave mode with angular frequency ω (not to be confused with vorticity $\boldsymbol{\omega}$) and wave number k, in which the horizontal fluid velocity oscillates in the x direction and (in accord with the Taylor-Proudman theorem; Sec. 14.5.3) is independent of z, so

$$v_x \text{ and } v_y \propto \exp[i(kx - \omega t)], \quad \frac{\partial v_x}{\partial z} = \frac{\partial v_y}{\partial z} = 0. \tag{16.41}$$

The only allowed vertical variation is in the vertical velocity v_z; differentiating $\nabla \cdot \mathbf{v} = 0$ with respect to z, we obtain

$$\frac{\partial^2 v_z}{\partial z^2} = 0. \tag{16.42}$$

The vertical velocity therefore varies linearly between the surface and the sea floor (Fig. 16.9). One boundary condition is that the vertical velocity must vanish at the sea's surface. The other is that at the sea floor $z = -h$, we must have $v_z(-h) = -\alpha v_y$, where α is the tangent of the angle of inclination of the sea floor. The solution to Eq. (16.42) satisfying these boundary conditions is

$$v_z = \frac{\alpha z}{h} v_y. \tag{16.43}$$

Taking the vertical component of Eq. (16.40) and evaluating $\omega_z = v_{y,x} - v_{x,y} = ikv_y$, we obtain

$$\omega k v_y = 2\Omega\frac{\partial v_z}{\partial z} = \frac{2\Omega\alpha v_y}{h}. \tag{16.44}$$

The dispersion relation therefore has the quite unusual form $\omega \propto 1/k$:

$$\boxed{\omega k = \frac{2\Omega\alpha}{h}.} \tag{16.45}$$

dispersion relation for
Rossby waves

Rossby waves have interesting properties. They can only propagate in one direction—parallel to the intersection of the sea floor with the horizontal (our $\mathbf{e}_x$ direction). Their phase velocity $\mathbf{V}_{ph}$ and group velocity $\mathbf{V}_g$ are equal in magnitude but opposite in direction:

$$\mathbf{V}_{ph} = -\mathbf{V}_g = \frac{2\Omega\alpha}{k^2 h}\mathbf{e}_x. \tag{16.46}$$

Using $\nabla \cdot \mathbf{v} = 0$, we discover that the two components of horizontal velocity are in quadrature: $v_x = i\alpha v_y/(kh)$. This means that the fluid circulates in the opposite sense to the angular velocity Ω.

Rossby waves play an important role in the circulation of Earth's oceans (see, e.g., Chelton and Schlax, 1996). A variant of these Rossby waves in air can be seen as undulations in the atmosphere's jet stream, produced when the stream goes over a sloping terrain, such as that of the Rocky Mountains. Another variant found in neutron stars, called "r modes," generates gravitational waves (ripples of spacetime curvature) that are a promising source for ground-based gravitational-wave detectors, such as LIGO.

EXERCISES

Exercise 16.12 *Example: Rossby Waves in a Cylindrical Tank with a Sloping Bottom*
In the film Fultz (1969), about 20 min 40 s into the film, an experiment is described in which Rossby waves are excited in a rotating cylindrical tank with inner and outer vertical walls and a sloping bottom. Figure 16.10a is a photograph of the tank from the side, showing its bottom, which slopes upward toward the center, and a hump on the bottom that generates the Rossby waves. The tank is filled with water, then set into rotation with an angular velocity Ω. The water is given time to settle down into rigid rotation with the cylinder. Then the cylinder's angular velocity is reduced by a small amount, so the water is rotating at angular velocity $\Delta\Omega \ll \Omega$ relative to the cylinder. As the water passes over the hump on the tank bottom, the hump generates Rossby waves. Those waves are made visible by injecting dye at a fixed radius through a syringe attached to the tank. Figure 16.10b is a photograph of the dye trace as seen looking down on the tank from above. If there were no Rossby waves present, the trace would be circular. The Rossby waves make it pentagonal. In this exercise you will work out the details of the Rossby waves, explore their physics, and explain the shape of the trace.

Because the slope of the bottom is cylindrical rather than planar, this is somewhat different from the situation discussed in the text (see Fig. 16.9). However, we can deduce the details of the waves in this cylindrical case from those for the planar case by using geometric-optics considerations (Sec. 7.3), making modest errors because the wavelength of the waves is not all that small compared to the circumference of the tank.

(a) Using geometric optics, show that the rays along which the waves propagate are circles centered on the tank's symmetry axis.

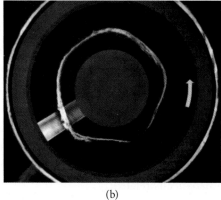

(a)   (b)

FIGURE 16.10 Rossby waves in a rotating cylinder with sloping bottom. (a) Side view. (b) Top view, showing dye trace. Images from *NCFMF Book of Film Notes*, 1972; The MIT Press with Education Development Center, Inc. © 2014 Education Development Center, Inc. Reprinted with permission with all other rights reserved.

(b) Focus on the ray that is halfway between the inner and outer walls of the tank. Let its radius be a, the depth of the water there be h, and the slope angle of the tank floor be α. Introduce quasi-Cartesian coordinates $x = a\phi$ and $y = -\varpi$, where $\{\varpi, \phi, z\}$ are cylindrical coordinates. By translating the Cartesian-coordinate waves of the text into quasi-Cartesian coordinates and noting from Fig. 16.10b that five wavelengths must fit into the circumference around the cylinder, show that the velocity field has the form $v_\varpi, v_\phi, v_z \propto e^{i(5\phi+\omega t)}$, and deduce the ratios of the three components of velocity to one another. The solution has nonzero radial velocity at the walls—a warning that edge effects will modify the waves somewhat. This analysis ignores those edge effects.

(c) Because the waves are generated by the ridge on the bottom of the tank, the wave pattern must remain at rest relative to that ridge, which means it must rotate relative to the fluid's frame with the angular velocity $d\phi/dt = -\Delta\Omega$. From the waves' dispersion relation, deduce $\Delta\Omega/\Omega$ (the fractional slowdown of the tank that had to be imposed to generate the observed pentagonal wave).

(d) Compute the displacement field $\delta\mathbf{x}(\varpi, \phi, z, t)$ of a fluid element whose undisplaced location (in the rigidly rotating cylindrical coordinates) is (ϖ, ϕ, z). Explain the pentagonal shape of the movie's dye lines in terms of this displacement field.

(e) Compute the wave's vertical vorticity field ω_z (relative to the rigidly rotating flow), and show that as a fluid element moves and the vertical vortex line through it shortens or lengthens due to the changing water depth, ω_z changes proportionally to the vortex line's length.

16.5 Sound Waves

So far our discussion of fluid dynamics has mostly been concerned with flows sufficiently slow that the density can be treated as constant. We now introduce the effects of compressibility in the context of sound waves (in a nonrotating reference frame). Sound waves are prototypical scalar waves and therefore are simpler in many respects than vector electromagnetic waves and tensor gravitational waves.

Consider a small-amplitude sound wave propagating through a homogeneous, time-independent fluid. The wave's oscillations are generally quick compared to the time for heat to diffuse across a wavelength, so the pressure and density perturbations are adiabatically related:

$$\delta P = C^2 \delta \rho, \tag{16.47}$$

where

adiabatic sound speed

$$\boxed{C \equiv \left[\left(\frac{\partial P}{\partial \rho} \right)_s \right]^{1/2}}, \tag{16.48}$$

which will turn out to be the wave's propagation speed—the speed of sound. The perturbation of the fluid velocity (which we denote $\mathbf{v}$, since the unperturbed fluid is static) is related to the pressure perturbation by the linearized Euler equation:

$$\frac{\partial \mathbf{v}}{\partial t} = -\frac{\nabla \delta P}{\rho}. \tag{16.49a}$$

A second relation between $\mathbf{v}$ and δP can be obtained by combining the linearized law of mass conservation, $\partial \rho / \partial t = -\rho \nabla \cdot \mathbf{v}$, with the adiabatic pressure-density relation (16.47):

$$\nabla \cdot \mathbf{v} = -\frac{1}{\rho C^2} \frac{\partial \, \delta P}{\partial t}. \tag{16.49b}$$

By equating the divergence of Eq. (16.49a) to the time derivative of Eq. (16.49b), we obtain a simple, dispersion-free wave equation for the pressure perturbation:

wave equation for sound waves

$$\left(\frac{\partial^2}{\partial t^2} - C^2 \nabla^2 \right) \delta P = 0. \tag{16.50}$$

Thus, as claimed, C is the wave's propagation speed.

For a perfect gas, this adiabatic sound speed is $C = (\gamma P / \rho)^{1/2}$, where γ is the ratio of specific heats (see Ex. 5.4). The sound speed in air at 20 °C is 343 m s^{-1}. In water under atmospheric conditions, it is about 1.5 km s^{-1} (not much different from sound speeds in solids).

Because the vorticity of the unperturbed fluid vanishes and the wave contains no vorticity-producing forces, the wave's vorticity vanishes: $\nabla \times \mathbf{v} = 0$. This permits us to express the wave's velocity perturbation as the gradient of a velocity potential: $\mathbf{v} = \nabla \psi$.

Inserting this expression into the perturbed Euler equation (16.49a), we express the pressure perturbation in terms of ψ:

$$\delta P = -\rho \frac{\partial \psi}{\partial t}, \quad \text{where} \quad \mathbf{v} = \boldsymbol{\nabla}\psi.$$

(16.51)

velocity potential for sound waves

The first of these relations guarantees that ψ satisfies the same wave equation as δP:

$$\left(\frac{\partial^2}{\partial t^2} - C^2\nabla^2 \right) \psi = 0.$$

(16.52)

It is sometimes useful to describe the wave by its oscillating pressure δP and sometimes by its oscillating potential ψ.

The general solution of the wave equation (16.52) for plane sound waves propagating in the $\pm x$ directions is

$$\psi = f_1(x - Ct) + f_2(x + Ct),$$

(16.53)

where f_1 and f_2 are arbitrary functions.

Exercise 16.13 *Problem: Sound Wave in an Inhomogeneous Fluid*

EXERCISES

Consider a sound wave propagating through a static, inhomogeneous fluid with no gravity. Explain why the unperturbed fluid has velocity $\mathbf{v} = 0$ and pressure $P_o = $ constant, but can have variable density and sound speed, $\rho_o(\mathbf{x})$ and $C(\mathbf{x}, t)$. By repeating the analysis in Eqs. (16.47)–(16.50), show that the wave equation is $\partial^2 \delta P/\partial t^2 = C^2 \rho_o \boldsymbol{\nabla}\cdot(\rho_o^{-1}\boldsymbol{\nabla}\delta P)$, which can be rewritten as

$$W\frac{\partial^2 \delta P}{\partial t^2} - \boldsymbol{\nabla}\cdot(WC^2\boldsymbol{\nabla}\delta P) = 0,$$

(16.54)

where $W = (C^2\rho_o)^{-1}$. [Hint: It may be helpful to employ the concept of Lagrangian versus Eulerian perturbations, as described by Eq. (19.44).] Equation (16.54) is an example of the prototypical wave equation (7.17) that we used in Sec. 7.3.1 to illustrate the geometric-optics formalism. The functional form of W and the placement of W and C^2 (inside versus outside the derivatives) have no influence on the wave's dispersion relation or its rays or phase in the geometric-optics limit, but they do influence the propagation of the wave's amplitude. See Sec. 7.3.1.

16.5.1 Wave Energy

16.5.1

In Sec. 7.3.1 and Ex. 7.4, we used formal mathematical techniques to derive the energy density U and energy flux $\mathbf{F}$ [Eqs. (7.18)] associated with waves satisfying the prototypical wave equation (16.54). In this section, we rederive U and $\mathbf{F}$ for sound waves using a physical, fluid dynamical analysis. We get precisely the same expressions up to a constant multiplicative factor ρ^2. Because of the formal nature of the arguments leading to Eqs. (7.18), we only had a right to expect the same answer up to some multiplicative constant.

The fluid's energy density is $U = (\frac{1}{2}v^2 + u)\rho$ (Table 13.1 with $\Phi = 0$). The first term is the fluid's kinetic energy density; the second is its internal energy density. The internal energy density can be evaluated by a Taylor expansion in the wave's density perturbation:

$$u\rho = [u\rho] + \left[\left(\frac{\partial(u\rho)}{\partial\rho}\right)_s\right]\delta\rho + \frac{1}{2}\left[\left(\frac{\partial^2(u\rho)}{\partial\rho^2}\right)_s\right]\delta\rho^2, \tag{16.55}$$

where the three coefficients in square brackets are evaluated at the equilibrium density. The first term in Eq. (16.55) is the energy of the background fluid, so we drop it. The second term averages to zero over a wave period, so we also drop it. The third term can be simplified using the first law of thermodynamics in the form $du = Tds - Pd(1/\rho)$ (which implies $[\partial(u\rho)/\partial\rho]_s = u + P/\rho$). We then apply the definition $h = u + P/\rho$ of enthalpy density, followed by the first law in the form $dh = Tds + dP/\rho$, and then followed by expression (16.48) for the speed of sound. The result is

$$\left(\frac{\partial^2(u\rho)}{\partial\rho^2}\right)_s = \left(\frac{\partial h}{\partial\rho}\right)_s = \frac{C^2}{\rho}. \tag{16.56}$$

Inserting this relation into the third term of Eq. (16.55) and averaging over a wave period and wavelength, we obtain for the wave energy per unit volume $U = \frac{1}{2}\rho\overline{v^2} + [C^2/(2\rho)]\overline{\delta\rho^2}$. Using $\mathbf{v} = \nabla\psi$ [the second of Eqs. (16.51)] and $\delta\rho = (\rho/C^2)\partial\psi/\partial t$ [from $\delta\rho = (\partial\rho/\partial P)_s\delta P = \delta P/C^2$ and the first of Eqs. (16.51)], we bring the equation for U into the form

$$U = \frac{1}{2}\rho\overline{\left[(\nabla\psi)^2 + \frac{1}{C^2}\left(\frac{\partial\psi}{\partial t}\right)^2\right]} = \rho\overline{(\nabla\psi)^2}. \tag{16.57}$$

The second equality can be deduced by multiplying the wave equation (16.52) by ψ and averaging. Thus, energy is equipartitioned between the kinetic and internal energy terms.

The energy flux is $\mathbf{F} = (\frac{1}{2}v^2 + h)\rho\mathbf{v}$ (Table 13.1 with $\Phi = 0$). The kinetic energy flux (first term) is third order in the velocity perturbation and therefore vanishes on average. For a sound wave, the internal energy flux (second term) can be brought into a more useful form by expanding the enthalpy per unit mass:

$$h = [h] + \left[\left(\frac{\partial h}{\partial P}\right)_s\right]\delta P = [h] + \frac{\delta P}{\rho}. \tag{16.58}$$

Here we have used the first law of thermodynamics $dh = Tds + (1/\rho)dP$ and adiabaticity of the perturbation ($s = $ const); the terms in square brackets are unperturbed quantities. Inserting Eq. (16.58) into $\mathbf{F} = h\rho\mathbf{v}$, expressing δP and $\mathbf{v}$ in terms of the velocity potential [Eqs. (16.51)], and averaging over a wave period and wavelength, we obtain for the energy flux $\mathbf{F} = \overline{\rho h \mathbf{v}} = \overline{\delta P \, \mathbf{v}}$, which becomes

$$\mathbf{F} = -\rho \overline{\left(\frac{\partial \psi}{\partial t}\right) \nabla \psi}. \tag{16.59}$$

energy flux for soundwaves

Aside from a multiplicative constant factor ρ^2, this equation and Eq. (16.57) agree with Eqs. (7.18) [with ψ there being this chapter's velocity potential ψ, and with $W = (C^2\rho)^{-1}$; Eq. (16.54)], which we derived by formal techniques in Sec. 7.3.1 and Ex. 7.4.

For a locally plane wave with $\psi = \psi_o \cos(\mathbf{k} \cdot \mathbf{x} - \omega t + \varphi)$ (where φ is an arbitrary phase), the energy density (16.57) is $U = \frac{1}{2}\rho\psi_o^2 k^2$, and the energy flux (16.59) is $\mathbf{F} = \frac{1}{2}\rho\psi_o^2 \omega \mathbf{k}$. Since for this dispersion-free wave, the phase and group velocities are both $\mathbf{V} = (\omega/k)\hat{\mathbf{k}} = C\hat{\mathbf{k}}$ (where $\hat{\mathbf{k}} = \mathbf{k}/k$ is the unit vector pointing in the wave-propagation direction), the energy density and flux are related by

$$\boxed{\mathbf{F} = U\mathbf{V} = UC\hat{\mathbf{k}}.} \tag{16.60}$$

The energy flux is therefore the product of the energy density and the wave velocity, as it must be [Eq. (7.31), where we see that, if the waves were to have dispersion, it would be the group velocity that appears in this expression].

The energy flux carried by sound is conventionally measured in dB (decibels). The flux in decibels, F_{dB}, is related to the flux F in W m^{-2} by

$$\boxed{F_{\mathrm{dB}} = 120 + 10 \log_{10}(F).} \tag{16.61}$$

Sound that is barely audible is about 1 dB. Normal conversation is about 50–60 dB. Jet aircraft, rock concerts, and volcanic eruptions can cause exposure to more than 120 dB, with consequent damage to the ear.

16.5.2 Sound Generation

16.5.2

So far in this book we have been concerned with describing how different types of waves propagate. It is also important to understand how they are generated. We now outline some aspects of the theory of sound generation.

The reader should be familiar with the theory of electromagnetic wave emission (e.g., Jackson, 1999, Chap. 9). For electromagnetic waves one considers a localized region containing moving charges and varying currents. The source can be described as a sum over electric and magnetic multipoles, and each multipole produces a characteristic angular variation of the distant radiation field. The radiation-field amplitude decays inversely with distance from the source, and so the Poynting flux varies with the inverse square of the distance. Integrating over a large sphere gives the total power radiated by the source, broken down into the power radiated by each multipolar component. The ratio of the power in successive multipole pairs [e.g., (magnetic dipole power)/(electric dipole power) $\sim$ (electric quadrupole power)/(electric dipole power)] is typically $\sim (b/\lambda)^2$, where b is the size of the source, and $\lambda = 1/k$ is the waves' reduced wavelength. When $\lambdabar$ is large compared to b (a situation referred to as *slow motion*, since the source's charges then generally move at speeds $\sim (b/\lambdabar)c$ small compared to the speed of light c), the most powerful radiating multipole is the electric dipole $\mathbf{d}(t)$, unless it happens to be suppressed. The dipole's average emitted power is given by the Larmor formula:

$$\mathcal{P} = \frac{\overline{\ddot{\mathbf{d}}^2}}{6\pi\epsilon_0 c^3},\tag{16.62}$$

where $\ddot{\mathbf{d}}$ is the second time derivative of $\mathbf{d}$, the bar denotes a time average, ϵ_0 is the permittivity of free space, and c is the speed of light.

This same procedure can be followed when describing sound generation. However, as we are dealing with a scalar wave, sound can have a monopolar source. As a pedagogical example, let us set a small, spherical, elastic ball, surrounded by fluid, into radial oscillation (not necessarily sinusoidal) with oscillation frequencies of order ω, so the emitted waves have reduced wavelengths of order $\lambdabar = C/\omega$. Let the surface of the ball have radius $a + \xi(t)$, and impose the slow-motion and small-amplitude conditions that

$$\lambdabar \gg a \gg |\xi|.\tag{16.63}$$

As the waves will be spherical, the relevant outgoing-wave solution of the wave equation (16.52) is

$$\psi = \frac{f(t - r/C)}{r},\tag{16.64}$$

where f is a function to be determined. Since the fluid's velocity at the ball's surface must match that of the ball, we have (to first order in $\mathbf{v}$ and ψ):

$$\dot{\xi}\mathbf{e}_r = \mathbf{v}(a, t) = \nabla\psi \simeq -\frac{f(t - a/C)}{a^2}\mathbf{e}_r \simeq -\frac{f(t)}{a^2}\mathbf{e}_r,\tag{16.65}$$

where in the third equality we have used the slow-motion condition $\lambdabar \gg a$. Solving for $f(t)$ and inserting into Eq. (16.64), we see that

$$\psi(r, t) = -\frac{a^2\dot{\xi}(t - r/C)}{r}.\tag{16.66}$$

It is customary to express the radial velocity perturbation v in terms of an oscillating fluid *monopole moment*

monopole moment for spherical sound waves from an oscillating ball

$$\boxed{q = 4\pi\rho a^2\dot{\xi}.}\tag{16.67}$$

Physically, this is the total radial discharge of air mass (i.e., mass per unit time) crossing an imaginary fixed spherical surface of radius slightly larger than that of the oscillating ball. In terms of q, we have $\dot{\xi}(t) = q(t)/[4\pi\rho a^2]$. Using this expression and Eq. (16.66), we compute for the power radiated as sound waves [Eq. (16.59) integrated over a sphere centered on the ball]:

$$\boxed{\mathcal{P} = \frac{\overline{\dot{q}^2}}{4\pi\rho C}.}$$

(16.68)

Note that the power is inversely proportional to the signal speed, which is characteristic of monopolar emission and is in contrast to the inverse-cube variation for dipolar emission [Eq. (16.62)].

The emission of monopolar waves requires that the volume of the emitting solid body oscillate. When the solid simply oscillates without changing its volume (e.g., the reed in a musical instrument), dipolar emission usually dominates. We can think of this as two monopoles of size a in antiphase separated by some displacement $b \sim a$. The velocity potential in the far field is then the sum of two monopolar contributions, which almost cancel. Making a Taylor expansion, we obtain

$$\frac{\psi_{\text{dipole}}}{\psi_{\text{monopole}}} \sim \frac{b}{\lambda} \sim \frac{\omega b}{C},$$

(16.69)

where ω and λ are the characteristic magnitudes of the angular frequency and reduced wavelength of the waves (which we have not assumed to be precisely sinusoidal).

This reduction of ψ by the slow-motion factor b/λ implies that the dipolar power emission is weaker than the monopolar power by a factor $\sim (b/\lambda)^2$ for similar frequencies and amplitudes of motion—the same factor as for electromagnetic waves (see the start of this subsection). However, to emit dipole radiation, momentum must be given to and removed from the fluid. In other words, the fluid must be forced by a solid body. In the absence of such a solid body, the lowest multipole that can be radiated effectively is quadrupolar radiation, which is weaker by yet one more factor of $(b/\lambda)^2$.

These considerations are important for understanding how noise is produced by the intense turbulence created by jet engines, especially close to airports. We expect that the sound emitted by the free turbulence in the wake just behind the engine will be quadrupolar and will be dominated by emission from the largest (and hence fastest) turbulent eddies. (See the discussion of turbulent eddies in Sec. 15.4.4.) Denote by ℓ and v_ℓ the size and turnover speed of these largest eddies. Then the characteristic size of the sound's source is $a \sim b \sim \ell$, the mass discharge is $q \sim \rho\ell^2 v_\ell$, the characteristic frequency is $\omega \sim v_\ell/\ell$, the reduced wavelength of the sound waves is $\lambda = C/\omega \sim \ell C/v_\ell$, and the slow-motion parameter is $b/\lambda \sim \omega b/C \sim v_\ell/C$. The quadrupolar power radiated per unit volume [Eq. (16.68) divided by the volume ℓ^3 of an eddy and reduced by $\sim (b/\lambda)^4$] is therefore

$$\frac{d\mathcal{P}}{d^3 x} \sim \rho \frac{v_\ell^3}{\ell} \left(\frac{v_\ell}{C}\right)^5,$$

(16.70)

and this power is concentrated around frequency $\omega \sim v_\ell/\ell$. For air of fixed sound speed and lengthscale ℓ of the largest eddies, and for which the largest eddy speed v_ℓ is proportional to some characteristic speed V (e.g., the average speed of the air leaving the engine), Eq. (16.70) says the sound generation increases proportional to the eighth power of the Mach number $M = V/C$. This is known as Lighthill's law (Lighthill 1952, 1954). The implications for the design of jet engines should be obvious.

EXERCISES

Exercise 16.14 *Problem: Attenuation of Sound Waves*
Viscosity and thermal conduction will attenuate sound waves. For the moment just consider a monatomic gas where the bulk viscosity can be neglected.

(a) Consider the entropy equation (13.75), and evaluate the influence of the heat flux on the relationship between the pressure and the density perturbations. [Hint: Assume that all quantities vary as $e^{i(kx-\omega t)}$, where k is real and ω complex.]

(b) Consider the momentum equation (13.69), and include the viscous term as a perturbation.

(c) Combine these two relations in parts (a) and (b), together with the equation of mass conservation, to solve for the imaginary part of ω in the linear regime.

(d) Substitute kinetic-theory expressions for the coefficient of shear viscosity and the coefficient of thermal conductivity (Secs. 13.7.3 and 3.7) to obtain a simple expression for the attenuation length involving the wave's wavelength and the atoms' collisional mean free path.

(e) How do you think the wave attenuation will be affected if the fluid is air or is turbulent, or both?

See Faber (1995) for more details.

Exercise 16.15 *Example: Plucked Violin String*
Consider the G string (196 Hz) of a violin. It is ~30 cm from bridge to nut (the fixed endpoints), and the tension in the string is ~40 N.

(a) Infer the mass per unit length in the string and estimate its diameter. Hence estimate the strain in the string before being plucked. Estimate the strain's increase if its midpoint is displaced through 3 mm.

(b) Now suppose that the string is released. Estimate the speed with which it moves as it oscillates back and forth.

(c) Estimate the dipolar sound power emitted and the distance out to which the note can be heard (when its intensity is a few decibels). Do your answers seem reasonable? What factors, omitted from this calculation, might change your answers?

Exercise 16.16 *Example: Trumpet*
Idealize the trumpet as a bent pipe of length 1.2 m from the mouthpiece (a node of the air's displacement) to the bell (an antinode). The lowest note is a first overtone and should correspond to B flat (233 Hz). Does it?

Exercise 16.17 *Problem: Aerodynamic Sound Generation*

Consider the emission of quadrupolar sound waves by a Kolmogorov spectrum of free turbulence (Sec. 15.4.4). Show that the power radiated per unit frequency interval has a spectrum

$$\mathcal{P}_\omega \propto \omega^{-7/2}.$$

Also show that the total power radiated is roughly a fraction M^5 of the power dissipated in the turbulence, where M is the Mach number.

16.5.3 Radiation Reaction, Runaway Solutions, and Matched Asymptotic Expansions [T2]

Let us return to our idealized example of sound waves produced by a radially oscillating, spherical ball. We use this example to illustrate several deep issues in theoretical physics: the *radiation-reaction force* that acts back on a source due to its emission of radiation, a spurious *runaway solution* to the source's equation of motion caused by the radiation-reaction force, and *matched asymptotic expansions*, a mathematical technique for solving field equations when there are two different regions of space in which the equations have rather different behaviors.[7] These issues also arise, in a rather more complicated way, in analyses of the electromagnetic radiation reaction force on an accelerated electron (the "Abraham-Lorentz force"), and the radiation-reaction force caused by emission of gravitational waves; see the derivation, by Burke (1971), of gravitational results quoted in Sec. 27.5.3.

sound waves from radially oscillating ball

For our oscillating ball, the two different regions of space that we match to each other are the *near zone, $r \ll \lambda$*, and the *wave zone, $r \gtrsim \lambda$*.

near zone and wave zone

We consider, first, the near zone, and we redo, from a new point of view, the analysis of the matching of the near-zone fluid velocity to the ball's surface velocity and the computation of the pressure perturbation. Because the region near the ball is small compared to λ and the fluid speeds are small compared to C, the flow is very nearly incompressible, $\nabla \cdot \mathbf{v} = \nabla^2 \psi = 0$; see the discussion of conditions for incompressibility in Sec. 13.6. (The near-zone equation $\nabla^2 \psi = 0$ is analogous to $\nabla^2 \Phi = 0$ for the Newtonian gravitational potential in the weak-gravity near zone of a gravitational-wave source; Sec. 27.5.)

The general monopolar (spherical) solution to $\nabla^2 \psi = 0$ is

$$\psi = \frac{A(t)}{r} + B(t). \tag{16.71}$$

Matching the fluid's radial velocity $v = \partial \psi / \partial r = -A/r^2$ at $r = a$ to the ball's radial velocity $\dot{\xi}$, we obtain

$$A(t) = -a^2 \dot{\xi}(t). \tag{16.72}$$

---

7. Our treatment is based on Burke (1970).

From the point of view of near-zone physics, no mechanism exists for generating a nonzero spatially constant term $B(t)$ in ψ [Eq. (16.71)], so if one were unaware of the emitted sound waves and their action back on the source, one would be inclined to set $B(t)$ to zero. [This line of reasoning is analogous to that of a Newtonian physicist, who would be inclined to write the quadrupolar contribution to an axisymmetric source's external gravitational field in the form $\Phi = P_2(\cos\theta)[A(t)r^{-3} + B(t)r^2]$ and then, being unaware of gravitational waves and their action back on the source, would set $B(t)$ to zero.] Taking this near-zone viewpoint, with $B = 0$, we infer that the fluid's pressure perturbation acting on the ball's surface is

$$\delta P = -\rho \frac{\partial \psi(a, t)}{\partial t} = -\rho \frac{\dot{A}}{a} = \rho a \ddot{\xi} \tag{16.73}$$

[Eqs. (16.51) and (16.72)].

The motion $\xi(t)$ of the ball's surface is controlled by the elastic restoring forces in its interior and the fluid pressure perturbation δP on its surface. In the absence of δP the surface would oscillate sinusoidally with some angular frequency ω_o, so $\ddot{\xi} + \omega_o^2 \xi = 0$. The pressure will modify this expression to

$$m(\ddot{\xi} + \omega_o^2 \xi) = -4\pi a^2 \delta P, \tag{16.74}$$

where m is an effective mass, roughly equal to the ball's true mass, and the right-hand side is the integral of the radial component of the pressure perturbation force over the sphere's surface. Inserting the near-zone viewpoint's pressure perturbation (16.73), we obtain

$$(m + 4\pi a^3 \rho)\ddot{\xi} + m\omega_o^2 \xi = 0. \tag{16.75}$$

Evidently, the fluid increases the ball's effective inertial mass (it *loads* additional mass on the ball) and thereby reduces its frequency of oscillation to

$$\omega = \frac{\omega_o}{\sqrt{1+\kappa}}, \quad \text{where } \kappa = \frac{4\pi a^3 \rho}{m} \tag{16.76}$$

is a measure of the coupling strength between the ball and the fluid. In terms of this loaded frequency, the equation of motion becomes

$$\ddot{\xi} + \omega^2 \xi = 0. \tag{16.77}$$

This near-zone viewpoint is not quite correct, just as the standard Newtonian viewpoint is not quite correct for the near-zone gravity of a gravitational-wave source (Sec. 27.5.3). To improve on this viewpoint, we temporarily move out into the wave zone and identify the general, outgoing-wave solution to the sound wave equation:

$$\psi = \frac{f(t - \epsilon r/C)}{r} \tag{16.78}$$

[Eq. (16.64)]. Here f is a function to be determined by matching to the near zone, and ϵ is a parameter that has been inserted to trace the influence of the outgoing-wave boundary condition. For outgoing waves (the real, physical, situation), $\epsilon = +1$; if the waves were ingoing, we would have $\epsilon = -1$.

This wave-zone solution remains valid into the near zone. In the near zone we can perform a slow-motion expansion to bring it into the same form as the near-zone velocity potential (16.71): slow-motion expansion

$$\psi = \frac{f(t)}{r} - \epsilon \frac{\dot{f}(t)}{C} + \ldots . \tag{16.79}$$

The second term is sensitive to whether the waves are outgoing or incoming and thus must ultimately be responsible for the radiation-reaction force that acts back on the oscillating ball; for this reason we call it the *radiation-reaction potential*. radiation-reaction potential

Equating the first term of this ψ to the first term of Eq. (16.71) and using the value in Eq. (16.72) of $A(t)$, which was obtained by matching the fluid velocity to the ball velocity, we obtain

$$f(t) = A(t) = -a^2 \dot{\xi}(t). \tag{16.80}$$

This equation tells us that the wave field $f(t - r/C)/r$ generated by the ball's surface displacement $\xi(t)$ is given by $\psi = -a^2 \dot{\xi}(t - r/C)/r$ [Eq. (16.66)]—the result we derived more quickly in the previous section. We can regard Eq. (16.80) as matching the near-zone solution outward onto the wave-zone solution to determine the wave field as a function of the source's motion.

Equating the second term of Eq. (16.79) to the second term of the near-zone velocity potential (16.71), we obtain

$$B(t) = -\epsilon \frac{\dot{f}(t)}{C} = \epsilon \frac{a^2}{C} \ddot{\xi}(t). \tag{16.81}$$

This is the term in the near-zone velocity potential $\psi = A/r + B$ that is responsible for radiation reaction. *We can regard this radiation-reaction potential $\psi^{RR} = B(t)$ as having been generated by matching the wave zone's outgoing ($\epsilon = +1$) or ingoing ($\epsilon = -1$) wave field back into the near zone.* [A similar matching analysis by Burke (1971) led him to the gravitational radiation-reaction potential (27.64).] how radiation-reaction potential is generated

This pair of matchings, outward then inward (Fig. 16.11), is a special, almost trivial example of the technique of matched asymptotic expansions—a technique developed by applied mathematicians to deal with much more complicated matching problems than this one (see, e.g., Cole, 1974). matched asymptotic expansions

The radiation-reaction potential $\psi^{RR} = B(t) = \epsilon(a^2/C)\ddot{\xi}(t)$ gives rise to a radiation-reaction contribution to the pressure on the ball's surface: $\delta P^{RR} = -\rho \dot{\psi}^{RR} = -\epsilon(\rho a^2/C)\dddot{\xi}$. Inserting this into the equation of motion (16.74) along with the loading pressure (16.73) and performing the same algebra as before, we get the following radiation-reaction-modified form of Eq. (16.77): radiation-reaction pressure on ball's surface

$$\boxed{\ddot{\xi} + \omega^2 \xi = \epsilon \tau \dddot{\xi},} \quad \text{where} \quad \tau = \frac{\kappa}{1 + \kappa} \frac{a}{C} \tag{16.82}$$

ball's equation of motion with radiation reaction

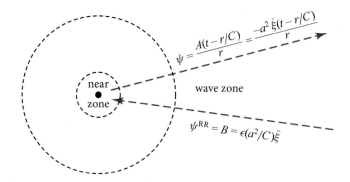

FIGURE 16.11 Matched asymptotic expansions for an oscillating ball emitting sound waves. The near-zone expansion feeds the radiation field $\psi = \frac{1}{r}A(t - r/C) = -\frac{1}{r}a^2\dot\xi(t - r/C)$ into the wave zone. The wave-zone expansion then feeds the radiation-reaction field $\psi^{\rm RR} = B = \epsilon(a^2/C)\dddot\xi$ back into the near zone, where it produces the radiation-reaction pressure $\delta P^{\rm RR} = -\rho\dot\psi^{\rm RR}$ on the ball's surface.

is less than the fluid's sound travel time to cross the ball's radius, a/C. The term $\epsilon\tau\dddot\xi$ in the equation of motion is the ball's *radiation-reaction acceleration,* as we see from the fact that it would change sign if we switched from outgoing waves ($\epsilon = +1$) to incoming waves ($\epsilon = -1$).

radiation-reaction induced damping

In the absence of radiation reaction, the ball's surface oscillates sinusoidally in time: $\xi = e^{\pm i\omega t}$. The radiation-reaction term produces a weak damping of these oscillations:

$$\xi \propto e^{\pm i\omega t}e^{-\sigma t}, \quad \sigma = \frac{1}{2}\epsilon(\omega\tau)\omega, \tag{16.83}$$

where σ is the radiation-reaction-induced damping rate (with $\epsilon = +1$). Note that in order of magnitude the ratio of the damping rate to the oscillation frequency is $\sigma/\omega \sim \omega\tau \lesssim \omega a/C = a/\lambda$, which is small compared to unity by virtue of the slow-motion assumption. If the waves were incoming rather than outgoing, $\epsilon = -1$, the fluid's oscillations would grow. In either case, outgoing waves or ingoing waves, the radiation-reaction force removes energy from the ball or adds it at the same rate as the sound waves carry energy off or bring it in. The total energy, wave plus ball, is conserved.

Expression (16.83) is two linearly independent solutions to the equation of motion (16.82), one with the plus sign and the other with the minus sign. Since this equation of motion has been made third order by the radiation-reaction term, there must be a third independent solution. It is easy to see that, up to a tiny fractional correction, that third solution is

$$\xi \propto e^{\epsilon t/\tau}. \tag{16.84}$$

runaway solution to ball's equation of motion

For outgoing waves, $\epsilon = +1$, this solution grows exponentially in time on an extremely rapid timescale, $\tau \lesssim a/C$; it is called a *runaway solution.*

Such runaway solutions are ubiquitous in equations of motion with radiation reaction. For example, a computation of the electromagnetic radiation reaction on a small, classical, electrically charged, spherical particle gives the Abraham-Lorentz equation of motion:

runaway solutions for an electrically charged particle

$$m(\ddot{\mathbf{x}} - \tau \dddot{\mathbf{x}}) = \mathbf{F}_{\text{ext}} \tag{16.85}$$

(Rohrlich, 1965; Jackson, 1999, Sec. 16.2). Here $\mathbf{x}(t)$ is the particle's world line, $\mathbf{F}_{\text{ext}}$ is the external force that causes the particle to accelerate, and the particle's inertial mass m includes an electrostatic contribution analogous to $4\pi a^3 \rho$ in our fluid problem. The timescale τ, like that in our fluid problem, is very short, and when the external force is absent, there is a runaway solution $\mathbf{x} \propto e^{t/\tau}$.

Much human heat and confusion were generated, in the early and mid-twentieth century, over these runaway solutions (see, e.g., Rohrlich, 1965). For our simple model problem, there need be little heat or confusion. One can easily verify that the runaway solution (16.84) violates the slow-motion assumption $a/\lambda \ll 1$ that underlies our derivation of the radiation-reaction acceleration. It therefore is a spurious solution.

Our model problem is sufficiently simple that one can dig deeper into it and learn that the runaway solution arises from the slow-motion approximation failing to reproduce a genuine, rapidly damped solution and getting the sign of the damping wrong (Ex. 16.19 and Burke, 1970).

origin of runaway solution

EXERCISES

Exercise 16.18 *Problem: Energy Conservation for Radially Oscillating Ball Plus Sound Waves*
For the radially oscillating ball as analyzed in Sec. 16.5.3, verify that the radiation-reaction acceleration removes energy from the ball, plus the fluid loaded onto it, at the same rate as the sound waves carry energy away. See Ex. 27.12 for the analogous gravitational-wave result.

Exercise 16.19 *Problem: Radiation Reaction without the Slow-Motion Approximation*
Redo the computation of radiation reaction for a radially oscillating ball immersed in a fluid without imposing the slow-motion assumption and approximation. Thereby obtain the following coupled equations for the radial displacement $\xi(t)$ of the ball's surface and the function $\Phi(t) \equiv a^{-2} f(t - \epsilon a/C)$, where $\psi = r^{-1} f(t - \epsilon r/C)$ is the sound-wave field:

$$\ddot{\xi} + \omega_o^2 \xi = \kappa \dot{\Phi}, \quad \dot{\xi} = -\Phi - \epsilon(a/C)\dot{\Phi}. \tag{16.86}$$

Show that in the slow-motion regime, this equation of motion has two weakly damped solutions of the same form as we derived using the slow-motion approximation [Eq. (16.83)], and one rapidly damped solution: $\xi \propto \exp(-\epsilon \kappa t/\tau)$. Burke (1970) shows that the runaway solution (16.84) obtained using the slow-motion approximation is caused by that approximation's futile attempt to reproduce this genuine, rapidly damped solution.

Exercise 16.20 *Problem: Sound Waves from a Ball Undergoing Quadrupolar Oscillations*

Repeat the analysis of sound-wave emission, radiation reaction, and energy conservation—as given in Sec. 16.5.3 and Ex. 16.18—for axisymmetric, quadrupolar oscillations of an elastic ball: $r_{\text{ball}} = a + \xi(t)\, P_2(\cos\theta)$.

Comment: Since the lowest multipolar order for gravitational waves is quadrupolar, this exercise is closer to the analogous problem of gravitational wave emission (Secs. 27.5.2 and 27.5.3) than is the monopolar analysis in the text.

[Hint: If ω is the frequency of the ball's oscillations, then the sound waves have the form

$$\psi = K\, \Re \left[e^{-i\omega t} \left(\frac{n_2(\omega r / C) - i\epsilon j_2(\omega r / C)}{r} \right) \right] P_2(\cos\theta), \qquad (16.87)$$

where K is a constant; $\Re(X)$ is the real part of X; ϵ is $+1$ for outgoing waves and -1 for ingoing waves; j_2 and n_2 are the spherical Bessel and spherical Neuman functions of order 2, and P_2 is the Legendre polynomial of order 2. In the distant wave zone, $x \equiv \omega r / C \gg 1$, we have

$$n_2(x) - i\epsilon j_2(x) = \frac{e^{i\epsilon x}}{x}; \qquad (16.88)$$

in the near zone $x = \omega r / C \ll 1$, we have

$$n_2(x) = -\frac{3}{x^3}\left(1\, \&\, x^2\, \&\, x^4\, \&\, \ldots\right), \qquad j_2(x) = \frac{x^2}{15}\left(1\, \&\, x^2\, \&\, x^4\, \&\, \ldots\right). \qquad (16.89)$$

Here "$\&\, x^n$" means "+ (some constant) x^n".]

Bibliographic Note

For physical insight into waves in fluids, we recommend the movies discussed in Box 16.2. Among fluid-dynamics textbooks, those that we most like for their treatment of waves are Acheson (1990), Lautrup (2005), and Kundu, Cohen, and Dowling (2012). For greater depth and detail, we recommend two books solely devoted to waves: Lighthill (2001) and Whitham (1974).

For Rossby waves (which are omitted from most fluid-dynamics texts), we recommend the very physical descriptions and analyses in Tritton (1987). For solitons, we like Whitham (1974, Chap. 17); also Drazin and Johnson (1989), which focuses on the mathematics of the Korteweg–de Vries equation; Ablowitz (2011), which treats the mathematics of solitons plus applications to fluids and nonlinear optics; and Dauxois and Peyrard (2010), which treats the mathematics and applications to plasmas and condensed-matter physics.

The mathematics of matched asymptotic expansions is nicely developed by Cole (1974) and Lagerstrom (1988). Radiation reaction in wave emission is not treated pedagogically in any textbook that we know of except ours; for pedagogical original literature, we like Burke (1970).

Compressible and Supersonic Flow

Rocket science is tough, and rockets have a way of failing.

PHYSICIST, ASTRONAUT, AND EDUCATOR SALLY RIDE (2012)

17.1 Overview

So far we have mainly been concerned with flows that are slow enough that they may be treated as incompressible. We now consider flows in which the velocity approaches or even exceeds the speed of sound and in which changes of density along streamlines cannot be ignored. Such flows are common in aeronautics and astrophysics. For example, the motion of a rocket through the atmosphere is faster than the speed of sound in air. In other words, it is *supersonic*. Therefore, if we transform into the frame of the rocket, the flow of air past the rocket is also supersonic.

When the flow speed exceeds the speed of sound in some reference frame, it is not possible for a pressure pulse to travel upstream in that frame and change the direction of the flow. However, if there is a solid body in the way (e.g., a rocket or aircraft), the flow direction must change (Fig. 17.1). In a supersonic flow, this change happens nearly discontinuously, through the formation of *shock fronts,* at which the flow suddenly decelerates from supersonic to subsonic. Shock fronts are an inevitable feature of supersonic flows.

In another example of supersonic flow, a rocket itself is propelled by the thrust created by its nozzle's escaping hot gases. These hot gases move through the rocket nozzle at supersonic speeds, expanding and cooling as they accelerate. In this manner, the random thermal motion of the gas molecules is converted into an organized bulk motion that carries negative momentum away from the rocket and pushes it forward.

The solar wind furnishes yet another example of a supersonic flow. This high-speed flow of ionized gas is accelerated in the solar corona and removes a fraction of $\sim 10^{-14}$ of the Sun's mass every year. Its own pressure accelerates it to supersonic speeds of ~ 400 km s^{-1}. When the outflowing solar wind encounters a planet, it is rapidly decelerated to subsonic speed by passing through a strong discontinuity, known as a *bow shock,* that surrounds the planet (Fig. 17.2). The bulk kinetic energy in the solar wind, built up during acceleration, is rapidly and irreversibly transformed into heat as it passes through this shock front.

FIGURE 17.1 Complex pattern of shock fronts formed around a model aircraft in a wind tunnel with air moving 10% faster than the speed of sound (i.e., with Mach number $M = 1.1$). Image from NASA/Ames Imaging Library System.

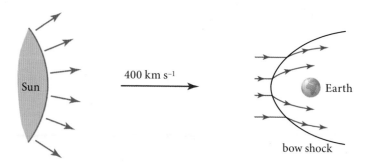

FIGURE 17.2 The supersonic solar wind forms a type of shock front known as a bow shock when it passes by a planet. Earth is $\sim$200 solar radii from the Sun.

In this chapter, we study some properties of supersonic flows. After restating the basic equations of compressible fluid dynamics (Sec. 17.2), we analyze three important, simple cases: stationary, quasi-1-dimensional flow (Sec. 17.3); time-dependent, 1-dimensional flow (Sec. 17.4); and normal adiabatic shock fronts (Sec. 17.5). In these sections, we apply the results of our analyses to some contemporary examples, including the Space Shuttle (Box 17.4); rocket engines; shock tubes; and the Mach cone, N-wave, and sonic booms produced by supersonic projectiles and aircraft. In Sec. 17.6, we develop similarity-solution techniques for supersonic flows and apply them to supernovae, underwater depth charges, and nuclear-bomb explosions in Earth's atmosphere.

As in our previous fluid-dynamics chapters, we strongly encourage readers to view relevant movies in parallel with reading this chapter. See Box 17.2.

17.2 Equations of Compressible Flow

In Chap. 13, we derived the equations of fluid dynamics, allowing for compressibility. We expressed them as laws of mass conservation [Eq. (13.29)], momentum conservation [$\partial(\rho\mathbf{v})/\partial t + \nabla \cdot \mathbf{T} = 0$, with $\mathbf{T}$ as given in Table 13.3], energy conservation [$\partial U/\partial t + \nabla \cdot \mathbf{F} = 0$, with U and $\mathbf{F}$ as given in Table 13.3], and also an evolution law for entropy [Eq. (13.76)]. When, as in this chapter, heat conduction is negligible ($\kappa \to 0$) and the gravitational field is a time-independent, external one (not generated by the flowing fluid), these equations become

$$\frac{\partial \rho}{\partial t} + \nabla \cdot (\rho\mathbf{v}) = 0,$$

(17.1a)   **mass conservation**

$$\frac{\partial (\rho\mathbf{v})}{\partial t} + \nabla \cdot (P\mathbf{g} + \rho\mathbf{v} \otimes \mathbf{v} - 2\eta\boldsymbol{\sigma} - \zeta\theta\mathbf{g}) = \rho\mathbf{g},$$

(17.1b)   **momentum conservation**

17.2

equations of compressible flow

$$\frac{\partial}{\partial t}\left[\left(\tfrac{1}{2}v^2 + u + \Phi\right)\rho\right] + \nabla \cdot \left[\left(\tfrac{1}{2}v^2 + h + \Phi\right)\rho\mathbf{v} - 2\eta\boldsymbol{\sigma}\cdot\mathbf{v} - \zeta\theta\mathbf{v}\right] = 0, \quad (17.1c)$$

$$\frac{\partial(\rho s)}{\partial t} + \nabla \cdot (\rho s \mathbf{v}) = \frac{1}{T}\left(2\eta\boldsymbol{\sigma} : \boldsymbol{\sigma} + \zeta\theta^2\right). \quad (17.1d)$$

Here $\boldsymbol{\sigma} : \boldsymbol{\sigma}$ is index-free notation for $\sigma_{ij}\sigma_{ij}$.

Some comments are in order. Equation (17.1a) is the complete mass-conservation equation (continuity equation), assuming that matter is neither added to nor removed from the flow, for example, no electron-positron pair creation. Equation (17.1b) expresses the conservation of momentum allowing for one external force, gravity. Other external forces (e.g., electromagnetic) can be added. Equation (17.1c), expressing energy conservation, includes a viscous contribution to the energy flux. If there are sources or sinks of fluid energy, then these must be included on the right-hand side of this equation. Possible sources of energy include chemical or nuclear reactions; possible energy sinks include cooling by emission of radiation. Equation (17.1d) expresses the evolution of entropy and will also need modification if there are additional contributions to the energy equation. The right-hand side of this equation is the rate of increase of entropy due to viscous heating. This equation is not independent of the preceding equations and the laws of thermodynamics, but it is often convenient to use. In particular, one often uses it (together with the first law of thermodynamics) in place of energy conservation (17.1c).

These equations must be supplemented with an equation of state in the form $P(\rho, T)$ or $P(\rho, s)$. For simplicity, we often focus on an ideal gas (one with $P \propto \rho k_B T$) that undergoes adiabatic evolution with constant specific-heat ratio (adiabatic index γ; Ex. 5.4), so the equation of state has the simple polytropic form (Box 13.2)

$$P = K(s)\rho^\gamma. \quad (17.2a)$$

Here $K(s)$ is a function of the entropy per unit mass s and is thus constant during adiabatic evolution, but it will change across shocks, because the entropy increases in a shock (Sec. 17.5). The value of γ depends on the number of thermalized internal degrees of freedom of the gas's constituent particles (Ex. 17.1). For a gas of free particles (e.g., fully ionized hydrogen), the value is $\gamma = 5/3$; for Earth's atmosphere at temperatures between about 10 K and 400 K, it is $\gamma = 7/5 = 1.4$ (Ex. 17.1).

For a polytropic gas with $P = K(s)\rho^\gamma$, we can integrate the first law of thermodynamics (Box 13.2) to obtain a formula for the internal energy per unit mass:

$$u = \frac{P}{(\gamma - 1)\rho}, \quad (17.2b)$$

where we have assumed that the internal energy vanishes as the temperature $T \to 0$ and thence $P \to 0$. It will prove convenient to express the density ρ, the internal energy per unit mass u, and the enthalpy per unit mass h in terms of the sound speed:

$$C = \sqrt{\left(\frac{\partial P}{\partial \rho}\right)_s} = \sqrt{\frac{\gamma P}{\rho}} \qquad (17.2c)$$

[Eq. (16.48)]. A little algebra gives

$$\rho = \left(\frac{C^2}{\gamma K}\right)^{1/(\gamma-1)}, \quad u = \frac{C^2}{\gamma(\gamma-1)}, \quad h = u + \frac{P}{\rho} = \frac{C^2}{\gamma-1} = \frac{\gamma P}{(\gamma-1)\rho}. \qquad (17.2d)$$

Exercise 17.1 **Example: Values of γ**

Consider an ideal gas consisting of several different particle species (e.g., diatomic oxygen molecules and nitrogen molecules in the case of Earth's atmosphere). Consider a sample of this gas with volume V, containing N_A particles of various species A, all in thermodynamic equilibrium at a temperature T sufficiently low that we can ignore the effects of special relativity. Let species A have ν_A internal degrees of freedom with the hamiltonian quadratic in their generalized coordinates (e.g., rotation and vibration), and assume that those degrees of freedom are sufficiently thermally excited to have reached energy equipartition. Then the equipartition theorem (Sec. 4.4.4) dictates that each such particle has $\frac{3}{2}k_B T$ of translational energy plus $\frac{1}{2}\nu_A k_B T$ of internal energy, and because the gas is ideal, each particle contributes $k_B T/V$ to the pressure. Correspondingly, the sample's total energy E and pressure P are

$$E = \sum_A \left(\frac{3}{2} + \frac{\nu_A}{2}\right) N_A k_B T, \quad P = \frac{1}{V}\sum_A N_A k_B T. \qquad (17.3a)$$

(a) Use the laws of thermodynamics to show that the specific heats at fixed volume and pressure are

$$C_V \equiv \left(T\frac{\partial S}{\partial T}\right)_{V,N_A} = \frac{E}{T} = \sum_A \left(\frac{3}{2} + \frac{\nu_A}{2}\right) N_A k_B,$$

$$C_P = \left(T\frac{\partial S}{\partial T}\right)_{P,N_A} = C_V + \frac{PV}{T}, \qquad (17.3b)$$

so the ratio of specific heats is

$$\gamma = \frac{C_P}{C_V} = 1 + \frac{\sum_A N_A}{\sum_A N_A \left(\frac{3}{2} + \frac{\nu_A}{2}\right)} \qquad (17.3c)$$

(cf. Ex. 5.4).

(b) If there are no thermalized internal degrees of freedom, $\nu_A = 0$ (e.g., for a fully ionized, nonrelativistic gas), then $\gamma = 5/3$. For Earth's atmosphere, at temperatures between about 10 K and 400 K, the rotational degrees of freedom of the O_2 and N_2 molecules are thermally excited, but the temperature is too low to excite their vibrational degrees of freedom. Explain why this means that $\nu_{O_2} = \nu_{N_2} = 2$,

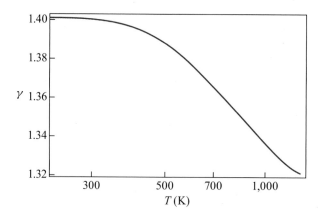

FIGURE 17.3 The ratio of specific heats γ for air as a function of temperature.

which implies $\gamma = 7/5 = 1.4$. [Hint: There are just two orthogonal axes around which the diatomic molecule can rotate.]

(c) Between about 1,300 K and roughly 10,000 K the vibrational degrees of freedom are thermalized, but the molecules have not dissociated substantially into individual atoms, nor have they become substantially ionized. Explain why this means that $\nu_{O_2} = \nu_{N_2} = 4$ in this temperature range, which implies $\gamma = 9/7 \simeq 1.29$. [Hint: An oscillator has kinetic energy and potential energy.]

(d) At roughly 10,000 K the two oxygen atoms in O_2 dissociate, the two nitrogen atoms in N_2 dissociate, and electrons begin to ionize. Explain why this drives γ up toward $5/3 \simeq 1.67$.

The actual value of γ as a function of temperature for the range 200 K to 1,300 K is shown in Fig. 17.3. Evidently, as stated, $\gamma = 1.4$ is a good approximation only up to about 400 K, and the transition toward $\gamma = 1.29$ occurs gradually between about 400 K and 1,400 K as the vibrational degrees of freedom gradually become thermalized and begin to obey the equipartition theorem (Sec. 4.4.4).

17.3

17.3 Stationary, Irrotational, Quasi-1-Dimensional Flow

17.3.1

17.3.1 Basic Equations; Transition from Subsonic to Supersonic Flow

In their full generality, the fluid dynamic equations (17.1) are quite unwieldy. To demonstrate some of the novel features of supersonic flow, we proceed as in earlier chapters: we specialize to a simple type of flow in which the physical effects of interest are strong, and extraneous effects are negligible.

In particular, in this section, we seek insight into smooth transitions between subsonic and supersonic flow by restricting ourselves to a stationary ($\partial/\partial t = 0$),

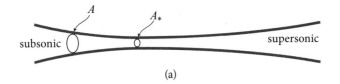

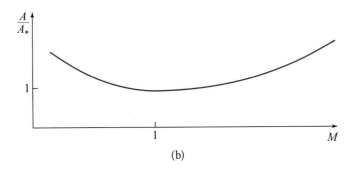

FIGURE 17.4 Stationary, transonic flow in a converging and then diverging streamtube. (a) The streamtube. (b) The flow's Mach number $M = v/C$ (horizontal axis) as a function of the streamtube's area A (vertical axis). The flow is subsonic to the left of the streamtube's throat $A = A_*$, sonic at the throat, and supersonic to the right.

irrotational ($\nabla \times \mathbf{v} = 0$) flow in which gravity and viscosity are negligible ($\Phi = \mathbf{g} = \eta = \zeta = 0$), as are various effects not included in our general equations: chemical reactions, thermal conductivity, and radiative losses. (We explore some effects of gravity in Ex. 17.5.) The vanishing viscosity implies [from the entropy evolution equation (17.1d)] that the entropy per baryon s is constant along each flow line. We assume that s is the same on all flow lines, so the flow is fully isentropic (s is constant everywhere), and the pressure $P = P(\rho, s)$ can thus be regarded as a function only of the density: $P = P(\rho)$. When we need a specific form for $P(\rho)$, we will use the polytropic form $P = K(s)\rho^{\gamma}$ for an ideal gas with constant specific-heat ratio γ [Eqs. (17.2); Ex. 17.1], but much of our analysis is done for a general isentropic $P(\rho)$. We make one further approximation: the flow is almost 1-dimensional. In other words, the velocity vectors all make small angles with one another in the region of interest.

stationary, irrotational, isentropic flow without gravity or viscosity

These drastic simplifications are actually appropriate for many cases of practical interest. Granted these simplifications, we can consider a narrow bundle of streamlines—which we call a *streamtube*—and introduce as a tool for analysis its cross sectional area A normal to the flow (Fig. 17.4a).

streamtube

As the flow is stationary, the equation of mass conservation (17.1a) states that the rate $\dot{m}$ at which mass passes through the streamtube's cross section must be independent of position along the tube:

$$\rho v A = \dot{m} = \text{const};$$ (17.4a)

here v is the speed of the fluid in the streamtube. Rewriting this in differential form, we obtain

$$\frac{dA}{A} + \frac{d\rho}{\rho} + \frac{dv}{v} = 0. \tag{17.4b}$$

Because the flow is stationary and inviscid, the law of energy conservation (17.1c) reduces to Bernoulli's theorem [Eqs. (13.51), (13.50)]:

$$h + \frac{1}{2}v^2 = h_1 = \text{const} \tag{17.4c}$$

along each streamline and thus along our narrow streamtube. Here h_1 is the specific enthalpy at a location where the flow velocity v vanishes (e.g., in chamber 1 of Fig. 17.5a below). Since the flow is adiabatic, we can use the first law of thermodynamics (Box 13.2) $dh = dP/\rho + Tds = dP/\rho = C^2 d\rho/\rho$ [where C is the speed of sound; Eq. (17.2c)] to write Eq. (17.4c) in the differential form

$$\frac{d\rho}{\rho} + \frac{vdv}{C^2} = 0. \tag{17.4d}$$

Finally and most importantly, we combine Eqs. (17.4b) and (17.4d) to obtain

$$\boxed{\frac{dv}{v} = \frac{dA/A}{M^2 - 1}, \qquad \frac{d\rho}{\rho} = \frac{dA/A}{M^{-2} - 1},} \tag{17.5}$$

where

Mach number

$$\boxed{M \equiv v/C} \tag{17.6}$$

is the *Mach number*. This Mach number is an important dimensionless number that is used to characterize compressible flows. When the Mach number is less than 1, the

subsonic and supersonic flow

flow is *subsonic*; when $M > 1$, it is *supersonic*. By contrast with the Reynolds, Rossby, and Ekman numbers, which are usually defined using a single set of (characteristic) values of the flow parameters (V, ν, Ω, and L) and thus have a single value for any given flow, the Mach number by convention is defined at each point in the flow and thus is a flow variable, $M(\mathbf{x})$, similar to $v(\mathbf{x})$ and $\rho(\mathbf{x})$.

properties of subsonic and supersonic flow

Equations (17.5) make remarkable predictions, which we illustrate in Fig. 17.4 for a particular flow called "transonic":

1. The only locations along a streamtube at which M can be unity ($v = C$) are those where A is an extremum—for example, for the streamtube in Fig. 17.4, the minimum $A = A_*$ (the tube's throat).

2. At points along a streamtube where the flow is subsonic, $M < 1$ (left side of the streamtube in Fig. 17.4), v increases when A decreases, in accord with everyday experience.

3. At points where the flow is supersonic, $M > 1$ (right side of Fig. 17.4), v increases when A increases—just the opposite of everyday experience.

These conclusions are useful when analyzing stationary, high-speed flows.

17.3.2 Setting up a Stationary, Transonic Flow

At this point the reader may wonder whether it is easy to set up a transonic flow in which the speed of the fluid changes continuously from subsonic to supersonic, as in Fig. 17.4. The answer is quite illuminating. We can illustrate the answer using two chambers maintained at different pressures, P_1 and P_2, and connected through a narrow channel, along which the cross sectional area passes smoothly through a minimum $A = A_*$, the channel's throat (Fig. 17.5a). When $P_2 = P_1$, no flow occurs between the two chambers. When we decrease P_2 slightly below P_1, a slow subsonic flow moves through the channel (curves 1 in Fig. 17.5b,c). As we decrease P_2 further, there comes a point ($P = P_2^{\text{crit}}$) at which the flow becomes transonic at the channel's throat $A = A_*$ (curves 2). For all pressures $P_2 < P_2^{\text{crit}}$, the flow is also transonic at the throat and has a universal form to the left of and near the throat, independent of the value of P_2 (curves 2)—including a universal value $\dot{m}_{\text{crit}}$ for the rate of mass flow through the throat! This universal flow is supersonic to the right of the throat (curves 2b), but it must be brought to rest in chamber 2, since there is a hard wall at the chamber's end. How is it brought to rest? Through a shock front, where it is driven subsonic almost discontinuously (curves 3 and 4; see Sec. 17.5).

inducing transonic flow by increasing pressure difference

How, physically, is it possible for the transonic flow to have a universal form to the left of the shock? The key is that, in any supersonic region of the flow, disturbances are unable to propagate upstream, so the upstream fluid has no way of knowing what the pressure P_2 is in chamber 2. Although the flow to the left of the shock is universal, the location of the shock and the nature of the subsonic, post-shock flow are affected by P_2, since information can propagate upstream through that subsonic flow, from chamber 2 to the shock.

The reader might now begin to suspect that the throat, in the transonic case, is a special location. It is, and that location is known as a *critical point* of the stationary flow. From a mathematical point of view, critical points are singular points of the equations (17.4) and (17.5) of stationary flow. This singularity shows up in the solutions to the equations, as depicted in Fig. 17.5c. The universal solution that passes transonically through the critical point (solution 2) joins onto two different solutions to the right of the throat: solution 2a, which is supersonic, and solution 2b, which is subsonic. Which solution occurs in practice depends on conditions downstream. Other solutions that are arbitrarily near this universal solution (dashed curves in Fig. 17.5c) are either double valued and consequently unphysical, or are everywhere subsonic or everywhere supersonic (in the absence of shocks); see Box 17.3 and Ex. 17.2.

critical point of a stationary flow

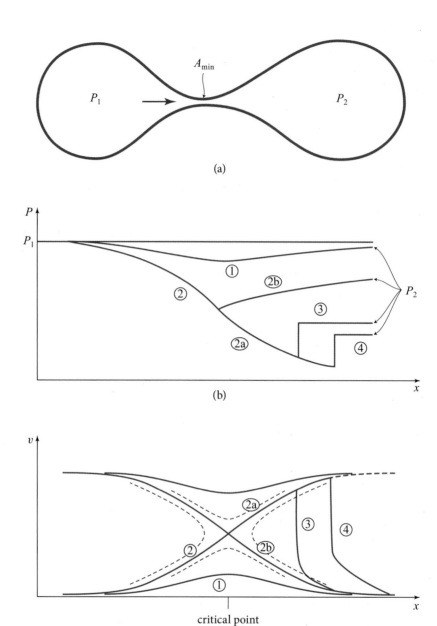

FIGURE 17.5 Stationary flow through a channel between two chambers maintained at different pressures P_1 and P_2. (a) Setup of the chambers. When the pressure difference $P_1 - P_2$ is large enough [see curves in panel (b)], the flow is subsonic to the left of the channel's throat and supersonic to the right [see curves in panel (c)]. As it nears or enters the second chamber, the supersonic flow must encounter a strong shock, where it decelerates abruptly to subsonic speed [panels (b) and (c)]. The forms of the various velocity profiles $v(x)$ in panel (c) are explained in Box 17.3.

BOX 17.3. VELOCITY PROFILES FOR 1-DIMENSIONAL FLOW
BETWEEN CHAMBERS

Consider the adiabatic, stationary flow of an isentropic, polytropic fluid $P = K\rho^\gamma$ between the two chambers shown in Fig. 17.5. Describe the channel between chambers by its cross sectional area $A(x)$ as a function of distance x, and describe the flow by the fluid's velocity $v(x)$ and its sound speed $C(x)$. There are two coupled algebraic equations for $v(x)$ and $C(x)$: mass conservation $\rho v A = \dot{m}$ [Eq. (17.4a)] and the Bernoulli theorem $h + \frac{1}{2}v^2 = h_1$ [Eq. (17.4c)], which, for our polytropic fluid, become [see Eqs. (17.2d)]:

$$C^{2/(\gamma-1)}v = (\gamma K)^{1/(\gamma-1)}\dot{m}/A, \qquad \frac{C^2}{\gamma-1} + \frac{v^2}{2} = \frac{C_1^2}{\gamma-1}. \qquad (1)$$

These equations are graphed in diagrams below for three different mass flow rates $\dot{m}$. Mass conservation [the first of Eqs. (1)] is a set of generalized hyperbolas, one for each value of the channel's area $A_* < A_a < A_b < A_c$. The Bernoulli theorem [the second of Eqs. (1)] is a single ellipse. On a chosen diagram (for a chosen $\dot{m}$), the dot at the intersection of the ellipse with a hyperbola tells us the flow velocity v and speed of sound C at each of the two points in the channel where the area A has the hyperbola's value.

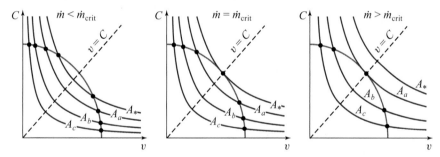

There is a critical mass-flow rate $\dot{m}_{crit}$ (central diagram above), such that the hyperbola $A = A_*$ (the channel's throat) is tangent to the ellipse at the point $v = C$, so the flow is Mach 1 ($v = C$). For this $\dot{m}$, the sequence of dots along the ellipse, moving from lower right to upper left, represents the transonic flow, which begins as subsonic and becomes supersonic (upward swooping solid curve in the drawing below); and the same sequence of dots, moving in the opposite direction from upper left to lower right, represents a flow that begins as supersonic and smoothly transitions to subsonic (downward swooping solid curve, below). When $\dot{m} < \dot{m}_{crit}$ (left diagram above; top and bottom quadrants below), the sequence of dots beginning at lower

(continued)

BOX 17.3.   (continued)

right reaches the throat $A = A_*$ at a subsonic velocity, so the solution climbs up along the ellipse to that point and then descends back down, mapping out a fully subsonic solution $v(x)$ below—and similarly for the dots on the upper branch of the ellipse, which map out a fully supersonic solution below. When $\dot{m} > \dot{M}_{crit}$ (right diagram above), the dots map out curves in the left and right quadrants below that never reach the sonic point and are double valued for $v(x)$—and are thus unphysical. Therefore, the mass flow rate $\dot{m}$ can never exceed $\dot{m}_{crit}$.

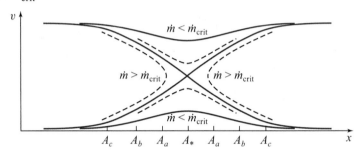

The existence of critical points is a price we must pay, mathematically, for not allowing our equations to be time dependent. If we were to solve the time-dependent equations (which would then be partial differential equations), we would find that they change from elliptic to hyperbolic as the flow passes through a critical point.

properties of critical points

From a physical point of view, critical points are the places where a sound wave propagating upstream remains at rest in the flow. They are therefore the one type of place from which time-dependent transients, associated with setting up the flow in the first place, cannot decay away (if the equations are dissipation-free, i.e., inviscid). Thus, even the time-dependent equations can display peculiar behaviors at a critical point. However, when dissipation, for example due to viscosity, is introduced, these

dissipation smears out critical points

peculiarities get smeared out.

EXERCISES

Exercise 17.2 *Problem: Explicit Solution for Flow between Chambers When $\gamma = 3$*
For $\gamma = 3$ and for a channel with $A = A_*(1 + x^2)$, solve the flow equations (1) of Box 17.3 analytically and explicitly for $v(x)$, and verify that the solutions have the qualitative forms depicted in the last figure of Box 17.3.

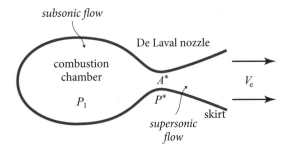

FIGURE 17.6 Schematic illustration of a rocket engine. Note the skirt, which increases the thrust produced by the escaping exhaust gases.

17.3.3 Rocket Engines

We have shown that, to push a quasi-1-dimensional flow from subsonic to supersonic, one must send it through a throat. This result is exploited in the design of rocket engines and jet engines.

In a rocket engine, hot gas is produced by controlled burning of fuel in a large chamber, and the gas then escapes through a *converging-diverging* (also known as a *De Laval*[1]) nozzle, as shown in Fig. 17.6. The nozzle is designed with a skirt so the flow becomes supersonic smoothly when it passes through the nozzle's throat.

De Laval nozzle

To analyze this flow in some detail, let us approximate it as precisely steady, isentropic, and quasi-1-dimensional, and the gas as ideal and inviscid with a constant ratio of specific heats γ. In this case, the enthalpy is $h = C^2/(\gamma - 1)$ [Eqs. (17.2d)], so Bernoulli's theorem (17.4c) reduces to

$$\frac{C^2}{(\gamma - 1)} + \frac{1}{2}v^2 = \frac{C_1^2}{(\gamma - 1)}. \tag{17.7}$$

Here C is the sound speed in the flow and C_1 is the *stagnation* sound speed (i.e., the sound speed evaluated in the rocket chamber where $v = 0$). Dividing this Bernoulli theorem by C^2 and manipulating, we learn how the sound speed varies with Mach number $M = v/C$:

$$C = C_1 \left[1 + \frac{\gamma - 1}{2} M^2 \right]^{-1/2}. \tag{17.8}$$

how sound speed varies with Mach number

From mass conservation [Eq. (17.4a)], we know that the cross sectional area A varies as $A \propto \rho^{-1} v^{-1} \propto \rho^{-1} M^{-1} C^{-1} \propto M^{-1} C^{(\gamma+1)/(1-\gamma)}$, where we have used $\rho \propto C^{2/(\gamma-1)}$ [Eqs. (17.2d)]. Combining with Eq. (17.8), and noting that $M = 1$ where $A = A_*$ (i.e.,

---

1. First used in a steam turbine in 1882 by the Swedish engineer Gustaf de Laval. De Laval is most famous for developing a device to separate cream from milk, centrifugally!

the flow is sonic at the throat), we find that

$$\frac{A}{A_*} = \frac{1}{M}\left[\frac{2}{\gamma+1} + \left(\frac{\gamma-1}{\gamma+1}\right)M^2\right]^{\frac{(\gamma+1)}{2(\gamma-1)}}.$$
(17.9)

The pressure P_* at the throat can be deduced from $P \propto \rho^\gamma \propto C^{2\gamma/(\gamma-1)}$ [Eqs. (17.2a) and (17.2d)] together with Eq. (17.8) with $M = 0$ and $P = P_1 =$ (stagnation pressure) in the chamber and $M = 1$ at the throat:

$$P_* = P_1\left(\frac{2}{\gamma+1}\right)^{\frac{\gamma}{\gamma-1}}.$$
(17.10)

We use these formulas in Box 17.4 and Ex. 17.3 to evaluate, numerically, some features of the Space Shuttle and its rocket engines.

Bernoulli's theorem is a statement that the fluid's energy is conserved along a streamtube. (For conceptual simplicity we regard the entire interior of the nozzle as a single streamtube.) By contrast with energy, the fluid's momentum is not conserved, since it pushes against the nozzle wall as it flows. As the subsonic flow accelerates down the nozzle's converging region, the area of its streamtube diminishes, and the momentum flowing per second in the streamtube, $(P + \rho v^2)A$, decreases; momentum is being transferred to the nozzle wall. If the rocket did not have a skirt but instead opened up completely to the outside world at its throat, the rocket thrust would be

$$T_* = (\rho_* v_*^2 + P_*)A_* = (\gamma+1)P_*A_*.$$
(17.11)

This is much less than if momentum had been conserved along the subsonic, accelerating streamtubes.

Much of the "lost" momentum is regained, and the thrust is made significantly larger than T_*, by the force of the skirt on the stream tube in the diverging part of the nozzle (Fig. 17.6). The nozzle's skirt keeps the flow quasi-1-dimensional well beyond the throat, driving it to be more and more strongly supersonic. In this accelerating, supersonic flow the tube's rate of momentum flow $(\rho v^2 + P)A$ increases downstream, with a compensating increase of the rocket's forward thrust. This skirt-induced force accounts for a significant fraction of the thrust of a well-designed rocket engine.

Rockets work most efficiently when the exit pressure of the gas, as it leaves the base of the skirt, matches the external pressure in the surrounding air. When the pressure in the exhaust is larger than the external pressure, the flow is termed *underexpanded* and a pulse of low pressure, known as a *rarefaction*, will be driven into the escaping gases, causing them to expand and increasing their speed. However, the exhaust will now be pushing on the surrounding air, rather than on the rocket. More thrust could have been exerted on the rocket if the flow had not been underexpanded. By contrast, when the exhaust has a smaller pressure than the surrounding air (i.e., is *overexpanded*), shock fronts will form near the exit of the nozzle, affecting the fluid flow and sometimes causing separation of the flow from the nozzle's walls. It is important that the nozzle's skirt be shaped so that the exit flow is neither seriously over- nor underexpanded.

BOX 17.4.   SPACE SHUTTLE

NASA's (now retired) Space Shuttle provides many nice examples of the behavior of supersonic flows. At launch, the shuttle and fuel had a mass $\sim 2 \times 10^6$ kg. The maximum thrust, $T \sim 3 \times 10^7$ N, occurred at lift-off and gave the rocket an initial acceleration relative to the ground of $\sim 0.5g$. This increased to $\sim 3g$ as the fuel was burned and the total mass diminished. Most of the thrust was produced by two solid-fuel boosters that burned fuel at a combined rate of $\dot{m} \sim 10{,}000$ kg s^{-1} over a 2-min period. Their combined thrust was $T \sim 2 \times 10^7$ N averaged over the 2 minutes, from which we can estimate the speed of the escaping gases as they left the nozzles' skirts. Assuming this speed was quite supersonic (so $P_e \ll \rho_e v_e^2$), we estimate that $v_e \sim T/\dot{m} \sim 2$ km s^{-1}. The combined exit areas of the two skirts was $A_e \sim 20$ m^2, roughly four times the combined throat area, A_*. Using Eq. (17.9) with $\gamma \sim 1.29$, we deduce that the exit Mach number was $M_e \sim 3$.

From $T \sim \rho_e v_e^2 A_e$ and $P_e = C_e^2 \rho_e / \gamma$, we deduce the exit pressure, $P_e \sim T/(\gamma M_e^2 A_e) \sim 8 \times 10^4$ N m^{-2}, about atmospheric. The stagnation pressure in the combustion region was [combine Eqs. (17.2a), (17.2d), and (17.8)]

$$P_1 \sim P_e \left[1 + \frac{(\gamma - 1)M_e^2}{2} \right]^{\frac{\gamma}{\gamma - 1}} \sim 35 \text{ atmospheres.} \tag{1}$$

Of course, the actual operation was far more complex than this. For example, to optimize the final altitude, one must allow for the decreasing mass and atmospheric pressure as well as the 2-dimensional gas flow through the nozzle.

The Space Shuttle can also be used to illustrate the properties of shock waves (Sec. 17.5). When the shuttle reentered the atmosphere, it was highly supersonic, and therefore was preceded by a strong shock front that heated the onrushing air and consequently heated the shuttle. The shuttle continued moving supersonically down to an altitude of 15 km, and until this time it created a shock-front pattern that could be heard on the ground as a sonic boom. The maximum heating rate occurred at 70 km. Here, the shuttle moved at $V \sim 7$ km s^{-1}, and the sound speed is about 280 m s^{-1}, giving a Mach number of 25. For the specific-heat ratio $\gamma \sim 1.5$ and mean molecular weight $\mu \sim 10$ appropriate to dissociated air, the strong-shock Rankine-Hugoniot relations (17.37) (see Sec. 17.5.2), together with $P = [\rho/(\mu m_p)] k_B T$ and $C^2 = \gamma P/\rho$, predict a post-shock temperature of

$$T \sim \frac{2(\gamma - 1)\mu m_p V^2}{(\gamma + 1)^2 k_B} \sim 9{,}000 \text{ K.} \tag{2}$$

Exposure to gas at this high temperature heated the shuttle's nose to $\sim 1{,}800$ K.

(continued)

There is a second, well-known consequence of this high temperature: it is sufficient to ionize the air partially as well as dissociate it. As a result, during reentry the shuttle was surrounded by a sheath of plasma, which, as we shall discover in Chap. 19, prevented radio communication. The blackout lasted for about 12 minutes.

EXERCISES

Exercise 17.3 *Problem: Space Shuttle's Solid-Fuel Boosters*
Use the rough figures in Box 17.4 to estimate the energy released per unit mass in burning the fuel. Does your answer seem reasonable?

Exercise 17.4 *Problem: Relativistic 1-Dimensional Flow* **T2**
Use the development of relativistic gas dynamics in Sec. 13.8.2 to show that the cross sectional area of a relativistic 1-dimensional flow tube is also minimized when the flow is transonic. Assume that the equation of state is $P = \frac{1}{3}\rho c^2$. For details see Blandford and Rees (1974).

Exercise 17.5 ***Example: Adiabatic, Spherical Accretion of Gas
onto a Black Hole or Neutron Star*
Consider a black hole or neutron star with mass $\mathcal{M}$ at rest in interstellar gas that has constant ratio of specific heats γ. In this exercise you will derive some features of the adiabatic, spherical accretion of the gas onto the hole or star, a problem first solved by Bondi (1952). This exercise shows how gravity can play a role analogous to a De Laval nozzle: it can trigger a transition of the flow from subsonic to supersonic. Although, near the black hole or neutron star, spacetime is significantly curved and the flow becomes relativistic, we shall confine ourselves to a Newtonian treatment.

(a) Let ρ_∞ and C_∞ be the density and sound speed in the gas far from the hole (at radius $r = \infty$). Use dimensional analysis to estimate the rate of accretion of mass $\dot{\mathcal{M}}$ onto the star or hole in terms of the parameters of the system: $\mathcal{M}, \gamma, \rho_\infty, C_\infty$, and Newton's gravitation constant G. [Hint: Dimensional considerations alone cannot give the answer. Why? Augment your dimensional considerations by a knowledge of how the answer should scale with one of the parameters (e.g., the density ρ_∞).]

(b) Give a simple physical argument, devoid of dimensional considerations, that produces the same answer for $\dot{\mathcal{M}}$, to within a multiplicative factor of order unity, as you deduced in part (a).

(c) Because the neutron star and black hole are both very compact with intense gravity near their surfaces, the inflowing gas is guaranteed to accelerate to supersonic speeds as it falls in. Explain why the speed will remain supersonic in the case of the hole, but must transition through a shock to subsonic flow near the surface of the neutron star. If the star has the same mass $\mathcal{M}$ as the hole, will the details of its accretion flow $[\rho(r), C(r), v(r)]$ be the same as or different from those for the hole, outside the star's shock? Will the mass accretion rates $\dot{\mathcal{M}}$ be the same or different? Justify your answers physically.

(d) By combining the Euler equation for $v(r)$ with the equation of mass conservation, $\dot{\mathcal{M}} = 4\pi r^2 \rho v$, and with the sound-speed equation $C^2 = (\partial P/\partial\rho)_s$, show that

$$(v^2 - C^2)\frac{1}{\rho}\frac{d\rho}{dr} = \frac{GM}{r^2} - \frac{2v^2}{r}, \tag{17.12}$$

and the flow speed v_s, sound speed C_s, and radius r_s at the *sonic point* (the transition from subsonic to supersonic; the flow's critical point) are related by

$$v_s^2 = C_s^2 = \frac{GM}{2r_s}. \tag{17.13}$$

(e) By combining with the Bernoulli equation (with the effects of gravity included), deduce that the sound speed at the sonic point is related to that at infinity by

$$C_s^2 = \frac{2C_\infty^2}{5 - 3\gamma} \tag{17.14}$$

and that the radius of the sonic point is

$$r_s = \frac{(5 - 3\gamma)}{4}\frac{GM}{C_\infty^2}. \tag{17.15}$$

Thence also deduce a precise value for the mass accretion rate $\dot{\mathcal{M}}$ in terms of the parameters of the problem. Compare with your estimate of $\dot{\mathcal{M}}$ in parts (a) and (b). [Comment: For $\gamma = 5/3$, which is the value for hot, ionized gas, this analysis places the sonic point at an arbitrarily small radius. In this limiting case, (i) general relativistic effects strengthen the gravitational field (Sec. 26.2), thereby moving the sonic point well outside the star or hole, and (ii) the answer for $\dot{\mathcal{M}}$ has a finite value close to the general relativistic prediction.]

(f) Much of the interstellar medium is hot and ionized, with a density of about 1 proton per cubic centimeter and temperature of about 10^4 K. In such a medium, what is the mass accretion rate onto a 10-solar-mass hole, and approximately how long does it take for the hole's mass to double?

17.4 1-Dimensional, Time-Dependent Flow

17.4.1 Riemann Invariants

Let us turn now to time-dependent flows. Again we confine our attention to the simplest situation that illustrates the physics—in this case, truly 1-dimensional motion of an isentropic fluid in the absence of viscosity, thermal conductivity, and gravity, so the flow is adiabatic as well as isentropic (entropy constant in time as well as space). The motion of the gas in such a flow is described by the equation of continuity (17.1a) and the Euler equation (17.1b) specialized to 1 dimension:

$$\frac{d\rho}{dt} = -\rho \frac{\partial v}{\partial x}, \qquad \frac{dv}{dt} = -\frac{1}{\rho}\frac{\partial P}{\partial x}, \tag{17.16}$$

where

$$\frac{d}{dt} = \frac{\partial}{\partial t} + v\frac{\partial}{\partial x} \tag{17.17}$$

is the convective (advective) time derivative—the time derivative moving with the fluid.

Given an isentropic equation of state $P = P(\rho)$ that relates the pressure to the density, these two nonlinear equations can be combined into a single second-order differential equation in the velocity. However, it is more illuminating to work with the first-order set. As the gas is isentropic, the density ρ and sound speed $C = (dP/d\rho)^{1/2}$ can both be regarded as functions of a single thermodynamic variable, which we choose to be the pressure.

Taking linear combinations of Eqs. (17.16), we obtain two partial differential equations:

$$\frac{\partial v}{\partial t} \pm \frac{1}{\rho C}\frac{\partial P}{\partial t} + (v \pm C)\left(\frac{\partial v}{\partial x} \pm \frac{1}{\rho C}\frac{\partial P}{\partial x}\right) = 0, \tag{17.18}$$

which together are equivalent to Eqs. (17.16). We can rewrite these equations in terms of *Riemann invariants*

Riemann invariants

$$\boxed{J_\pm \equiv v \pm \int \frac{dP}{\rho C}} \tag{17.19}$$

and *characteristic speeds*

characteristic speeds

$$\boxed{V_\pm \equiv v \pm C} \tag{17.20}$$

in the following way:

flow equations

$$\boxed{\left(\frac{\partial}{\partial t} + V_\pm \frac{\partial}{\partial x}\right)J_\pm = 0.} \tag{17.21}$$

Equation (17.21) tells us that the convective derivative of each Riemann invariant $J_\pm$ vanishes for an observer who moves, not with the fluid speed, but, instead, with the speed $V_\pm$. We say that each Riemann invariant is conserved along its characteristic (denoted by $\mathcal{C}_\pm$), which is a path through spacetime satisfying

Riemann invariant conserved along its characteristic

$$\boxed{\mathcal{C}_\pm: \quad \frac{dx}{dt} = v \pm C.} \tag{17.22}$$

Note that in these equations, both v and C are functions of x and t.

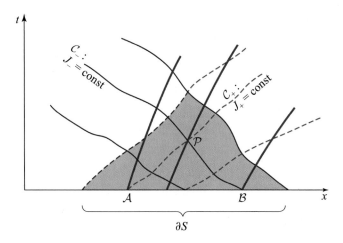

FIGURE 17.7 Spacetime diagram showing the characteristics (thin solid and dashed lines) for a 1-dimensional adiabatic flow of an isentropic gas. The paths of the fluid elements are shown as thick solid lines. Initial data are presumed to be specified over some interval ∂S of x at time $t = 0$. The Riemann invariant J_+ is constant along each characteristic $\mathcal{C}_+$ (thin dashed line) and thus at point $\mathcal{P}$ it has the same value, unchanged, as at point $\mathcal{A}$ in the initial data. Similarly J_- is invariant along each characteristic $\mathcal{C}_-$ (thin solid line) and thus at $\mathcal{P}$ it has the same value as at $\mathcal{B}$. The shaded area of spacetime is the domain of dependence S of ∂S.

The characteristics have a natural interpretation. They describe the motion of small disturbances traveling backward and forward relative to the fluid at the local sound speed. As seen in the fluid's local rest frame $v = 0$, two neighboring events in the flow, separated by a small time interval Δt and a space interval $\Delta x = +C\,\Delta t$—so that they lie on the same $\mathcal{C}_+$ characteristic—will have small velocity and pressure differences satisfying $\Delta v = -\Delta P/(\rho C)$ [as one can deduce from Eqs. (17.16) with $v = 0$, $d/dt = \partial/\partial t$, and $C^2 = dP/d\rho$]. Now, for a linear sound wave propagating along the positive x direction, Δv and ΔP will separately vanish. However, in a nonlinear wave, only the combination $\Delta J_+ = \Delta v + \Delta P/(\rho C)$ will vanish along $\mathcal{C}_+$. Integrating over a finite interval of time, we recover the constancy of J_+ along the characteristic $\mathcal{C}_+$ [Eq. (17.21)].

physical interpretation of characteristics

The Riemann invariants provide a general method for deriving the details of the flow from initial conditions. Suppose that the fluid velocity and the thermodynamic variables are specified over an interval of x, designated ∂S, at an initial time $t = 0$ (Fig. 17.7). This means that $J_\pm$ are also specified over this interval. We can then determine $J_\pm$ at any point $\mathcal{P}$ in the *domain of dependence S* of ∂S (i.e., at any point linked to ∂S by two characteristics $\mathcal{C}_\pm$) by simply propagating each of $J_\pm$ unchanged along its characteristic. From these values of $J_\pm$ at $\mathcal{P}$, we can solve algebraically for all the other flow variables (v, P, ρ, etc.) at $\mathcal{P}$. To learn the evolution outside the domain of dependence S, we must specify the initial conditions outside ∂S.

using characteristics and Riemann invariants to solve for details of flow in domain of dependence

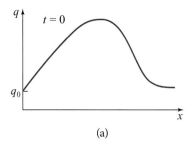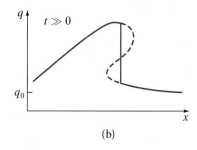

(a)　　　　　　　　　　　　　　　(b)

FIGURE 17.8 Evolution of a nonlinear sound wave. (a) The fluid at the crest of the wave moves faster than the fluid in the trough. (b) Mathematically, the flow eventually becomes triple-valued (dashed curve). Physically, a shock wave develops (vertical solid line).

In practice, we do not actually know the characteristics $\mathcal{C}_\pm$ until we have solved for the flow variables, so we must solve for the characteristics as part of the solution process. This means, in practice, that the solution involves algebraic manipulations of (i) the equation of state and the relations $J_\pm = v \pm \int dP/(\rho C)$, which give $J_\pm$ in terms of v and C; and (ii) the conservation laws that $J_\pm$ are constant along $\mathcal{C}_\pm$ (i.e., along curves $dx/dt = v \pm C$). These algebraic manipulations have the goal of deducing $C(x, t)$ and $v(x, t)$ from the initial conditions on ∂S. We exhibit a specific example in Sec. 17.4.2.

how a nonlinear sound wave evolves to form a shock wave

We can use Riemann invariants to understand qualitatively how a nonlinear sound wave evolves with time. If the wave propagates in the positive x direction into previously undisturbed fluid (fluid with $v = 0$), then the J_- invariant, propagating backward along $\mathcal{C}_-$, is constant everywhere, so $v = \int dP/(\rho C) + \text{const}$. Let us use $q \equiv \int dP/(\rho C)$ as our wave variable. For an ideal gas with a constant ratio of specific heats γ, we have $q = 2C/(\gamma - 1)$, so our oscillating wave variable is essentially the oscillating sound speed. Constancy of J_- then says that $v = q - q_0$, where q_0 is the stagnation value of q (i.e., the value of q in the undisturbed fluid in front of the wave).

Now, $J_+ = v + q$ is conserved on each rightward characteristic $\mathcal{C}_+$, and so both v and q are separately conserved on each $\mathcal{C}_+$. If we sketch a profile of the wave pulse as in Fig. 17.8 and measure its amplitude using the quantity q, then the relation $v = q - q_0$ requires that the fluid at the crest of the wave moves faster than the fluid in a trough. This causes the leading edge of the wave to steepen, a process we have already encountered in our discussion of shallow-water solitons (Fig. 16.4). Now, sound waves, by contrast with shallow-water waves (where dispersion counteracts the steepening), are nondispersive so the steepening will continue until $|dv/dx| \to \infty$ (Fig. 17.8). When the velocity gradient becomes sufficiently large, viscosity and dissipation become strong, producing an increase of entropy and a breakdown of our isentropic flow. This breakdown and entropy increase will occur in an extremely thin region—a shock wave, which we study in Sec. 17.5.

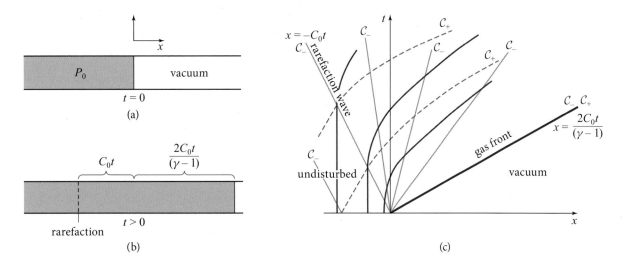

FIGURE 17.9 Shock tube. (a) At $t \leq 0$ gas is held at rest at high pressure P_0 in the left half of the tube. (b) At $t > 0$ the high-pressure gas moves rightward down the tube at high speed, and a rarefaction wave propagates leftward at the sound speed. (c) Space-time diagram showing the flow's characteristics ($\mathcal{C}_+$: thin dashed lines; $\mathcal{C}_-$: thin solid lines) and fluid paths (thick solid lines). To the left of the rarefaction wave, $x < -C_0 t$, the fluid is undisturbed. To the right of the gas front, $x > [2/(\gamma - 1)]C_0 t$, resides undisturbed (near) vacuum.

17.4.2 Shock Tube

We have shown how 1-dimensional isentropic flows can be completely analyzed by propagating the Riemann invariants along characteristics. Let us illustrate this in more detail by analyzing a shock tube, a laboratory device for creating supersonic flows and studying the behavior of shock waves. In a shock tube, high-pressure gas is retained at rest in the left half of a long tube by a thin membrane (Fig. 17.9a). At time $t = 0$, the membrane is ruptured by a laser beam, and the gas rushes into the tube's right half, which has usually been evacuated. Diagnostic photographs and velocity and pressure measurements are synchronized with the onset of the flow.

Let us idealize the operation of a shock tube by assuming, once more, that the gas is ideal with constant γ, so that $P \propto \rho^\gamma$. For times $t \leq 0$, we suppose that the gas has uniform density ρ_0 and pressure P_0 (and consequently, uniform sound speed C_0) at $x \leq 0$, and that $\rho = P = 0$ at $x \geq 0$. At time $t = 0$, the barrier is ruptured, and the gas flows toward positive x. The first Riemann invariant J_+ is conserved on $\mathcal{C}_+$, which originates in the static gas, so it has the value

$$J_+ = v + \frac{2C}{\gamma - 1} = \frac{2C_0}{\gamma - 1}. \tag{17.23}$$

Note that in this case, the invariant is the same on all rightward characteristics (i.e., throughout the flow), so that

$$v = \frac{2(C_0 - C)}{\gamma - 1} \quad \text{everywhere.} \tag{17.24}$$

The second invariant is

$$J_- = v - \frac{2C}{\gamma - 1}.$$ (17.25)

Its constant values are not so easy to identify, because those characteristics C_- that travel through the perturbed flow all emerge from the origin, where v and C are indeterminate (cf. Fig. 17.9c). However, by combining Eq. (17.24) with Eq. (17.25), we deduce that v and C are separately constant on each characteristic C_-. This enables us, trivially, to solve the differential equation $dx/dt = v - C$ for the leftward characteristics C_-, obtaining

$$C_-: \quad x = (v - C)t.$$ (17.26)

Here we have set the constant of integration equal to zero to obtain all the characteristics that propagate through the perturbed fluid. (For those in the unperturbed fluid, $v = 0$ and $C = C_0$, so $x = x_0 - C_0 t$, with $x_0 < 0$ the characteristic's initial location.)

Now Eq. (17.26) is true on each characteristic in the perturbed fluid. Therefore, it is true throughout the perturbed fluid. We can then combine Eqs. (17.24) and (17.26) to solve for $v(x, t)$ and $C(x, t)$ throughout the perturbed fluid. That solution, together with the obvious solution (same as initial data) to the left and right of the perturbed fluid, is:

$$v = 0, \quad C = C_0 \quad \text{at } x < -C_0 t,$$

$$v = \frac{2}{\gamma + 1}\left(C_0 + \frac{x}{t}\right), \quad C = \frac{2C_0}{\gamma + 1} - \left(\frac{\gamma - 1}{\gamma + 1}\right)\frac{x}{t} \quad \text{at } -C_0 t < x < \frac{2C_0}{\gamma - 1}t,$$

vacuum prevails at $x > \frac{2C_0}{\gamma - 1}t.$ (17.27)

In this solution, notice that the gas at $x < 0$ remains at rest until a *rarefaction wave* from the origin reaches it. Thereafter, it is accelerated rightward by the local pressure gradient, and as it accelerates it expands and cools, so its sound speed C goes down; asymptotically it reaches zero temperature, as exhibited by $C = 0$ and an asymptotic speed $v = 2C_0/(\gamma - 1)$ [cf. Eq. (17.23)]; see Fig. 17.9b,c. In the expansion, the internal random velocity of the gas molecules is transformed into an ordered velocity, just as in a rocket's exhaust. However, the total energy per unit mass in the stationary gas is $u = C_0^2/[\gamma(\gamma - 1)]$ [Eq. (17.2d)], which is less than the asymptotic kinetic energy per unit mass of $2C_0^2/(\gamma - 1)^2$. The additional energy has come from the gas in the rarefaction wave, which is pushing the asymptotic gas rightward.

In the more realistic case, where there initially is some low-density gas in the evacuated half of the tube, the expanding driver gas creates a strong shock as it plows into the low-density gas; hence the name "shock tube." In the next section we explore the structure of this and other shock fronts.

flow in a shock tube

rarefaction wave as flow initiator

Exercise 17.6 *Problem: Fluid Paths in Free Expansion*

We have computed the velocity field for a freely expanding gas in 1 dimension, Eqs. (17.27). Use this result to show that the path of an individual fluid element, which begins at $x = x_0 < 0$, is

$$x = \frac{2C_0 t}{\gamma - 1} + \left(\frac{\gamma + 1}{\gamma - 1}\right) x_0 \left(\frac{-C_0 t}{x_0}\right)^{\frac{2}{\gamma + 1}} \quad \text{at} \quad 0 < -\frac{x_0}{C_0} < t.$$

Exercise 17.7 *Problem: Riemann Invariants for Shallow-Water Flow; Breaking of a Dam*

Consider the 1-dimensional flow of shallow water in a straight, narrow channel, neglecting dispersion and boundary layers. The equations governing the flow, as derived and discussed in Box 16.3 and Eqs. (16.23), are

$$\frac{\partial h}{\partial t} + \frac{\partial (hv)}{\partial x} = 0, \quad \frac{\partial v}{\partial t} + v \frac{\partial v}{\partial x} + g \frac{\partial h}{\partial x} = 0. \tag{17.28}$$

Here $h(x, t)$ is the height of the water, and $v(x, t)$ is its depth-independent velocity.

(a) Find two Riemann invariants $J_{\pm}$ for these equations, and find two conservation laws for these $J_{\pm}$ that are equivalent to the shallow-water equations (17.28).

(b) Use these Riemann invariants to demonstrate that shallow-water waves steepen in the manner depicted in Fig. 16.4, a manner analogous to the peaking of the nonlinear sound wave in Fig. 17.8.

(c) Use these Riemann invariants to solve for the flow of water $h(x, t)$ and $v(x, t)$ after a dam breaks (the problem posed in Ex. 16.9, and there solved via similarity methods). The initial conditions (at $t = 0$) are $v = 0$ everywhere, and $h = h_o$ at $x < 0$, $h = 0$ (no water) at $x > 0$.

17.5 Shock Fronts

We have just demonstrated that in an ideal gas with constant adiabatic index γ, large perturbations to fluid dynamical variables inevitably evolve to form a divergently large velocity gradient—a "shock front," or a "shock wave," or simply a "shock." When the velocity gradient becomes large, we can no longer ignore the viscous stress, because the viscous terms in the Navier-Stokes equation involve second derivatives of **v** in space, whereas the inertial term involves only first derivatives. As in turbulence and in boundary layers, so also in a shock front, the viscous stresses convert the fluid's ordered, bulk kinetic energy into microscopic kinetic energy (i.e., thermal energy). The ordered fluid velocity **v** thereby is rapidly—almost discontinuously—reduced from supersonic to subsonic, and the fluid is heated.

inevitability of shock fronts

The cooler, supersonic region of incoming fluid is said to be *ahead of* or *upstream from* the shock, and it hits the shock's *front side;* the hotter, subsonic region of outgoing fluid is said to be *behind* or *downstream from* the shock, and it emerges from the shock's *back side;* see Fig. 17.10 below.

17.5.1 Junction Conditions across a Shock; Rankine-Hugoniot Relations

Viscosity is crucial to the internal structure of the shock, but it is just as negligible in the downstream flow behind the shock as in the upstream flow ahead of the shock, since there velocity gradients are modest again. Remarkably, if (as is usually the case) the shock front is very thin compared to the length scales in the upstream and downstream flows, and the time for the fluid to pass through the shock is short compared to the upstream and downstream timescales, then we can deduce the net influence of the shock on the flow without any reference to the viscous processes that operate in the shock and without reference to the shock's detailed internal structure. We do so by treating the shock as a discontinuity across which certain junction conditions must be satisfied. This is similar to electromagnetic theory, where the junction conditions for the electric and magnetic fields across a material interface are independent of the detailed structure of the interface.

The keys to the shock's junction conditions are the conservation laws for mass, momentum, and energy: The fluxes of mass, momentum, and energy must usually be the same in the downstream flow, emerging from the shock, as in the upstream flow, entering it. To understand this, we first note that, because the time to pass through the shock is so short, mass, momentum, and energy cannot accumulate in the shock, so the flow can be regarded as stationary. In a stationary flow, the mass flux is always constant, as there is no way to create new mass or destroy old mass. Its continuity across the shock can be written as

$$\boxed{[\rho \mathbf{v} \cdot \mathbf{n}] = 0,} \tag{17.29a}$$

where $\mathbf{n}$ is the unit normal to the shock front, and the square bracket means the difference in the values on the downstream and upstream sides of the shock. Similarly, the total momentum flux $\mathbf{T} \cdot \mathbf{n}$ must be conserved in the absence of external forces. Now $\mathbf{T}$ has both a mechanical component, $P\mathbf{g} + \rho \mathbf{v} \otimes \mathbf{v}$, and a viscous component, $-\zeta\theta\mathbf{g} - 2\eta\boldsymbol{\sigma}$. However, the viscous component is negligible in the upstream and downstream flows, which are being matched to each other, so the mechanical component by itself must be conserved across the shock front:

$$\boxed{[(P\mathbf{g} + \rho\mathbf{v} \otimes \mathbf{v}) \cdot \mathbf{n}] = 0.} \tag{17.29b}$$

Similar remarks apply to the energy flux, though here we must be slightly more restrictive. There are three ways that a change in the energy flux could occur. First, energy may be added to the flow by chemical or nuclear reactions that occur in the shock front. Second, the gas may be heated to such a high temperature that it will lose energy in the shock front through the emission of radiation. Third, energy may be conducted far upstream by suprathermal particles so as to preheat the incoming gas. Any of these processes will thicken the shock front and may make it so thick that it can no longer sensibly be approximated as a discontinuity. If any of these processes occurs, we must check to see whether it is strong enough to significantly influence energy conservation across the shock. What such a check often reveals is that preheating is negligible, and the lengthscales over which the chemical and nuclear reactions and radiation emission operate are much greater than the length over which viscosity acts. In this case we can conserve energy flux across the viscous shock and then follow the evolutionary effects of reactions and radiation (if significant) in the downstream flow.

A shock with negligible preheating—and with negligible radiation emission and chemical and nuclear reactions inside the shock—will have the same energy flux in the departing, downstream flow as in the entering, upstream flow, so it will satisfy

$$\left[\!\left[\left(\frac{1}{2}v^2 + h\right)\rho\mathbf{v}\cdot\mathbf{n}\right]\!\right] = 0.$$

(17.29c)   **energy conservation**

Shocks that satisfy the conservation laws of mass, momentum, and energy, Eqs. (17.29), are said to be *adiabatic*.

By contrast to mass, momentum, and energy, the entropy will not be conserved across a shock front, since viscosity and other dissipative processes increase the entropy as the fluid flows through the shock. So far, the only type of dissipation that we have discussed is viscosity, and this is sufficient by itself to produce a shock front and keep it thin. However, heat conduction (Sec. 18.2) and electrical resistivity, which is important in magnetic shocks (Chap. 19), can also contribute to the dissipation and can influence the detailed structure of the shock front.

For an adiabatic shock, the three requirements of mass, momentum, and energy conservation [Eqs. (17.29)], known collectively as the *Rankine-Hugoniot relations*, enable us to relate the downstream flow and its thermodynamic variables to their upstream counterparts.[2]

Let us work in a reference frame where the incoming flow is normal to the shock front and the shock is at rest, so the flow is stationary in the shock's vicinity. Then the conservation of tangential momentum—the tangential component of Eq. (17.29b)— tells us that the outgoing flow is also normal to the shock in our chosen reference frame. We say that the shock is *normal*, not *oblique*.

We use the subscripts 1, 2 to denote quantities measured ahead of and behind the shock, respectively (i.e., 1 denotes the incoming flow, and 2 denotes the outgoing flow;

2. The existence of shocks was actually understood quite early on, more or less in this way, by Stokes. However, he was persuaded by his former student Rayleigh and others that such discontinuities were impossible, because they would violate energy conservation. With a deference that professors traditionally show their students, Stokes believed Rayleigh. They were both making an error in their analysis of energy conservation, due to an inadequate understanding of thermodynamics in that era.

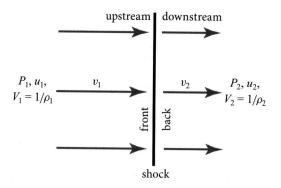

upstream | downstream

$P_1, u_1,$
$V_1 = 1/\rho_1$

v_1

front | back

v_2

$P_2, u_2,$
$V_2 = 1/\rho_2$

shock

FIGURE 17.10 Terminology and notation for a shock front and the flow into and out of it.

Fig. 17.10). The Rankine-Hugoniot relations (17.29) then take the forms:

$$\rho_2 v_2 = \rho_1 v_1 = j, \tag{17.30a}$$

$$P_2 + \rho_2 v_2^2 = P_1 + \rho_1 v_1^2, \tag{17.30b}$$

$$h_2 + \frac{1}{2}v_2^2 = h_1 + \frac{1}{2}v_1^2, \tag{17.30c}$$

where j is the mass flux, which is determined by the upstream flow.

These equations can be brought into a more useful form by replacing the density ρ with the specific volume $V \equiv 1/\rho$, replacing the specific enthalpy h by its value in terms of P and V, $h = u + P/\rho = u + PV$, and performing some algebra. The result is

Rankine-Hugoniot jump conditions across a shock

$$u_2 - u_1 = \frac{1}{2}(P_1 + P_2)(V_1 - V_2), \tag{17.31a}$$

$$j^2 = \frac{P_2 - P_1}{V_1 - V_2}, \tag{17.31b}$$

$$v_1 - v_2 = [(P_2 - P_1)(V_1 - V_2)]^{1/2}. \tag{17.31c}$$

This is the most widely used form of the Rankine-Hugoniot relations. It must be augmented by an equation of state in the form

equation of state

$$u = u(P, V). \tag{17.32}$$

shock adiabat

Some of the physical content of these Rankine-Hugoniot relations is depicted in Fig. 17.11. The thermodynamic state of the upstream (incoming) fluid is the point (V_1, P_1) in this volume-pressure diagram. The thick solid curve, called the *shock adiabat*, is the set of all possible downstream (outgoing) fluid states. This shock adiabat can be computed by combining Eq. (17.31a) with the equation of state (17.32). Those

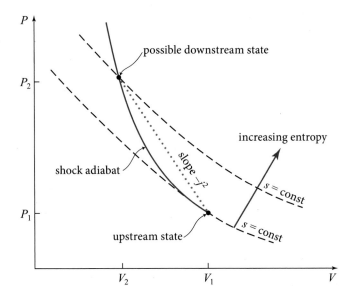

FIGURE 17.11 Shock adiabat. The pressure and specific volume $V = 1/\rho$ in the upstream flow are P_1 and V_1, and in the downstream flow they are P_2 and V_2. The dashed curves are ordinary adiabats (curves of constant entropy per unit mass s). The thick curve is the shock adiabat, the curve of allowed downstream states (V_2, P_2) for a given upstream state (V_1, P_1). The actual location of the downstream state on this adiabat is determined by the mass flux j flowing through the shock: the slope of the dotted line connecting the upstream and downstream states is $-j^2$.

equations will actually give a curve that extends away from (V_1, P_1) in both directions, up-leftward and down-rightward. Only the up-leftward portion is compatible with an increase of entropy across the shock; the down-rightward portion requires an entropy decrease, which is forbidden by the second law of thermodynamics and therefore is not drawn in Fig. 17.11. The actual location (V_2, P_2) of the downstream state along the shock adiabat is determined by Eq. (17.31b) in a simple way: the slope of the dotted line connecting the upstream and downstream states is $-j^2$, where j is the mass flux passing through the shock. When one thereby has learned (V_2, P_2), one can compute the downstream speed v_2 from Eq. (17.31c).

It can be shown that the pressure and density always increase across a shock (as is the case in Fig. 17.11), and the fluid always decelerates:

$$P_2 > P_1, \quad V_2 < V_1, \quad v_2 < v_1; \tag{17.33}$$

see Ex. 17.8. It also can be demonstrated in general—and will be verified in a particular case below, that the Rankine-Hugoniot relations require the flow to be supersonic with respect to the shock front upstream $v_1 > C_1$ and subsonic downstream, $v_2 < C_2$. Physically, this requirement is sensible (as we have seen): When the fluid approaches the shock supersonically, it is not possible to communicate a pressure pulse upstream from the shock (via a Riemann invariant moving at the speed of sound relative to the

pressure and density increase, and velocity decrease across shock

flow) and thereby cause the flow to decelerate; therefore, to slow the flow, a shock must develop.[3] By contrast, the shock front can and does respond to changes in the downstream conditions, since it is in causal contact with the downstream flow; sound waves and a Riemann invariant can propagate upstream, through the downstream flow, to the shock.

summary of shocks

To summarize, shocks are machines that decelerate a normally incident upstream flow to a subsonic speed, so it can be in causal contact with conditions downstream. In the process, bulk momentum flux ρv^2 is converted into pressure, bulk kinetic energy is converted into internal energy, and entropy is manufactured by the dissipative processes at work in the shock front. For given upstream conditions, the downstream conditions are fixed by the conservation laws of mass, momentum, and energy and are independent of the detailed dissipation mechanisms.

EXERCISES

Exercise 17.8 *Derivation and Challenge: Signs of Change across a Shock*

(a) Almost all equations of state satisfy the condition $(\partial^2 V / \partial P^2)_s > 0$. Show that, when this condition is satisfied, the Rankine-Hugoniot relations and the law of entropy increase imply that the pressure and density must increase across a shock and the fluid must decelerate: $P_2 > P_1$, $V_2 < V_1$, and $v_2 < v_1$.

(b) Show that in a fluid that violates $(\partial^2 V / \partial P^2)_s > 0$, the pressure and density must still increase and the fluid decelerate across a shock, as otherwise the shock would be unstable.

For a solution to this exercise, see Landau and Lifshitz (1959, Sec. 84).

Exercise 17.9 *Problem: Relativistic Shock* **T2**

In astrophysics (e.g., in supernova explosions and in jets emerging from the vicinities of black holes), one sometimes encounters shock fronts for which the flow speeds relative to the shock approach the speed of light, and the internal energy density is comparable to the fluid's rest-mass density.

(a) Show that the relativistic Rankine-Hugoniot equations for such a shock take the following form:

$$\boxed{\eta_2^2 - \eta_1^2 = (P_2 - P_1)(\eta_1 V_1 + \eta_2 V_2),} \tag{17.34a}$$

$$\boxed{j^2 = \frac{P_2 - P_1}{\eta_1 V_1 - \eta_2 V_2},} \tag{17.34b}$$

$$\boxed{v_2 \gamma_2 = j V_2, \qquad v_1 \gamma_1 = j V_1.} \tag{17.34c}$$

---

3. Of course, if there is some faster means of communication, for example, photons or, in an astrophysical context, cosmic rays or neutrinos, then there may be a causal contact between the shock and the inflowing gas, which can either prevent shock formation or lead to a more complex shock structure.

Here,

(i) We use units in which the speed of light is 1 (as in Chap. 2).

(ii) The volume per unit rest mass is $V \equiv 1/\rho_o$, and ρ_o is the rest-mass density (equal to some standard rest mass per baryon times the number density of baryons; cf. Sec. 2.12.3).

(iii) We denote the total density of mass-energy including rest mass by ρ_R (it was denoted ρ in Chap. 2) and the internal energy per unit rest mass by u, so $\rho_R = \rho_o(1 + u)$. In terms of these the quantity $\eta \equiv (\rho_R + P)/\rho_o = 1 + u + P/\rho_o = 1 + h$ is the relativistic enthalpy per unit rest mass (i.e., the enthalpy per unit rest mass, including the rest-mass contribution to the energy) as measured in the fluid rest frame.

(iv) The pressure as measured in the fluid rest frame is P.

(v) The flow velocity in the shock's rest frame is v, and $\gamma \equiv 1/\sqrt{1 - v^2}$ (*not* the adiabatic index!), so $v\gamma$ is the spatial part of the flow 4-velocity.

(vi) The rest-mass flux is j (rest mass per unit area per unit time) entering and leaving the shock.

(b) Use a pressure-volume diagram to discuss these relativistic Rankine-Hugoniot equations in a manner analogous to Fig. 17.11.

(c) Show that in the nonrelativistic limit, the relativistic Rankine-Hugoniot equations (17.34) reduce to the nonrelativistic ones (17.31).

(d) It can be shown (Thorne, 1973) that relativistically, just as for nonrelativistic shocks, in general $P_2 > P_1$, $V_2 < V_1$, and $v_2 < v_1$. Consider, as an example, a relativistic shock propagating through a fluid in which the mass density due to radiation greatly exceeds that due to matter (a *radiation-dominated fluid*), so $P = \rho_R/3$ (Sec. 3.5.5). Show that $v_1 v_2 = 1/3$, which implies $v_1 > 1/\sqrt{3}$ and $v_2 < 1/\sqrt{3}$. Show further that $P_2/P_1 = (9v_1^2 - 1)/[3(1 - v_1^2)]$.

Exercise 17.10 **Problem: *Hydraulic Jumps and Breaking Ocean Waves*
Run water at a high flow rate from a kitchen faucet onto a dinner plate (Fig. 17.12). What you see is called a hydraulic jump. It is the kitchen analog of a breaking ocean wave, and the shallow-water-wave analog of a shock front in a compressible gas. In this exercise you will develop the theory of hydraulic jumps (and breaking ocean waves) using the same tools as those used for shock fronts.

(a) Recall that for shallow-water waves, the water motion, below the water's surface, is nearly horizontal, with speed independent of depth z (Ex. 16.1). The same is true of the water in front of and behind a hydraulic jump. Apply the conservation of mass and momentum to a hydraulic jump, in the jump's rest frame, to obtain equations for the height of the water h_2 and water speed v_2 behind the jump (emerging from it) in terms of those in front of the jump, h_1 and v_1. These are the analog of the Rankine-Hugoniot relations for a shock front. [Hint: For

FIGURE 17.12 Hydraulic jump on a dinner plate under a kitchen tap.

momentum conservation you will need to use the pressure P as a function of height in front of and behind the jump.]

(b) You did not use energy conservation across the jump in your derivation, but it was needed in the analysis of a shock front. Why?

(c) Show that the upstream speed v_1 is greater than the speed $\sqrt{gh_1}$ of small-amplitude, upstream gravity waves [shallow-water waves; Eq. (16.10) and associated discussion]; thus the upstream flow is supersonic. Similarly, show that the downstream flow speed v_2 is slower than the speed $\sqrt{gh_2}$ of small-amplitude, downstream gravity waves (i.e., the downstream flow is subsonic).

(d) We normally view a breaking ocean wave in the rest frame of the quiescent upstream water. Use your hydraulic-jump equations to show that the speed of the breaking wave as seen in this frame is related to the depths h_1 and h_2 in front of and behind the breaking wave by

$$v_{\text{break}} = \left[\frac{g(h_1 + h_2)h_2}{2h_1} \right]^{1/2};$$

see Fig. 17.13.

---

shock Mach number

17.5.2 Junction Conditions for Ideal Gas with Constant γ

To make the shock junction conditions more explicit, let us again specialize to an ideal gas with constant specific-heat ratio γ (a polytropic gas), so the equation of state is

Chapter 17. Compressible and Supersonic Flow

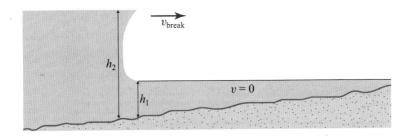

FIGURE 17.13 An ocean wave breaking on a gradually sloping beach. The depth of water ahead of the wave is h_1, and the depth behind the wave is h_2.

$u = PV/(\gamma - 1)$, and the sound speed is $C = \sqrt{\gamma P/\rho} = \sqrt{\gamma PV}$ [Eqs. (17.2)]. We measure the strength of the shock using the *shock Mach number* M, which is defined to be the Mach number in the upstream flow relative to the shock:

$$M \equiv M_1 = v_1/C_1 = \sqrt{v_1^2/(\gamma P_1 V_1)}. \tag{17.35}$$

With the aid of this equation of state and Mach number, we can bring the Rankine-Hugoniot relations (17.31) into the form:

$$\boxed{\frac{\rho_1}{\rho_2} = \frac{V_2}{V_1} = \frac{v_2}{v_1} = \frac{\gamma - 1}{\gamma + 1} + \frac{2}{(\gamma + 1)M^2},} \tag{17.36a}$$

Rankine-Hugoniot relations for polytropic equation of state

$$\boxed{\frac{P_2}{P_1} = \frac{2\gamma M^2}{\gamma + 1} - \frac{\gamma - 1}{\gamma + 1},} \tag{17.36b}$$

$$\boxed{M_2^2 = \frac{2 + (\gamma - 1)M^2}{2\gamma M^2 - (\gamma - 1)}.} \tag{17.36c}$$

Here $M_2 \equiv v_2/c_2$ is the downstream Mach number.

The results for this equation of state illustrate a number of general features of shocks. The density and pressure increase across the shock, the flow speed decreases, and the downstream flow is subsonic—all discussed previously—and one important new feature: a shock weakens as its Mach number M decreases. In the limit that $M \to 1$, the jumps in pressure, density, and speed vanish, and the shock disappears.

In the *strong-shock limit*, $M \gg 1$, the jumps are

$$\boxed{\frac{\rho_1}{\rho_2} = \frac{V_2}{V_1} = \frac{v_2}{v_1} \simeq \frac{\gamma - 1}{\gamma + 1},} \tag{17.37a}$$

Rankine-Hugoniot relations for strong polytropic shock

$$\boxed{\frac{P_2}{P_1} \simeq \frac{2\gamma M^2}{\gamma + 1}.} \tag{17.37b}$$

Thus the density jump is always of order unity, but the pressure jump grows ever larger as M increases. Air has $\gamma \simeq 1.4$ (Ex. 17.1), so the density compression ratio for a strong shock in air is $\rho_2/\rho_1 = 6$, and the pressure ratio is $P_2/P_1 = 1.2M^2$. The Space Shuttle's reentry provides a nice example of these strong-shock Rankine-Hugoniot relations; see the bottom half of Box 17.4.

EXERCISES

Exercise 17.11 *Problem: Shock Tube*
Consider a shock tube as discussed in Sec. 17.4.2 and Fig. 17.9. High density "driver" gas with sound speed C_0 and specific heat ratio γ is held in place by a membrane that separates it from target gas with very low density, sound speed C_1, and the same specific-heat ratio γ. When the membrane is ruptured, a strong shock propagates into the target gas. Show that the Mach number of this shock is given approximately by

$$M = \left(\frac{\gamma + 1}{\gamma - 1}\right)\left(\frac{C_0}{C_1}\right) . \tag{17.38}$$

[Hint: Think carefully about which side of the shock is supersonic and which side subsonic.]

17.5.3

17.5.3 Internal Structure of a Shock

Although they are often regarded as discontinuities, shocks, like boundary layers, do have structure. The simplest case is that of a gas in which the shear-viscosity coefficient is molecular in origin and is given by $\eta = \rho\nu \sim \rho\lambda v_{th}/3$, where λ is the molecular mean free path, and $v_{th} \sim C$ is the mean thermal speed of the molecules (Ex. 3.19). In this case for 1-dimensional flow $v = v_x(x)$, the viscous stress $T_{xx} = -\zeta\theta - 2\eta\sigma_{xx}$ is $-(\zeta + 4\eta/3)dv/dx$, where ζ is the coefficient of bulk viscosity, which can be of the same order as the coefficient of shear viscosity. In the shock, this viscous stress must roughly balance the total kinetic momentum flux $\sim\rho v^2$. If we estimate the velocity gradient dv/dx by v_1/δ_S, where δ_S is a measure of the shock thickness, and we estimate the sound speed in the shock front by $C \sim v_1$, then we deduce that the shock thickness δ_S is roughly equal to λ, the collision mean free path in the gas. For air at standard temperature and pressure, the mean free path is $\lambda \sim (\sqrt{2}n\pi\sigma^2)^{-1} \sim 70$ nm, where n is the molecular density, and σ is the molecular diameter. This is very small! Microscopically, it makes sense that $\delta_S \sim \lambda$, as an individual molecule only needs a few collisions to randomize its ordered motion perpendicular to the shock front. However, this estimate raises a problem, as it brings into question our use of the continuum approximation (cf. Sec. 13.2). It turns out that, when a more careful calculation of the shock structure is carried out incorporating heat conduction, the shock thickness is several mean free paths, fluid dynamics is acceptable for an approximate theory, and the results are in rough accord with measurements of the velocity profiles of shocks with modest Mach numbers. Despite this, for an accurate description of the shock structure, a kinetic treatment is usually necessary.

molecular collisions as dissipation mechanism in some shocks

shock thickness: several mean free paths

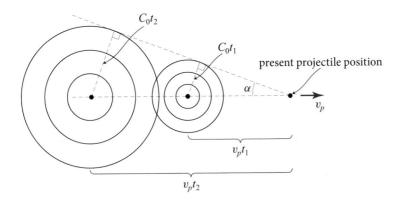

FIGURE 17.14 Construction for the Mach cone formed by a supersonic projectile. The cone angle is $\alpha = \sin^{-1}(1/M)$, where $M = v_p/C_0$ is the Mach number of the projectile.

So far we have assumed that the shocked fluid is made of uncharged molecules. A more complicated type of shock can arise in an ionized gas (i.e., a plasma; Part VI). Shocks in the solar wind are examples. In this case the collision mean free paths are enormous; in fact, they are comparable to the transverse size of the shock, and therefore one might expect the shocks to be so thick that the Rankine-Hugoniot relations fail. However, spacecraft measurements reveal solar-wind shocks that are relatively thin—far thinner than the collisional mean free paths of the plasma's electrons and ions. In this case, it turns out that collisionless, collective electromagnetic and charged-particle interactions in the plasma are responsible for the viscosity and dissipation. (The particles create plasma waves, which in turn deflect the particles.) These processes are so efficient that thin shock fronts can occur without individual particles having to hit one another. Since the shocks are thin, they must satisfy the Rankine-Hugoniot relations. We discuss these collisionless shocks further in Sec. 23.6.

dissipation mechanisms for shocks in collisionless plasmas

17.5.4 Mach Cone

The shock waves formed by a supersonically moving body are quite complex close to the body and depend on its detailed shape, Reynolds' number, and so forth (see, e.g., Fig. 17.1). However, far from the body, the leading shock has the form of the *Mach cone* shown in Fig. 17.14. We can understand this cone by the construction shown in the figure. The shock is the boundary between the fluid that is in sound-based causal contact with the projectile and the fluid that is not. This boundary is mapped out by (conceptual) sound waves that propagate into the fluid from the projectile at the ambient sound speed C_0. When the projectile is at the indicated position, the envelope of the circles is the shock front and has the shape of the Mach cone, with opening angle (the *Mach angle*)

17.5.4

Mach cone

$$\boxed{\alpha = \sin^{-1}(1/M).}$$

(17.39)

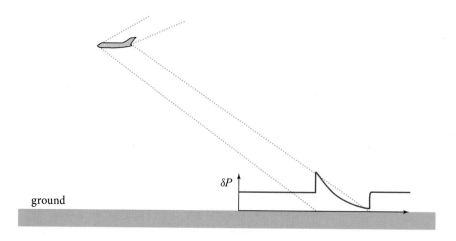

FIGURE 17.15 Double shock created by a supersonic body and its associated N-wave pressure distribution on the ground.

Usually, there are two such shock cones, one attached to the projectile's bow shock and the other formed out of the complex shock structure in its tail region. The pressure must jump twice, once across each of these shocks, and therefore forms an *N wave,* which propagates cylindrically away from the projectile, as shown in Fig. 17.15. Behind the first shock, the density and pressure drop off gradually by more than the first shock's compression. As a result, the fluid flowing into the second shock has a lower pressure, density, and sound speed than that flowing into the first (Fig. 17.15). This causes the Mach number of the second shock to be higher than that of the first, and its Mach angle thus to be smaller. As a result, the separation between the shocks increases as they travel as $\varpi^{1/2}$, it turns out. The pressure jumps across the shocks decrease as $\varpi^{-3/4}$. Here ϖ is the perpendicular distance of the point of observation from the projectile's trajectory. Often a double boom can be heard on the ground. For a detailed analysis, see Whitham (1974, Sec. 9.3).

EXERCISES

Exercise 17.12 *Problem: Sonic Boom from the Space Shuttle*
Use the quoted scaling of N-wave amplitude with cylindrical radius ϖ to make an order-of-magnitude estimate of the flux of acoustic energy produced by the Space Shuttle flying at Mach 2 at an altitude of 20 km. Give your answer in decibels [Eq. (16.61)].

17.6 17.6 Self-Similar Solutions—Sedov-Taylor Blast Wave

Strong explosions can generate shock waves. Examples include atmospheric nuclear explosions, supernova explosions, and depth charges. The debris from a strong explosion is at much higher pressure than the surrounding gas and therefore drives a strong spherical shock into its surroundings. Initially, this shock wave travels at roughly the radial speed of the expanding debris. However, the mass of fluid swept up by the shock eventually exceeds that of the explosion debris. The shock then decelerates and the energy of the explosion is transferred to the swept-up fluid. It is of obvious importance to be able to calculate how fast and how far the shock front will travel.

17.6.1 The Sedov-Taylor Solution

We first make an order-of-magnitude estimate. Let the total energy of the explosion be E and the density of the surrounding fluid (assumed uniform) be ρ_0. Then after time t, when the shock radius is $R(t)$, the mass of swept-up fluid will be $m \sim \rho_0 R^3$. The fluid velocity behind the shock will be roughly the radial velocity of the shock front, $v \sim \dot{R} \sim R/t$, and so the kinetic energy of the swept-up gas will be $\sim m v^2 \sim \rho_0 R^5/t^2$. There will also be internal energy in the post-shock flow, with energy density roughly equal to the post-shock pressure: $\rho u \sim P \sim \rho_0 \dot{R}^2$ [cf. the strong-shock jump condition (17.37b) with $P_1 \sim \rho_0 C_0^2$, so $P_1 M^2 \sim \rho_0 v^2 \sim \rho_0 \dot{R}^2$]. The total internal energy behind the expanding shock will then be $\sim \rho \dot{R}^2 R^3 \sim \rho_0 R^5/t^2$, equal in order of magnitude to the kinetic energy. Equating this to the total energy E of the explosion, we obtain the estimate

$$E = \kappa \rho_0 R^5 t^{-2}, \tag{17.40}$$

where κ is a numerical constant of order unity. This expression implies that at time t the shock front has reached the radius

$$R = \left(\frac{E}{\kappa \rho_0}\right)^{1/5} t^{2/5}. \tag{17.41}$$

radius of a strong, spherical shock as a function of time

This scaling should hold roughly from the time that the mass of the swept-up gas is of order that of the exploding debris, to the time that the shock weakens to a Mach number of order unity so we can no longer use the strong-shock value $\sim \rho_0 \dot{R}^2$ for the post-shock pressure.

Note that we could have obtained Eq. (17.41) by a purely dimensional argument: E and ρ_0 are the only significant controlling parameters in the problem, and $E^{1/5}\rho_0^{-1/5}t^{2/5}$ is the only quantity with dimensions of length that can be constructed from E, ρ_0, and t. However, it is usually possible and always desirable to justify any such dimensional argument on the basis of the governing equations.

If, as we shall assume, the motion remains radial and the gas is ideal with constant specific-heat ratio γ, then we can solve for the details of the flow behind the shock front by integrating the radial flow equations:

$$\frac{\partial \rho}{\partial t} + \frac{1}{r^2}\frac{\partial}{\partial r}(r^2 \rho v) = 0, \tag{17.42a}$$

$$\frac{\partial v}{\partial t} + v\frac{\partial v}{\partial r} + \frac{1}{\rho}\frac{\partial P}{\partial r} = 0, \tag{17.42b}$$

$$\frac{\partial}{\partial t}\left(\frac{P}{\rho^\gamma}\right) + v\frac{\partial}{\partial r}\left(\frac{P}{\rho^\gamma}\right) = 0. \tag{17.42c}$$

The first two equations are the familiar continuity equation and Euler equation written for a spherical flow. The third equation is energy conservation expressed as the adiabatic-expansion relation, $P/\rho^\gamma = $ const moving with a fluid element. Although P/ρ^γ is time independent for each fluid element, its value will change from element to element. Gas that has passed through the shock more recently will be given a smaller entropy than gas that was swept up when the shock was stronger, and thus it will have a smaller value of P/ρ^γ.

Given suitable initial conditions, the partial differential equations (17.42) can be integrated numerically. However, there is a practical problem: it is not easy to determine the initial conditions in an explosion! Fortunately, at late times, when the initial debris mass is far less than the swept-up mass, the fluid evolution is independent of the details of the initial expansion and in fact can be understood analytically as a *similarity solution*. By this term we mean that the shape of the radial profiles of pressure, density, and velocity are independent of time.

examples of similarity solutions

We have already met three examples of similarity solutions: the Blasius structure of a laminar boundary layer (Sec. 14.4.1), the structure of a turbulent jet (Ex. 15.3), and the flow of water following the sudden rupture of a dam (Ex. 16.9). The one we explored in greatest detail was the Blasius boundary layer (Sec. 14.4.1). There we argued on the basis of mass and momentum conservation (or, equally well, by dimensional analysis) that the thickness of the boundary layer as a function of distance x downstream would be $\sim\delta = (\nu x/V)^{1/2}$, where V is the speed of the flow above the boundary layer. This motivated us to introduce the dimensionless variable $\xi = y/\delta$ and argue that the boundary layer's speed $v_x(x, y)$ would be equal to the free-stream velocity V times some universal function $f'(\xi)$. This ansatz converted the fluid's partial differential equations into an ordinary differential equation for $f(\xi)$, which we solved numerically.

Our explosion problem is somewhat similar. The characteristic scaling length in the explosion is the radius $R(t) = [E/(\kappa\rho_0)]^{1/5}t^{2/5}$ of the shock [Eq. (17.41)], with κ an as-yet-unknown constant, so the fluid and thermodynamic variables should be expressible as some characteristic values multiplying universal functions of

$$\xi \equiv r/R(t). \tag{17.43}$$

similarity solution behind strong, spherical shock

Our thermodynamic variables are P, ρ, and u, and natural choices for their characteristic values are the values P_2, ρ_2, and v_2 immediately behind the shock. If we assume the shock is strong, then we can use the strong-shock jump conditions (17.37) to determine those values, and then write

$$P = \frac{2}{\gamma + 1}\rho_0\dot{R}^2\tilde{P}(\xi), \tag{17.44a}$$

$$\rho = \frac{\gamma + 1}{\gamma - 1}\rho_0\,\tilde{\rho}(\xi), \tag{17.44b}$$

$$v = \frac{2}{\gamma + 1}\dot{R}\,\tilde{v}(\xi), \tag{17.44c}$$

with $\tilde{P}(1) = \tilde{\rho}(1) = \tilde{v}(1) = 1$, since $\xi = 1$ is the shock's location. Note that the velocity v is scaled to the post-shock velocity v_2 measured in the inertial frame in which the upstream fluid is at rest, rather than in the noninertial frame in which the decelerating shock is at rest. The self-similarity ansatz (17.44) and resulting similarity solution for the flow are called the *Sedov-Taylor blast-wave solution,* since L. I. Sedov and G. I. Taylor independently developed it (Sedov, 1946, 1993; Taylor, 1950).

The partial differential equations (17.42) can now be transformed into ordinary differential equations by inserting the ansatz (17.44) together with expression (17.41) for $R(t)$, changing the independent variables from r and t to R and ξ and using

$$\left(\frac{\partial}{\partial t}\right)_r = -\left(\frac{\xi \dot{R}}{R}\right)\left(\frac{\partial}{\partial \xi}\right)_R + \dot{R}\left(\frac{\partial}{\partial R}\right)_\xi = -\left(\frac{2\xi}{5t}\right)\left(\frac{\partial}{\partial \xi}\right)_R + \frac{2R}{5t}\left(\frac{\partial}{\partial R}\right)_\xi,$$

(17.45)

$$\left(\frac{\partial}{\partial r}\right)_t = \left(\frac{1}{R}\right)\left(\frac{\partial}{\partial \xi}\right)_R.$$

(17.46)

Mass conservation, the Euler equation, and the equation of adiabatic expansion become, in that order:

$$0 = 2\tilde{\rho}\tilde{v}' - (\gamma + 1)\xi\tilde{\rho}' + \tilde{v}\left(2\tilde{\rho}' + \frac{4}{\xi}\tilde{\rho}\right),$$

(17.47a)

$$0 = \tilde{\rho}\tilde{v}[3(\gamma + 1) - 4\tilde{v}'] + 2(\gamma + 1)\xi\tilde{\rho}\tilde{v}' - 2(\gamma - 1)\tilde{P}',$$

(17.47b)

$$3 = \left(\frac{2\tilde{v}}{\gamma + 1} - \xi\right)\left(\frac{\tilde{P}'}{\tilde{P}} - \gamma\frac{\tilde{\rho}'}{\tilde{\rho}}\right).$$

(17.47c)

differential equations for similarity solution

These self-similarity equations can be solved numerically, subject to the boundary conditions that $\tilde{v}$, $\tilde{\rho}$, and $\tilde{P}$ are all zero at $\xi = 0$ and 1 at $\xi = 1$. Remarkably, Sedov and Taylor independently found an analytic solution (also given in Landau and Lifshitz, 1959, Sec. 99). The solutions for an explosion in air ($\gamma = 1.4$) are exhibited in Fig. 17.16.

Armed with the solution for $\tilde{v}(\xi)$, $\tilde{\rho}(\xi)$, and $\tilde{P}(\xi)$ (numerical or analytic), we can evaluate the flow's energy E, which is equal to the explosion's total energy during the time interval when this similarity solution is accurate. The energy E is given by the integral

$$E = \int_0^R 4\pi r^2 dr\rho\left(\frac{1}{2}v^2 + u\right)$$

explosion's total energy

$$= \frac{4\pi\rho_0 R^3\dot{R}^2(\gamma + 1)}{(\gamma - 1)}\int_0^1 d\xi\xi^2\tilde{\rho}\left(\frac{2\tilde{v}^2}{(\gamma + 1)^2} + \frac{2\tilde{P}}{(\gamma + 1)^2\tilde{\rho}}\right).$$

(17.48)

Here we have used Eqs. (17.44) and substituted $u = P/[\rho(\gamma - 1)]$ for the internal energy [Eq. (17.2b)]. The energy E appears not only on the left-hand side of this equation but also on the right, in the terms $\rho_o R^3\dot{R}^2 = (4/25)E/\kappa$. Thus, E cancels

17.6 Self-Similar Solutions—Sedov-Taylor Blast Wave   **911**

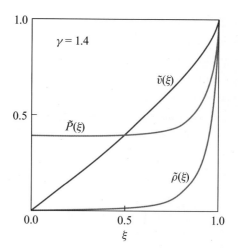

FIGURE 17.16 Scaled pressure, density, and velocity as a function of scaled radius behind a Sedov-Taylor blast wave in air with $\gamma = 1.4$.

out, and Eq. (17.48) becomes an equation for the unknown constant κ. Evaluating that equation numerically, we find that κ varies from $\kappa = 0.85$ for $\gamma = 1.4$ (air) to $\kappa = 0.49$ for $\gamma = 1.67$ (monatomic gas or fully ionized plasma).

It is enlightening to see how the fluid behaves in this blast-wave solution. The fluid that passes through the shock is compressed, so that it mostly occupies a fairly thin spherical shell immediately behind the shock [see the spike in $\tilde{\rho}(\xi)$ in Fig. 17.16]. The fluid in this shell moves somewhat more slowly than the shock [$v = 2\dot{R}/(\gamma + 1)$; Eq. (17.44c) and Fig. 17.16]; it flows from the shock front through the high-density shell (fairly slowly relative to the shell, which remains attached to the shock), and on into the lower density post-shock region. Since the post-shock flow is subsonic, the pressure in the blast wave is fairly uniform [see the curve $\tilde{P}(\xi)$ in Fig. 17.16]; in fact the central pressure is typically about half the maximum pressure immediately behind the shock. This pressure pushes on the spherical shell, thereby accelerating the freshly swept-up fluid.

17.6.2 Atomic Bomb

The first atomic bomb was exploded in New Mexico in 1945, and photographs released in 1947 (Fig. 17.17) showed the radius of the blast wave as a function of time. The pictures were well fit by $R \sim 37(t/1\,\mathrm{ms})^{0.4}$ m up to about $t = 100$ ms, when the shock Mach number fell to about unity (Fig. 17.17). Combining this information with the similarity solution (Sec. 17.6.1), that they had earlier derived independently, the Russian physicist L. I. Sedov and the British physicist G. I. Taylor were both able to infer the total energy released, which was an official American secret at the time. Their analyses of the data were published later (Taylor, 1950; Sedov, 1957, 1993).

Adopting the specific-heat ratio $\gamma = 1.4$ of air, the corresponding value $\kappa = 0.85$, and the measured $R \sim 37(t/1\,\mathrm{ms})^{0.4}$ m, we obtain from Eq. (17.48) the estimate

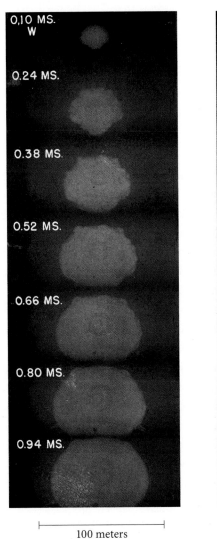
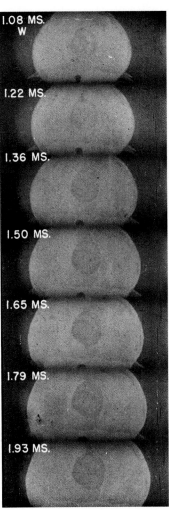

100 meters

FIGURE 17.17 Photographs of the fireball (very hot post-shock gas) from the first atomic bomb explosion, at Almagordo, New Mexico, on July 16, 1945. Courtesy of the Los Alamos National Laboratory Archives.

$E \sim 7.2 \times 10^{13}$ J, which is about the same energy release as 17 kilotons of TNT. This estimate is close to the Los Alamos scientists' estimate of 18–20 kilotons. (Hydrogen bombs have been manufactured that are more than a thousand times more energetic than this—as much as 57 megatons—but such awesome weapons have not been deemed militarily useful, so today's arsenals contain bombs that are typically *only* $\sim$1 megaton!)

We can use the Sedov-Taylor solution to infer some further features of the explosion. The post-shock gas is at density $\sim(\gamma + 1)/(\gamma - 1) \sim 5$ times the ambient density $\rho_0 \sim 1$ kg m^{-3}. Similarly, using the ideal gas law $P = [\rho/(m_p\mu)]k_B T$ with a mean molecular weight $\mu \sim 10$ and the strong-shock jump conditions (17.37), the post-shock temperature can be computed:

$$T_2 = \frac{m_p \mu}{\rho_2 k_B} P_2 \sim 4 \times 10^4 \left(\frac{t}{1 \text{ ms}}\right)^{-1.2} \text{K}. \tag{17.49}$$

At early times, this temperature is high enough to ionize the gas.

17.6.3 Supernovae

The evolution of most massive stars ends in a supernova explosion (like the one observed in 1987 in the Large Magellanic Cloud), in which a neutron star of mass $m \sim 3 \times 10^{30}$ kg is usually formed. This neutron star has a gravitational binding energy of about $0.1 mc^2 \sim 3 \times 10^{46}$ J. Most of this binding energy is released in the form of neutrinos in the collapse that forms the neutron star, but an energy $E \sim 10^{44}$ J drives off the outer envelope of the pre-supernova star, a mass $M_0 \sim 10^{31}$ kg. This stellar material escapes with an rms speed $V_0 \sim (2E/M_0)^{1/2} \sim 5{,}000$ km s^{-1}. The expanding debris drives a blast wave into the surrounding interstellar medium, which has density $\rho_0 \sim 10^{-21}$ kg m^{-3}. The expansion of the blast wave can be modeled using the Sedov-Taylor solution after the mass of the swept-up interstellar gas has become large enough to dominate the blast debris, so the star-dominated initial conditions are no longer important—after a time $\sim (3M_0/4\pi\rho_0)^{1/3}/V_0 \sim 1{,}000$ yr. The blast wave then decelerates as for a Sedov-Taylor similarity solution until the shock speed nears the sound speed in the surrounding gas; this takes about 100,000 yr. *Supernova remnants* of this sort are efficient emitters of radio waves and X-rays, and several hundred have been observed in our Milky Way galaxy.

In some of the younger examples, like Tycho's remnant (Fig. 17.18), it is possible to determine the expansion speed, and the effects of deceleration can be measured. The observations from remnants like this are close to the prediction of the Sedov-Taylor solution; namely, the radius variation satisfies $d \ln R/d \ln t = 0.4$. (For Tycho's remnant, a value of 0.54 ± 0.05 is found, and using the estimated density of the interstellar medium plus the kinematics, an explosion energy of $E \sim 5 \times 10^{43}$ J is inferred.)

EXERCISES

Exercise 17.13 *Problem: Underwater Explosions*

A simple analytical solution to the Sedov-Taylor similarity equations can be found for the particular case $\gamma = 7$. This is a fair approximation to the behavior of water under explosive conditions, as it will be almost incompressible.

(a) Make the ansatz (whose self-consistency we will check later) that the velocity in the post-shock flow varies linearly with radius from the origin to the shock: $\tilde{v}(\xi) = \xi$. Use Eq. (17.45) to transform the equation of continuity into an ordinary differential equation and hence solve for the density function $\tilde{\rho}(\xi)$.

FIGURE 17.18 This remnant of the supernova explosion observed by the astronomer Tycho Brahe in 1572 is roughly 7,000 light-years from us. (Recent observations of a *light echo* have demonstrated that the supernova was of the same type used to discover the acceleration of the universe, which is discussed in Ex. 28.6.) This image was made by the Chandra X-ray Observatory and shows the emission from the hot (in red) ($\sim 1 - 10 \times 10^7$ K) gas heated by the outer and inner shock fronts as well as the even more energetic (in blue) X-rays that delineate the position of the outer shock front. Shock fronts like this are believed to be the sites for acceleration of galactic cosmic rays (cf. Sec. 19.7, Ex. 23.8). NASA/CXC/Rutgers/J. Warren and J. Hughes, et al.

(b) Next use the equation of motion to discover that $\tilde{P}(\xi) = \xi^3$.

(c) Verify that your solutions for the functions $\tilde{P}$, $\tilde{\rho}$, and $\tilde{v}$ satisfy the remaining entropy equation, thereby vindicating the original ansatz.

(d) Substitute your results from parts (a) and (b) into Eq. (17.48) to show that

$$E = \frac{2\pi R^5 \rho_0}{225 t^2}.$$

(e) An explosive charge weighing 100 kg with an energy release of 10^8 J kg^{-1} is detonated underwater. For what range of shock radii do you expect the Sedov-Taylor similarity solution to be valid?

Exercise 17.14 *Problem: Stellar Winds*

Many stars possess powerful stellar winds that drive strong spherical shock waves into the surrounding interstellar medium. If the strength of the wind remains constant, the kinetic and internal energy of the swept-up interstellar medium will increase linearly with time.

(a) Modify the text's analysis of a point explosion to show that in this case the speed of the shock wave at time t is $\frac{3}{5}R(t)/t$, where R is the associated shock radius. What is the speed of the post-shock gas?

(b) Now suppose that the star explodes as a supernova, and the blast wave expands into the relatively slowly moving stellar wind. Suppose that before the explosion the rate at which mass left the star and the speed of the wind were constant for a long time. How do you expect the density of gas in the wind to vary with radius? Modify the Sedov-Taylor analysis again to show that the expected speed of the shock wave at time t is now $\frac{2}{3}R(t)/t$.

Exercise 17.15 *Problem: Similarity Solution for a Shock Tube*
Use a similarity analysis to derive the solution (17.27) for the shock-tube flow depicted in Fig. 17.9.

Bibliographic Note

For physical insight into compressible flows and shock waves, we recommend the movies cited in Box 17.2. For textbook treatments, we recommend the relevant sections of Landau and Lifshitz (1959); Thompson (1984); Liepmann and Roshko (2002); Anderson (2003); and, at a more elementary and sometimes cursory level, Lautrup (2005). Whitham (1974) is superb on all aspects, from an applied mathematician's viewpoint.

Engineering-oriented textbooks on fluid mechanics generally contain detailed treatments of quasi-1-dimensional flows, shocks, hydraulic jumps, and their real-world applications. We like Munson, Young, and Okiishi (2006), White (2006), and Potter, Wiggert, and Ramadan (2012).

The two-volume treatise Zel'dovich and Raizer (2002) is a compendium of insights into shock waves and high-temperature hydrodynamics by an author (Yakov Borisovich Zel'dovich) who had a huge influence on the design of atomic and thermonuclear weapons in the Soviet Union and later on astrophysics and cosmology. Stanyukovich (1960) is a classic treatise on nonstationary flows, shocks and detonation waves. Sedov (1993)—the tenth edition of a book whose first was in 1943—is a classic and insightful treatise on similarity methods in physics.

18

Convection

Approximation methods derived from physical intuition are frequently more reliable than rigorous mathematical methods, because in the case of the latter it is easier for errors to creep into the fundamental assumptions.

WERNER HEISENBERG (1969)

18.1 Overview

In Chaps. 13 and 14, we demonstrated that viscosity can exert a major influence on subsonic fluid flows. When the viscosity ν is large and the Reynolds number ($\text{Re} = LV/\nu$) is low, viscous stresses transport momentum directly, and the fluid's behavior can be characterized by the diffusion of the vorticity ($\boldsymbol{\omega} = \boldsymbol{\nabla} \times \mathbf{v}$) through the fluid [cf. Eq. (14.3)]. As the Reynolds number increases, the advection of the vorticity becomes more important. In the limit of large Reynolds number, we think of the vortex lines as being frozen into the flow. However, as we learned in Chap. 15, this insight is only qualitatively helpful, because high-Reynolds-number flows are invariably turbulent. Large, irregular, turbulent eddies transport shear stress very efficiently. This is particularly in evidence in turbulent boundary layers.

When viewed microscopically, heat conduction is a similar transport process to viscosity, and it is responsible for analogous physical effects. If a viscous fluid has high viscosity, then vorticity diffuses through it rapidly; similarly, if a fluid has high thermal conductivity, then heat diffuses through it rapidly. In the other extreme, when viscosity is low (i.e., when the Reynolds number is high), instabilities produce turbulence, which transports vorticity far more rapidly than diffusion could possibly do. Analogously, in heated fluids with low conductivity, the local accumulation of heat drives the fluid into convective motion, and the heat is transported much more efficiently by this motion than by thermal diffusion. As the convective heat transport increases, the fluid motion becomes more vigorous, and if the viscosity is sufficiently low, the thermally driven flow can also become turbulent. These effects are very much in evidence near solid boundaries, where thermal boundary layers can be formed, analogous to viscous boundary layers.

In addition to thermal effects that resemble the effects of viscosity, there are also unique thermal effects—particularly the novel and subtle combined effects of gravity and heat. Heat, unlike vorticity, causes a fluid to expand and thus, in the presence of gravity, to become buoyant; this buoyancy can drive thermal circulation or *free convection* in an otherwise stationary fluid. (Free convection should be distinguished from *forced convection,* in which heat is carried passively in a flow driven by externally imposed pressure gradients, e.g., when you blow on hot food to cool it, or stir soup on a hot stove.)

The transport of heat is a fundamental characteristic of many flows. It dictates the form of global weather patterns and ocean currents. It is also of great technological importance and is studied in detail, for example, in the cooling of nuclear reactors and the design of automobile engines. From a more fundamental perspective, as we have already discussed, the analysis and experimental studies of convection have led to major insights into the route to chaos (Sec. 15.6).

In this chapter, we describe some flows where thermal effects are predominant. We begin in Sec. 18.2 by writing down and then simplifying the equations of fluid mechanics with heat conduction. Then in Sec. 18.3, we discuss the *Boussinesq approximation*, which is appropriate for modest-scale flows where buoyancy is important. This allows us in Sec. 18.4 to derive the conditions under which convection is initiated. Unfortunately, this Boussinesq approximation sometimes breaks down. In particular, as we discuss in Sec. 18.5, it is inappropriate for convection in stars and planets, where circulation takes place over several gravitational scale heights. For such cases we have to use alternative, more heuristic arguments to derive the relevant criterion for convective instability, known as the *Schwarzschild criterion*, and to quantify the associated heat flux. We shall apply this theory to the solar convection zone.

Finally, in Sec. 18.6 we return to simple buoyancy-driven convection in a stratified fluid to consider *double diffusion*, a quite general type of instability that can arise when the diffusion of two physical quantities (in our case heat and the concentration of salt) render a fluid unstable even though the fluid would be stably stratified if there were only concentration gradients of one of these quantities.

18.2   18.2 Diffusive Heat Conduction—Cooling a Nuclear Reactor; Thermal Boundary Layers

So long as the mean free path of heat-carrying particles is small compared to the fluid's inhomogeneity lengthscales (as is almost always the case), and the fractional temperature change in one mean free path is small (as is also almost always true), the energy flux of heat flow takes the thermal-diffusion form

$$\mathbf{F}_{\text{cond}} = -\kappa \boldsymbol{\nabla} T;$$ (18.1)

see Secs. 3.7 and 13.7.4. Here κ is the thermal conductivity.

For a viscous, heat-conducting fluid flowing in an external gravitational field, the most general governing equations are the fundamental thermodynamic potential $u(\rho, s)$; the first law of thermodynamics, Eq. (2) or (3) of Box 13.2; the law of mass conservation [Eq. (13.29) or (13.31)]; the Navier-Stokes equation (13.69); and the law of dissipative entropy production (13.75):

> evolution equations for a viscous, heat-conducting fluid in an external gravitational field

$$u = u(\rho, s),$$ (18.2a)

$$\frac{du}{dt} = T\frac{ds}{dt} - P\frac{d(1/\rho)}{dt},$$ (18.2b)

$$\frac{d\rho}{dt} = -\rho \boldsymbol{\nabla} \cdot \mathbf{v},$$ (18.2c)

$$\rho\frac{d\mathbf{v}}{dt} = -\boldsymbol{\nabla} P + \rho \mathbf{g} + \boldsymbol{\nabla}(\zeta\theta) + 2\boldsymbol{\nabla} \cdot (\eta\boldsymbol{\sigma}),$$ (18.2d)

$$T\left[\rho\left(\frac{ds}{dt}\right) + \boldsymbol{\nabla} \cdot \left(\frac{-\kappa\boldsymbol{\nabla} T}{T}\right)\right] = \zeta\theta^2 + 2\eta\boldsymbol{\sigma} : \boldsymbol{\sigma} + \frac{\kappa}{T}(\boldsymbol{\nabla} T)^2.$$ (18.2e)

These are four scalar equations and one vector equation for four scalar and one vector variables: the density ρ, internal energy per unit mass u, entropy per unit mass s, pressure P, and velocity $\mathbf{v}$. The thermal conductivity κ and coefficients of shear viscosity ζ and bulk viscosity $\eta = \rho\nu$ are presumed to be functions of ρ and s (or equally well, ρ and T).

This set of equations is far too complicated to solve, except via massive numerical simulations, unless some strong simplifications are imposed. We therefore introduce approximations. Our first approximation (already implicit in the above equations) is that the thermal conductivity κ is constant, as are the coefficients of viscosity. For most real applications this approximation is close to true, and no significant physical effects are missed by assuming it. Our second approximation, which does limit somewhat the type of problem we can address, is that the fluid motions are very slow—slow enough that, not only can the flow be regarded as incompressible ($\theta = \boldsymbol{\nabla} \cdot \mathbf{v} = 0$), but also the squares of the shear $\boldsymbol{\sigma}$ and expansion θ (which are quadratic in the fluid speed) are negligibly small, and we thus can ignore viscous dissipation. These approximations bring the last three of the fluid evolution equations (18.2) into the simplified form

> approximations

$$\boldsymbol{\nabla} \cdot \mathbf{v} \simeq 0, \quad d\rho/dt \simeq 0,$$ (18.3a)

$$\frac{d\mathbf{v}}{dt} = -\frac{\boldsymbol{\nabla} P}{\rho} + \mathbf{g} + \nu\nabla^2\mathbf{v},$$ (18.3b)

$$\rho T\frac{ds}{dt} = \kappa\nabla^2 T.$$ (18.3c)

[Our reasons for using "≃" in Eqs. (18.3a) become clear in Sec. 18.3, in connection with buoyancy.] Note that Eq. (18.3b) is the standard form of the Navier-Stokes equation for incompressible flows, which we have used extensively in the past several chapters. Equation (18.3c) is an elementary law of energy conservation: the temperature times the rate of increase of entropy density moving with the fluid is equal to minus the divergence of the conductive energy flux, $\mathbf{F}_{\text{heat}} = -\kappa \boldsymbol{\nabla} T$.

We can convert the entropy evolution equation (18.3c) into an evolution equation for temperature by expressing the changes ds/dt of entropy per unit mass in terms of changes dT/dt of temperature. The usual way to do this is to note that $T\,ds$ (the amount of heat deposited in a unit mass of fluid) is given by $c\,dT$, where c is the fluid's specific heat per unit mass. However, the specific heat depends on what one holds fixed during the energy deposition: the fluid element's volume or its pressure. As we have assumed that the fluid motions are slow, the fractional pressure fluctuations will be correspondingly small. (This assumption does not preclude significant temperature perturbations, provided they are compensated by density fluctuations of opposite sign.) Therefore, the relevant specific heat for a slowly moving fluid is the one at constant pressure, c_P, and we must write $T\,ds = c_P\,dT$.[1] Equation (18.3c) then becomes a linear partial differential equation for the temperature:

<table>
<tr><td>temperature diffusion
equation</td><td>

$$\frac{dT}{dt} \equiv \frac{\partial T}{\partial t} + \mathbf{v} \cdot \boldsymbol{\nabla} T = \chi \nabla^2 T,$$

</td><td>(18.4)</td></tr>
</table>

where

<table>
<tr><td>thermal diffusivity</td><td>

$$\chi = \kappa/(\rho\, c_P)$$

</td><td>(18.5)</td></tr>
</table>

is known as the *thermal diffusivity,* and we have again taken the easiest route in treating c_P and ρ as constant. When the fluid moves so slowly that the advective term $\mathbf{v} \cdot \boldsymbol{\nabla} T$ is negligible, then Eq. (18.4) shows that the heat simply diffuses through the fluid, with the thermal diffusivity χ being the diffusion coefficient for temperature.

Prandtl number: vorticity diffusion over heat diffusion

The diffusive transport of heat by thermal conduction is similar to the diffusive transport of vorticity by viscous stress [Eq. (14.3)], and the thermal diffusivity χ is the direct analog of the kinematic viscosity ν. This observation motivates us to introduce a new dimensionless number known as the *Prandtl number,* which measures the relative importance of viscosity and heat conduction (in the sense of their relative abilities to diffuse vorticity and heat):

$$\mathrm{Pr} = \frac{\nu}{\chi}. \tag{18.6}$$

---

1. See, e.g., Turner (1973) for a more formal justification of the use of the specific heat at constant pressure rather than at constant volume.

Fluid	ν (m^2 s^{-1})	χ (m^2 s^{-1})	Pr
Earth's mantle	10^{17}	10^{-6}	10^{23}
Solar interior	10^{-2}	10^{2}	10^{-4}
Atmosphere	10^{-5}	10^{-5}	1
Ocean	10^{-6}	10^{-7}	10

For gases, both ν and χ are given to order of magnitude by the product of the mean molecular speed and the mean free path, and so Prandtl numbers are typically of order unity. (For air, $\mathrm{Pr} \sim 0.7$.) By contrast, in liquid metals the free electrons carry heat very efficiently compared with the transport of momentum (and vorticity) by diffusing ions, and so their Prandtl numbers are small. This is why liquid sodium is used as a coolant in nuclear power reactors. At the other end of the spectrum, water is a relatively poor thermal conductor with $\mathrm{Pr} \sim 6$, and Prandtl numbers for oils, which are quite viscous and are poor conductors, measure in the thousands. Other Prandtl numbers are given in Table 18.1.

One might think that, when the Prandtl number is small (so κ is large compared to ν), one should necessarily include heat flow in the fluid equations and pay attention to thermally induced buoyancy (Sec. 18.3). Not so. In some low-Prandtl-number flows the heat conduction is so effective that the fluid becomes essentially isothermal, and buoyancy effects are minimized. Conversely, in some large-Prandtl-number flows the large viscous stress reduces the velocity gradient so that slow, thermally driven circulation takes place, and thermal effects are very important. In general the kinematic viscosity is of direct importance in controlling the transport of momentum, and hence in establishing the velocity field, whereas heat conduction affects the velocity field only indirectly (Sec. 18.3). We must therefore examine each flow on its individual merits.

Another dimensionless number is commonly introduced when discussing thermal effects: the *Péclet number*. It is defined, by analogy with the Reynolds number, as

$$\mathrm{Pe} = \frac{LV}{\chi}, \qquad (18.7)$$

where L is a characteristic length scale of the flow, and V is a characteristic speed. The Péclet number measures the relative importance of advection and heat conduction.

typical Prandtl numbers

Péclet number: heat advection over heat conduction

Exercise 18.1 *Problem: Fukushima-Daiichi Power Plant*

After the earthquake that triggered the catastrophic failure of the Fukushima-Daiichi nuclear power plant on March 11, 2011, reactor operation was immediately stopped. However, the subsequent tsunami disabled the cooling system needed to remove the decay heat, which was still being generated. This system failure led to a meltdown of three reactors and escape of radioactive material.

The *boiling water reactors* that were in use each generated $\sim$500 MW of heat, under normal operation, and the decay-heat production amounted to $\sim$30 MW. Suppose that this decay heat was being carried off equally by a large number of cylindrical pipes of length $L \sim 10$ m and inner radius $R \sim 10$ mm, taking it from the reactor core, where the water temperature was $T_0 \sim 550$ K and the pressure was $P_0 \sim 10^7$ N m^{-2}, to a heat exchanger. Suppose, initially, that the flow was laminar, so that the fluid velocity had the parabolic Poiseuille: profile

$$v = 2\bar{v}\left(1 - \frac{\varpi^2}{R^2}\right) \tag{18.8}$$

[Eq. (13.80) and associated discussion]. Here, ϖ is the cylindrical radial coordinate measured from the axis of the pipe, and $\bar{v}$ is the mean speed along the pipe. As the goal was to carry the heat away efficiently during normal operation, the pipe was thermally well insulated, so its inner wall was at nearly the same temperature as the core of the fluid (at $\varpi = 0$). The total temperature drop ΔT down the length L was then $\Delta T \ll T_0$, and the longitudinal temperature gradient was constant, so the temperature distribution in the pipe had the form:

$$T = T_0 - \Delta T \frac{z}{L} + f(\varpi). \tag{18.9}$$

(a) Use Eq. (18.4) to show that

$$f = \frac{\bar{v} R^2 \Delta T}{2\chi L}\left[\frac{3}{4} - \frac{\varpi^2}{R^2} + \frac{1}{4}\frac{\varpi^4}{R^4}\right]. \tag{18.10}$$

(b) Derive an expression for the conductive heat flux through the walls of the pipe and show that the ratio of the heat escaping through the walls to that advected by the fluid was $\Delta T / T$. (Ignore the influence of the temperature gradient on the velocity field, and treat the thermal diffusivity and specific heat as constant throughout the flow.)

(c) The flow would only remain laminar so long as the Reynolds number was Re $\lesssim$ Re$_{\rm crit} \sim 2{,}000$, the critical value for transition to turbulence. Show that the maximum power that could be carried by a laminar flow was:

$$\dot{Q}_{\rm max} \sim \frac{\pi}{2}\rho R v \text{Re}_{\rm crit} c_p T_0 \sim \frac{\pi}{4}\left(\frac{\rho P_0 R^5 \text{Re}_{\rm crit}}{L}\right)^{1/2} c_p T_0. \tag{18.11}$$

Estimate the mean velocity $\bar{v}$, the temperature drop ΔT, and the number of cooling pipes needed in normal operation, when the flow was about to transition to turbulence. Assume that $\chi = 10^{-7}\,\mathrm{m^2\,s^{-1}}$, $c_p = 6\,\mathrm{kJ\,kg^{-1}\,K^{-1}}$, and $\rho v = \eta = 1 \times 10^{-4}\,\mathrm{Pa\,s^{-1}}$.

(d) Describe qualitatively what would happen if the flow became turbulent (as it did under normal operation).

Exercise 18.2 *Problem: Thermal Boundary Layers*

In Sec. 14.4, we introduced the notion of a laminar boundary layer by analyzing flow past a thin plate. Now suppose that this same plate is maintained at a different temperature from the free flow. A thermal boundary layer will form, in addition to the viscous boundary layer, which we presume to be laminar. These two boundary layers both extend outward from the wall but (usually) have different thicknesses.

(a) Explain why their relative thicknesses depend on the Prandtl number.

(b) Using Eq. (18.4), show that in order of magnitude the thickness δ_T of the thermal boundary layer is given by

$$v(\delta_T)\delta_T^2 = \ell\chi,$$

where $v(\delta_T)$ is the fluid velocity parallel to the plate at the outer edge of the thermal boundary layer, and ℓ is the distance downstream from the leading edge.

(c) Let V be the free-stream fluid velocity and ΔT be the temperature difference between the plate and the body of the flow. Estimate δ_T in the limits of large and small Prandtl numbers.

(d) What will be the boundary layer's temperature profile when the Prandtl number is exactly unity? [Hint: Seek a self-similar solution to the relevant equations. For the solution, see Lautrup (2005, Sec. 31.1).]

18.3 Boussinesq Approximation

When heat fluxes are sufficiently small, we can use Eq. (18.4) to solve for the temperature distribution in a given velocity field, ignoring the feedback of thermal effects on the velocity. However, if we imagine increasing the flow's temperature differences, so the heat fluxes also increase, at some point thermal feedback effects begin to influence the velocity significantly. Typically, the first feedback effect to occur is *buoyancy*, the tendency of the hotter (and hence lower-density) fluid to rise in a gravitational field and the colder (and hence denser) fluid to descend.[2] In this section, we describe the

---

2. This effect is put to good use in a domestic "gravity-fed" warm-air circulation system. The furnace generally resides in the basement, not the attic!

effects of buoyancy as simply as possible. The minimal approach, which is adequate surprisingly often, is called the *Boussinesq approximation*. Leading to Eqs. (18.12), (18.18), and (18.19), it can be used to describe many heat-driven laboratory flows and atmospheric flows, and some geophysical flows.

Boussinesq approximation for describing heat-driven buoyancy effects

The types of flows for which the Boussinesq approximation is appropriate are those in which the fractional density changes are small ($|\Delta\rho| \ll \rho$). By contrast, the velocity can undergo large changes, though it remains constrained by the incompressibility relation (18.3a):

$$\boxed{\nabla \cdot \mathbf{v} = 0 \quad \text{Boussinesq (1)}.} \tag{18.12}$$

One might think that this constraint implies constant density moving with a fluid element, since mass conservation requires $d\rho/dt = -\rho\nabla \cdot v$. However, thermal expansion causes small density changes, with tiny corresponding violations of Eq. (18.12); this explains the "$\simeq$" that we used in Eqs. (18.3a). The key point is that, for these types of flows, the density is controlled to high accuracy by thermal expansion, and the velocity field is divergence free to high accuracy.

When discussing thermal expansion, it is convenient to introduce a *reference density* ρ_0 and *reference temperature* T_0, equal to some mean of the density and temperature in the region of fluid that one is studying. We shall denote by

$$\boxed{\tau \equiv T - T_0} \tag{18.13}$$

the perturbation of the temperature away from its reference value. The thermally perturbed density can then be written as

thermal expansion

$$\boxed{\rho = \rho_0(1 - \alpha\tau),} \tag{18.14}$$

where α is the thermal expansion coefficient for volume[3] [evaluated at constant pressure for the same reason as the specific heat was chosen as at constant pressure in the paragraph following Eq. (18.3c)]:

$$\boxed{\alpha = -\left(\frac{\partial \ln \rho}{\partial T}\right)_P.} \tag{18.15}$$

Equation (18.14) enables us to eliminate density perturbations as an explicit variable and replace them by temperature perturbations.

Now turn to the Navier-Stokes equation (18.3b) in a uniform external gravitational field. We expand the pressure-gradient term as

$$-\frac{\nabla P}{\rho} \simeq -\frac{\nabla P}{\rho_0}(1 + \alpha\tau) = \frac{\nabla P}{\rho_0} - \alpha\tau\mathbf{g}, \tag{18.16}$$

---

3. Note that α is three times larger than the thermal expansion coefficient for the linear dimensions of the fluid.

where we have used hydrostatic equilibrium for the unperturbed flow. As in our analysis of rotating flows [Eq. (14.55)], we introduce an *effective pressure* designed to compensate for the first-order effects of the uniform gravitational field:

$$P' = P + \rho_0 \Phi = P - \rho_0 \mathbf{g} \cdot \mathbf{x}. \qquad (18.17)$$

(Notice that P' measures the amount the pressure differs from the value it would have in supporting a hydrostatic atmosphere of the fluid at the reference density.) The Navier-Stokes equation (18.3b) then becomes

$$\frac{d\mathbf{v}}{dt} = -\frac{\nabla P'}{\rho_0} - \alpha \tau \mathbf{g} + \nu \nabla^2 \mathbf{v} \quad \text{Boussinesq (2),} \qquad (18.18)$$

dropping the small term $O(\alpha P')$. In words, a fluid element accelerates in response to a buoyancy force [the sum of the first and second terms on the right-hand side of Eq. (18.18)] and a viscous force.

To solve this equation, we must be able to solve for the temperature perturbation τ. This evolves according to the standard equation of heat diffusion [Eq. (18.4)]:

$$\frac{d\tau}{dt} = \chi \nabla^2 \tau \quad \text{Boussinesq (3).} \qquad (18.19)$$

Equations (18.12), (18.18), and (18.19) are the equations of fluid flow in the Boussinesq approximation; they control the coupled evolution of the velocity $\mathbf{v}$ and the temperature perturbation τ. We now use them to discuss free convection in a laboratory apparatus.

Boussinesq equations

18.4 Rayleigh-Bénard Convection

18.4

In a relatively simple laboratory experiment to demonstrate convection, a fluid is confined between two rigid plates a distance d apart, each maintained at a fixed temperature, with the upper plate cooler than the lower by ΔT. When ΔT is small, viscous stresses, together with the no-slip boundary conditions at the plates, inhibit circulation; so, despite the upward buoyancy force on the hotter, less-dense fluid near the bottom plate, the fluid remains stably at rest with heat being conducted diffusively upward. If the plates' temperature difference ΔT is gradually increased, the buoyancy becomes gradually stronger. At some critical ΔT it will overcome the restraining viscous forces, and the fluid starts to circulate (convect) between the two plates. Our goal is to determine the critical temperature difference ΔT_{crit} for the onset of this *Rayleigh-Bénard* convection.

Rayleigh-Bénard convection

We now make some physical arguments to simplify the calculation of ΔT_{crit}. From our experience with earlier instability calculations, especially those involving elastic bifurcations (Secs. 11.6.1 and 12.3.5), we anticipate that for $\Delta T < \Delta T_{\text{crit}}$ the response of the equilibrium to small perturbations will be oscillatory (i.e., will have positive squared eigenfrequency ω^2), while for $\Delta T > \Delta T_{\text{crit}}$, perturbations will grow

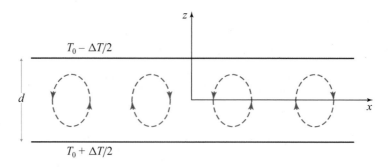

FIGURE 18.1 Rayleigh-Bénard convection. A fluid is confined between two horizontal surfaces separated by a vertical distance d. When the temperature difference between the two plates ΔT is increased sufficiently, the fluid starts to convect heat vertically. The reference effective pressure P_0' and reference temperature T_0 are the values of P' and T measured at the midplane $z = 0$.

exponentially (i.e., will have negative ω^2). Correspondingly, at $\Delta T = \Delta T_{\text{crit}}$, ω^2 for some mode will be zero. This zero-frequency mode marks the bifurcation of equilibria from one with no fluid motions to one with slow, convective motions. We search for ΔT_{crit} by searching for a solution to the Boussinesq equations (18.12), (18.18), and (18.19) that represents this zero-frequency mode. In those equations, we choose for the reference temperature T_0, density ρ_0, and effective pressure P_0 the values at the midplane between the plates, $z = 0$ (cf. Fig. 18.1).

The unperturbed equilibrium, when $\Delta T = \Delta T_{\text{crit}}$, is a solution of the Boussinesq equations (18.12), (18.18), and (18.19) with vanishing velocity, a time-independent vertical temperature gradient $dT/dz = -\Delta T/d$, and a compensating, time-independent, vertical pressure gradient:

$$\mathbf{v} = 0, \quad \tau = T - T_0 = -\frac{\Delta T}{d} z, \quad P' = P_0' + g\rho_0 \alpha \frac{\Delta T}{d} \frac{z^2}{2}. \tag{18.20}$$

When the zero-frequency mode is present, the velocity $\mathbf{v}$ is nonzero, and the temperature and effective pressure have additional perturbations $\delta\tau$ and $\delta P'$:

$$\mathbf{v} \neq 0, \quad \tau = T - T_0 = -\frac{\Delta T}{d} z + \delta\tau, \quad P' = P_0' + g\rho_0 \alpha \frac{\Delta T}{d} \frac{z^2}{2} + \delta P'. \tag{18.21}$$

The perturbations $\mathbf{v}$, $\delta\tau$, and $\delta P'$ are governed by the Boussinesq equations and the boundary conditions at the plates ($z = \pm d/2$): $\mathbf{v} = 0$ (no-slip) and $\delta\tau = 0$. We manipulate these equations and boundary conditions in such a way as to get a partial differential equation for the scalar temperature perturbation $\delta\tau$ by itself, decoupled from the velocity and the pressure perturbation.

First consider the result of inserting expressions (18.21) into the Boussinesq-approximated Navier-Stokes equation (18.18). Because the perturbation mode has zero frequency, $\partial \mathbf{v}/\partial t$ vanishes; and because $\mathbf{v}$ is extremely small, we can neglect the quadratic advective term $\mathbf{v} \cdot \nabla \mathbf{v}$, thereby bringing Eq. (18.18) into the form:

$$\frac{\nabla \delta P'}{\rho_0} = \nu \nabla^2 \mathbf{v} - \mathbf{g}\alpha\delta\tau. \tag{18.22}$$

We want to eliminate $\delta P'$ from this equation. The other Boussinesq equations are of no help for this, since $\delta P'$ is absent from them. One might be tempted to eliminate δP using the equation of state $P = P(\rho, T)$; but in the present analysis our Boussinesq approximation insists that the only significant changes of density are those due to thermal expansion (i.e., it neglects the influence of pressure on density), so the equation of state cannot help us. Lacking any other way to eliminate $\delta P'$, we employ a common trick: we take the curl of Eq. (18.22). As the curl of a gradient vanishes, $\delta P'$ drops out. We then take the curl one more time and use $\nabla \cdot \mathbf{v} = 0$ to obtain

$$\nu \nabla^2 (\nabla^2 \mathbf{v}) = \alpha \mathbf{g} \nabla^2 \delta\tau - \alpha(\mathbf{g} \cdot \nabla)\nabla\delta\tau. \tag{18.23}$$

Next turn to the Boussinesq version of the equation of heat transport [Eq. (18.19)]. Inserting into it Eqs. (18.21) for τ and $\mathbf{v}$, setting $\partial \delta\tau/\partial t$ to zero (because our perturbation has zero frequency), linearizing in the perturbation, and using $\mathbf{g} = -g\mathbf{e}_z$, we obtain

$$\frac{v_z \Delta T}{d} = -\chi \nabla^2 \delta\tau. \tag{18.24}$$

This is an equation for the vertical velocity v_z in terms of the temperature perturbation $\delta\tau$. By inserting this v_z into the z component of Eq. (18.23), we achieve our goal of a scalar equation for $\delta\tau$ alone:

$$\boxed{\nu\chi \nabla^2\nabla^2\nabla^2\delta\tau = \frac{\alpha g \Delta T}{d}\left(\frac{\partial^2\delta\tau}{\partial x^2} + \frac{\partial^2\delta\tau}{\partial y^2}\right).} \tag{18.25}$$

linearized perturbation equation for Rayleigh-Bénard convection

This is a sixth-order differential equation, even more formidable than the fourth-order equations that arise in the elasticity calculations of Chaps. 11 and 12. We now see how prudent it was to make simplifying assumptions at the outset!

The differential equation (18.25) is, however, linear, so we can seek solutions using separation of variables. As the equilibrium is unbounded horizontally, we look for a single horizontal Fourier component with some wave number k; that is, we seek a solution of the form

$$\delta\tau \propto \exp(ikx)f(z), \tag{18.26}$$

where $f(z)$ is some unknown function. Such a $\delta\tau$ will be accompanied by motions $\mathbf{v}$ in the x and z directions (i.e., $v_y = 0$) that also have the form $v_j \propto \exp(ikx)f_j(z)$ for some other functions $f_j(z)$.

The ansatz (18.26) converts the partial differential equation (18.25) into the single ordinary differential equation

$$\left(\frac{d^2}{dz^2} - k^2\right)^3 f + \frac{\text{Ra}\,k^2 f}{d^4} = 0, \tag{18.27}$$

where we have introduced yet another dimensionless number

$$\boxed{\text{Ra} = \frac{\alpha g \Delta T d^3}{\nu \chi}} \tag{18.28}$$

called the *Rayleigh number*. By virtue of relation (18.24) between v_z and $\delta\tau$, the Rayleigh number is a measure of the ratio of the strength of the buoyancy term $-\alpha\delta\tau\mathbf{g}$ to the viscous term $\nu\nabla^2\mathbf{v}$ in the Boussinesq version [Eq. (18.18)] of the Navier-Stokes equation:

$$\boxed{\text{Ra} \sim \frac{\text{buoyancy force}}{\text{viscous force}}.} \tag{18.29}$$

The general solution of Eq. (18.27) is an arbitrary, linear combination of three sine functions and three cosine functions:

$$f = \sum_{n=1}^{3} A_n \cos(\mu_n k z) + B_n \sin(\mu_n k z), \tag{18.30}$$

where the dimensionless numbers μ_n are given by

$$\mu_n = \left[\left(\frac{\text{Ra}}{k^4 d^4}\right)^{1/3} e^{2\pi n i/3} - 1\right]^{1/2} ; \quad n = 1, 2, 3, \tag{18.31}$$

which involves the three cube roots of unity, $e^{2\pi n i/3}$. The values of five of the coefficients A_n, B_n are fixed in terms of the sixth (an overall arbitrary amplitude) by five boundary conditions at the bounding plates. A sixth boundary condition then determines the critical temperature difference ΔT_{crit} (or equivalently, the critical Rayleigh number Ra_{crit}) at which convection sets in.

The six boundary conditions are paired as follows:

1. The requirement that the fluid temperature be the same as the plate temperature at each plate, so $\delta\tau = 0$ at $z = \pm d/2$.

2. The no-slip boundary condition $v_z = 0$ at each plate, which by virtue of Eq. (18.24) and $\delta\tau = 0$ at the plates, translates to $\delta\tau_{,zz} = 0$ at $z = \pm d/2$ (where the indices after the comma are partial derivatives).

3. The no-slip boundary condition $v_x = 0$, which by virtue of incompressibility $\nabla \cdot \mathbf{v} = 0$ implies $v_{z,z} = 0$ at the plates, which in turn by Eq. (18.24) implies $\delta\tau_{,zzz} + \delta\tau_{,xxz} = 0$ at $z = \pm d/2$.

It is straightforward but computationally complex to impose these six boundary conditions and from them deduce the critical Rayleigh number for onset of convection (Pellew and Southwell, 1940). Rather than present the nasty details, we switch to a toy problem in which the boundary conditions are adjusted to give a simpler solution but one with the same qualitative features as for the real problem. Specifically, we replace the no-slip condition (3) ($v_x = 0$ at the plates) by a condition of no shear:

toy problem for Rayleigh-Bénard convection

3′. $v_{x,z} = 0$ at the plates. By virtue of incompressibility $\nabla \cdot \mathbf{v} = 0$, the x derivative of this condition translates to $v_{z,zz} = 0$, which by Eq. (18.24) becomes $\delta\tau_{,zzxx} + \delta\tau_{,zzzz} = 0$.

To recapitulate, we seek a solution of the form (18.30) and (18.31) that satisfies the boundary conditions (1), (2), and (3′).

The terms in Eq. (18.30) with $n = 1, 2$ always have complex arguments and thus always have z dependences that are products of hyperbolic and trigonometric functions with real arguments. For $n = 3$ and a large enough Rayleigh number, μ_3 is positive, and the solutions are pure sines and cosines. Let us just consider the $n = 3$ terms alone, in this regime, and impose boundary condition (1): $\delta\tau = 0$ at the plates. The cosine term by itself,

$$\delta\tau = \text{const} \times \cos(\mu_3 kz)\, e^{ikx}, \tag{18.32}$$

satisfies this boundary condition, if we set

$$\frac{\mu_3 kd}{2} \equiv \left[\left(\frac{\text{Ra}}{k^4 d^4}\right)^{1/3} - 1\right]^{1/2} \frac{kd}{2} = \left(m + \frac{1}{2}\right)\pi, \tag{18.33}$$

where m is an integer. It is straightforward to show, remarkably, that Eqs. (18.32) and (18.33) also satisfy boundary conditions (2) and (3′), so they solve the toy version of our problem.

As ΔT is gradually increased from zero, the Rayleigh number gradually grows, passing through the sequence of values given by Eq. (18.33) with $m = 0, 1, 2, \ldots$ (for any chosen k). At each of these values there is a zero-frequency, circulatory mode of fluid motion with horizontal wave number k, which is passing from stability to instability. The first of these, $m = 0$, represents the onset of circulation for the chosen k, and the Rayleigh number at this onset [Eq. (18.33) with $m = 0$] is

$$\text{Ra} = \frac{(k^2 d^2 + \pi^2)^3}{k^2 d^2}. \tag{18.34}$$

This Ra(k) relation is plotted as a thick curve in Fig. 18.2.

Notice in Fig. 18.2 that there is a critical Rayleigh number Ra$_{\text{crit}}$ below which all modes are stable, independent of their wave numbers, and above which modes in some range $k_{\min} < k < k_{\max}$ are unstable. From Eq. (18.34) we deduce that for our toy problem, Ra$_{\text{crit}} = 27\pi^4/4 \simeq 658$.

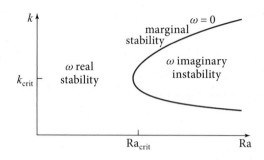

FIGURE 18.2 Horizontal wave number k of the first mode to go unstable, as a function of Rayleigh number. Along the solid curve the mode has zero frequency; to the left of the curve it is stable, to the right it is unstable. Ra_{crit} is the minimum Rayleigh number for convective instability.

When one imposes the correct boundary conditions (1), (2), and (3) [instead of our toy choice (1), (2), and (3′)] and works through the nasty details of the computation, one obtains a relation for $Ra(k)$ that looks qualitatively the same as Fig. 18.2. One deduces that convection should set in at $Ra_{crit} = 1{,}708$, which agrees reasonably well with experiment. One can carry out the same computation with the fluid's upper surface free to move (e.g., due to placing air rather than a solid plate at $z = d/2$). Such a computation predicts that convection begins at $Ra_{crit} \simeq 1{,}100$, though in practice surface tension is usually important, and its effect must be included.

One feature of these critical Rayleigh numbers is very striking. Because the Rayleigh number is an estimate of the ratio of buoyancy forces to viscous forces [Eq. (18.29)], an order-of-magnitude analysis suggests that convection should set in at $Ra \sim 1$—which is wrong by three orders of magnitude! This provides a vivid reminder that order-of-magnitude estimates can be quite inaccurate. In this case, the main reason for the discrepancy is that the convective onset is governed by a sixth-order differential equation (18.25) and thus is highly sensitive to the lengthscale d used in the order-of-magnitude analysis. If we choose d/π rather than d as the length scale, then an order-of-magnitude estimate could give $Ra \sim \pi^6 \sim 1{,}000$, a much more satisfactory value.

Once convection has set in, the unstable modes grow until viscosity and non-linearities stabilize them, at which point they carry far more heat upward between the plates than does conduction. The convection's velocity pattern depends in practice on the manner in which the heat is applied, the temperature dependence of the viscosity, and the fluid's boundaries. For a limited range of Rayleigh numbers near Ra_{crit}, it is possible to excite a hexagonal pattern of *convection cells* as is largely but not entirely the case in Fig. 18.3; however other patterns can also be excited.

critical Rayleigh number for onset of Rayleigh-Bénard convection

how order-of-magnitude analyses can fail

patterns of convection cells in Rayleigh-Bénard convection

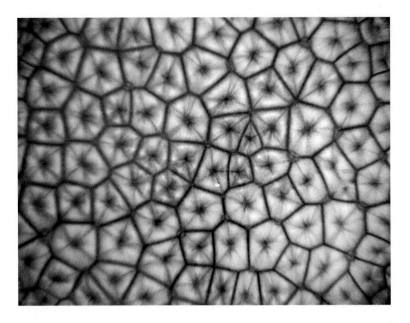

FIGURE 18.3 Convection cells in Rayleigh-Bénard convection. The fluid, which is visualized using aluminum powder, rises at the center of each cell and falls around its edges. From Maroto, Perez-Munuzuri, and Romero-Cano (2007).

Drazin and Reid (2004) suggest a kitchen experiment for observing convection cells. Place a 2-mm layer of corn or canola oil on the bottom of a skillet, and sprinkle cocoa or Ovaltine or other powder over it. Heat the skillet bottom gently and uniformly. The motion of the powder particles will reveal the convection cells, with upwelling at the cell centers and surface powder collecting and falling at the edges.

In Rayleigh-Bénard convection experiments, as the Rayleigh number is increased beyond the onset of convection, one or another sequences of equilibrium bifurcations leads to weak turbulence (see Secs. 15.6.2 and 15.6.3). When the Rayleigh number becomes very large, the convection becomes strongly turbulent.

Free convection, like that in these laboratory experiments, also occurs in meteorological and geophysical flows. For example, for air in a room, the relevant parameter values are $\alpha = 1/T \sim 0.003 \text{ K}^{-1}$ (Charles' Law), and $\nu \sim \chi \sim 10^{-5} \text{ m}^2 \text{ s}^{-1}$, so the Rayleigh number is $\text{Ra} \sim 3 \times 10^8 (\Delta T / 1 \text{ K})(d/1 \text{ m})^3$. Convection in a room thus occurs extremely readily, even for small temperature differences. In fact, so many modes of convective motion can be excited that heat-driven air flow is invariably turbulent. It is therefore common in everyday situations to describe heat transport using a phenomenological turbulent thermal conductivity (Sec. 15.4.2; White, 2008, Sec. 6.10.1).

A second example, convection in Earth's mantle, is described in Box 18.2.

kitchen experiment

bifurcation of Rayleigh-Bénard equilibria leading to convective turbulence

BOX 18.2.   MANTLE CONVECTION AND CONTINENTAL DRIFT

As is now well known, the continents drift over the surface of the globe on a timescale of roughly 100 million years. Despite the clear geographical evidence that the continents fit together, some geophysicists were, for a long while, skeptical that this occurred, because they were unable to identify the forces responsible for overcoming the visco-elastic resilience of the crust. It is now known that these motions are in fact slow convective circulation of the mantle driven by internally generated heat from the radioactive decay of unstable isotopes, principally uranium, thorium, and potassium.

When the heat is generated in the convective layer (which has radial thickness d), rather than passively transported from below, we must modify our definition of the Rayleigh number. Let the heat generated per unit mass per unit time be Q. In the analog of our laboratory analysis, where the fluid is assumed marginally unstable to convective motions, this Q will generate a heat flux $\sim \rho Q d$, which must be carried diffusively. Equating this flux to $\kappa \Delta T / d$, we can solve for the temperature difference ΔT between the lower and upper edges of the convective mantle: $\Delta T \sim \rho Q d^2 / \kappa$. Inserting this ΔT into Eq. (18.28), we obtain a modified expression for the Rayleigh number

$$\mathrm{Ra}' = \frac{\alpha \rho g Q d^5}{\kappa \chi \nu}. \tag{1}$$

Let us now estimate the value of Ra' for Earth's mantle. The mantle's kinematic viscosity can be measured by post-glacial rebound studies (cf. Ex. 14.13) to be $\nu \sim 10^{17}\,\mathrm{m^2\,s^{-1}}$. We can use the rate of attenuation of diurnal and annual temperature variation with depth in surface rock to estimate a thermal diffusivity $\chi \sim 10^{-6}\,\mathrm{m^2\,s^{-1}}$. Direct experiment furnishes an expansion coefficient, $\alpha \sim 3 \times 10^{-5}\,\mathrm{K^{-1}}$ and thermal conductivity $\kappa \sim 4\,\mathrm{W\,m^{-1}\,K^{-1}}$. The thickness of the upper mantle is $d \sim 700$ km, and the rock density is $\rho \sim 4{,}000\,\mathrm{kg\,m^{-3}}$. The rate of heat generation can be estimated both by chemical analysis and direct measurement at Earth's surface and turns out to be $Q \sim 10^{-11}\,\mathrm{W\,kg^{-1}}$. Combining these quantities, we obtain an estimated Rayleigh number $\mathrm{Ra}' \sim 10^7$, well in excess of the critical value for convection under free slip conditions, which evaluates to $\mathrm{Ra}'_{\mathrm{crit}} = 868$ (Turcotte and Schubert, 1982). For this reason, it is now believed that continental drift is driven primarily by mantle convection.

Exercise 18.3 *Problem: Critical Rayleigh Number*
Use the Rayleigh criterion to estimate the temperature difference that would have to be maintained for 2 mm of corn/canola oil, or water, or mercury in a skillet to start convecting. Look up the relevant physical properties and comment on your answers. Do not perform this experiment with mercury.

Exercise 18.4 *Problem: Width of a Thermal Plume*
Consider a knife on its back, so its sharp edge points in the upward, z direction. The edge (idealized as extending infinitely far in the y direction) is hot, and by heating adjacent fluid, it creates a rising thermal plume. Introduce a temperature deficit $\Delta T(z)$ that measures the typical difference in temperature between the plume and the surrounding, ambient fluid at height z above the knife edge, and let $\delta_p(z)$ be the width of the plume at height z.

(a) Show that energy conservation implies the constancy of $\delta_p \Delta T \bar{v}_z$, where $\bar{v}_z(z)$ is the plume's mean vertical speed at height z.

(b) Make an estimate of the buoyancy acceleration, and use it to estimate $\bar{v}_z$.

(c) Use Eq. (18.19) to relate the width of the plume to the speed. Hence, show that the width of the plume scales as $\delta_p \propto z^{2/5}$ and the temperature deficit as $\Delta T \propto z^{-3/5}$.

(d) Repeat this exercise for a 3-dimensional plume above a hot spot.

18.5 Convection in Stars

18.5

The Sun and other stars generate heat in their interiors by nuclear reactions. In most stars the internal energy is predominantly in the form of hot hydrogen and helium ions and their electrons, while the thermal conductivity is due primarily to diffusing photons (Sec. 3.7.1), which have much longer mean free paths than the ions and electrons. When the photon mean free path becomes small due to high opacity (as happens in the outer 30% of the Sun; Fig. 18.4), the thermal conductivity goes down, so to transport the heat from nuclear burning, the star develops an increasingly steep

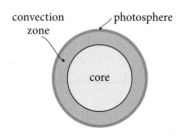

FIGURE 18.4 A convection zone occupies the outer 30% of a solar-type star.

temperature gradient. The star may then become convectively unstable and transport its energy far more efficiently by circulating its hot gas than it could have done by photon diffusion. Describing this convection is a key step in understanding the interiors of the Sun and other stars.

A heuristic argument provides the basis for a surprisingly simple description of this convection. As a foundation for our argument, let us identify the relevant physics:

1. The pressure in stars varies through many orders of magnitude (a factor $\sim 10^{12}$ for the Sun). Therefore, we cannot use the Boussinesq approximation; instead, as a fluid element rises or descends, we must allow for its density to change in response to large changes of the surrounding pressure.

2. The convection involves circulatory motions on such large scales that the attendant shears are small, and viscosity is thus unimportant.

3. Because the convection is driven by the ineffectiveness of conduction, we can idealize each fluid element as retaining its heat as it moves, so the flow is adiabatic.

4. The convection is usually well below sonic, as subsonic motions are easily sufficient to transport the nuclear-generated heat, except very close to the solar surface.

heuristic analysis of the onset of convection in stars

Our heuristic argument, then, focuses on convecting fluid blobs that move through the star's interior very subsonically, adiabatically, and without viscosity. As the motion is subsonic, each blob remains in pressure equilibrium with its surroundings. Now, suppose we make a virtual interchange between two blobs at different heights (Fig. 18.5). The blob that rises (blob B in the figure) experiences a decreased pressure and thus expands, so its density diminishes. If its density after rising is lower than that of its surroundings, then it is buoyant and continues to rise. Conversely, if the raised blob is denser than its surroundings, then it will sink back to its original location. Therefore, a criterion for convective instability is that the raised blob has lower density than its surroundings. Since the blob and its surroundings have the same pressure, and since the larger is the entropy s per unit mass of gas, the lower is its density (there being more phase space available to its particles), the fluid is convectively unstable if the raised blob has a higher entropy than its surroundings. Now, the blob's motion was adiabatic, so its entropy per unit mass s is the same after it rises as before. Therefore, the fluid is convectively unstable if the entropy per unit mass s at the location where the blob began (lower in the star) is greater than that at the location to which it rose (higher in the star); that is, *the star is convectively unstable if its entropy per unit mass decreases outward: $ds/dr < 0$.* For small blobs, this instability will be counteracted by both viscosity and heat conduction. But for large blobs, viscosity and conduction are ineffective, and the convection proceeds.

When building stellar models, astrophysicists find it convenient to determine whether a region of a model is convectively unstable by computing what its struc-

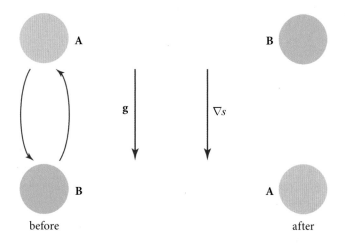

before　　　　　　　　　　　　　　　after

FIGURE 18.5 Convectively unstable interchange of two blobs in a star whose entropy per unit mass increases downward. Blob B rises to the former position of blob A and expands adiabatically to match the surrounding pressure. The entropy per unit mass of the blob is higher than that of the surrounding gas, and so the blob has a lower density. It will therefore be buoyant and continue to rise. Similarly, blob A will continue to sink.

ture would be without convection (i.e., with all its heat carried radiatively). That computation gives some temperature gradient dT/dr. If this computed dT/dr is *superadiabatic,* that is, if

Schwarzschild criterion for convection in stars

$$-\frac{d\ln T}{d\ln r} > \left(\frac{\partial \ln T}{\partial \ln P}\right)_s \left(-\frac{d\ln P}{d\ln r}\right) \equiv -\left(\frac{d\ln T}{d\ln r}\right)_s, \qquad (18.35)$$

then the entropy s decreases outward, and the star is convectively unstable. This is known as the *Schwarzschild criterion for convection,* since it was formulated by the same Karl Schwarzschild who discovered the Schwarzschild solution to Einstein's equations (which describes a nonrotating black hole; Chap. 26).

In practice, if the star is convective, then the convection is usually so efficient at transporting heat that the actual temperature gradient is only slightly superadiabatic, that is, the entropy s is nearly independent of radius—it decreases outward only very slightly. (Of course, the entropy can increase significantly outward in a convectively stable zone, where radiative diffusion is adequate to transport heat.)

actual temperature gradient in convective region of a star

We can demonstrate the efficiency of convection by estimating the convective heat flux when the temperature gradient is slightly superadiabatic, that is, when $\Delta|\nabla T| \equiv |(dT/dr)| - |(dT/dr)_s|$ is slightly positive. As a tool in our estimate, we introduce the concept of the *mixing length,* denoted by l—the typical distance a blob travels before breaking up. As the blob is in pressure equilibrium, we can estimate its fractional density difference from its surroundings by $\Delta\rho/\rho \sim \Delta T/T \sim \Delta|\nabla T| l/T$. Invoking Archimedes' law, we estimate the blob's acceleration to be $\sim g\Delta\rho/\rho \sim g\Delta|\nabla T|l/T$ (where g is the local acceleration of gravity); hence the average speed with which a

mixing length

blob rises or sinks is $\bar{v} \sim (g\Delta|\nabla T|/T)^{1/2}l$. The convective heat flux is then given by

$$F_{\text{conv}} \sim c_P \rho \bar{v} l \Delta |\nabla T|$$
$$\sim c_P \rho (g/T)^{1/2} (\Delta |\nabla T|)^{3/2} l^2. \tag{18.36}$$

We can bring this expression into a more useful form, accurate to within factors of order unity, by (i) setting the mixing length equal to the pressure scale height $l \sim H = |dr/d \ln P|$ (as is usually the case in the outer parts of a star); (ii) setting $c_P \sim h/T$, where h is the enthalpy per unit mass [cf. the first law of thermodynamics, Eq. (3) of Box 13.2]; (iii) setting $g = -(P/\rho)d \ln P/dr \sim C^2 |d \ln P/dr|$ [cf. the equation of hydrostatic equilibrium (13.13) and Eq. (16.48) for the speed of sound C]; and (iv) setting $|\nabla T| \equiv |dT/dr| \sim T |d \ln P/dr|$. The resulting expression for F_{conv} can then be inverted to give

$$\frac{|\Delta \nabla T|}{|\nabla T|} \sim \left(\frac{F_{\text{conv}}}{h\rho C} \right)^{2/3} \sim \left(\frac{F_{\text{conv}}}{\frac{5}{2} P \sqrt{k_B T/m_p}} \right)^{2/3}. \tag{18.37}$$

Here the last expression is obtained from the fact that the gas is fully ionized, so its enthalpy is $h = \frac{5}{2} P/\rho$, and its speed of sound is about the thermal speed of its protons (the most numerous massive particle), $C \sim \sqrt{k_B T/m_p}$ (with k_B Boltzmann's constant and m_p the proton rest mass).

It is informative to apply this estimate to the convection zone of the Sun (the outer $\sim$30% of its radius; Fig. 18.4). The luminosity of the Sun is $\sim 4 \times 10^{26}$ W, and its radius is 7×10^5 km, so its convective energy flux is $F_{\text{conv}} \sim 10^8$ W m^{-2}. First consider the convection zone's base. The pressure there is $P \sim 1$ TPa, and the temperature is $T \sim 10^6$ K, so Eq. (18.37) predicts $|\Delta \nabla T|/|\nabla T| \sim 3 \times 10^{-7}$, so the temperature gradient at the base of the convection zone need only be superadiabatic by a few parts in 10 million to carry the solar energy flux.

By contrast, at the top of the convection zone (which is nearly at the solar surface), the gas pressure is only $\sim$10 kPa, and the sound speed is $\sim$10 km s^{-1}, so $h\rho c \sim 10^8$ W m^{-2}, and $|\Delta \nabla T|/|\nabla T| \sim 1$; that is, the temperature gradient must depart significantly from the adiabatic gradient to carry the heat. Moreover, the convective elements, in their struggle to carry the heat, move with a significant fraction of the sound speed, so it is no longer true that they are in pressure equilibrium with their surroundings. A more sophisticated theory of convection is therefore necessary near the solar surface.

Convection is important in some other types of stars. It is the primary means of heat transport in the cores of stars with high mass and high luminosity, and throughout very young stars before they start to burn their hydrogen in nuclear reactions.

Exercise 18.5 *Problem: Radiative Transport*

The density and temperature in the deep interior of the Sun are roughly $0.1\,\mathrm{kg\,m^{-3}}$ and $1.5 \times 10^7\,\mathrm{K}$.

(a) Estimate the central gas pressure and radiation pressure and their ratio.

(b) The mean free path of the radiation is determined almost equally by Thomson scattering, bound-free absorption, and free-free absorption. Estimate numerically the photon mean free path and hence estimate the photon escape time and the luminosity. How well do your estimates compare with the known values for the Sun?

Exercise 18.6 *Problem: Bubbles*

Consider a small bubble of air rising slowly in a large expanse of water. If the bubble is large enough for surface tension to be ignored, then it will form an irregular cap of radius r. Show that the speed with which the bubble rises is roughly $(gr)^{1/2}$. (A more refined estimate gives a numerical coefficient of $2/3$.)

18.6 Double Diffusion—Salt Fingers T2

18.6

As we have described it so far, convection is driven by the presence of an unbalanced buoyancy force in an equilibrium distribution of fluid. However, it can also arise as a higher-order effect, even if the fluid initially is stably stratified (i.e., if the density gradient is in the same direction as gravity). An example is *salt fingering*, a rapid mixing that can occur when warm, salty water lies at rest above colder fresh water. The higher temperature of the upper fluid outbalances the weight of its salt, making it more buoyant than the fresh water below. However, in a small, localized, downward perturbation of the warm, salty water, heat diffuses laterally into the colder surrounding water faster than salt diffuses, increasing the perturbation's density, so it will continue to sink.

salt fingering and the mechanism behind it

It is possible to describe this instability using a local perturbation analysis. The setup is somewhat similar to the one we used in Sec. 18.4 to analyze Rayleigh-Bénard convection. We consider a stratified fluid in an equilibrium state, in which there is a vertical gradient of the temperature, and as before, we measure its departure from a reference temperature T_0 at a midplane ($z = 0$) by $\tau \equiv T - T_0$. We presume that in the equilibrium state τ varies linearly with z, so $\nabla\tau = (d\tau/dz)\mathbf{e}_z$ is constant. Similarly, we characterize the salt concentration by $\mathcal{C} \equiv$ (concentration) − (equilibrium concentration at the midplane); and we assume that in the equilibrium state, $\mathcal{C}$, like τ, varies linearly with height, so $\nabla\mathcal{C} = (d\mathcal{C}/dz)\mathbf{e}_z$ is constant. The density ρ is equal to the equilibrium density at the midplane plus corrections due to thermal expansion and salt concentration:

$$\rho = \rho_0 - \alpha\rho_0\tau + \beta\rho_0\mathcal{C} \qquad (18.38)$$

[cf. Eq. (18.14)]. Here β is a constant for concentration analogous to the thermal expansion coefficient α for temperature. In this problem, by contrast with Rayleigh-Bénard convection, it is easier to work directly with the pressure than with the modified pressure. In equilibrium, hydrostatic equilibrium dictates that its gradient be $\nabla P = -\rho\mathbf{g}$.

Now, let us perturb about this equilibrium state and write down the linearized equations for the evolution of the perturbations. We denote the perturbation of temperature (relative to the reference temperature) by $\delta\tau$, of salt concentration by δC, of density by $\delta\rho$, of pressure by δP, and of velocity by simply $\mathbf{v}$ (since the unperturbed state has $\mathbf{v} = 0$). We do not ask about the onset of instability, but rather (because we expect our situation to be generically unstable) we seek a dispersion relation $\omega(\mathbf{k})$ for the perturbations. Correspondingly, in all our perturbation equations we replace $\partial/\partial t$ with $-i\omega$ and ∇ with $i\mathbf{k}$, except for the equilibrium ∇C and $\nabla\tau$, which are constants.

perturbation equations

The first of our perturbation equations is the linearized Navier-Stokes equation (18.3b):

$$-i\omega\rho_0\mathbf{v} = -i\mathbf{k}\delta P + \mathbf{g}\delta\rho - \nu k^2\rho_0\mathbf{v}, \tag{18.39a}$$

where we have kept the viscous term, because we expect the Prandtl number to be of order unity (for water, Pr $\sim$ 6). Low velocity implies incompressibity $\nabla \cdot \mathbf{v} = 0$, which becomes

$$\mathbf{k} \cdot \mathbf{v} = 0. \tag{18.39b}$$

The density perturbation follows from the perturbed form of Eq. (18.38):

$$\delta\rho = -\alpha\rho_0\delta\tau + \beta\rho_0\delta C. \tag{18.39c}$$

The temperature perturbation is governed by Eq. (18.19), which linearizes to

$$-i\omega\delta\tau + (\mathbf{v} \cdot \nabla)\tau = -\chi k^2\delta\tau. \tag{18.39d}$$

Assuming that the timescale for the salt to diffuse is much longer than that for the temperature to diffuse, we can ignore salt diffusion altogether, so that $d\delta C/dt = 0$:

$$-i\omega\delta C + (\mathbf{v} \cdot \nabla)C = 0. \tag{18.39e}$$

Equations (18.39) are five equations for the five unknowns δP, $\delta\rho$, δC, $\delta\tau$, and $\mathbf{v}$, one of which is a three-component vector! Unless we are careful, we will end up with a seventh-order algebraic equation. Fortunately, there is a way to keep the algebra manageable. First, we eliminate the pressure perturbation by taking the curl of Eq. (18.39a) [or equivalently, by crossing $\mathbf{k}$ into Eq. (18.39a)]:

$$(-i\omega + \nu k^2)\rho_0\mathbf{k} \times \mathbf{v} = \mathbf{k} \times \mathbf{g}\delta\rho. \tag{18.40a}$$

Taking the curl of this equation again allows us to incorporate incompressibility (18.39b). Then dotting into $\mathbf{g}$, we obtain

$$(i\omega - \nu k^2)\rho_0 k^2\mathbf{g} \cdot \mathbf{v} = [(\mathbf{k} \cdot \mathbf{g})^2 - k^2 g^2]\delta\rho. \tag{18.40b}$$

Since **g** points vertically, Eq. (18.40b) is one equation for the density perturbation in terms of the vertical velocity perturbation v_z. We can obtain a second equation of this sort by inserting Eq. (18.39d) for $\delta\tau$ and Eq. (18.39e) for $\delta\mathcal{C}$ into Eq. (18.39c); the result is

$$\delta\rho = -\left(\frac{\alpha\rho_0}{i\omega - \chi k^2}\right)(\mathbf{v}\cdot\mathbf{\nabla})\tau + \frac{\beta\rho_0}{i\omega}(\mathbf{v}\cdot\mathbf{\nabla})\mathcal{C}. \qquad (18.40c)$$

Since the unperturbed gradients of temperature and salt concentration are both vertical, Eq. (18.40c), like Eq. (18.40b), involves only v_z and not v_x or v_y. Solving both Eqs. (18.40b) and (18.40c) for the ratio $\delta\rho/v_z$ and equating these two expressions, we obtain the following dispersion relation for our perturbations:

$$\omega(\omega + i\nu k^2)(\omega + i\chi k^2) + \left[1 - \frac{(\mathbf{k}\cdot\mathbf{g})^2}{k^2 g^2}\right][\omega\alpha(\mathbf{g}\cdot\mathbf{\nabla})\tau - (\omega + i\chi k^2)\beta(\mathbf{g}\cdot\mathbf{\nabla})\mathcal{C}] = 0.$$

dispersion relation for salt-fingering perturbations

$$(18.41)$$

When **k** is real, as we shall assume, we can write this dispersion relation as a cubic equation for $p = -i\omega$ with real coefficients. The roots for p are either all real or one real and two complex conjugates, and growing modes have the real part of p positive. When the constant term in the cubic is negative, that is, when

$$(\mathbf{g}\cdot\mathbf{\nabla})\mathcal{C} < 0, \qquad (18.42)$$

we are guaranteed that there is at least one positive, real root p and this root corresponds to an unstable, growing mode. Therefore, *a sufficient condition for instability is that the concentration of salt increase with height!*

sufficient condition for salt-fingering instability

By inspecting the dispersion relation, we conclude that the growth rate will be maximal when $\mathbf{k}\cdot\mathbf{g} = 0$ (i.e., when the wave vector is horizontal). What is the direction of the velocity **v** for these fastest-growing modes? The incompressibility equation (18.39b) shows that **v** is orthogonal to the horizontal **k**; Eq. (18.40a) states that $\mathbf{k}\times\mathbf{v}$ points in the same direction as $\mathbf{k}\times\mathbf{g}$, which is horizontal, since **g** is vertical. These two conditions imply that **v** points vertically. Therefore, these fastest modes represent *fingers* of salty water descending past rising fingers of fresh water (Fig. 18.6). For large k (narrow fingers), the dispersion relation (18.41) predicts a growth rate given approximately by

fastest salt-fingering modes

$$p = -i\omega \sim \frac{\beta(-\mathbf{g}\cdot\mathbf{\nabla})\mathcal{C}}{\nu k^2}. \qquad (18.43)$$

Thus the growth of narrow fingers is driven by the concentration gradient and retarded by viscosity. For larger fingers, the temperature gradient participates in the retardation, since the heat must diffuse to break the buoyant stability.

Now let us turn to the nonlinear development of this instability. Although we have just considered a single Fourier mode, the fingers that grow are roughly cylindrical rather than sheet-like. They lengthen at a rate that is slow enough for the heat to diffuse horizontally, though not so slow that the salt can diffuse. Let the diffusion coefficient for the salt be χ_C by analogy with χ for temperature. If the length of the fingers is L

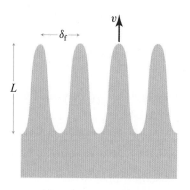

FIGURE 18.6 Salt fingers in a fluid in which warm, salty water lies on top of cold, fresh water.

and their width is δ_f, then to facilitate heat diffusion and prevent salt diffusion, the vertical speed v must satisfy

$$\frac{\chi_C L}{\delta_f^2} \ll v \ll \frac{\chi L}{\delta_f^2}. \tag{18.44}$$

Balancing the viscous acceleration $v\nu/\delta_f^2$ by the buoyancy acceleration $g\beta\delta C$, we obtain

$$v \sim \frac{g\beta\delta C \delta_f^2}{\nu}. \tag{18.45}$$

We can therefore rewrite Eq. (18.44) as

width of nonlinear salt-fingering modes

$$\left(\frac{\chi_C \nu L}{g\beta\delta C}\right)^{1/4} \ll \delta_f \ll \left(\frac{\chi \nu L}{g\beta\delta C}\right)^{1/4}. \tag{18.46}$$

Typically, $\chi_C \sim 0.01\chi$, so Eq. (18.46) implies that the widths of the fingers lie in a narrow range, as is verified by laboratory experiments.

Salt fingering can occur naturally, for example, in an estuary where cold river water flows beneath sea water warmed by the Sun. However, the development of salt fingers is quite slow, and in practice it only leads to mixing when the equilibrium velocity field is very small.

This instability is one example of a quite general type of instability known as *double diffusion*, which can arise when two physical quantities can diffuse through a fluid at different rates. Other examples include the diffusion of two different solutes and the diffusion of vorticity and heat in a rotating flow.

EXERCISES

Exercise 18.7 *Problem: Laboratory Experiment with Salt Fingers* T2
Make an order-of-magnitude estimate of the size of the fingers and the time it takes for them to grow in a small transparent jar. You might like to try an experiment.

Exercise 18.8 *Problem: Internal Waves* `T2`

Consider a stably stratified fluid at rest with a small (negative) vertical density gradient $d\rho/dz$.

(a) By modifying the analysis in this section, ignoring the effects of viscosity, heat conduction, and concentration gradients, show that small-amplitude linear waves, which propagate in a direction making an angle θ to the vertical, have an angular frequency given by $\omega = N|\sin\theta|$, where $N \equiv [(\mathbf{g}\cdot\nabla)\ln\rho]^{1/2}$ is known as the *Brunt-Väisälä frequency*. These waves are called *internal waves*. They can also be found at abrupt discontinuities as well as in the presence of a slow variation in the background medium. They are analogous to the Love and Rayleigh waves we have already met in our discussion of seismology (Sec. 12.4.2). Another type of internal wave is the Kelvin-Helmholtz wave (Sec. 14.6.1).

(b) Show that the group velocity of these waves is orthogonal to the phase velocity, and interpret this result physically.

Bibliographic Note

For pedagogical treatments of almost all the topics in this chapter plus much more related material, we particularly like Tritton (1987), whose phenomenological approach is lucid and appealing; and also Turner (1973), which is a thorough treatise on the influence of buoyancy (thermally induced and otherwise) on fluid motions.

Lautrup (2005) treats very nicely all this chapter's topics except convection in stars, salt fingers, and double diffusion. Landau and Lifshitz (1959, Chaps. 5 and 6) give a fairly succinct treatment of diffusive heat flow in fluids, the onset of convection in several different physical situations, and the concepts underlying double diffusion. Chandrasekhar (1961, Chaps. 2–6) gives a thorough and rich treatment of the influence of a wide variety of phenomena on the onset of convection, and on the types of fluid motions that can occur near the onset of convection. For a few pages on strongly turbulent convective heat transfer, see White (2006, Sec. 6-10).

Engineering-oriented textbooks typically say little about convection. For an engineer's viewpoint and engineering issues in convection, we recommend more specialized texts, such as Bejan (2013). For an applied mathematician's viewpoint, we suggest the treatise Pop and Ingham (2001).

A

Newtonian Physics: Geometric Viewpoint

(Chapter 1 of *Modern Classical Physics*)

> Geometry postulates the solution of these problems from mechanics and teaches the use of the problems thus solved. And geometry can boast that with so few principles obtained from other fields, it can do so much.
>
> ISAAC NEWTON, 1687

1.1 Introduction

1.1.1 The Geometric Viewpoint on the Laws of Physics

In this book, we adopt a different viewpoint on the laws of physics than that in many elementary and intermediate texts. In most textbooks, physical laws are expressed in terms of quantities (locations in space, momenta of particles, etc.) that are measured in some coordinate system. For example, Newtonian vectorial quantities are expressed as triplets of numbers [e.g., $\mathbf{p} = (p_x, p_y, p_z) = (1, 9, -4)$], representing the components of a particle's momentum on the axes of a Cartesian coordinate system; and tensors are expressed as arrays of numbers (e.g.,

$$\mathbf{I} = \begin{bmatrix} I_{xx} & I_{xy} & I_{xz} \\ I_{yx} & I_{yy} & I_{yz} \\ I_{zx} & I_{zy} & I_{zz} \end{bmatrix} \tag{1.1}$$

for the moment of inertia tensor).

By contrast, in this book we express all physical quantities and laws in *geometric forms,* i.e., in forms that are *independent of any coordinate system or basis vectors.* For example, a particle's velocity $\mathbf{v}$ and the electric and magnetic fields $\mathbf{E}$ and $\mathbf{B}$ that it encounters will be vectors described as arrows that live in the 3-dimensional, flat Euclidean space of everyday experience.[1] They require no coordinate system or basis vectors for their existence or description—though often coordinates will be useful. In other words, $\mathbf{v}$ represents the vector itself and is not just shorthand for an ordered list of numbers.

---

1. This interpretation of a vector is close to the ideas of Newton and Faraday. Lagrange, Hamilton, Maxwell, and many others saw vectors in terms of Cartesian components. The vector notation was streamlined by Gibbs, Heaviside, and others, but the underlying coordinate system was still implicit, and $\mathbf{v}$ was usually regarded as shorthand for (v_x, v_y, v_z).

- This chapter is a foundation for almost all of this book.

- Many readers already know the material in this chapter, but from a viewpoint different from our *geometric* one. Such readers will be able to understand almost all of Parts II–VI of this book without learning our viewpoint. Nevertheless, that geometric viewpoint has such power that we encourage them to learn it by browsing this chapter and focusing especially on Secs. 1.1.1, 1.2, 1.3, 1.5, 1.7, and 1.8.

- The stress tensor, introduced and discussed in Sec. 1.9, plays an important role in kinetic theory (Chap. 3) and a crucial role in elasticity (Part IV), fluid dynamics (Part V), and plasma physics (Part VI).

- The integral and differential conservation laws derived and discussed in Secs. 1.8 and 1.9 play major roles throughout this book.

- The Box labeled **T2** is advanced material (Track Two) that can be skipped in a time-limited course or on a first reading of this book.

We insist that the Newtonian laws of physics all obey a *Geometric Principle:* they are all geometric relationships among geometric objects (primarily scalars, vectors, and tensors), expressible without the aid of any coordinates or bases. An example is the Lorentz force law $md\mathbf{v}/dt = q(\mathbf{E} + \mathbf{v} \times \mathbf{B})$—a (coordinate-free) relationship between the geometric (coordinate-independent) vectors $\mathbf{v}$, $\mathbf{E}$, and $\mathbf{B}$ and the particle's scalar mass m and charge q. As another example, a body's moment of inertia tensor $\mathbf{I}$ can be viewed as a vector-valued linear function of vectors (a coordinate-independent, basis-independent geometric object). Insert into the tensor $\mathbf{I}$ the body's angular velocity vector $\boldsymbol{\Omega}$, and you get out the body's angular momentum vector: $\mathbf{J} = \mathbf{I}(\boldsymbol{\Omega})$. No coordinates or basis vectors are needed for this law of physics, nor is any description of $\mathbf{I}$ as a matrix-like entity with components I_{ij} required. Components are secondary; they only exist after one has chosen a set of basis vectors. Components (we claim) are an impediment to a clear and deep understanding of the laws of classical physics. The coordinate-free, component-free description is deeper, and—once one becomes accustomed to it—much more clear and understandable.[2]

---

2. This philosophy is also appropriate for quantum mechanics (see Box 1.2) and, especially, quantum field theory, where it is the invariance of the description under gauge and other symmetry operations that is the powerful principle. However, its implementation there is less direct, simply because the spaces in which these symmetries lie are more abstract and harder to conceptualize.

By adopting this geometric viewpoint, we gain great conceptual power and often also computational power. For example, when we ignore experiment and simply ask what forms the laws of physics can possibly take (what forms are allowed by the requirement that the laws be geometric), we shall find that there is remarkably little freedom. Coordinate independence and basis independence strongly constrain the laws of physics.[3]

This power, together with the elegance of the geometric formulation, suggests that in some deep sense, Nature's physical laws are geometric and have nothing whatsoever to do with coordinates or components or vector bases.

1.1.2 Purposes of This Chapter

The principal purpose of this foundational chapter is to teach the reader this geometric viewpoint.

The mathematical foundation for our geometric viewpoint is *differential geometry* (also called "tensor analysis" by physicists). Differential geometry can be thought of as an extension of the vector analysis with which all readers should be familiar. *A second purpose of this chapter is to develop key parts of differential geometry in a simple form well adapted to Newtonian physics.*

1.1.3 Overview of This Chapter

In this chapter, we lay the geometric foundations for the Newtonian laws of physics in flat Euclidean space. We begin in Sec. 1.2 by introducing some foundational geometric concepts: points, scalars, vectors, inner products of vectors, and the distance between points. Then in Sec. 1.3, we introduce the concept of a tensor as a linear function of vectors, and we develop a number of geometric tools: the tools of coordinate-free tensor algebra. In Sec. 1.4, we illustrate our tensor-algebra tools by using them to describe—without any coordinate system—the kinematics of a charged point particle that moves through Euclidean space, driven by electric and magnetic forces.

In Sec. 1.5, we introduce, for the first time, Cartesian coordinate systems and their basis vectors, and also the components of vectors and tensors on those basis vectors; and we explore how to express geometric relationships in the language of components. In Sec. 1.6, we deduce how the components of vectors and tensors transform when one rotates the chosen Cartesian coordinate axes. (These are the transformation laws that most physics textbooks use to define vectors and tensors.)

In Sec. 1.7, we introduce directional derivatives and gradients of vectors and tensors, thereby moving from tensor algebra to true differential geometry (in Euclidean space). We also introduce the Levi-Civita tensor and use it to define curls and cross

---

3. Examples are the equation of elastodynamics (12.4b) and the Navier-Stokes equation of fluid mechanics (13.69), which are both dictated by momentum conservation plus the form of the stress tensor [Eqs. (11.18), (13.43), and (13.68)]—forms that are dictated by the irreducible tensorial parts (Box 11.2) of the strain and rate of strain.

products, and we learn how to use *index gymnastics* to derive, quickly, formulas for multiple cross products. In Sec. 1.8, we use the Levi-Civita tensor to define vectorial areas, scalar volumes, and integration over surfaces. These concepts then enable us to formulate, in geometric, coordinate-free ways, integral and differential conservation laws. In Sec. 1.9, we discuss, in particular, the law of momentum conservation, formulating it in a geometric way with the aid of a geometric object called the *stress tensor*. As important examples, we use this geometric conservation law to derive and discuss the equations of Newtonian fluid dynamics, and the interaction between a charged medium and an electromagnetic field. We conclude in Sec. 1.10 with some concepts from special relativity that we shall need in our discussions of Newtonian physics.

1.2

1.2 Foundational Concepts

In this section, we sketch the foundational concepts of Newtonian physics without using any coordinate system or basis vectors. This is the geometric viewpoint that we advocate.

space and time

The arena for the Newtonian laws of physics is a spacetime composed of the familiar 3-dimensional Euclidean space of everyday experience (which we call *3-space*) and a universal time t. We denote points (locations) in 3-space by capital script letters, such as $\mathcal{P}$ and $\mathcal{Q}$. These points and the 3-space in which they live require no coordinates for their definition.

scalar

A *scalar* is a single number. We are most interested in scalars that directly represent physical quantities (e.g., temperature T). As such, they are real numbers, and when they are functions of location $\mathcal{P}$ in space [e.g., $T(\mathcal{P})$], we call them *scalar fields*. However, sometimes we will work with complex numbers—most importantly in quantum mechanics, but also in various Fourier representations of classical physics.

vector

A *vector* in Euclidean 3-space can be thought of as a straight arrow (or more formally a directed line segment) that reaches from one point, $\mathcal{P}$, to another, $\mathcal{Q}$ (e.g., the arrow $\Delta \mathbf{x}$ in Fig. 1.1a). Equivalently, $\Delta \mathbf{x}$ can be thought of as a direction at $\mathcal{P}$ and a number, the vector's length. Sometimes we shall select one point $\mathcal{O}$ in 3-space as an "origin" and identify all other points, say, $\mathcal{Q}$ and $\mathcal{P}$, by their vectorial separations $\mathbf{x}_{\mathcal{Q}}$ and $\mathbf{x}_{\mathcal{P}}$ from that origin.

distance and length

The Euclidean distance $\Delta \sigma$ between two points $\mathcal{P}$ and $\mathcal{Q}$ in 3-space can be measured with a ruler and so, of course, requires no coordinate system for its definition. (If one does have a Cartesian coordinate system, then $\Delta \sigma$ can be computed by the Pythagorean formula, a precursor to the invariant interval of flat spacetime; Sec. 2.2.3.) This distance $\Delta \sigma$ is also the *length* $|\Delta \mathbf{x}|$ of the vector $\Delta \mathbf{x}$ that reaches from $\mathcal{P}$ to $\mathcal{Q}$, and the square of that length is denoted

$$|\Delta \mathbf{x}|^2 \equiv (\Delta \mathbf{x})^2 \equiv (\Delta \sigma)^2. \qquad (1.2)$$

Of particular importance is the case when $\mathcal{P}$ and $\mathcal{Q}$ are neighboring points and $\Delta \mathbf{x}$ is a differential (infinitesimal) quantity $d\mathbf{x}$. This *infinitesimal displacement* is a more fundamental physical quantity than the finite $\Delta \mathbf{x}$. To create a finite vector out

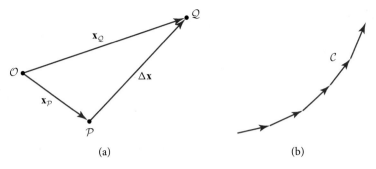

FIGURE 1.1 (a) A Euclidean 3-space diagram depicting two points $\mathcal{P}$ and $\mathcal{Q}$, their respective vectorial separations $\mathbf{x}_\mathcal{P}$ and $\mathbf{x}_\mathcal{Q}$ from the (arbitrarily chosen) origin $\mathcal{O}$, and the vector $\Delta\mathbf{x} = \mathbf{x}_\mathcal{Q} - \mathbf{x}_\mathcal{P}$ connecting them. (b) A curve $\mathcal{P}(\lambda)$ generated by laying out a sequence of infinitesimal vectors, tail-to-tip.

of infinitesimal vectors, one has to add several infinitesimal vectors head to tail, head to tail, and so on, and then take a limit. This involves *translating* a vector from one point to the next. There is no ambiguity about doing this in flat Euclidean space using the geometric notion of parallelism.[4] This simple property of Euclidean space enables us to add (and subtract) vectors at a point. We attach the tail of a second vector to the head of the first vector and then construct the sum as the vector from the tail of the first to the head of the second, or vice versa, as should be quite familiar. The point is that we do not need to add the Cartesian components to sum vectors.

We can also rotate vectors about their tails by pointing them along a different direction in space. Such a rotation can be specified by two angles. The space that is defined by all possible changes of length and direction at a point is called that point's *tangent space*. Again, we generally view the rotation as being that of a physical vector in space, and not, as it is often useful to imagine, the rotation of some coordinate system's basis vectors, with the chosen vector itself kept fixed.

tangent space

We can also construct a path through space by laying down a sequence of infinitesimal $d\mathbf{x}$s, tail to head, one after another. The resulting path is a *curve* to which these $d\mathbf{x}$s are tangent (Fig. 1.1b). The curve can be denoted $\mathcal{P}(\lambda)$, with λ a parameter along the curve and $\mathcal{P}(\lambda)$ the point on the curve whose parameter value is λ, or $\mathbf{x}(\lambda)$ where $\mathbf{x}$ is the vector separation of $\mathcal{P}$ from the arbitrary origin $\mathcal{O}$. The infinitesimal vectors that map the curve out are $d\mathbf{x} = (d\mathcal{P}/d\lambda)\,d\lambda = (d\mathbf{x}/d\lambda)\,d\lambda$, and $d\mathcal{P}/d\lambda = d\mathbf{x}/d\lambda$ is the tangent vector to the curve.

curve

tangent vector

If the curve followed is that of a particle, and the parameter λ is time t, then we have defined the *velocity* $\mathbf{v} \equiv d\mathbf{x}/dt$. In effect we are multiplying the vector $d\mathbf{x}$ by the scalar $1/dt$ and taking the limit. Performing this operation at every point $\mathcal{P}$ in the space occupied by a fluid defines the fluid's *velocity field* $\mathbf{v}(\mathbf{x})$. Multiplying a particle's velocity $\mathbf{v}$ by its scalar mass gives its *momentum* $\mathbf{p} = m\mathbf{v}$. Similarly, the difference $d\mathbf{v}$

---

4. The statement that there is just one choice of line parallel to a given line, through a point not lying on the line, is the famous fifth axiom of Euclid.

of two velocity measurements during a time interval dt, multiplied by $1/dt$, generates the particle's *acceleration* $\mathbf{a} = d\mathbf{v}/dt$. Multiplying by the particle's mass gives the force $\mathbf{F} = m\mathbf{a}$ that produced the acceleration; dividing an electrically produced force by the particle's charge q gives the electric field $\mathbf{E} = \mathbf{F}/q$. And so on.

We can define inner products [see Eq. (1.4a) below] and cross products [Eq. (1.22a)] of pairs of vectors at the same point geometrically; then using those vectors we can define, for example, the rate that work is done by a force and a particle's angular momentum about a point.

These two products can be expressed geometrically as follows. If we allow the two vectors to define a parallelogram, then their cross product is the vector orthogonal to the parallelogram with length equal to the parallelogram's area. If we first rotate one vector through a right angle in a plane containing the other, and then define the parallelogram, its area is the vectors' inner product.

derivatives of scalars and vectors

We can also define spatial derivatives. We associate the difference of a scalar between two points separated by $d\mathbf{x}$ at the same time with a *gradient* and, likewise, go on to define the scalar *divergence* and the vector *curl*. The freedom to translate vectors from one point to the next also underlies the association of a single vector (e.g., momentum) with a group of particles or an extended body. One simply adds all the individual momenta, taking a limit when necessary.

In this fashion (which should be familiar to the reader and will be elucidated, formalized, and generalized below), we can construct all the standard scalars and vectors of Newtonian physics. What is important is that *these physical quantities require no coordinate system for their definition.* They are geometric (coordinate-independent) objects residing in Euclidean 3-space at a particular time.

Geometric Principle

It is a fundamental (though often ignored) principle of physics that *the Newtonian physical laws are all expressible as geometric relationships among these types of geometric objects, and these relationships do not depend on any coordinate system or orientation of axes, nor on any reference frame* (i.e., on any purported velocity of the Euclidean space in which the measurements are made).[5] We call this the *Geometric Principle* for the laws of physics, and we use it throughout this book. It is the Newtonian analog of Einstein's Principle of Relativity (Sec. 2.2.2).

1.3

1.3 Tensor Algebra without a Coordinate System

In preparation for developing our geometric view of physical laws, we now introduce, in a coordinate-free way, some fundamental concepts of differential geometry: tensors, the inner product, the metric tensor, the tensor product, and contraction of tensors.

We have already defined a vector $\mathbf{A}$ as a straight arrow from one point, say $\mathcal{P}$, in our space to another, say $\mathcal{Q}$. Because our space is flat, there is a unique and obvious way to

---

5. By changing the velocity of Euclidean space, one adds a constant velocity to all particles, but this leaves the laws (e.g., Newton's $\mathbf{F} = m\mathbf{a}$) unchanged.

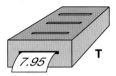

FIGURE 1.2 A rank-3 tensor **T**.

transport such an arrow from one location to another, keeping its length and direction unchanged.[6] Accordingly, we shall regard vectors as unchanged by such transport. This enables us to ignore the issue of where in space a vector actually resides; it is completely determined by its direction and its length.

A *rank-n tensor* **T** is, by definition, a real-valued linear function of n vectors.[7] Pictorially we regard **T** as a box (Fig. 1.2) with n slots in its top, into which are inserted n vectors, and one slot in its end, which prints out a single real number: the value that the tensor **T** has when evaluated as a function of the n inserted vectors. Notationally we denote the tensor by a boldfaced sans-serif character **T**:

tensor

$$\mathsf{T}(\underbrace{\_\_, \_\_, \_\_, \_\_}_{})$$

(1.3a)

↖ n slots in which to put the vectors.

This definition of a tensor is very different (and far simpler) than the one found in most standard physics textbooks (e.g., Marion and Thornton, 1995; Jackson, 1999; Griffiths, 1999). There, a tensor is an array of numbers that transform in a particular way under rotations. We shall learn the connection between these definitions in Sec. 1.6 below.

To illustrate this approach, if **T** is a rank-3 tensor (has 3 slots) as in Fig. 1.2, then its value on the vectors **A**, **B**, **C** is denoted **T**(**A**, **B**, **C**). Linearity of this function can be expressed as

$$\mathsf{T}(e\mathbf{E} + f\mathbf{F}, \mathbf{B}, \mathbf{C}) = e\mathsf{T}(\mathbf{E}, \mathbf{B}, \mathbf{C}) + f\mathsf{T}(\mathbf{F}, \mathbf{B}, \mathbf{C}),$$

(1.3b)

where e and f are real numbers, and similarly for the second and third slots.

We have already defined the squared length $(\mathbf{A})^2 \equiv \mathbf{A}^2$ of a vector **A** as the squared distance between the points at its tail and its tip. The *inner product* (also called the dot product) $\mathbf{A} \cdot \mathbf{B}$ of two vectors is defined in terms of this squared length by

inner product

$$\boxed{\mathbf{A} \cdot \mathbf{B} \equiv \frac{1}{4}\left[(\mathbf{A} + \mathbf{B})^2 - (\mathbf{A} - \mathbf{B})^2\right].}$$

(1.4a)

In Euclidean space, this is the standard inner product, familiar from elementary geometry and discussed above in terms of the area of a parallelogram.

---

6. This is not so in curved spaces, as we shall see in Sec. 24.3.4.
7. This is a different use of the word *rank* than for a matrix, whose rank is its number of linearly independent rows or columns.

One can show that the inner product (1.4a) is a real-valued linear function of each of its vectors. Therefore, we can regard it as a tensor of rank 2. When so regarded, the inner product is denoted $\mathbf{g}(\_\_, \_\_)$ and is called the *metric tensor*. In other words, the metric tensor $\mathbf{g}$ is that linear function of two vectors whose value is given by

$$\boxed{\mathbf{g}(\mathbf{A}, \mathbf{B}) \equiv \mathbf{A} \cdot \mathbf{B}.} \tag{1.4b}$$

Notice that, because $\mathbf{A} \cdot \mathbf{B} = \mathbf{B} \cdot \mathbf{A}$, the metric tensor is *symmetric* in its two slots—one gets the same real number independently of the order in which one inserts the two vectors into the slots:

$$\mathbf{g}(\mathbf{A}, \mathbf{B}) = \mathbf{g}(\mathbf{B}, \mathbf{A}). \tag{1.4c}$$

With the aid of the inner product, we can regard any vector $\mathbf{A}$ as a tensor of rank one: the real number that is produced when an arbitrary vector $\mathbf{C}$ is inserted into $\mathbf{A}$'s single slot is

$$\boxed{\mathbf{A}(\mathbf{C}) \equiv \mathbf{A} \cdot \mathbf{C}.} \tag{1.4d}$$

In Newtonian physics, we rarely meet tensors of rank higher than two. However, second-rank tensors appear frequently—often in roles where one sticks a single vector into the second slot and leaves the first slot empty, thereby producing a single-slotted entity, a vector. An example that we met in Sec. 1.1.1 is a rigid body's moment-of-inertia tensor $\mathbf{I}(\_\_, \_\_)$, which gives us the body's angular momentum $\mathbf{J}(\_\_) = \mathbf{I}(\_\_, \boldsymbol{\Omega})$ when its angular velocity $\boldsymbol{\Omega}$ is inserted into its second slot.[8] Another example is the stress tensor of a solid, a fluid, a plasma, or a field (Sec. 1.9 below).

From three vectors $\mathbf{A}$, $\mathbf{B}$, $\mathbf{C}$, we can construct a tensor, their *tensor product* (also called *outer product* in contradistinction to the inner product $\mathbf{A} \cdot \mathbf{B}$), defined as follows:

$$\boxed{\mathbf{A} \otimes \mathbf{B} \otimes \mathbf{C}(\mathbf{E}, \mathbf{F}, \mathbf{G}) \equiv \mathbf{A}(\mathbf{E})\mathbf{B}(\mathbf{F})\mathbf{C}(\mathbf{G}) = (\mathbf{A} \cdot \mathbf{E})(\mathbf{B} \cdot \mathbf{F})(\mathbf{C} \cdot \mathbf{G}).} \tag{1.5a}$$

Here the first expression is the notation for the value of the new tensor, $\mathbf{A} \otimes \mathbf{B} \otimes \mathbf{C}$ evaluated on the three vectors $\mathbf{E}$, $\mathbf{F}$, $\mathbf{G}$; the middle expression is the ordinary product of three real numbers, the value of $\mathbf{A}$ on $\mathbf{E}$, the value of $\mathbf{B}$ on $\mathbf{F}$, and the value of $\mathbf{C}$ on $\mathbf{G}$; and the third expression is that same product with the three numbers rewritten as scalar products. Similar definitions can be given (and should be obvious) for the tensor product of any number of vectors, and of any two or more tensors of any rank; for example, if $\mathbf{T}$ has rank 2 and $\mathbf{S}$ has rank 3, then

$$\mathbf{T} \otimes \mathbf{S}(\mathbf{E}, \mathbf{F}, \mathbf{G}, \mathbf{H}, \mathbf{J}) \equiv \mathbf{T}(\mathbf{E}, \mathbf{F})\mathbf{S}(\mathbf{G}, \mathbf{H}, \mathbf{J}). \tag{1.5b}$$

One last geometric (i.e., frame-independent) concept we shall need is *contraction*. We illustrate this concept first by a simple example, then give the general definition.

---

8. Actually, it doesn't matter which slot, since $\mathbf{I}$ is symmetric.

metric tensor (margin note)

tensor product (margin note)

contraction (margin note)

From two vectors **A** and **B** we can construct the tensor product $\mathbf{A} \otimes \mathbf{B}$ (a second-rank tensor), and we can also construct the scalar product $\mathbf{A} \cdot \mathbf{B}$ (a real number, i.e., a *scalar*, also known as a *rank-0 tensor*). The process of contraction is the construction of $\mathbf{A} \cdot \mathbf{B}$ from $\mathbf{A} \otimes \mathbf{B}$:

$$\boxed{\text{contraction}(\mathbf{A} \otimes \mathbf{B}) \equiv \mathbf{A} \cdot \mathbf{B}.}$$ (1.6a)

One can show fairly easily using component techniques (Sec. 1.5 below) that any second-rank tensor **T** can be expressed as a sum of tensor products of vectors, $\mathbf{T} = \mathbf{A} \otimes \mathbf{B} + \mathbf{C} \otimes \mathbf{D} + \ldots$. Correspondingly, it is natural to define the contraction of **T** to be contraction$(\mathbf{T}) = \mathbf{A} \cdot \mathbf{B} + \mathbf{C} \cdot \mathbf{D} + \ldots$. Note that this contraction process lowers the rank of the tensor by two, from 2 to 0. Similarly, for a tensor of rank n one can construct a tensor of rank $n - 2$ by contraction, but in this case one must specify which slots are to be contracted. For example, if **T** is a third-rank tensor, expressible as $\mathbf{T} = \mathbf{A} \otimes \mathbf{B} \otimes \mathbf{C} + \mathbf{E} \otimes \mathbf{F} \otimes \mathbf{G} + \ldots$, then the contraction of **T** on its first and third slots is the rank-1 tensor (vector)

$$1\&3\text{contraction}(\mathbf{A} \otimes \mathbf{B} \otimes \mathbf{C} + \mathbf{E} \otimes \mathbf{F} \otimes \mathbf{G} + \ldots) \equiv (\mathbf{A} \cdot \mathbf{C})\mathbf{B} + (\mathbf{E} \cdot \mathbf{G})\mathbf{F} + \ldots.$$

(1.6b)

Unfortunately, there is no simple index-free notation for contraction in common use.

All the concepts developed in this section (vector, tensor, metric tensor, inner product, tensor product, and contraction of a tensor) can be carried over, with no change whatsoever, into any vector space[9] that is endowed with a concept of squared length—for example, to the 4-dimensional spacetime of special relativity (next chapter).

1.4 Particle Kinetics and Lorentz Force in Geometric Language

In this section, we illustrate our geometric viewpoint by formulating Newton's laws of motion for particles.

In Newtonian physics, a classical particle moves through Euclidean 3-space as universal time t passes. At time t it is located at some point $\mathbf{x}(t)$ (its *position*). The function $\mathbf{x}(t)$ represents a curve in 3-space, the particle's *trajectory*. The particle's *velocity* $\mathbf{v}(t)$ is the time derivative of its position, its *momentum* $\mathbf{p}(t)$ is the product of its mass m and velocity, its *acceleration* $\mathbf{a}(t)$ is the time derivative of its velocity, and its *kinetic energy* $E(t)$ is half its mass times velocity squared:

trajectory, velocity, momentum, acceleration, and energy

$$\mathbf{v}(t) = \frac{d\mathbf{x}}{dt}, \quad \mathbf{p}(t) = m\mathbf{v}(t), \quad \mathbf{a}(t) = \frac{d\mathbf{v}}{dt} = \frac{d^2\mathbf{x}}{dt^2}, \quad E(t) = \frac{1}{2}m\mathbf{v}^2.$$ (1.7a)

9. Or, more precisely, any vector space over the real numbers. If the vector space's scalars are complex numbers, as in quantum mechanics, then slight changes are needed.

Since points in 3-space are geometric objects (defined independently of any coordinate system), so also are the trajectory $\mathbf{x}(t)$, the velocity, the momentum, the acceleration, and the energy. (Physically, of course, the velocity has an ambiguity; it depends on one's standard of rest.)

Newton's second law of motion states that the particle's momentum can change only if a force $\mathbf{F}$ acts on it, and that its change is given by

$$d\mathbf{p}/dt = m\mathbf{a} = \mathbf{F}. \tag{1.7b}$$

If the force is produced by an electric field $\mathbf{E}$ and magnetic field $\mathbf{B}$, then this law of motion in SI units takes the familiar Lorentz-force form

$$d\mathbf{p}/dt = q(\mathbf{E} + \mathbf{v} \times \mathbf{B}). \tag{1.7c}$$

(Here we have used the vector cross product, with which the reader should be familiar, and which will be discussed formally in Sec. 1.7.)

laws of motion

The laws of motion (1.7) are geometric relationships among geometric objects. Let us illustrate this using something very familiar, planetary motion. Consider a light planet orbiting a heavy star. If there were no gravitational force, the planet would continue in a straight line with constant velocity $\mathbf{v}$ and speed $v = |\mathbf{v}|$, sweeping out area A at a rate $dA/dt = rv_t/2$, where r is the radius, and v_t is the tangential speed. Elementary geometry equates this to the constant $vb/2$, where b is the impact parameter—the smallest separation from the star. Now add a gravitational force $\mathbf{F}$ and let it cause a small radial impulse. A second application of geometry showed Newton that the product $rv_t/2$ is unchanged to first order in the impulse, and he recovered Kepler's second law ($dA/dt = \text{const}$) without introducing coordinates.[10]

Contrast this approach with one relying on coordinates. For example, one introduces an (r, ϕ) coordinate system, constructs a lagrangian and observes that the coordinate ϕ is ignorable; then the Euler-Lagrange equations immediately imply the conservation of angular momentum, which is equivalent to Kepler's second law. So, which of these two approaches is preferable? The answer is surely "both!" Newton wrote the *Principia* in the language of geometry at least partly for a reason that remains valid today: it brought him a quick understanding of fundamental laws of physics. Lagrange followed his coordinate-based path to the function that bears his name, because he wanted to solve problems in celestial mechanics that would not yield to

10. Continuing in this vein, when the force is inverse square, as it is for gravity and electrostatics, we can use Kepler's second law to argue that when the orbit turns through a succession of equal angles $d\theta$, its successive changes in velocity $d\mathbf{v} = \mathbf{a}dt$ (with $\mathbf{a}$ the gravitational acceleration) all have the same magnitude $|d\mathbf{v}|$ and have the same angles $d\theta$ from one to another. So, if we trace the head of the velocity vector in velocity space, it follows a circle. The circle is not centered on zero velocity when the eccentricity is nonzero but there exists a reference frame in which the speed of the planet is constant. This graphical representation is known as a *hodograph,* and similar geometrical approaches are used in fluid mechanics. For Richard Feynman's masterful presentation of these ideas to first-year undergraduates, see Goodstein and Goodstein (1996).

Newton's approach. So it is today. Geometry and analysis are both indispensible. In the domain of classical physics, the geometry is of greater importance in deriving and understanding fundamental laws and has arguably been underappreciated; coordinates hold sway when we apply these laws to solve real problems. Today, both old and new laws of physics are commonly expressed geometrically, using lagrangians, hamiltonians, and actions, for example Hamilton's action principle $\delta \int L\,dt = 0$ where L is the coordinate-independent lagrangian. Indeed, being able to do this without introducing coordinates is a powerful guide to deriving these laws and a tool for comprehending their implications.

A comment is needed on the famous connection between *symmetry* and *conservation laws*. In our example above, angular momentum conservation followed from axial symmetry which was embodied in the lagrangian's independence of the angle ϕ; but we also deduced it geometrically. This is usually the case in classical physics; typically, we do not need to introduce a specific coordinate system to understand symmetry and to express the associated conservation laws. However, symmetries are sometimes well hidden, for example with a nutating top, and coordinate transformations are then usually the best approach to uncover them.

symmetry and conservation laws

Often in classical physics, real-world factors invalidate or complicate Lagrange's and Hamilton's coordinate-based analytical dynamics, and so one is driven to geometric considerations. As an example, consider a spherical marble rolling on a flat horizontal table. The analytical dynamics approach is to express the height of the marble's center of mass and the angle of its rotation as constraints and align the basis vectors so there is a single horizontal coordinate defined by the initial condition. It is then deduced that linear and angular momenta are conserved. Of course that result is trivial and just as easily gotten without this formalism. However, this model is also used for many idealized problems where the outcome is far from obvious and the approach is brilliantly effective. But consider the real world in which tables are warped and bumpy, marbles are ellipsoidal and scratched, air imposes a resistance, and wood and glass comprise polymers that attract one another. And so on. When one includes these factors, it is to geometry that one quickly turns to understand the real marble's actual dynamics. Even ignoring these effects and just asking what happens when the marble rolls off the edge of a table introduces a *nonholonomic* constraint, and figuring out where it lands and how fast it is spinning are best addressed not by the methods of Lagrange and Hamilton, but instead by considering the geometry of the gravitational and reaction forces. In the following chapters, we shall encounter many examples where we have to deal with messy complications like these.

EXERCISES

Exercise 1.1 *Practice: Energy Change for Charged Particle*
Without introducing any coordinates or basis vectors, show that when a particle with charge q interacts with electric and magnetic fields, its kinetic energy changes at a rate

$$dE/dt = q\,\mathbf{v} \cdot \mathbf{E}. \tag{1.8}$$

1.4 Particle Kinetics and Lorentz Force in Geometric Language   **953**

Exercise 1.2 *Practice: Particle Moving in a Circular Orbit*

Consider a particle moving in a circle with uniform speed $v = |\mathbf{v}|$ and uniform magnitude $a = |\mathbf{a}|$ of acceleration. Without introducing any coordinates or basis vectors, do the following.

(a) At any moment of time, let $\mathbf{n} = \mathbf{v}/v$ be the unit vector pointing along the velocity, and let s denote distance that the particle travels in its orbit. By drawing a picture, show that $d\mathbf{n}/ds$ is a unit vector that points to the center of the particle's circular orbit, divided by the radius of the orbit.

(b) Show that the vector (not unit vector) pointing from the particle's location to the center of its orbit is $(v/a)^2 \mathbf{a}$.

1.5 Component Representation of Tensor Algebra

Cartesian coordinates and orthonormal basis vectors

In the Euclidean 3-space of Newtonian physics, there is a unique set of *orthonormal basis vectors* $\{\mathbf{e}_x, \mathbf{e}_y, \mathbf{e}_z\} \equiv \{\mathbf{e}_1, \mathbf{e}_2, \mathbf{e}_3\}$ associated with any *Cartesian coordinate system* $\{x, y, z\} \equiv \{x^1, x^2, x^3\} \equiv \{x_1, x_2, x_3\}$. (In Cartesian coordinates in Euclidean space, we usually place indices down, but occasionally we place them up. It doesn't matter. By definition, in Cartesian coordinates a quantity is the same whether its index is down or up.) The basis vector $\mathbf{e}_j$ points along the x_j coordinate direction, which is orthogonal to all the other coordinate directions, and it has unit length (Fig. 1.3), so

$$\mathbf{e}_j \cdot \mathbf{e}_k = \delta_{jk}, \tag{1.9a}$$

where δ_{jk} is the Kronecker delta.

Any vector $\mathbf{A}$ in 3-space can be expanded in terms of this basis:

$$\mathbf{A} = A_j \mathbf{e}_j. \tag{1.9b}$$

Einstein summation convention

Cartesian components of a vector

Here and throughout this book, we adopt the *Einstein summation convention*: repeated indices (in this case j) are to be summed (in this 3-space case over $j = 1, 2, 3$), unless otherwise instructed. By virtue of the orthonormality of the basis, the components A_j of $\mathbf{A}$ can be computed as the scalar product

$$A_j = \mathbf{A} \cdot \mathbf{e}_j. \tag{1.9c}$$

[The proof of this is straightforward: $\mathbf{A} \cdot \mathbf{e}_j = (A_k \mathbf{e}_k) \cdot \mathbf{e}_j = A_k (\mathbf{e}_k \cdot \mathbf{e}_j) = A_k \delta_{kj} = A_j$.]

Any tensor, say, the third-rank tensor $\mathbf{T}(\_, \_, \_)$, can be expanded in terms of tensor products of the basis vectors:

$$\mathbf{T} = T_{ijk} \mathbf{e}_i \otimes \mathbf{e}_j \otimes \mathbf{e}_k. \tag{1.9d}$$

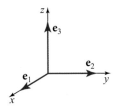

FIGURE 1.3 The orthonormal basis vectors $\mathbf{e}_j$ associated with a Euclidean coordinate system in Euclidean 3-space.

The components T_{ijk} of $\mathbf{T}$ can be computed from $\mathbf{T}$ and the basis vectors by the generalization of Eq. (1.9c):

Cartesian components of a tensor

$$T_{ijk} = \mathbf{T}(\mathbf{e}_i, \mathbf{e}_j, \mathbf{e}_k). \qquad (1.9e)$$

[This equation can be derived using the orthonormality of the basis in the same way as Eq. (1.9c) was derived.] As an important example, the components of the metric tensor are $g_{jk} = \mathbf{g}(\mathbf{e}_j, \mathbf{e}_k) = \mathbf{e}_j \cdot \mathbf{e}_k = \delta_{jk}$ [where the first equality is the method (1.9e) of computing tensor components, the second is the definition (1.4b) of the metric, and the third is the orthonormality relation (1.9a)]:

$$g_{jk} = \delta_{jk}. \qquad (1.9f)$$

The components of a tensor product [e.g., $\mathbf{T}(\_, \_, \_) \otimes \mathbf{S}(\_, \_)$] are easily deduced by inserting the basis vectors into the slots [Eq. (1.9e)]; they are $\mathbf{T}(\mathbf{e}_i, \mathbf{e}_j, \mathbf{e}_k) \otimes \mathbf{S}(\mathbf{e}_l, \mathbf{e}_m) = T_{ijk}S_{lm}$ [cf. Eq. (1.5a)]. In words, the components of a tensor product are equal to the ordinary arithmetic product of the components of the individual tensors.

In component notation, the inner product of two vectors and the value of a tensor when vectors are inserted into its slots are given by

$$\mathbf{A} \cdot \mathbf{B} = A_j B_j, \qquad \mathbf{T}(\mathbf{A}, \mathbf{B}, \mathbf{C}) = T_{ijk} A_i B_j C_k, \qquad (1.9g)$$

as one can easily show using previous equations. Finally, the contraction of a tensor [say, the fourth-rank tensor $\mathbf{R}(\_, \_, \_, \_)$] on two of its slots (say, the first and third) has components that are easily computed from the tensor's own components:

$$\text{components of } [1\&3\text{contraction of } \mathbf{R}] = R_{ijik}. \qquad (1.9h)$$

Note that R_{ijik} is summed on the i index, so it has only two free indices, j and k, and thus is the component of a second-rank tensor, as it must be if it is to represent the contraction of a fourth-rank tensor.

1.5.1 Slot-Naming Index Notation

We now pause in our development of the component version of tensor algebra to introduce a very important new viewpoint.

BOX 1.2. VECTORS AND TENSORS IN QUANTUM THEORY T2

The laws of quantum theory, like all other laws of Nature, can be expressed as geometric relationships among geometric objects. Most of quantum theory's geometric objects, like those of classical theory, are vectors and tensors: the quantum state $|\psi\rangle$ of a physical system (e.g., a particle in a harmonic-oscillator potential) is a Hilbert-space vector—a generalization of a Euclidean-space vector **A**. There is an inner product, denoted $\langle\phi|\psi\rangle$, between any two states $|\phi\rangle$ and $|\psi\rangle$, analogous to **B** · **A**; but **B** · **A** is a real number, whereas $\langle\phi|\psi\rangle$ is a complex number (and we add and subtract quantum states with complex-number coefficients). The Hermitian operators that represent observables (e.g., the hamiltonian $\hat{H}$ for the particle in the potential) are two-slotted (second-rank), complex-valued functions of vectors; $\langle\phi|\hat{H}|\psi\rangle$ is the complex number that one gets when one inserts ϕ and ψ into the first and second slots of $\hat{H}$. Just as, in Euclidean space, we get a new vector (first-rank tensor) **T**(__, **A**) when we insert the vector **A** into the second slot of **T**, so in quantum theory we get a new vector (physical state) $\hat{H}|\psi\rangle$ (the result of letting $\hat{H}$ "act on" $|\psi\rangle$) when we insert $|\psi\rangle$ into the second slot of $\hat{H}$. In these senses, we can regard **T** as a linear map of Euclidean vectors into Euclidean vectors and $\hat{H}$ as a linear map of states (Hilbert-space vectors) into states.

For the electron in the hydrogen atom, we can introduce a set of orthonormal basis vectors $\{|1\rangle, |2\rangle, |3\rangle, \ldots\}$, that is, the atom's energy eigenstates, with $\langle m|n\rangle = \delta_{mn}$. But by contrast with Newtonian physics, where we only need three basis vectors (because our Euclidean space is 3-dimensional), for the particle in a harmonic-oscillator potential, we need an infinite number of basis vectors (since the Hilbert space of all states is infinite-dimensional). In the particle's quantum-state basis, any observable (e.g., the particle's position $\hat{x}$ or momentum $\hat{p}$) has components computed by inserting the basis vectors into its two slots: $x_{mn} = \langle m|\hat{x}|n\rangle$, and $p_{mn} = \langle m|\hat{p}|n\rangle$. In this basis, the operator $\hat{x}\hat{p}$ (which maps states into states) has components $x_{jk}p_{km}$ (a matrix product), and the noncommutation of position and momentum $[\hat{x}, \hat{p}] = i\hbar$ (an important physical law) is expressible in terms of components as $x_{jk}p_{km} - p_{jk}x_{km} = i\hbar\delta_{jm}$.

Consider the rank-2 tensor **F**(__, __). We can define a new tensor **G**(__, __) to be the same as **F**, but with the slots interchanged: i.e., for any two vectors **A** and **B**, it is true that **G**(**A**, **B**) = **F**(**B**, **A**). We need a simple, compact way to indicate that **F** and **G** are equal except for an interchange of slots. The best way is to give the slots names, say a and b—i.e., to rewrite **F**(__, __) as **F**(__$_a$, __$_b$) or more conveniently as F_{ab}, and then to write the relationship between **G** and **F** as $G_{ab} = F_{ba}$. "NO!" some readers might object. This notation is indistinguishable from our notation for components on a particular basis. "GOOD!" a more astute reader will exclaim. The relation $G_{ab} = F_{ba}$ in a particular basis is a true statement if and only if "**G** = **F** with slots interchanged" is true, so why not use the same notation to symbolize both? In fact, we shall do this. We ask our readers to look at any "index equation," such as $G_{ab} = F_{ba}$, like they would look at an Escher drawing: momentarily think of it as a relationship between components of tensors in a specific basis; then do a quick mind-flip and regard it quite differently, as a relationship between geometric, basis-independent tensors with the indices playing the roles of slot names. This mind-flip approach to tensor algebra will pay substantial dividends.

As an example of the power of this *slot-naming index notation,* consider the contraction of the first and third slots of a third-rank tensor **T**. In any basis the components of 1&3contraction(**T**) are T_{aba}; cf. Eq. (1.9h). Correspondingly, in slot-naming index notation we denote 1&3contraction(**T**) by the simple expression T_{aba}. We can think of the first and third slots as annihilating each other by the contraction, leaving free only the second slot (named b) and therefore producing a rank-1 tensor (a vector).

slot-naming index notation

We should caution that the phrase "slot-naming index notation" is unconventional. You are unlikely to find it in any other textbooks. However, we like it. It says precisely what we want it to say.

1.5.2 Particle Kinetics in Index Notation

1.5.2

As an example of slot-naming index notation, we can rewrite the equations of particle kinetics (1.7) as follows:

$$v_i = \frac{dx_i}{dt}, \quad p_i = mv_i, \quad a_i = \frac{dv_i}{dt} = \frac{d^2x_i}{dt^2},$$

$$E = \frac{1}{2}mv_jv_j, \quad \frac{dp_i}{dt} = q(E_i + \epsilon_{ijk}v_jB_k). \tag{1.10}$$

(In the last equation ϵ_{ijk} is the so-called Levi-Civita tensor, which is used to produce the cross product; we shall learn about it in Sec. 1.7. And note that the scalar energy E must not be confused with the electric field vector E_i.)

Equations (1.10) can be viewed in either of two ways: (i) as the basis-independent geometric laws $\mathbf{v} = d\mathbf{x}/dt$, $\mathbf{p} = m\mathbf{v}$, $\mathbf{a} = d\mathbf{v}/dt = d^2\mathbf{x}/dt^2$, $E = \frac{1}{2}m\mathbf{v}^2$, and $d\mathbf{p}/dt = q(\mathbf{E} + \mathbf{v} \times \mathbf{B})$ written in slot-naming index notation; or (ii) as equations for the components of $\mathbf{v}$, $\mathbf{p}$, $\mathbf{a}$, $\mathbf{E}$, and $\mathbf{B}$ in some particular Cartesian coordinate system.

EXERCISES

Exercise 1.3 *Derivation: Component Manipulation Rules*
Derive the component manipulation rules (1.9g) and (1.9h).

Exercise 1.4 *Example and Practice: Numerics of Component Manipulations*
The third-rank tensor **S**(__, __, __) and vectors **A** and **B** have as their only nonzero components $S_{123} = S_{231} = S_{312} = +1$, $A_1 = 3$, $B_1 = 4$, $B_2 = 5$. What are the

components of the vector $\mathbf{C} = \mathbf{S}(\mathbf{A}, \mathbf{B}, \_\_)$, the vector $\mathbf{D} = \mathbf{S}(\mathbf{A}, \_\_, \mathbf{B})$, and the tensor $\mathbf{W} = \mathbf{A} \otimes \mathbf{B}$?

[Partial solution: In component notation, $C_k = S_{ijk} A_i B_j$, where (of course) we sum over the repeated indices i and j. This tells us that $C_1 = S_{231} A_2 B_3$, because S_{231} is the only component of $\mathbf{S}$ whose last index is a 1; this in turn implies that $C_1 = 0$, since $A_2 = 0$. Similarly, $C_2 = S_{312} A_3 B_1 = 0$ (because $A_3 = 0$). Finally, $C_3 = S_{123} A_1 B_2 = +1 \times 3 \times 5 = 15$. Also, in component notation $W_{ij} = A_i B_j$, so $W_{11} = A_1 \times B_1 = 3 \times 4 = 12$, and $W_{12} = A_1 \times B_2 = 3 \times 5 = 15$. Here the $\times$ stands for numerical multiplication, not the vector cross product.]

Exercise 1.5 *Practice: Meaning of Slot-Naming Index Notation*
(a) The following expressions and equations are written in slot-naming index notation. Convert them to geometric, index-free notation: $A_i B_{jk}$, $A_i B_{ji}$, $S_{ijk} = S_{kji}$, $A_i B_i = A_i B_j g_{ij}$.

(b) The following expressions are written in geometric, index-free notation. Convert them to slot-naming index notation: $\mathbf{T}(\_\_, \_\_, \mathbf{A})$, $\mathbf{T}(\_\_, \mathbf{S}(\mathbf{B}, \_\_), \_\_)$.

1.6 Orthogonal Transformations of Bases

Consider two different Cartesian coordinate systems $\{x, y, z\} \equiv \{x_1, x_2, x_3\}$, and $\{\bar{x}, \bar{y}, \bar{z}\} \equiv \{x_{\bar{1}}, x_{\bar{2}}, x_{\bar{3}}\}$. Denote by $\{\mathbf{e}_i\}$ and $\{\mathbf{e}_{\bar{p}}\}$ the corresponding bases. It is possible to expand the basis vectors of one basis in terms of those of the other. We denote the expansion coefficients by the letter R and write

$$\mathbf{e}_i = \mathbf{e}_{\bar{p}} R_{\bar{p}i}, \qquad \mathbf{e}_{\bar{p}} = \mathbf{e}_i R_{i\bar{p}}. \tag{1.11}$$

The quantities $R_{\bar{p}i}$ and $R_{i\bar{p}}$ are not the components of a tensor; rather, they are the elements of transformation matrices

$$[R_{\bar{p}i}] = \begin{bmatrix} R_{\bar{1}1} & R_{\bar{1}2} & R_{\bar{1}3} \\ R_{\bar{2}1} & R_{\bar{2}2} & R_{\bar{2}3} \\ R_{\bar{3}1} & R_{\bar{3}2} & R_{\bar{3}3} \end{bmatrix}, \qquad [R_{i\bar{p}}] = \begin{bmatrix} R_{1\bar{1}} & R_{1\bar{2}} & R_{1\bar{3}} \\ R_{2\bar{1}} & R_{2\bar{2}} & R_{2\bar{3}} \\ R_{3\bar{1}} & R_{3\bar{2}} & R_{3\bar{3}} \end{bmatrix}. \tag{1.12a}$$

(Here and throughout this book we use square brackets to denote matrices.) These two matrices must be the inverse of each other, since one takes us from the barred basis to the unbarred, and the other in the reverse direction, from unbarred to barred:

$$R_{\bar{p}i} R_{i\bar{q}} = \delta_{\bar{p}\bar{q}}, \qquad R_{i\bar{p}} R_{\bar{p}j} = \delta_{ij}. \tag{1.12b}$$

The orthonormality requirement for the two bases implies that $\delta_{ij} = \mathbf{e}_i \cdot \mathbf{e}_j = (\mathbf{e}_{\bar{p}} R_{\bar{p}i}) \cdot (\mathbf{e}_{\bar{q}} R_{\bar{q}j}) = R_{\bar{p}i} R_{\bar{q}j} (\mathbf{e}_{\bar{p}} \cdot \mathbf{e}_{\bar{q}}) = R_{\bar{p}i} R_{\bar{q}j} \delta_{\bar{p}\bar{q}} = R_{\bar{p}i} R_{\bar{p}j}$. This says that the transpose of $[R_{\bar{p}i}]$ is its inverse—which we have already denoted by $[R_{i\bar{p}}]$:

$$[R_{i\bar{p}}] \equiv \text{inverse} \left([R_{\bar{p}i}]\right) = \text{transpose} \left([R_{\bar{p}i}]\right). \tag{1.12c}$$

This property implies that the transformation matrix is orthogonal, so the transformation is a reflection or a rotation (see, e.g., Goldstein, Poole, and Safko, 2002). Thus (as should be obvious and familiar), the bases associated with any two Euclidean coordinate systems are related by a reflection or rotation, and the matrices (1.12a) are called *rotation matrices*. Note that Eq. (1.12c) does not say that $[R_{i\bar{p}}]$ is a symmetric matrix. In fact, most rotation matrices are not symmetric [see, e.g., Eq. (1.14)].

The fact that a vector $\mathbf{A}$ is a geometric, basis-independent object implies that $\mathbf{A} = A_i \mathbf{e}_i = A_i (\mathbf{e}_{\bar{p}} R_{\bar{p}i}) = (R_{\bar{p}i} A_i) \mathbf{e}_{\bar{p}} = A_{\bar{p}} \mathbf{e}_{\bar{p}}$:

$$A_{\bar{p}} = R_{\bar{p}i} A_i, \quad \text{and similarly,} \quad A_i = R_{i\bar{p}} A_{\bar{p}}; \tag{1.13a}$$

and correspondingly for the components of a tensor:

$$T_{\bar{p}\bar{q}\bar{r}} = R_{\bar{p}i} R_{\bar{q}j} R_{\bar{r}k} T_{ijk}, \quad T_{ijk} = R_{i\bar{p}} R_{j\bar{q}} R_{k\bar{r}} T_{\bar{p}\bar{q}\bar{r}}. \tag{1.13b}$$

It is instructive to compare the transformation law (1.13a) for the components of a vector with Eqs. (1.11) for the bases. To make these laws look natural, we have placed the transformation matrix on the left in the former and on the right in the latter. In Minkowski spacetime (Chap. 2), the placement of indices, up or down, will automatically tell us the order.

If we choose the origins of our two coordinate systems to coincide, then the vector $\mathbf{x}$ reaching from the common origin to some point $\mathcal{P}$, whose coordinates are x_j and $x_{\bar{p}}$, has components equal to those coordinates; and as a result, the coordinates themselves obey the same transformation law as any other vector:

$$x_{\bar{p}} = R_{\bar{p}i} x_i, \quad x_i = R_{i\bar{p}} x_{\bar{p}}. \tag{1.13c}$$

The product of two rotation matrices $[R_{i\bar{p}} R_{\bar{p}\bar{s}}]$ is another rotation matrix $[R_{i\bar{s}}]$, which transforms the Cartesian bases $\mathbf{e}_{\bar{s}}$ to $\mathbf{e}_i$. Under this product rule, the rotation matrices form a mathematical *group*: the *rotation group*, whose *group representations* play an important role in quantum theory.

Exercise 1.6 **Example and Practice: Rotation in* x-y *Plane*

Consider two Cartesian coordinate systems rotated with respect to each other in the x-y plane as shown in Fig. 1.4.

(a) Show that the rotation matrix that takes the barred basis vectors to the unbarred basis vectors is

$$[R_{\bar{p}i}] = \begin{bmatrix} \cos\phi & \sin\phi & 0 \\ -\sin\phi & \cos\phi & 0 \\ 0 & 0 & 1 \end{bmatrix}, \tag{1.14}$$

and show that the inverse of this rotation matrix is, indeed, its transpose, as it must be if this is to represent a rotation.

(b) Verify that the two coordinate systems are related by Eq. (1.13c).

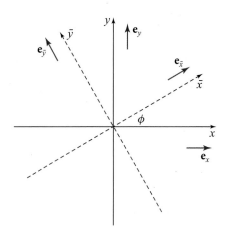

FIGURE 1.4 Two Cartesian coordinate systems $\{x, y, z\}$ and $\{\bar{x}, \bar{y}, \bar{z}\}$ and their basis vectors in Euclidean space, rotated by an angle ϕ relative to each other in the x-y plane. The z- and $\bar{z}$-axes point out of the paper or screen and are not shown.

(c) Let A_j be the components of the electromagnetic vector potential that lies in the x-y plane, so that $A_z = 0$. The two nonzero components A_x and A_y can be regarded as describing the two polarizations of an electromagnetic wave propagating in the z direction. Show that $A_{\bar{x}} + i A_{\bar{y}} = (A_x + i A_y)e^{-i\phi}$. One can show (cf. Sec. 27.3.3) that the factor $e^{-i\phi}$ implies that the quantum particle associated with the wave—the photon—has spin one [i.e., spin angular momentum $\hbar = $ (Planck's constant)$/2\pi$].

(d) Let h_{jk} be the components of a symmetric tensor that is *trace-free* (its contraction h_{jj} vanishes) and is confined to the x-y plane (so $h_{zk} = h_{kz} = 0$ for all k). Then the only nonzero components of this tensor are $h_{xx} = -h_{yy}$ and $h_{xy} = h_{yx}$. As we shall see in Sec. 27.3.1, this tensor can be regarded as describing the two polarizations of a gravitational wave propagating in the z direction. Show that $h_{\bar{x}\bar{x}} + i h_{\bar{x}\bar{y}} = (h_{xx} + i h_{xy})e^{-2i\phi}$. The factor $e^{-2i\phi}$ implies that the quantum particle associated with the gravitational wave (the graviton) has spin two (spin angular momentum $2\hbar$); cf. Eq. (27.31) and Sec. 27.3.3.

1.7

1.7 Differentiation of Scalars, Vectors, and Tensors; Cross Product and Curl

Consider a tensor field $\mathbf{T}(\mathcal{P})$ in Euclidean 3-space and a vector $\mathbf{A}$. We define the

directional derivative

directional derivative of $\mathbf{T}$ along $\mathbf{A}$ by the obvious limiting procedure

$$\nabla_{\mathbf{A}}\mathbf{T} \equiv \lim_{\epsilon \to 0} \frac{1}{\epsilon}[\mathbf{T}(\mathbf{x}_{\mathcal{P}} + \epsilon \mathbf{A}) - \mathbf{T}(\mathbf{x}_{\mathcal{P}})] \tag{1.15a}$$

and similarly for the directional derivative of a vector field $\mathbf{B}(\mathcal{P})$ and a scalar field $\psi(\mathcal{P})$. [Here we have denoted points, e.g., $\mathcal{P}$, by the vector $\mathbf{x}_{\mathcal{P}}$ that reaches from some arbitrary origin to the point, and $\mathbf{T}(\mathbf{x}_{\mathcal{P}})$ denotes the field's dependence on location in space; $\mathbf{T}$'s slots and dependence on what goes into the slots are suppressed; and the units of ϵ are chosen to ensure that $\epsilon\mathbf{A}$ has the same units as $\mathbf{x}_{\mathcal{P}}$. There is no other appearance of vectors in this chapter.] In definition (1.15a), the quantity in square brackets is simply the difference between two linear functions of vectors (two tensors), so the quantity on the left-hand side is also a tensor with the same rank as $\mathbf{T}$.

It should not be hard to convince oneself that this directional derivative $\nabla_{\mathbf{A}}\mathbf{T}$ of any tensor field $\mathbf{T}$ is linear in the vector $\mathbf{A}$ along which one differentiates. Correspondingly, if $\mathbf{T}$ has rank n (n slots), then there is another tensor field, denoted $\nabla\mathbf{T}$, with rank $n+1$, such that

$$\boxed{\nabla_{\mathbf{A}}\mathbf{T} = \nabla\mathbf{T}(\_,\_,\_,\mathbf{A}).} \qquad (1.15b)$$

Here on the right-hand side the first n slots (3 in the case shown) are left empty, and $\mathbf{A}$ is put into the last slot (the "differentiation slot"). The quantity $\nabla\mathbf{T}$ is called the *gradient* of $\mathbf{T}$. In slot-naming index notation, it is conventional to denote this gradient by $T_{abc;d}$, where in general the number of indices preceding the semicolon is the rank of $\mathbf{T}$. Using this notation, the directional derivative of $\mathbf{T}$ along $\mathbf{A}$ reads [cf. Eq. (1.15b)] $T_{abc;j}A_j$.

gradient

It is not hard to show that in any Cartesian coordinate system, the components of the gradient are nothing but the partial derivatives of the components of the original tensor, which we denote by a comma:

$$\boxed{T_{abc;j} = \frac{\partial T_{abc}}{\partial x_j} \equiv T_{abc,j}.} \qquad (1.15c)$$

In a non-Cartesian basis (e.g., the spherical and cylindrical bases often used in electromagnetic theory), the components of the gradient typically are not obtained by simple partial differentiation [Eq. (1.15c) fails] because of turning and/or length changes of the basis vectors as we go from one location to another. In Sec. 11.8, we shall learn how to deal with this by using objects called *connection coefficients*. Until then, we confine ourselves to Cartesian bases, so subscript semicolons and subscript commas (partial derivatives) can be used interchangeably.

Because the gradient and the directional derivative are defined by the same standard limiting process as one uses when defining elementary derivatives, they obey the standard (Leibniz) rule for differentiating products:

$$\nabla_{\mathbf{A}}(\mathbf{S} \otimes \mathbf{T}) = (\nabla_{\mathbf{A}}\mathbf{S}) \otimes \mathbf{T} + \mathbf{S} \otimes \nabla_{\mathbf{A}}\mathbf{T},$$
$$\text{or} \quad (S_{ab}T_{cde})_{;j}A_j = (S_{ab;j}A_j)T_{cde} + S_{ab}(T_{cde;j}A_j); \qquad (1.16a)$$

and

$$\nabla_{\mathbf{A}}(f\mathbf{T}) = (\nabla_{\mathbf{A}}f)\mathbf{T} + f\nabla_{\mathbf{A}}\mathbf{T}, \quad \text{or} \quad (fT_{abc})_{;j}A_j = (f_{;j}A_j)T_{abc} + fT_{abc;j}A_j. \qquad (1.16b)$$

In an orthonormal basis these relations should be obvious: they follow from the Leibniz rule for partial derivatives.

Because the components g_{ab} of the metric tensor are constant in any Cartesian coordinate system, Eq. (1.15c) (which is valid in such coordinates) guarantees that $g_{ab;j} = 0$; i.e., the metric has vanishing gradient:

$$\nabla \mathbf{g} = 0, \quad \text{or} \quad g_{ab;j} = 0. \tag{1.17}$$

From the gradient of any vector or tensor we can construct several other important derivatives by contracting on slots:

1. Since the gradient $\nabla \mathbf{A}$ of a vector field $\mathbf{A}$ has two slots, $\nabla \mathbf{A}(\_, \_)$, we can contract its slots on each other to obtain a scalar field. That scalar field is the *divergence* of $\mathbf{A}$ and is denoted

$$\nabla \cdot \mathbf{A} \equiv (\text{contraction of } \nabla \mathbf{A}) = A_{a;a}. \tag{1.18}$$

2. Similarly, if $\mathbf{T}$ is a tensor field of rank 3, then $T_{abc;c}$ is its divergence on its third slot, and $T_{abc;b}$ is its divergence on its second slot.

3. By taking the double gradient and then contracting on the two gradient slots we obtain, from any tensor field $\mathbf{T}$, a new tensor field with the same rank,

$$\nabla^2 \mathbf{T} \equiv (\nabla \cdot \nabla) \mathbf{T}, \quad \text{or} \quad T_{abc;jj}. \tag{1.19}$$

Here and henceforth, all indices following a semicolon (or comma) represent gradients (or partial derivatives): $T_{abc;jj} \equiv T_{abc;j;j}$, $T_{abc,jk} \equiv \partial^2 T_{abc}/\partial x_j \partial x_k$. The operator ∇^2 is called the *laplacian*.

The metric tensor is a fundamental property of the space in which it lives; it embodies the inner product and hence the space's notion of distance. In addition to the metric, there is one (and only one) other fundamental tensor that describes a piece of Euclidean space's geometry: the *Levi-Civita tensor* $\boldsymbol{\epsilon}$, which embodies the space's notion of volume.

In a Euclidean space with dimension n, the Levi-Civita tensor $\boldsymbol{\epsilon}$ is a completely antisymmetric tensor with rank n (with n slots). A parallelepiped whose edges are the n vectors $\mathbf{A}, \mathbf{B}, \ldots, \mathbf{F}$ is said to have the *volume*

$$\boxed{\text{volume} = \boldsymbol{\epsilon}(\mathbf{A}, \mathbf{B}, \ldots, \mathbf{F}).} \tag{1.20}$$

(We justify this definition in Sec. 1.8.) Notice that this volume can be positive or negative, and if we exchange the order of the parallelepiped's legs, the volume's sign changes: $\boldsymbol{\epsilon}(\mathbf{B}, \mathbf{A}, \ldots, \mathbf{F}) = -\boldsymbol{\epsilon}(\mathbf{A}, \mathbf{B}, \ldots, \mathbf{F})$ by antisymmetry of $\boldsymbol{\epsilon}$.

It is easy to see (Ex. 1.7) that (i) the volume vanishes unless the legs are all linearly independent, (ii) once the volume has been specified for one parallelepiped (one set of linearly independent legs), it is thereby determined for all parallelepipeds, and therefore, (iii) we require only one number plus antisymmetry to determine $\boldsymbol{\epsilon}$

fully. If the chosen parallelepiped has legs that are orthonormal (all are orthogonal to one another and all have unit length—properties determined by the metric **g**), then it must have unit volume, or more precisely volume ± 1. This is a compatibility relation between **g** and ϵ. It is easy to see (Ex. 1.7) that (iv) ϵ is fully determined by its antisymmetry, compatibility with the metric, and a single sign: the choice of which parallelepipeds have positive volume and which have negative. It is conventional in Euclidean 3-space to give right-handed parallelepipeds positive volume and left-handed ones negative volume: $\epsilon(\mathbf{A}, \mathbf{B}, \mathbf{C})$ is positive if, when we place our right thumb along **C** and the fingers of our right hand along **A**, then bend our fingers, they sweep toward **B** and not $-\mathbf{B}$.

These considerations dictate that in a right-handed orthonormal basis of Euclidean 3-space, the only nonzero components of ϵ are

$$\epsilon_{123} = +1,$$

$$\epsilon_{abc} = \begin{cases} +1 & \text{if } a, b, c \text{ is an even permutation of } 1, 2, 3 \\ -1 & \text{if } a, b, c \text{ is an odd permutation of } 1, 2, 3 \\ 0 & \text{if } a, b, c \text{ are not all different;} \end{cases} \tag{1.21}$$

and in a left-handed orthonormal basis, the signs of these components are reversed.

The Levi-Civita tensor is used to define the cross product and the curl:

cross product and curl

$$\mathbf{A} \times \mathbf{B} \equiv \epsilon(\_\_, \mathbf{A}, \mathbf{B}); \quad \text{in slot-naming index notation, } \epsilon_{ijk} A_j B_k; \tag{1.22a}$$

$$\boldsymbol{\nabla} \times \mathbf{A} \equiv (\text{the vector field whose slot-naming index form is } \epsilon_{ijk} A_{k;j}). \tag{1.22b}$$

[Equation (1.22b) is an example of an expression that is complicated if stated in index-free notation; it says that $\boldsymbol{\nabla} \times \mathbf{A}$ is the double contraction of the rank-5 tensor $\epsilon \otimes \boldsymbol{\nabla} \mathbf{A}$ on its second and fifth slots, and on its third and fourth slots.]

Although Eqs. (1.22a) and (1.22b) look like complicated ways to deal with concepts that most readers regard as familiar and elementary, they have great power. The power comes from the following property of the Levi-Civita tensor in Euclidean 3-space [readily derivable from its components (1.21)]:

$$\boxed{\epsilon_{ijm}\epsilon_{klm} = \delta^{ij}_{kl} \equiv \delta^i_k \delta^j_l - \delta^i_l \delta^j_k.} \tag{1.23}$$

Here δ^i_k is the Kronecker delta. Examine the 4-index delta function δ^{ij}_{kl} carefully; it says that either the indices above and below each other must be the same ($i = k$ and $j = l$) with a $+$ sign, or the diagonally related indices must be the same ($i = l$ and $j = k$) with a $-$ sign. [We have put the indices ij of δ^{ij}_{kl} up solely to facilitate remembering this rule. Recall (first paragraph of Sec. 1.5) that in Euclidean space and Cartesian coordinates, it does not matter whether indices are up or down.] With the aid of Eq. (1.23) and the index-notation expressions for the cross product and curl, one can quickly and easily derive a wide variety of useful vector identities; see the very important Ex. 1.8.

Exercise 1.7 *Derivation: Properties of the Levi-Civita Tensor*

From its complete antisymmetry, derive the four properties of the Levi-Civita tensor, in n-dimensional Euclidean space, that are claimed in the text following Eq. (1.20).

Exercise 1.8 ***Example and Practice: Vectorial Identities for the Cross Product and Curl*

Here is an example of how to use index notation to derive a vector identity for the double cross product $\mathbf{A} \times (\mathbf{B} \times \mathbf{C})$: in index notation this quantity is $\epsilon_{ijk} A_j (\epsilon_{klm} B_l C_m)$. By permuting the indices on the second ϵ and then invoking Eq. (1.23), we can write this as $\epsilon_{ijk}\epsilon_{lmk} A_j B_l C_m = \delta_{ij}^{lm} A_j B_l C_m$. By then invoking the meaning of the 4-index delta function [Eq. (1.23)], we bring this into the form $A_j B_i C_j - A_j B_j C_i$, which is the slot-naming index-notation form of $(\mathbf{A} \cdot \mathbf{C})\mathbf{B} - (\mathbf{A} \cdot \mathbf{B})\mathbf{C}$. Thus, it must be that $\mathbf{A} \times (\mathbf{B} \times \mathbf{C}) = (\mathbf{A} \cdot \mathbf{C})\mathbf{B} - (\mathbf{A} \cdot \mathbf{B})\mathbf{C}$. Use similar techniques to evaluate the following quantities.

(a) $\nabla \times (\nabla \times \mathbf{A})$.

(b) $(\mathbf{A} \times \mathbf{B}) \cdot (\mathbf{C} \times \mathbf{D})$.

(c) $(\mathbf{A} \times \mathbf{B}) \times (\mathbf{C} \times \mathbf{D})$.

Exercise 1.9 ***Example and Practice: Levi-Civita Tensor in 2-Dimensional Euclidean Space*

In Euclidean 2-space, let $\{\mathbf{e}_1, \mathbf{e}_2\}$ be an orthonormal basis with positive volume.

(a) Show that the components of $\boldsymbol{\epsilon}$ in this basis are

$$\epsilon_{12} = +1, \qquad \epsilon_{21} = -1, \qquad \epsilon_{11} = \epsilon_{22} = 0. \qquad (1.24a)$$

(b) Show that

$$\epsilon_{ik}\epsilon_{jk} = \delta_{ij}. \qquad (1.24b)$$

1.8 Volumes, Integration, and Integral Conservation Laws

In Cartesian coordinates of 2-dimensional Euclidean space, the basis vectors are orthonormal, so (with a conventional choice of sign) the components of the Levi-Civita tensor are given by Eqs. (1.24a). Correspondingly, the area (i.e., 2-dimensional volume) of a parallelogram whose sides are $\mathbf{A}$ and $\mathbf{B}$ is

$$2\text{-volume} = \boldsymbol{\epsilon}(\mathbf{A}, \mathbf{B}) = \epsilon_{ab} A_a B_b = A_1 B_2 - A_2 B_1 = \det \begin{bmatrix} A_1 & B_1 \\ A_2 & B_2 \end{bmatrix}, \qquad (1.25)$$

a relation that should be familiar from elementary geometry. Equally familiar should be the following expression for the 3-dimensional volume of a parallelepiped with legs

A, **B**, and **C** [which follows from the components (1.21) of the Levi-Civita tensor]:

$$3\text{-volume} = \epsilon(\mathbf{A}, \mathbf{B}, \mathbf{C}) = \epsilon_{ijk}A_i B_j C_k = \mathbf{A} \cdot (\mathbf{B} \times \mathbf{C}) = \det \begin{bmatrix} A_1 & B_1 & C_1 \\ A_2 & B_2 & C_2 \\ A_3 & B_3 & C_3 \end{bmatrix}. \quad (1.26)$$

Our formal definition (1.20) of volume is justified because it gives rise to these familiar equations.

Equations (1.25) and (1.26) are foundations from which one can derive the usual formulas $dA = dx\,dy$ and $dV = dx\,dy\,dz$ for the area and volume of elementary surface and volume elements with Cartesian side lengths dx, dy, and dz (Ex. 1.10).

In Euclidean 3-space, we define the vectorial surface area of a 2-dimensional parallelogram with legs **A** and **B** to be

$$\boxed{\mathbf{\Sigma} = \mathbf{A} \times \mathbf{B} = \epsilon(\_\_, \mathbf{A}, \mathbf{B}).} \quad (1.27)$$

This vectorial surface area has a magnitude equal to the area of the parallelogram and a direction perpendicular to it. Notice that this surface area $\epsilon(\_\_, \mathbf{A}, \mathbf{B})$ can be thought of as an object that is waiting for us to insert a third leg, **C**, so as to compute a 3-volume $\epsilon(\mathbf{C}, \mathbf{A}, \mathbf{B})$—the volume of the parallelepiped with legs **C**, **A**, and **B**.

A parallelogram's surface has two faces (two sides), called the *positive face* and the *negative face*. If the vector **C** sticks out of the positive face, then $\mathbf{\Sigma}(\mathbf{C}) = \epsilon(\mathbf{C}, \mathbf{A}, \mathbf{B})$ is positive; if **C** sticks out of the negative face, then $\mathbf{\Sigma}(\mathbf{C})$ is negative.

1.8.1 Gauss's and Stokes' Theorems

Such vectorial surface areas are the foundation for surface integrals in 3-dimensional space and for the familiar *Gauss's theorem,*

$$\boxed{\int_{\mathcal{V}_3} (\mathbf{\nabla} \cdot \mathbf{A}) dV = \int_{\partial \mathcal{V}_3} \mathbf{A} \cdot d\mathbf{\Sigma}} \quad (1.28a)$$

(where $\mathcal{V}_3$ is a compact 3-dimensional region, and $\partial \mathcal{V}_3$ is its closed 2-dimensional boundary) and *Stokes' theorem,*

$$\boxed{\int_{\mathcal{V}_2} \mathbf{\nabla} \times \mathbf{A} \cdot d\mathbf{\Sigma} = \int_{\partial \mathcal{V}_2} \mathbf{A} \cdot d\mathbf{l}} \quad (1.28b)$$

(where $\mathcal{V}_2$ is a compact 2-dimensional region, $\partial \mathcal{V}_2$ is the 1-dimensional closed curve that bounds it, and the last integral is a line integral around that curve); see, e.g., Arfken, Weber, and Harris (2013).

This mathematics is illustrated by the integral and differential conservation laws for electric charge and for particles: The total charge and the total number of particles inside a 3-dimensional region of space $\mathcal{V}_3$ are $\int_{\mathcal{V}_3} \rho_e\,dV$ and $\int_{\mathcal{V}_3} n\,dV$, where ρ_e is the charge density and n the number density of particles. The rates that charge and particles flow out of $\mathcal{V}_3$ are the integrals of the current density **j** and the particle flux vector **S** over its boundary $\partial \mathcal{V}_3$. Therefore, the *integral laws of charge conservation and particle conservation* are

integral conservation laws

$$\boxed{\frac{d}{dt}\int_{\mathcal{V}_3}\rho_e\,dV + \int_{\partial\mathcal{V}_3}\mathbf{j}\cdot d\mathbf{\Sigma}=0,} \qquad \boxed{\frac{d}{dt}\int_{\mathcal{V}_3}n\,dV + \int_{\partial\mathcal{V}_3}\mathbf{S}\cdot d\mathbf{\Sigma}=0.} \quad (1.29)$$

Pull the time derivative inside each volume integral (where it becomes a partial derivative), and apply Gauss's law to each surface integral; the results are $\int_{\mathcal{V}_3}(\partial\rho_e/\partial t + \nabla\cdot\mathbf{j})dV = 0$ and similarly for particles. The only way these equations can be true for all choices of $\mathcal{V}_3$ is for the integrands to vanish:

differential conservation laws

$$\boxed{\partial\rho_e/\partial t + \nabla\cdot\mathbf{j}=0,} \qquad \boxed{\partial n/\partial t + \nabla\cdot\mathbf{S}=0.} \qquad (1.30)$$

These are the *differential conservation laws for charge and for particles.* They have a standard, universal form: the time derivative of the density of a quantity plus the divergence of its flux vanishes.

Note that the integral conservation laws (1.29) and the differential conservation laws (1.30) require no coordinate system or basis for their description, and no coordinate system or basis was used in deriving the differential laws from the integral laws. This is an example of the fundamental principle that *the Newtonian physical laws are all expressible as geometric relationships among geometric objects.*

EXERCISES

Exercise 1.10 *Derivation and Practice: Volume Elements in Cartesian Coordinates*
Use Eqs. (1.25) and (1.26) to derive the usual formulas $dA = dxdy$ and $dV = dxdydz$ for the 2-dimensional and 3-dimensional integration elements, respectively, in right-handed Cartesian coordinates. [Hint: Use as the edges of the integration volumes $dx\,\mathbf{e}_x$, $dy\,\mathbf{e}_y$, and $dz\,\mathbf{e}_z$.]

Exercise 1.11 *Example and Practice: Integral of a Vector Field over a Sphere*
Integrate the vector field $\mathbf{A} = z\mathbf{e}_z$ over a sphere with radius a, centered at the origin of the Cartesian coordinate system (i.e., compute $\int\mathbf{A}\cdot d\mathbf{\Sigma}$). Hints:

(a) Introduce spherical polar coordinates on the sphere, and construct the vectorial integration element $d\mathbf{\Sigma}$ from the two legs $ad\theta\,\mathbf{e}_{\hat{\theta}}$ and $a\sin\theta d\phi\,\mathbf{e}_{\hat{\phi}}$. Here $\mathbf{e}_{\hat{\theta}}$ and $\mathbf{e}_{\hat{\phi}}$ are unit-length vectors along the θ and ϕ directions. (Here as in Sec. 1.6 and throughout this book, we use accents on indices to indicate which basis the index is associated with: hats here for the spherical orthonormal basis, bars in Sec. 1.6 for the barred Cartesian basis.) Explain the factors $ad\theta$ and $a\sin\theta d\phi$ in the definitions of the legs. Show that

$$d\mathbf{\Sigma} = \boldsymbol{\epsilon}(\_\_,\,\mathbf{e}_{\hat{\theta}},\,\mathbf{e}_{\hat{\phi}})a^2\sin\theta d\theta d\phi. \qquad (1.31)$$

(b) Using $z = a\cos\theta$ and $\mathbf{e}_z = \cos\theta\mathbf{e}_{\hat{r}} - \sin\theta\mathbf{e}_{\hat{\theta}}$ on the sphere (where $\mathbf{e}_{\hat{r}}$ is the unit vector pointing in the radial direction), show that

$$\mathbf{A}\cdot d\mathbf{\Sigma} = a\cos^2\theta\,\boldsymbol{\epsilon}(\mathbf{e}_{\hat{r}},\,\mathbf{e}_{\hat{\theta}},\,\mathbf{e}_{\hat{\phi}})\,a^2\sin\theta d\theta d\phi.$$

(c) Explain why $\epsilon(\mathbf{e}_{\hat{r}}, \mathbf{e}_{\hat{\theta}}, \mathbf{e}_{\hat{\phi}}) = 1$.

(d) Perform the integral $\int \mathbf{A} \cdot d\mathbf{\Sigma}$ over the sphere's surface to obtain your final answer $(4\pi/3)a^3$. This, of course, is the volume of the sphere. Explain pictorially why this had to be the answer.

Exercise 1.12 *Example: Faraday's Law of Induction*
One of Maxwell's equations says that $\mathbf{\nabla} \times \mathbf{E} = -\partial \mathbf{B}/\partial t$ (in SI units), where $\mathbf{E}$ and $\mathbf{B}$ are the electric and magnetic fields. This is a geometric relationship between geometric objects; it requires no coordinates or basis for its statement. By integrating this equation over a 2-dimensional surface $\mathcal{V}_2$ with boundary curve $\partial \mathcal{V}_2$ and applying Stokes' theorem, derive Faraday's law of induction—again, a geometric relationship between geometric objects.

1.9 The Stress Tensor and Momentum Conservation

Press your hands together in the y-z plane and feel the force that one hand exerts on the other across a tiny area A—say, one square millimeter of your hands' palms (Fig. 1.5). That force, of course, is a vector $\mathbf{F}$. It has a normal component (along the x direction). It also has a tangential component: if you try to slide your hands past each other, you feel a component of force along their surface, a "shear" force in the y and z directions. Not only is the force $\mathbf{F}$ vectorial; so is the 2-surface across which it acts, $\mathbf{\Sigma} = A\,\mathbf{e}_x$. (Here $\mathbf{e}_x$ is the unit vector orthogonal to the tiny area A, and we have chosen the negative side of the surface to be the $-x$ side and the positive side to be $+x$. With this choice, the force $\mathbf{F}$ is that which the negative hand, on the $-x$ side, exerts on the positive hand.)

force vector

Now, it should be obvious that the force $\mathbf{F}$ is a linear function of our chosen surface $\mathbf{\Sigma}$. Therefore, there must be a tensor, the *stress tensor*, that reports the force to us when we insert the surface into its second slot:

stress tensor

$$\boxed{\mathbf{F}(\underline{\ \ }) = \mathbf{T}(\underline{\ \ }, \mathbf{\Sigma}), \quad \text{or} \quad F_i = T_{ij}\Sigma_j.} \tag{1.32}$$

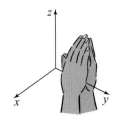

FIGURE 1.5 Hands, pressed together, exert a force on each other.

Newton's law of action and reaction tells us that the force that the positive hand exerts on the negative hand must be equal and opposite to that which the negative hand exerts on the positive. This shows up trivially in Eq. (1.32): by changing the sign of $\boldsymbol{\Sigma}$, one reverses which hand is regarded as negative and which positive, and since $\mathbf{T}$ is linear in $\boldsymbol{\Sigma}$, one also reverses the sign of the force.

The definition (1.32) of the stress tensor gives rise to the following physical meaning of its components:

$$T_{jk} = \left(\begin{array}{c} j \text{ component of force per unit area} \\ \text{across a surface perpendicular to } \mathbf{e}_k \end{array} \right)$$

meaning of components of stress tensor

$$= \left(\begin{array}{c} j \text{ component of momentum that crosses a unit} \\ \text{area that is perpendicular to } \mathbf{e}_k, \text{ per unit time,} \\ \text{with the crossing being from } -x_k \text{ to } +x_k \end{array} \right). \tag{1.33}$$

The stresses inside a table with a heavy weight on it are described by the stress tensor $\mathbf{T}$, as are the stresses in a flowing fluid or plasma, in the electromagnetic field, and in any other physical medium. Accordingly, we shall use the stress tensor as an important mathematical tool in our study of force balance in kinetic theory (Chap. 3), elasticity (Part IV), fluid dynamics (Part V), and plasma physics (Part VI).

symmetry of stress tensor

It is not obvious from its definition, but the stress tensor $\mathbf{T}$ is always symmetric in its two slots. To see this, consider a small cube with side L in any medium (or field) (Fig. 1.6). The medium outside the cube exerts forces, and hence also torques, on the cube's faces. The z-component of the torque is produced by the shear forces on the front and back faces and on the left and right. As shown in the figure, the shear forces on the front and back faces have magnitudes $T_{xy}L^2$ and point in opposite directions, so they exert identical torques on the cube, $N_z = T_{xy}L^2(L/2)$ (where $L/2$ is the distance of each face from the cube's center). Similarly, the shear forces on the left and right faces have magnitudes $T_{yx}L^2$ and point in opposite directions, thereby exerting identical torques on the cube, $N_z = -T_{yx}L^2(L/2)$. Adding the torques from all four faces and equating them to the rate of change of angular momentum, $\frac{1}{6}\rho L^5 d\Omega_z/dt$ (where ρ is the mass density, $\frac{1}{6}\rho L^5$ is the cube's moment of inertia, and Ω_z is the z component of its angular velocity), we obtain $(T_{xy} - T_{yx})L^3 = \frac{1}{6}\rho L^5 d\Omega_z/dt$. Now, let the cube's edge length become arbitrarily small, $L \to 0$. If $T_{xy} - T_{yx}$ does not vanish, then the cube will be set into rotation with an infinitely large angular acceleration, $d\Omega_z/dt \propto 1/L^2 \to \infty$— an obviously unphysical behavior. Therefore, $T_{yx} = T_{xy}$, and similarly for all other components: *the stress tensor is always symmetric under interchange of its two slots.*

1.9.1 Examples: Electromagnetic Field and Perfect Fluid

Two examples will make the concept of the stress tensor more concrete.

- **Electromagnetic field:** See Ex. 1.14.

perfect fluid

- **Perfect fluid:** A *perfect fluid* is a medium that can exert an isotropic pressure P but no shear stresses, so the only nonzero components of its stress tensor

Appendix A. Newtonian Physics: Geometric Viewpoint

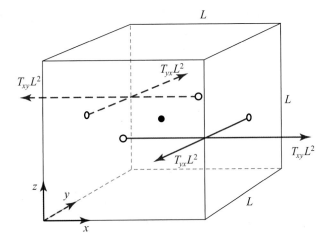

FIGURE 1.6 The shear forces exerted on the left, right, front, and back faces of a vanishingly small cube of side length L. The resulting torque about the z direction will set the cube into rotation with an arbitrarily large angular acceleration unless the stress tensor is symmetric.

in a Cartesian basis are $T_{xx} = T_{yy} = T_{zz} = P$. (Examples of nearly perfect fluids are air and water, but not molasses.) We can summarize this property by $T_{ij} = P\delta_{ij}$ or equivalently, since δ_{ij} are the components of the Euclidean metric, $T_{ij} = Pg_{ij}$. The frame-independent version of this is

$$\mathbf{T} = P\mathbf{g} \quad \text{or, in slot-naming index notation,} \quad T_{ij} = Pg_{ij}. \quad (1.34)$$

Note that, as always, the formula in slot-naming index notation looks identical to the formula $T_{ij} = Pg_{ij}$ for the components in our chosen Cartesian coordinate system. To check Eq. (1.34), consider a 2-surface $\mathbf{\Sigma} = A\mathbf{n}$ with area A oriented perpendicular to some arbitrary unit vector $\mathbf{n}$. The vectorial force that the fluid exerts across $\mathbf{\Sigma}$ is, in index notation, $F_j = T_{jk}\Sigma_k = Pg_{jk}An_k = PAn_j$ (i.e., it is a normal force with magnitude equal to the fluid pressure P times the surface area A). This is what it should be.

1.9.2 Conservation of Momentum

The stress tensor plays a central role in the Newtonian law of momentum conservation because (by definition) the force acting across a surface is the same as the rate of flow of momentum, per unit area, across the surface: *the stress tensor is the flux of momentum.*

Consider the 3-dimensional region of space $\mathcal{V}_3$ used above in formulating the integral laws of charge and particle conservation (1.29). The total momentum in $\mathcal{V}_3$ is $\int_{\mathcal{V}_3} \mathbf{G}\, dV$, where $\mathbf{G}$ is the momentum density. This quantity changes as a result of momentum flowing into and out of $\mathcal{V}_3$. The net rate at which momentum flows outward is the integral of the stress tensor over the surface $\partial\mathcal{V}_3$ of $\mathcal{V}_3$. Therefore, by analogy with charge and particle conservation (1.29), *the integral law of momentum conservation* says

integral conservation of momentum

$$\boxed{\frac{d}{dt}\int_{\mathcal{V}_3} \mathbf{G}\, dV + \int_{\partial\mathcal{V}_3} \mathbf{T}\cdot d\boldsymbol{\Sigma} = 0.}$$

(1.35)

By pulling the time derivative inside the volume integral (where it becomes a partial derivative) and applying the vectorial version of Gauss's law to the surface integral, we obtain $\int_{\mathcal{V}_3}(\partial\mathbf{G}/\partial t + \boldsymbol{\nabla}\cdot\mathbf{T})\, dV = 0$. This can be true for all choices of $\mathcal{V}_3$ only if the integrand vanishes:

differential conservation of momentum

$$\boxed{\frac{\partial\mathbf{G}}{\partial t} + \boldsymbol{\nabla}\cdot\mathbf{T} = 0, \quad \text{or} \quad \frac{\partial G_j}{\partial t} + T_{jk;k} = 0.}$$

(1.36)

(Because **T** is symmetric, it does not matter which of its slots the divergence acts on.) This is *the differential law of momentum conservation*. It has the standard form for any local conservation law: the time derivative of the density of some quantity (here momentum), plus the divergence of the flux of that quantity (here the momentum flux is the stress tensor), is zero. We shall make extensive use of this Newtonian law of momentum conservation in Part IV (elasticity), Part V (fluid dynamics), and Part VI (plasma physics).

EXERCISES

Exercise 1.13   **Example: Equations of Motion for a Perfect Fluid*
(a) Consider a perfect fluid with density ρ, pressure P, and velocity $\mathbf{v}$ that vary in time and space. Explain why the fluid's momentum density is $\mathbf{G} = \rho\mathbf{v}$, and explain why its momentum flux (stress tensor) is

$$\boxed{\mathbf{T} = P\mathbf{g} + \rho\mathbf{v}\otimes\mathbf{v}, \quad \text{or, in slot-naming index notation,} \quad T_{ij} = Pg_{ij} + \rho v_i v_j.}$$

(1.37a)

(b) Explain why the law of mass conservation for this fluid is

$$\frac{\partial\rho}{\partial t} + \boldsymbol{\nabla}\cdot(\rho\mathbf{v}) = 0.$$

(1.37b)

(c) Explain why the derivative operator

$$\frac{d}{dt} \equiv \frac{\partial}{\partial t} + \mathbf{v}\cdot\boldsymbol{\nabla}$$

(1.37c)

describes the rate of change as measured by somebody who moves locally with the fluid (i.e., with velocity $\mathbf{v}$). This is sometimes called the fluid's *advective time derivative* or *convective time derivative* or *material derivative*.

(d) Show that the fluid's law of mass conservation (1.37b) can be rewritten as

$$\frac{1}{\rho}\frac{d\rho}{dt} = -\boldsymbol{\nabla}\cdot\mathbf{v}, \qquad (1.37d)$$

which says that the divergence of the fluid's velocity field is minus the fractional rate of change of its density, as measured in the fluid's local rest frame.

(e) Show that the differential law of momentum conservation (1.36) for the fluid can be written as

$$\frac{d\mathbf{v}}{dt} = -\frac{\boldsymbol{\nabla}P}{\rho}. \qquad (1.37e)$$

This is called the fluid's *Euler equation*. Explain why this Euler equation is Newton's second law of motion, $\mathbf{F} = m\mathbf{a}$, written on a per unit mass basis.

In Part V of this book, we use Eqs. (1.37) to study the dynamical behaviors of fluids. For many applications, the Euler equation will need to be augmented by the force per unit mass exerted by the fluid's internal viscosity.

Exercise 1.14 **Problem: Electromagnetic Stress Tensor*
(a) An electric field $\mathbf{E}$ exerts (in SI units) a pressure $\epsilon_o\mathbf{E}^2/2$ orthogonal to itself and a tension of this same magnitude along itself. Similarly, a magnetic field $\mathbf{B}$ exerts a pressure $\mathbf{B}^2/2\mu_o = \epsilon_o c^2\mathbf{B}^2/2$ orthogonal to itself and a tension of this same magnitude along itself. Verify that the following stress tensor embodies these stresses:

$$\boxed{\mathbf{T} = \frac{\epsilon_o}{2}\left[(\mathbf{E}^2 + c^2\mathbf{B}^2)\mathbf{g} - 2(\mathbf{E}\otimes\mathbf{E} + c^2\mathbf{B}\otimes\mathbf{B})\right].} \qquad (1.38)$$

(b) Consider an electromagnetic field interacting with a material that has a charge density ρ_e and a current density $\mathbf{j}$. Compute the divergence of the electromagnetic stress tensor (1.38) and evaluate the derivatives using Maxwell's equations. Show that the result is the negative of the force density that the electromagnetic field exerts on the material. Use momentum conservation to explain why this has to be so.

1.10 Geometrized Units and Relativistic Particles for Newtonian Readers

Readers who are skipping the relativistic parts of this book will need to know two important pieces of relativity: (i) geometrized units and (ii) the relativistic energy and momentum of a moving particle.

1.10.1 Geometrized Units

The speed of light is independent of one's reference frame (i.e., independent of how fast one moves). This is a fundamental tenet of special relativity, and in the era before 1983, when the meter and the second were defined independently, it was tested and confirmed experimentally with very high precision. By 1983, this constancy had become so universally accepted that it was used to redefine the meter (which is hard to measure precisely) in terms of the second (which is much easier to measure with modern technology).[11] The meter is now related to the second in such a way that the speed of light is precisely $c = 299{,}792{,}458$ m s^{-1} (i.e., 1 meter is the distance traveled by light in $1/299{,}792{,}458$ seconds). Because of this constancy of the light speed, it is permissible when studying special relativity to set c to unity. Doing so is equivalent to the relationship

$$c = 2.99792458 \times 10^8 \text{ m s}^{-1} = 1 \qquad (1.39a)$$

between seconds and centimeters; i.e., equivalent to

$$1\,\text{s} = 2.99792458 \times 10^8 \text{ m}. \qquad (1.39b)$$

geometrized units

We refer to units in which $c = 1$ as *geometrized units,* and we adopt them throughout this book when dealing with relativistic physics, since they make equations look much simpler. Occasionally it will be useful to restore the factors of c to an equation, thereby converting it to ordinary (SI or cgs) units. This restoration is achieved easily using dimensional considerations. For example, the equivalence of mass m and relativistic energy $\mathcal{E}$ is written in geometrized units as $\mathcal{E} = m$. In SI units $\mathcal{E}$ has dimensions of joule = kg m^2 s^{-2}, while m has dimensions of kg, so to make $\mathcal{E} = m$ dimensionally correct we must multiply the right side by a power of c that has dimensions m^2 s^{-2} (i.e., by c^2); thereby we obtain $\mathcal{E} = mc^2$.

1.10.2

1.10.2 Energy and Momentum of a Moving Particle

A particle with rest mass m, moving with velocity $\mathbf{v} = d\mathbf{x}/dt$ and speed $v = |\mathbf{v}|$, has a

relativistic energy and momentum

relativistic energy $\mathcal{E}$ (including its rest mass), relativistic kinetic energy E (excluding its rest mass), and relativistic momentum $\mathbf{p}$ given by

$$\boxed{\mathcal{E} = \frac{m}{\sqrt{1 - v^2}} \equiv \frac{m}{\sqrt{1 - v^2/c^2}} \equiv E + m,} \qquad \boxed{\mathbf{p} = \mathcal{E}\mathbf{v} = \frac{m\mathbf{v}}{\sqrt{1 - v^2}};}$$

$$\qquad (1.40)$$

so $\boxed{\mathcal{E} = \sqrt{m^2 + \mathbf{p}^2}.}$

In the low-velocity (Newtonian) limit, the energy E with rest mass removed (kinetic energy) and the momentum $\mathbf{p}$ take their familiar Newtonian forms:

$$\text{When } v \ll c \equiv 1, \quad E \to \frac{1}{2}mv^2 \text{ and } \mathbf{p} \to m\mathbf{v}. \qquad (1.41)$$

---

11. The second is defined as the duration of 9,192,631,770 periods of the radiation produced by a certain hyperfine transition in the ground state of a ^{133}Cs atom that is at rest in empty space. Today (2016) all fundamental physical units except mass units (e.g., the kilogram) are defined similarly in terms of fundamental constants of Nature.

A particle with zero rest mass (a photon or a graviton)[12] always moves with the speed of light $v = c = 1$, and like other particles it has momentum $\mathbf{p} = \mathcal{E}\mathbf{v}$, so the magnitude of its momentum is equal to its energy: $|\mathbf{p}| = \mathcal{E}v = \mathcal{E}c = \mathcal{E}$.

When particles interact (e.g., in chemical reactions, nuclear reactions, and elementary-particle collisions) the sum of the particle energies $\mathcal{E}$ is conserved, as is the sum of the particle momenta $\mathbf{p}$.

For further details and explanations, see Chap. 2.

Exercise 1.15 *Practice: Geometrized Units*

Convert the following equations from the geometrized units in which they are written to SI units.

(a) The "Planck time" t_P expressed in terms of Newton's gravitation constant G and Planck's reduced constant $\hbar$, $t_P = \sqrt{G\hbar}$. What is the numerical value of t_P in seconds? in meters?

(b) The energy $\mathcal{E} = 2m$ obtained from the annihilation of an electron and a positron, each with rest mass m.

(c) The Lorentz force law $m d\mathbf{v}/dt = e(\mathbf{E} + \mathbf{v} \times \mathbf{B})$.

(d) The expression $\mathbf{p} = \hbar\omega\mathbf{n}$ for the momentum $\mathbf{p}$ of a photon in terms of its angular frequency ω and direction $\mathbf{n}$ of propagation.

How tall are you, in seconds? How old are you, in meters?

Bibliographic Note

Most of the concepts developed in this chapter are treated, though from rather different viewpoints, in intermediate and advanced textbooks on classical mechanics or electrodynamics, such as Marion and Thornton (1995); Jackson (1999); Griffiths (1999); Goldstein, Poole, and Safko (2002).

Landau and Lifshitz's (1976) advanced text *Mechanics* is famous for its concise and precise formulations; it lays heavy emphasis on symmetry principles and their implications. A similar approach is followed in the next volume in their Course of Theoretical Physics series, *The Classical Theory of Fields* (Landau and Lifshitz, 1975), which is rooted in special relativity and goes on to cover general relativity. We refer to other volumes in this remarkable series in subsequent chapters.

The three-volume *Feynman Lectures on Physics* (Feynman, Leighton, and Sands, 2013) had a big influence on several generations of physicists, and even more so on their teachers. Both of us (Blandford and Thorne) are immensely indebted to Richard Feynman for shaping our own approaches to physics. His insights on the foundations

---

12. We do not know for sure that photons and gravitons are massless, but the laws of physics as currently understood require them to be massless, and there are tight experimental limits on their rest masses.

of classical physics and its relationship to quantum mechanics, and on calculational techniques, are as relevant today as in 1963, when his course was first delivered.

The geometric viewpoint on the laws of physics, which we present and advocate in this chapter, is not common (but it should be because of its great power). For example, the vast majority of mechanics and electrodynamics textbooks, including all those listed above, define a tensor as a matrix-like entity whose components transform under rotations in the manner described by Eq. (1.13b). This is a complicated definition that hides the great simplicity of a tensor as nothing more than a linear function of vectors; it obscures thinking about tensors geometrically, without the aid of any coordinate system or basis.

The geometric viewpoint comes to the physics community from mathematicians, largely by way of relativity theory. By now, most relativity textbooks espouse it. See the Bibliographic Note to Chap. 2. Fortunately, this viewpoint is gradually seeping into the nonrelativistic physics curriculum (e.g., Kleppner and Kolenkow, 2013). We hope this chapter will accelerate that seepage.

REFERENCES

Abernathy, F. (1968). National Committee for Fluid Mechanics Films movie: Fundamentals of boundary layers.

Ablowitz, M. J. (2011). *Nonlinear Dispersive Waves: Asymptotic Analysis and Solitons*. Cambridge: Cambridge University Press.

Acheson, D. J. (1990). *Elementary Fluid Dynamics*. Oxford: Clarendon Press.

Adair, R. K. (1990). *The Physics of Baseball*. New York: Harper and Row.

Alligood, K. T., T. D. Sauer, and J. A. Yorke (1996). *Chaos, an Introduction to Dynamical Systems*. Berlin: Springer-Verlag.

Anderson, J. D. (2003). *Modern Compressible Flow: With Historical Perspective*. New York: McGraw-Hill.

Arfken, G. B., H. J. Weber, and F. E. Harris (2013). *Mathematical Methods for Physicists*. Amsterdam: Elsevier.

Armenti, A. J. (1992). *The Physics of Sports*. New York: American Institute of Physics.

Ashby, N., and J. Dreitlein (1975). Gravitational wave reception by a sphere. *Physical Review D* **12**, 336–349.

Bachman, H. E. (1994). *Vibration Problems in Structures*. Basel: Birkhauser.

Baker, G. L., and J. P. Gollub (1990). *Chaotic Dynamics, An Introduction*. Cambridge: Cambridge University Press.

Batchelor, G. K. (2000). *An Introduction to Fluid Dynamics*. Cambridge: Cambridge University Press.

Båth, M. (1966). Earthquake energy and magnitude. In L. H. Ahrens, F. Press, S. K. Runcorn, and H. C. Urey (eds.), *Physics and Chemistry of the Earth*, pp. 115–165. Oxford: Pergamon.

Bejan, A. (2013). *Convection Heat Transfer*. New York: Wiley.

Blandford R. D., and M. J. Rees (1974). A "twin-exhaust" for double radio sources. *Monthly Notices of the Royal Astronomical Society* **169**, 395–415.

Bondi, H. (1952). On spherically symmetric accretion. *Monthly Notices of the Royal Astronomical Society* **112**, 195–204.

Boresi, A. P., and K. P. Chong (1999). *Elasticity in Engineering Mechanics*. New York: Wiley.

Braginsky, V. B., M. L. Gorodetsky, and S. P. Vyatchanin (1999). Thermodynamical fluctuations and photo-thermal shot noise in gravitational wave antennae. *Physics Letters A* **264**, 1–10.

Brenner, M. P., S. Hilgenfeldt, and D. Lohse (2002). Single bubble sonoluminescence. *Reviews of Modern Physics* **74**, 425–484.

Bryson, A. (1964). National Committee for Fluid Mechanics Films movie: Waves in fluids.

Burke, W. L. (1970). Runaway solutions: Remarks on the asymptotic theory of radiation damping. *Physical Review A* **2**, 1501–1505.

Burke, W. L. (1971). Gravitational radiation damping of slowly moving systems calculated using matched asymptotic expansions. *Journal of Mathematical Physics* **12**, 402–418.

Canuto, C., Hussaini, M. Y., Quarteroni, A., and Zhang, T. A. (2014). *Spectral Methods: Evolution to Complex Geometries and Applications to Fluid Dynamics*. Berlin: Springer

Chandrasekhar, S. (1939). *Stellar Structure*. Chicago: University of Chicago Press.

Chandrasekhar, S. (1961). *Hydrodynamics and Hydromagnetic Stability*. Oxford: Oxford University Press.

Chandrasekhar, S. (1962). *Ellipsoidal Figures of Equilibrium*. New Haven, Conn.: Yale University Press.

Chelton, D., and M. Schlax (1996). Global observations of oceanic Rossby waves. *Science* **272**, 234–238.

Cohen-Tannoudji, C., B. Diu, and F. Laloë (1977). *Quantum Mechanics*. New York: Wiley.

Cole, J. (1974). *Perturbation Methods in Applied Mathematics*. New York: Blaisdell Publishing.

Coles, D. (1965). National Committee for Fluid Mechanics Films movie: Channel flow of a compressible fluid.

Constantinescu, A., and A. Korsunsky (2007). *Elasticity with Mathematica*. Cambridge: Cambridge University Press.

Cushman-Roisin, B., and J.-M. Beckers (2011). *Introduction to Geophysical Fluid Dynamics*. New York: Academic Press.

Dauxois, T., and M. Peyrard (2010). *Physics of Solitons*. Cambridge: Cambridge University Press.

Davidson, P. A. (2005). *Turbulence: An Introduction for Scientists and Engineers*. Oxford: Oxford University Press.

Davison, L. (2010). *Fundamentals of Shock Wave Propagation in Solids*. Berlin: Springer-Verlag.

Drazin, P. G., and R. S. Johnson (1989). *Solitons: An Introduction*. Cambridge: Cambridge University Press.

Drazin, P. G., and W. H. Reid (2004). *Hydrodynamic Stability*. Cambridge: Cambridge University Press.

Eddington, A. S. (1927). March 1927 Gifford lecture at the University of Edinburgh. As published in Arthur S. Eddington: *The Nature of the Physical World* Gifford Lectures of 1927: An Annotated Edition by H. G. Callaway. Cambridge: Cambridge Scholars Publishing (2014). Chapter 16 epigraph reprinted with permission of the publisher.

Eringen, A. C., and E. S. Suhubi (1975). *Elastodynamics, Vol. II: Linear Theory*. New York: Academic Press.

Faber, T. E. (1995). *Fluid Dynamics for Physicists*. Cambridge: Cambridge University Press.

Feigenbaum, M. (1978). Universal behavior in nonlinear systems. *Journal of Statistical Physics* **19**, 25–52.

Fenstermacher, P. R., H. L. Swinney, and J. P. Gollub (1979). Dynamical instabilities and the transition to chaotic Taylor vortex flow. *Journal of Fluid Mechanics* **94**, 103–128.

Ferziger, J. H. and Peric, M. (2001). *Computational Methods for Fluid Dynamics*. Berlin: Springer.

Feynman, R. P. (1972). *Statistical Mechanics*. New York: Benjamin.

Feynman, R. P., R. B. Leighton, and M. Sands (1964). *The Feynman Lectures on Physics*. Reading, Mass.: Addison-Wesley. Chapter 14 epigraph reprinted with permission of Caltech.

Fletcher, C. A. J. (1991). *Computational Techniques for Fluid Dynamics, Vol I: Fundamental and General Techniques*. Berlin: Springer-Verlag.

Fortere, Y., J. M. Skothelm, J. Dumals, and L. Mahadevan (2005). How the Venus flytrap snaps. *Nature* **433**, 421–425.

Fultz, D. (1969). National Committee for Fluid Mechanics Films movie: Rotating flows.

Gill, A. E. (1982). *Atmosphere-Ocean Dynamics*. New York: Academic Press.

Gladwell, G. M. L. (1980). *Contact Problems in the Classical Theory of Elasticity*. Alphen aan den Rijn: Sijthoff and Noordhoff.

Goldstein, H., C. Poole, and J. Safko (2002). *Classical Mechanics*. New York: Addison-Wesley.

Gollub, J. P., and S. V. Benson (1980). Many routes to turbulent convection. *Journal of Fluid Mechanics* **100**, 449–470.

Goodstein, D. L., and J. R. Goodstein (1996). *Feynman's Lost Lecture: The Motion of Planets around the Sun*. New York: W. W. Norton.

Gorman, M., and H. L. Swinney (1982). Spatial and temporal characteristics of modulated waves in the circular Couette system. *Journal of Fluid Mechanics* **117**, 123–142.

Greenspan, H. P. (1973). *The Theory of Rotating Fluids*. Cambridge: Cambridge University Press.

Griffiths, D. J. (1999). *Introduction to Electrodynamics*. Upper Saddle River, N.J.: Prentice-Hall.

Grossman, S. (2000). The onset of shear flow turbulence. *Reviews of Modern Physics* **72**, 603–618.

Gutzwiller, M. C. (1990). *Chaos in Classical and Quantum Mechanics*. New York: Springer Verlag.

Heisenberg, W. (1969). Significance of Sommerfeld's work today. In Bopp F., and H. Kleinpoppen (eds.), *Physics of the One and Two Electron Atoms*, p. 1. Amsterdam: North Holland. Chapter 18 epigraph reprinted with permission of the publisher.

Hooke, R. (1678). Answer to the anagram "ceiiinossssttuv," which he had previously published, to establish his priority on the linear law of elasticity. De Potentia, or of spring explaining the power of springing bodies, Hooke's Sixth Cutler Lecture, R. T. Gunther facsimile reprint. In *Early Science in Oxford*. Vol. 8. London: Dawsons of Pall Mall (1968).

Hosking, R. J. and Dewar, R. L. (2016). *Fundamental Fluid Mechanics and Magnetohydrodynamics*. Singapore: Springer.

Jackson, J. D. (1999). *Classical Electrodynamics*. New York: Wiley.

Johnson, K. L. (1985). *Contact Mechanics*. Cambridge: Cambridge University Press.

Johnson, L. R. (1974). Green's function for Lamb's problem. *Geophysical Journal of the Royal Astronomical Society* **37**, 99–131.

Kapner, D. J., T. S. Cook, E. G. Adelberger, J. H. Gundlach, et al. (2008). Tests of the gravitational inverse-square law below the dark-energy length scale. *Physical Review Letters* **98**, 021101.

Kausel, E. (2006). *Fundamental Solutions in Elastodynamics*. Cambridge: Cambridge University Press.

Kittel, C., and H. Kroemer (1980). *Thermal Physics*. London: Macmillan.

Kleppner, D., and R. K. Kolenkow (2013). *An Introduction to Mechanics*. Cambridge: Cambridge University Press.

Kolsky, H. (1963). *Stress Waves in Solids*. Mineola, N.Y.: Courier Dover Publications.

Kundu, P. K., I. M. Cohen, and D. R. Dowling (2012). *Fluid Mechanics*. New York: Academic Press.

Lagerstrom, P. (1988). *Matched Asymptotic Expansions: Ideas and Techniques*. Berlin: Springer-Verlag.

Lamb, H. (1882). On the vibrations of an elastic sphere. *Proceedings of the London Mathematical Society* **13**, 189–212.

Landau, L. D. (1944). On the problem of turbulence. *Doklady Akademii Nauk SSSR* **44**, 311–314.

Landau, L. D. and E. M. Lifshitz (1959). *Fluid Mechanics*. Oxford: Pergamon.

Landau, L. D. and E. M. Lifshitz (1975). *The Classical Theory of Fields*, fourth English edition. Oxford: Butterworth-Heinemann.

Landau, L. D. and E. M. Lifshitz (1976). *Mechanics*. Oxford: Butterworth-Heinemann.

Landau, L. D. and E. M. Lifshitz (1986). *Elasticity*. Oxford: Pergamon.

Lautrup, B. (2005). *Physics of Continuous Matter*. Bristol and Philadelphia: Institute of Physics Publishing.

Levin, Y. (1998). Internal thermal noise in the LIGO test masses: A direct approach. *Physical Review D* **57**, 659–663.

Libbrecht, K. G., and M. F. Woodard (1991). Advances in helioseismology. *Science* **253**, 152–157.

Libchaber, A., C. Laroche, and S. Fauve (1982). Period doubling cascade in mercury, a quantitative measurement. *Journal de Physique—Lettres* **43**, L211–L216.

Liepmann, H., and A. Roshko (2002). *Compressible Gas Dynamics*. Mineola, N.Y.: Courier Dover Publications.

Lighthill, M. J. (1952). On sound generated aerodynamically. I. General theory. *Proceedings of the Royal Society A* **211**, 564–587.

Lighthill, M. J. (1954). On sound generated aerodynamically. II. Turbulence as a source of sound. *Proceedings of the Royal Society A* **222**, 1–32.

Lighthill, M. J. (1986). *An Informal Introduction to Theoretical Fluid Mechanics*. Oxford: Oxford University Press.

Lighthill, M. J. (2001). *Waves in Fluids*. Cambridge: Cambridge University Press.

LIGO Scientific Collaboration (2015). Advanced LIGO. *Classical and Quantum Gravity* **32**, 074001.

Liu, Y. T., and K. S. Thorne (2000). Thermoelastic noise and thermal noise in finite-sized gravitational-wave test masses. *Physical Review D* **62**, 122002–122011.

Lorenz, E. N. (1963). Deterministic nonperiodic flow. *Journal of Atmospheric Sciences* **20**, 130–141.

Love, A. E. H. (1927). *A Treatise on the Mathematical Theory of Elasticity*. Mineola, N.Y.: Courier Dover Publications.

Majda, A. J., and A. L. Bertozzi (2002). *Vorticity and Incompressible Flow*. Cambridge: Cambridge University Press.

Marion, J. B., and S. T. Thornton (1995). *Classical Dynamics of Particles and Systems*. Philadelphia: Saunders College Publishing.

Marko, J. F., and S. Cocco (2003). The micro mechanics of DNA. *Physics World* **16**, 37–41.

Maroto, J. A., V. Perez-Munuzuri, and M. S. Romero-Cano (2007). Introductory analysis of Benard-Marangoni convection. *European Journal of Physics* **28**, 311–320.

Marsden, J. E., and T. J. Hughes (1986). *Mathematical Foundations of Elasticity*. Upper Saddle River, N.J.: Prentice-Hall.

Mathews, J., and R. L. Walker (1970). *Mathematical Methods of Physics*. New York: Benjamin.

Miles, J. (1993). Surface-wave generation revisited. *Journal of Fluid Mechanics* **256**, 427–441.

Misner, C. W., K. S. Thorne, and J. A. Wheeler (1973). *Gravitation*. San Francisco: Freeman.

Munson, B. R., D. F. Young, and T. H. Okiishi (2006). *Fundamentals of Fluid Mechanics*. New York: Wiley.

NIST (2005). *Final Report on the Collapse of the World Trade Center Towers*. National Institute of Standards and Technology Report Number NIST NCSTAR 1. Washington, D.C.: U.S. Government Printing Office.

NIST (2008). *Final Report on the Collapse of the World Trade Center Building 7*. National Institute of Standards and Technology Report Number NIST NCSTAR 1A. Washington, D.C.: U.S. Government Printing Office.

Nelson, P. (2008). *Biological Physics*. San Francisco: Freeman.

Newton, I. (1687). *Philosophiae Naturalis Principia Mathematica*. London: Royal Society. English translation by I. B. Cohen and A. Whitman. Berkeley: University of California Press (1999).

Ott, E. (1982). Strange attractors and chaotic motions of dynamical systems. *Reviews of Modern Physics* **53**, 655–671.

Ott, E. (1993). *Chaos in Dynamical Systems*. Cambridge: Cambridge University Press.

Panton, R. L. (2005). *Incompressible Flow*. New York: Wiley.

Pedlosky, J. (1987). *Geophysical Fluid Dynamics*. Berlin: Springer-Verlag.

Pellew, A., and R. V. Southwell (1940). On maintained convective motion in a fluid heated from below. *Proceedings of the Royal Society A* **176**, 312–343.

Phillips, O. M. (1957). On the generation of waves by turbulent wind. *Journal of Fluid Mechanics* **2**, 417–445.

Pop, I., and D. B. Ingham (2001). *Convective Heat Transfer: Mathematical Computational Modelling of Viscous Fluids and Porous Media*. Amsterdam: Elsevier.

Pope, S. B. (2000). *Turbulent Flows*. Cambridge: Cambridge University Press.

Poruchikov, V. B., V. A. Khokhryakov, and G. P. Groshev (1993). *Methods of the Classical Theory of Elastodynamics*. Berlin: Springer-Verlag.

Potter, M. C., D. C. Wiggert, and B. H. Ramadan (2012). *Mechanics of Fluids*. Stamford, Conn.: Cengage Learning.

Purcell, E. M. (1983). The back of the envelope. *American Journal of Physics* **51**, 205.

Rezzolla, L., and O. Zanotti (2013). *Relativistic Hydrodynamics*. Oxford: Oxford University Press.

Richardson, L. (1922). *Weather Prediction by Numerical Process*. Cambridge: Cambridge at the University Press.

Richter, C. F. (1980). Interview with Henry Spall. *Earthquake Information Bulletin*, January–February. Chapter 12 epigraph reprinted with permission of the publisher.

Ride, S. (2012). Interview with Jim Clash. Available at http://www.askmen.com/entertainment/right-stuff/sally-ride-interview.html. Reprinted by permission of Jim Clash.

Rohrlich, F. (1965). *Classical Charged Particles*. New York: Addison-Wesley.

Rouse, H. (1963a). University of Iowa movie: Introduction to the study of fluid motion. Available at
http://www.iihr.uiowa.edu/research/publications-and-media/films-by-hunter-rouse/.

Rouse, H. (1963b). University of Iowa movie: Fundamental principles of flow. Available at http://www.iihr.uiowa.edu/research/publications-and-media/films-by-hunter-rouse/.

Rouse, H. (1963c). University of Iowa movie: Fluid motion in a gravitational field. Available at http://www.iihr.uiowa.edu/research/publications-and-media/films-by-hunter-rouse/.

Rouse, H. (1963d). University of Iowa movie: Characteristics of laminar and turbulent flow. Available at
http://www.iihr.uiowa.edu/research/publications-and-media/films-by-hunter-rouse/.

Rouse, H. (1963e). University of Iowa movie: Form, drag, lift, and propulsion. Available at http://www.iihr.uiowa.edu/research/publications-and-media/films-by-hunter-rouse/.

Rouse, H. (1963f). University of Iowa movie: Effects of fluid compressibility. Available at http://www.iihr.uiowa.edu/research/publications-and-media/films-by-hunter-rouse/.

Ruelle, D. (1989). *Chaotic Evolution and Strange Attractors*. Cambridge: Cambridge University Press.

Sagdeev, R. Z., D. A. Usikov, and G. M. Zaslovsky (1988). *Non-linear Physics from the Pendulum to Turbulence and Chaos*. Newark, N.J.: Harwood Academic Publishers.

Scott-Russell, J. (1844). Report on waves. *British Association for the Advancement of Science* **14**, 311–390, Plates XLVII–LVII.

Sedov, L. I. (1946). Propagation of strong blast waves. *Prikhladnaya Matematika i Mekhanika* **10**, 241–250.

Sedov, L. I. (1957). Russian Language Fourth Edtion of Sedov (1993): *Metody podobiya i razmernosti v mekhanike*. Moskva: Gostekhizdat.

Sedov, L. I. (1993). *Similarity and Dimensional Methods in Mechanics*. Boca Raton, Fla.: CRC Press.

Sethna, J. P. (2006). *Statistical Mechanics: Entropy, Order Parameters, and Complexity*. Oxford: Oxford University Press.

Shapiro, A. (1961a). National Committee for Fluid Mechanics Films. Available at web.mit.edu/hml/ncfmf.html.

Shapiro, A. (1961b). National Committee for Fluid Mechanics Films movie: Vorticity.

Shapiro, S. L., and S. A. Teukolsky (1983). *Black Holes, White Dwarfs and Neutron Stars: The Physics of Compact Objects*. New York: Wiley.

Shearer, P. M. (2009). *Introduction to Seismology*. Cambridge: Cambridge University Press.

Slaughter, W. S. (2002). *The Linearized Theory of Elasticity*. Boston: Birkhäuser.

Southwell, R. V. (1941). *An Introduction to the Theory of Elasticity for Engineers and Physicists*. Oxford: Clarendon Press.

Stacey, F. D. (1977). *Physics of the Earth*. New York: Wiley.

Stanyukovich, K. P. (1960). *Unsteady Motion of Continuous Media*. Oxford: Pergamon.

Stein, S., and M. Wysession (2003). *An Introduction to Seismology, Earthquakes and Earth Structure*. Oxford: Blackwell.

Stewart, R. W. (1968). National Committee for Fluid Mechanics Films movie: Turbulence.

Strogatz, S. H. (2008). *Nonlinear Dynamics and Chaos: With Applications to Physics, Biology, Chemistry and Engineering*. Boulder: Westview Press.

Tanimoto, T., and J. Um (1999). Cause of continuous oscillations of the earth. *Journal of Geophysical Research* **104**, 28723–28739.

Taylor, E. (1968). National Committee for Fluid Mechanics Films movie: Secondary flow.

Taylor, G. (1950). The formation of a blast wave by a very intense explosion. II. The atomic explosion of 1945. *Proceedings of the Royal Society A* **201**, 175–186.

Taylor, G. (1964). National Committee for Fluid Mechanics Films movie: Low Reynolds number flows.

Tennekes, H., and J. L. Lumley (1972). *A First Course on Turbulence*. Cambridge, Mass.: MIT Press.

Thompson, P. A. (1984). *Compressible Fluid Dynamics*. Boulder: Maple Press.

Thorne, K. S. (1973). Relativistic shocks: The Taub adiabat. *Astrophysical Journal* **179**, 897–907.

Thorne, K. S. (1980). Multipole expansions of gravitational radiation. *Reviews of Modern Physics* **52**, 299–340.

Timoshenko, S., and J. N. Goodier (1970). *Theory of Elasticity*. New York: McGraw-Hill.

Todhunter, I., and K. Pearson (1886). *A History of the Theory of Elasticity and of the Strength of Materials, from Galilei to the Present Time*. Cambridge: Cambridge University Press.

Toro, E. F. (2010). *Riemann Solvers and Numerical Methods for Fluid Dynamics: A Practical Introduction*. Berlin: Springer-Verlag.

Townsend, A. A. (1949). The fully developed turbulent wake of a circular cylinder. *Australian Journal of Scientific Research* **2**, 451–468.

Tritton, D. J. (1987). *Physical Fluid Dynamics*. Oxford: Oxford University Press.

Turco, R. P., O. B. Toon, T. B. Ackerman, J. B. Pollack, and C. Sagan (1986). Nuclear winter: Global consequences of multiple nuclear explosions. *Science* **222**, 1283–1292.

Turcotte, D. L., and G. Schubert (1982). *Geodynamics*. New York: Wiley.

Turner, J. S. (1973). *Buoyancy Effects in Fluids*. Cambridge: Cambridge University Press.

Ugural, A. C., and S. K. Fenster (2012). *Advanced Mechanics of Materials and Applied Elasticity*. Upper Saddle River, N.J.: Prentice-Hall.

Vallis, G. K. (2006). *Atmospheric and Oceanic Fluid Dynamics*. Cambridge: Cambridge University Press.

Van Dyke, M. (1982). *An Album of Fluid Flow*. Stanford, Calif.: Parabolic Press.

Verhulst, P. F. (1838). Notice sur la loi que la population poursuit dans son accroissement. *Correspondance Mathématique et Physique* **10**, 113–121.

Vogel, S. (1994). *Life In Moving Fluids: The Physical Biology Of Flow,* 2nd Edition, Revised and Expanded by Steven Vogel, Illustrated by Susan Tanner Beety and the Author.

Wagner, T., S. Schlamminger, J. Gundlach, and E. Adelberger (2012). Torsion-balance tests of the weak equivalence principle. *Classical and Quantum Gravity* **29**, 1–15.

Weld, D. M., J. Xia, B. Cabrera, and A. Kapitulnik (2008). A new apparatus for detecting micron-scale deviations from Newtonian gravity. *Physical Review D* **77**, 062006.

White, F. M. (2006). *Viscous Fluid Flow*. New York: McGraw-Hill.

White, F. M. (2008). *Fluid Mechanics*. New York: McGraw-Hill.

Whitham, G. B. (1974). *Linear and Non-linear Waves*. New York: Wiley.

Wolgemuth, C. W., T. R. Powers, and R. E. Goldstein (2000). Twirling and whirling: Viscous dynamics of rotating elastic filaments. *Physical Review Letters* **84**, 1623–1626.

Xing, X., P. M. Goldbart, and L. Radzihovsky (2007). Thermal fluctuations and rubber elasticity. *Physical Review Letters* **98**, 075502.

Yeganeh-Haeri, A., D. J. Weidner, and J. B. Parise (1992). Elasticity of α-Cristobalite: A silicon dioxide with a negative Poisson's ratio. *Science* **257**, 650.

Zel'dovich, Ya. B., and Yu. P. Raizer (2002). *Physics of Shock Waves and High Temperature Hydrodynamic Phenomena*. Mineola, N.Y.: Courier Dover Publications.

NAME INDEX

Page numbers for entries in boxes are followed by "b," those for epigraphs at the beginning of a chapter by "e," those for figures by "f," and those for notes by "n."

SUBJECT INDEX

Second and third level entries are not ordered alphabetically. Instead, the most important or general entries come first, followed by less important or less general ones, with specific applications last.

Page numbers for entries in boxes are followed by "b," those for epigraphs at the beginning of a chapter by "e," those for figures by "f," for notes by "n," and for tables by "t."

nonlinear sound wave, steepening to form shock, 894, 894f

in shock tube. *See* shock tube, fluid flow in

transonic, quasi-1-dimensional, steady flow, 884f, 880–891

    equations in a stream tube, 880–882

    properties of, 882–883

    relativistic, 890

conductivity, thermal, κ

    energy flux for, 714

connection coefficients

    for orthonormal bases in Euclidean space, 615

        pictorial evaluation of, 616f

        used to compute components of gradient, 617

        for cylindrical orthonormal basis, 615

        for spherical orthonormal basis, 616

conservation laws

    differential and integral, 966

continental drift, 932b

contraction of tensors

    formal definition, 950–951

    in slot-naming index notation, 957

    component representation, 955

convection

    onset of convection and of convective turbulence, 830–831, 931

    Boussinesq approximation for, 924–925

    between two horizontal plates at different temperatures: Rayleigh-Bénard convection, 925–933

        Boussinesq-approximation analysis, 925–928, 930

        critical Rayleigh number for onset, 930, 930f, 933

        pattern of convection cells, 930–931, 931f

        toy model, 929

    in a room, 931

    in Earth's mantle, 932

    in a star, 933–937

    in the solar convection zone, 936

coordinate independence. *See* geometric principle

Coriolis acceleration, 735, 767–768

    as restoring force for Rossby waves, 858

critical point of transonic fluid flow, 883, 886, 891

Crocco's theorem, 702, 742

cross product, 963–964

curl, 963–964

curve, 947

cylindrical coordinates

    related to Cartesian coordinates, 614

    orthonormal basis and connection coefficients for, 614–615

        expansion and shear tensor in, 617, 618

d'Alembert's paradox, for potential flow around a cylinder, 765

dam, water flow after breaking, 857–858, 897

De Laval nozzle, 887

decibel, 865

derivatives of scalars, vectors, and tensors

    directional derivatives, 960–961

    gradients, 617, 961

    Lie derivative, 735n

diffusion coefficient

    for temperature, in thermally conducting fluid, 920

    for vorticity, in viscous fluid, 741

diffusion equation

    for temperature in homogenous medium, 920

    for vorticity, in viscous fluid, 741

dimensional analysis for functional form of a fluid flow, 790–791

dimensional reduction in elasticity theory, 590b

    for bent beam, 592–595

    for bent plate, 609–613

directional derivative, 960–961

displacement vector, in elasticity, 570

    gradient of, decomposed into expansion, shear, and rotation, 570–571

    Navier-Cauchy equation for, 587. *See also* Navier-Cauchy equation for elastostatic equilibrium

dissipation, 724b

divergence, 962

DNA molecule, elastostatics of, 599–600

double diffusion, 937–940

drag force and drag coefficient, 792

    at low Reynolds number (Stokes flow), 753–754

    influence of turbulence on, 792f, 794, 820–821

    on a flat plate, 763–764

    on a cylinder, 792–794, 792f

    on an airplane wing, 820–821

    on sports balls, 825

    on fish of various shapes, 797–798

Earth. *See also* atmosphere of Earth; elastodynamic waves in Earth

    internal structure of, 651t

    pressure at center of, 649

    Moho discontinuity, 650

    mantle viscosity, 755–756

    mantle convection, 932

    continental drift, 932

    normal modes excited by atmospheric turbulence, 816

eddies

    in flow past a cylinder, 791f, 793–794

    in turbulence, 798–800, 802, 804–807, 811–814

gravity waves on water *(continued)*
   shallow water, 840–843
     dam breaking: water flow after, 857
     nonlinear, 840–841, 843, 850–858, 897
     solitary waves (solitons) and KdV equation, 850–858
   deep water, 840
     viscous damping of, 842
   capillary (with surface tension), 844–848
Green's functions
   for elasticity theory, 590b
   for elastostatic displacement, 626–627
   for elastodynamic waves, 658–661, 660f
greenhouse effect, 748. *See also* climate change
gyre, 773, 775–776, 805

heat conduction, diffusive. *See also* diffusion equation
   in the sun, 937
   in a flowing fluid, 920
Heaviside Green's functions, 658–660, 660f
helioseismology, 848–850
high-Reynolds-number flow, 757–766
   boundary layers in. *See* boundary layers
Hilbert space, 956b
Hooke's law, 568f, 591
   realm of validity and breakdown of, 580, 581f
hydraulic jump, 903–904, 904f
hydrostatic equilibrium
   in uniform gravitational field
     equation of, 681
     theorems about, 682–683
   of nonrotating stars and planets, 686–689
   of rotating stars and planets, 689–691
     barotropic: von Zeipel's theorem, 702
     centrifugal flattening, 690, 691
hydrostatics, 681–691

ideal gas. *See* gas, ideal
impedance
   acoustic, 654
incompressible approximation for fluid dynamics, 709–710, 725b
index gymnastics. *See* component manipulation rules
index of refraction
   for Earth's atmosphere, 814–815
   for seismic waves in Earth, 652
inner product
   in Euclidean space, 948–950, 955
   in quantum theory, 956b
instabilities in fluid flows. *See* fluid-flow instabilities
integrals in Euclidean space
   over 2-surface, 965

   over 3-volume, 965
     Gauss's theorem, 965
intermittency in turbulence, 798–799, 807
internal waves in a stratified fluid, 941
interstellar medium, 891, 914–916
inviscid, 725b
irreducible tensorial parts of second-rank tensor, 572b–574b, 577, 711
irrotational flow (vorticity-free), 701, 725b, 837
isentropic, 725b
isobar, 725b

jets
   laminar, 796–797
   turbulent, 809f, 810
Joukowski's (Kutta-Joukowski's) theorem, 743
Joule-Kelvin cooling, 708, 708f
junction conditions
   elastodynamic, 588–589, 651, 654
   hydraulic jump, 903–904
   shock front. *See* Rankine-Hugoniot relations for a shock wave
Jupiter, 687, 689, 702, 801b

Kármán vortex street, 791f, 794
Keck telescopes, 609–611
Kelvin-Helmholtz instability in shear flow, 778–782
   influence of gravity on, 782–783
   onset of turbulence in, 801b
Kelvin's theorem for circulation, 740, 746, 824
Kepler's laws, 691, 784, 952
Knudsen number, 755n
Kolmogorov spectrum for turbulence, 810–815
   phenomena missed by, 814
   derivation of, 810–812
   for transported quantities, 814–816
Korteweg–de Vries equation and soliton solutions, 850–856, 858
Kutta-Joukowski's theorem, 743

Lagrangian changes, 725b
lagrangian methods for dynamics
   lagrangian density
     energy density and flux in terms of, 642
     for elastodynamic waves, 642
Lamé coefficients, 582
laminar flow, 716–717. *See also under* boundary layers; jets; wakes
laplacian, 665, 962
Levi-Civita tensor in Euclidean space, 962–964
   product of two, 963

perfect fluid (ideal fluid), 675, 675n, 968
    Euler equation for, 697, 971
    stress tensor for, 968–969, 970
phonons for modes of an elastic solid, 667–670
physical laws, geometric formulation of. *See* geometric
        principle
piezoelectric fields, 586
pipe
    stressed, elastostatics of, 619–621
    fluid flow in, 716–717, 766, 787
Pitot tube, 700, 701f
Planck energy, 580
Planck length, 579, 580
Poiseuille flow (confined laminar, viscous flow)
    between two plates, 718–719
    down a pipe, 717, 922
Poiseuille's law, for laminar fluid flow in a pipe, 717
Poisson's equation, 686, 705
Poisson's ratio, 591–592, 586t
polytrope, 687–689
potential flow (irrotational flow), 701
Prandtl number, 920, 921t
pressure, 968
    as component of stress tensor, 968–969
pressure self-adjustment in fluid dynamics, 742
proportionality limit, in elasticity, 580, 581f

radiation reaction, theory of
    slow-motion approximation, 871
    matched asymptotic expansion, 871
    radiation-reaction potential, 871
    damping and energy conservation, 872, 873
    runaway solutions, their origin and invalidity, 872–873
    examples
        electromagnetic waves from accelerated, charged
            particle, 873
        sound waves from oscillating ball, 869–874
radiative processes
    Thomson scattering, 937
rank of tensor, 949
Rankine-Hugoniot relations for a shock wave, 900
    derivation from conservation laws, 898–900
    physical implications of, 900–902
    for polytropic equation of state, 905
    for strong polytropic shock, 905
    relativistic, 902–903
rarefaction wave, 895f, 896
Rayleigh criterion for instability of rotating flows, 784
Rayleigh number, 928
Rayleigh waves at surface of a homogeneous solid, 654–657,
        659, 661, 839–840, 941

Rayleigh-Taylor instability, 783–784
renormalization group
    applied to the onset of chaos in the logistic equation, 831
resistance in Stokes fluid flow, 753
rest frame, local, 677, 719–720
rest mass, 972
Reynolds number, 716, 726b
    as ratio of inertial to viscous acceleration, 746
Reynolds stress for turbulence, 802
    and turbulent viscosity, 804
Richardson criterion for instability of shear flows, 785–786
Richardson number, 785–786
Riemann invariants, 852, 891–897, 901–902
rocket engines, fluid flow through, 887–890
rod. *See* bent beam, elastostatics of
Rossby number, 768
Rossby waves in rotating fluid, 858–861
rotating reference frame, fluid dynamics in, 766–777
rotation, rate of, in fluid mechanics, 711, 726b
    as vorticity in disguise, 711
rotation group, 572b–574b, 959
rotation matrix, 959
rotation tensor and vector, in elasticity theory, 571,
        573b–574b, 575–576, 577
rupture point, in elasticity, 580, 581f

Saint-Venant's principle, for elastostatic equilibrium, 590b
salt fingers due to double diffusion of salt and heat, 937–940
scaling relations in fluid flows
    between similar flows, 791–792
    for drag force on an object, 765
    for Kolmogorov turbulence spectrum, 789, 810–814
Schrödinger equation
    energy eigenstates (modes) of, 848–849
    nonlinear variant of, and solitons, 856–857
Schwarzschild criterion for onset of convection in a star, 935,
        935f
secondary fluid flows, 775–776
sedimentation, 749, 754–755
Sedov-Taylor blast wave, 909–912
seismic waves. *See* elastodynamic waves in Earth
self-gravity, in fluid dynamics, 705b–706b, 709
self-similar flows
    boundary layer near flat plate: Blasius profile, 758–763
    Sedov-Taylor blast wave, 909–912
    underwater blast wave, 914–915
    flow in shock tube, 916
    stellar wind, 915–916
    water flow when dam breaks, 857–858
separation of variables for Navier-Cauchy equation, 590b,
        624–625

superfluid, rotating, 733–734

supernovae

neutron stars produced in, 914

Sedov-Taylor blast wave from, 914–915

surface tension, 844b–845b

force balance at interface between two fluids, 846

swimming mechanisms, 744, 747b–748b, 756–757

tangent space, 947

tangent vector, 947

Taylor rolls, in rotating Couette flow, 826, 826f

Taylor-Proudman theorem for geostrophic flow, 771

tea cup: circulating flow and Ekman boundary layer, 776–777

temperature diffusion equation, 920

tensor in Euclidean space

definition and rank, 949

algebra of without coordinates or bases, 949–951

expanded in basis, 954

component representation, 955–957

tensor in quantum theory, 956b

tensor product, 950

thermal diffusivity, 920, 921t

thermal plume, 933

thermodynamics. *See also* equations of state; fundamental thermodynamic potentials

first law of for fluid, 679b–680b

thermoelastic noise in mirrors, 623, 626

thermoelasticity, 584–585

Thomson scattering of photons by electrons, 937

time derivative

advective (convective), 692, 724b, 892, 970

fluid, 736

Tollmien-Schlichting waves, 823

tomography, seismic, 663

tornado, 738, 739f

pressure differential in, 702, 738

torsion pendulum, elastostatics of, 621–622

transformation matrices between orthogonal bases, 958–959

trumpet, sound generation by, 868

tsunamis, 841b, 843, 922

turbulence, 787–834

weak and strong, 800

characteristics of

3-dimensional, 794

disorder, 798

irregularly distributed vorticity, 799

wide range of interacting scales, 798–799

eddies, 798–800, 802, 804–807, 811–814

efficient mixing and transport, 799

large dissipation, 799

intermittency, 798–799, 807, 814, 831

onset of. *See* turbulence, onset of

vorticity in, 799–800

drives energy from large scales to small, 799f

semiquantitative analysis of, 800–817

Kolmogorov spectrum, 810–813, 813f, 815. *See also* Kolmogorov spectrum for turbulence

weak turbulence formalism, 800–810. *See also* weak turbulence formalism

generation of sound by, 869

turbulence, 2-dimensional analog of, 801b

inverse cascade of energy, 799n

transition to (3-dimensional) turbulence, 800

turbulence, onset of. *See also* chaos, onset of in dynamical systems

critical Reynolds number for, 787, 794, 822, 826

in convection, 830, 831

in flow past a cylinder, 789–794, 800

in rotating Couette flow, 825–828

routes to turbulence

one frequency, two frequencies, turbulence, 825–828

one frequency, two frequencies, phase locking, turbulence, 831

one frequency, two frequencies, three frequencies, turbulence, 831

period doubling sequence, 830–831

intermittency, 831

ultrasound, 663–664

vector

as arrow, 946

as derivative of a point, 947

vector components, 954

vector in quantum theory, 956b

velocity, 947

velocity potential for irrotational flow, 701

violin string, sound generation by, 868

viscosity, bulk, coefficient of, 712, 724b

viscosity, molecular origin of, 713–714

viscosity, shear, coefficient of, 712, 726b

dynamic, η, 713, 724b

kinematic, ν, 713, 725b

values of, for various fluids, 713t, 921t

for monatomic gas, 714

volcanic explosions, 748, 755

volume in Euclidean space

2–volume (area), 964

vectorial surface area in 3-space, 965

3–volume, 965

n-volume, 962

differential volume elements, 966

CONTENTS OF THE UNIFIED WORK, *MODERN CLASSICAL PHYSICS*

T2   Track Two; see page xvii

PREFACE TO *MODERN CLASSICAL PHYSICS*

The study of physics (including astronomy) is one of the oldest academic enterprises. Remarkable surges in inquiry occurred in equally remarkable societies—in Greece and Egypt, in Mesopotamia, India and China—and especially in Western Europe from the late sixteenth century onward. Independent, rational inquiry flourished at the expense of ignorance, superstition, and obeisance to authority.

Physics is a constructive and progressive discipline, so these surges left behind layers of understanding derived from careful observation and experiment, organized by fundamental principles and laws that provide the foundation of the discipline today. Meanwhile the detritus of bad data and wrong ideas has washed away. The laws themselves were so general and reliable that they provided foundations for investigation far beyond the traditional frontiers of physics, and for the growth of technology.

The start of the twentieth century marked a watershed in the history of physics, when attention turned to the small and the fast. Although rightly associated with the names of Planck and Einstein, this turning point was only reached through the curiosity and industry of their many forerunners. The resulting quantum mechanics and relativity occupied physicists for much of the succeeding century and today are viewed very differently from each other. Quantum mechanics is perceived as an abrupt departure from the tacit assumptions of the past, while relativity—though no less radical conceptually—is seen as a logical continuation of the physics of Galileo, Newton, and Maxwell. There is no better illustration of this than Einstein's growing special relativity into the general theory and his famous resistance to the quantum mechanics of the 1920s, which others were developing.

This is a book about classical physics—a name intended to capture the pre-quantum scientific ideas, augmented by general relativity. Operationally, it is physics in the limit that Planck's constant $h \rightarrow 0$. Classical physics is sometimes used, pejoratively, to suggest that "classical" ideas were discarded and replaced by new principles and laws. Nothing could be further from the truth. The majority of applications of physics today are still essentially classical. This does not imply that physicists or others working in these areas are ignorant or dismissive of quantum physics. It is simply that the issues with which they are confronted are mostly addressed classically. Furthermore, classical physics has not stood still while the quantum world was being explored. In scope and in practice, it has exploded on many fronts and would now be quite unrecognizable to a Helmholtz, a Rayleigh, or a Gibbs. In this book, we have tried to emphasize these contemporary developments and applications at the expense of historical choices, and this is the reason for our seemingly oxymoronic title, *Modern Classical Physics.*

This book is ambitious in scope, but to make it bindable and portable (and so the authors could spend some time with their families), we do not develop classical mechanics, electromagnetic theory, or elementary thermodynamics. We assume the reader has already learned these topics elsewhere, perhaps as part of an undergraduate curriculum. We also assume a normal undergraduate facility with applied mathematics. This allows us to focus on those topics that are less frequently taught in undergraduate and graduate courses.

Another important exclusion is numerical methods and simulation. High-performance computing has transformed modern research and enabled investigations that were formerly hamstrung by the limitations of special functions and artificially imposed symmetries. To do justice to the range of numerical techniques that have been developed—partial differential equation solvers, finite element methods, Monte Carlo approaches, graphics, and so on—would have more than doubled the scope and size of the book. Nonetheless, because numerical evaluations are crucial for physical insight, the book includes many applications and exercises in which user-friendly numerical packages (such as Maple, Mathematica, and Matlab) can be used to produce interesting numerical results without too much effort. We hope that, via this pathway from fundamental principle to computable outcome, our book will bring readers not only physical insight but also enthusiasm for computational physics.

Classical physics as we develop it emphasizes physical phenomena on macroscopic scales: scales where the particulate natures of matter and radiation are secondary to their behavior in bulk; scales where particles' statistical—as opposed to individual—properties are important, and where matter's inherent graininess can be smoothed over.

In this book, we take a journey through spacetime and phase space; through statistical and continuum mechanics (including solids, fluids, and plasmas); and through optics and relativity, both special and general. In our journey, we seek to comprehend the fundamental laws of classical physics in their own terms, and also in relation to quantum physics. And, using carefully chosen examples, we show how the classical laws are applied to important, contemporary, twenty-first-century problems and to everyday phenomena; and we also uncover some deep relationships among the various fundamental laws and connections among the practical techniques that are used in different subfields of physics.

Geometry is a deep theme throughout this book and a very important connector. We shall see how a few geometrical considerations dictate or strongly limit the basic principles of classical physics. Geometry illuminates the character of the classical principles and also helps relate them to the corresponding principles of quantum physics. Geometrical methods can also obviate lengthy analytical calculations. Despite this, long, routine algebraic manipulations are sometimes unavoidable; in such cases, we occasionally save space by invoking modern computational symbol manipulation programs, such as Maple, Mathematica, and Matlab.

This book is the outgrowth of courses that the authors have taught at Caltech and Stanford beginning 37 years ago. Our goal was then and remains now to fill what we saw as a large hole in the traditional physics curriculum, at least in the United States:

- We believe that every masters-level or PhD physicist should be familiar with the basic concepts of all the major branches of classical physics and should have had some experience in applying them to real-world phenomena; this book is designed to facilitate this goal.

- Many physics, astronomy, and engineering graduate students in the United States and around the world use classical physics extensively in their research, and even more of them go on to careers in which classical physics is an essential component; this book is designed to expedite their efforts.

- Many professional physicists and engineers discover, in mid-career, that they need an understanding of areas of classical physics that they had not previously mastered. This book is designed to help them fill in the gaps and see the relationship to already familiar topics.

In pursuit of this goal, we seek, in this book, to *give the reader a clear understanding of the basic concepts and principles of classical physics.* We present these principles in the language of modern physics (not nineteenth-century applied mathematics), and we present them primarily for physicists—though we have tried hard to make the content interesting, useful, and accessible to a much larger community including engineers, mathematicians, chemists, biologists, and so on. As far as possible, we emphasize theory that involves general principles which extend well beyond the particular topics we use to illustrate them.

In this book, we also seek to *teach the reader how to apply the ideas of classical physics.* We do so by presenting contemporary applications from a variety of fields, such as

- fundamental physics, experimental physics, and applied physics;
- astrophysics and cosmology;
- geophysics, oceanography, and meteorology;
- biophysics and chemical physics; and

- engineering, optical science and technology, radio science and technology, and information science and technology.

Why is the range of applications so wide? Because we believe that physicists should have enough understanding of general principles to attack problems that arise in unfamiliar environments. In the modern era, a large fraction of physics students will go on to careers outside the core of fundamental physics. For such students, a broad exposure to non-core applications can be of great value. For those who wind up in the core, such an exposure is of value culturally, and also because ideas from other fields often turn out to have impact back in the core of physics. Our examples illustrate how basic concepts and problem-solving techniques are freely interchanged across disciplines.

We strongly believe that classical physics should *not* be studied in isolation from quantum mechanics and its modern applications. Our reasons are simple:

- Quantum mechanics has primacy over classical physics. Classical physics is an approximation—often excellent, sometimes poor—to quantum mechanics.

- In recent decades, many concepts and mathematical techniques developed for quantum mechanics have been imported into classical physics and there used to enlarge our classical understanding and enhance our computational capability. An example that we shall study is nonlinearly interacting plasma waves, which are best treated as quanta ("plasmons"), despite their being solutions of classical field equations.

- Ideas developed initially for classical problems are frequently adapted for application to avowedly quantum mechanical subjects; examples (not discussed in this book) are found in supersymmetric string theory and in the liquid drop model of the atomic nucleus.

Because of these intimate connections between quantum and classical physics, quantum physics appears frequently in this book.

The amount and variety of material covered in this book may seem overwhelming. If so, keep in mind the key goals of the book: to teach the fundamental concepts, which are not so extensive that they should overwhelm, and to illustrate those concepts. Our goal is not to provide a mastery of the many illustrative applications contained in the book, but rather to convey the spirit of how to apply the basic concepts of classical physics. To help students and readers who feel overwhelmed, we have labeled as "Track Two" sections that can be skipped on a first reading, or skipped entirely— but are sufficiently interesting that many readers may choose to browse or study them. Track-Two sections are labeled by the symbol **T2** . To keep Track One manageable for a one-year course, the Track-One portion of each chapter is rarely longer than 40 pages (including many pages of exercises) and is often somewhat shorter. Track One is designed for a full-year course at the first-year graduate level; that is how we have mostly used it. (Many final-year undergraduates have taken our course successfully, but rarely easily.)

The book is divided into seven parts:

I. **Foundations**—which introduces our book's powerful *geometric* point of view on the laws of physics and brings readers up to speed on some concepts and mathematical tools that we shall need. Many readers will already have mastered most or all of the material in Part I and might find that they can understand most of the rest of the book without adopting our avowedly geometric viewpoint. Nevertheless, we encourage such readers to browse Part I, at least briefly, before moving on, so as to become familiar with this viewpoint. We believe the investment will be repaid. Part I is split into two chapters, Chap. 1 on Newtonian physics and Chap. 2 on special relativity. Since nearly all of Parts II–VI is Newtonian, readers may choose to skip Chap. 2 and the occasional special relativity sections of subsequent chapters, until they are ready to launch into Part VII, General Relativity. Accordingly, Chap. 2 is labeled Track Two, though it becomes Track One when readers embark on Part VII.

II. **Statistical Physics**—including kinetic theory, statistical mechanics, statistical thermodynamics, and the theory of random processes. These subjects underlie some portions of the rest of the book, especially plasma physics and fluid mechanics.

III. **Optics**—by which we mean classical waves of all sorts: light waves, radio waves, sound waves, water waves, waves in plasmas, and gravitational waves. The major concepts we develop for dealing with all these waves include geometric optics, diffraction, interference, and nonlinear wave-wave mixing.

IV. **Elasticity**—elastic deformations, both static and dynamic, of solids. Here we develop the use of tensors to describe continuum mechanics.

V. **Fluid Dynamics**—with flows ranging from the traditional ones of air and water to more modern cosmic and biological environments. We introduce vorticity, viscosity, turbulence, boundary layers, heat transport, sound waves, shock waves, magnetohydrodynamics, and more.

VI. **Plasma Physics**—including plasmas in Earth-bound laboratories and in technological (e.g., controlled-fusion) devices, Earth's ionosphere, and cosmic environments. In addition to magnetohydrodynamics (treated in Part V), we develop two-fluid and kinetic approaches, and techniques of nonlinear plasma physics.

VII. **General Relativity**—the physics of curved spacetime. Here we show how the physical laws that we have discussed in flat spacetime are modified to account for curvature. We also explain how energy and momentum generate this curvature. These ideas are developed for their principal classical applications to neutron stars, black holes, gravitational radiation, and cosmology.

It should be possible to read and teach these parts independently, provided one is prepared to use the cross-references to access some concepts, tools, and results developed in earlier parts.

Five of the seven parts (II, III, V, VI, and VII) conclude with chapters that focus on applications where there is much current research activity and, consequently, there are many opportunities for physicists.

Exercises are a major component of this book. There are five types of exercises:

1. *Practice.* Exercises that provide practice at mathematical manipulations (e.g., of tensors).

2. *Derivation.* Exercises that fill in details of arguments skipped over in the text.

3. *Example.* Exercises that lead the reader step by step through the details of some important extension or application of the material in the text.

4. *Problem.* Exercises with few, if any, hints, in which the task of figuring out how to set up the calculation and get started on it often is as difficult as doing the calculation itself.

5. *Challenge.* Especially difficult exercises whose solution may require reading other books or articles as a foundation for getting started.

We urge readers to try working many of the exercises, especially the examples, which should be regarded as continuations of the text and which contain many of the most illuminating applications. Exercises that we regard as especially important are designated by **.

A few words on units and conventions. In this book we deal with practical matters and frequently need to have a quantitative understanding of the magnitudes of various physical quantities. This requires us to adopt a particular unit system. Physicists use both Gaussian and SI units; units that lie outside both formal systems are also commonly used in many subdisciplines. Both Gaussian and SI units provide a complete and internally consistent set for all of physics, and it is an often-debated issue as to which system is more convenient or aesthetically appealing. We will not enter this debate! One's choice of units should not matter, and a mature physicist should be able to change from one system to another with little thought. However, when learning new concepts, having to figure out "where the 2πs and 4πs go" is a genuine impediment to progress. Our solution to this problem is as follows. For each physics subfield that we study, we consistently use the set of units that seem most natural or that, we judge, constitute the majority usage by researchers in that subfield. We do not pedantically convert cm to m or vice versa at every juncture; we trust that the reader can easily make whatever translation is necessary. However, where the equations are actually different—primarily in electromagnetic theory—we occasionally provide, in brackets or footnotes, the equivalent equations in the other unit system and enough information for the reader to proceed in his or her preferred scheme.

We encourage readers to consult this book's website, http://press.princeton.edu/titles/MCP.html, for information, errata, and various resources relevant to the book.

A large number of people have influenced this book and our viewpoint on the material in it. We list many of them and express our thanks in the Acknowledgments. Many misconceptions and errors have been caught and corrected. However, in a book of this size and scope, others will remain, and for these we take full responsibility. We would be delighted to learn of these from readers and will post corrections and explanations on this book's website when we judge them to be especially important and helpful.

Above all, we are grateful for the support of our wives, Carolee and Liz—and especially for their forbearance in epochs when our enterprise seemed like a mad and vain pursuit of an unreachable goal, a pursuit that we juggled with huge numbers of other obligations, while Liz and Carolee, in the midst of their own careers, gave us the love and encouragement that were crucial in keeping us going.

ACKNOWLEDGMENTS FOR *MODERN CLASSICAL PHYSICS*

This book evolved gradually from notes written in 1980–81, through improved notes, then sparse prose, and on into text that ultimately morphed into what you see today. Over these three decades and more, courses based on our evolving notes and text were taught by us and by many of our colleagues at Caltech, Stanford, and elsewhere. From those teachers and their students, and from readers who found our evolving text on the web and dove into it, we have received an extraordinary volume of feedback,[1] and also patient correction of errors and misconceptions as well as help with translating passages that were correct but impenetrable into more lucid and accessible treatments. For all this feedback and to all who gave it, we are extremely grateful. We wish that we had kept better records; the heartfelt thanks that we offer all these colleagues, students, and readers, named and unnamed, are deeply sincere.

Teachers who taught courses based on our evolving notes and text, and gave invaluable feedback, include Professors Richard Blade, Yanbei Chen, Michael Cross, Steven Frautschi, Peter Goldreich, Steve Koonin, Christian Ott, Sterl Phinney, David Politzer, John Preskill, John Schwarz, and David Stevenson at Caltech; Professors Tom Abel, Seb Doniach, Bob Wagoner, and the late Shoucheng Zhang at Stanford; and Professor Sandor Kovacs at Washington University in St. Louis.

Our teaching assistants, who gave us invaluable feedback on the text, improvements of exercises, and insights into the difficulty of the material for the students, include Jeffrey Atwell, Nate Bode, Yu Cao, Yi-Yuh Chen, Jane Dai, Alexei Dvoretsky, Fernando Echeverria, Jiyu Feng, Eanna Flanagan, Marc Goroff, Dan Grin, Arun Gupta, Alexandr Ikriannikov, Anton Kapustin, Kihong Kim, Hee-Won Lee, Geoffrey Lovelace, Miloje Makivic, Draza Markovic, Keith Matthews, Eric Morganson, Mike Morris, Chung-Yi Mou, Rob Owen, Yi Pan, Jaemo Park, Apoorva Patel, Alexander Putilin, Shuyan Qi, Soo Jong Rey, Fintan Ryan, Bonnie Shoemaker, Paul Simeon,

---

1. Specific applications that were originated by others, to the best of our memory, are acknowledged in the text.

Hidenori Sinoda, Matthew Stevenson, Wai Mo Suen, Marcus Teague, Guodang Wang, Xinkai Wu, Huan Yang, Jimmy Yee, Piljin Yi, Chen Zheng, and perhaps others of whom we have lost track!

Among the students and readers of our notes and text, who have corresponded with us, sending important suggestions and errata, are Bram Achterberg, Mustafa Amin, Richard Anantua, Alborz Bejnood, Edward Blandford, Jonathan Blandford, Dick Bond, Phil Bucksbaum, James Camparo, Conrado Cano, U Lei Chan, Vernon Chaplin, Mina Cho, Ann Marie Cody, Sandro Commandè, Kevin Fiedler, Krzysztof Findeisen, Jeff Graham, Casey Handmer, John Hannay, Ted Jacobson, Matt Kellner, Deepak Kumar, Andrew McClung, Yuki Moon, Evan O'Connor, Jeffrey Oishi, Keith Olive, Zhen Pan, Eric Peterson, Laurence Perreault Levasseur, Rob Phillips, Vahbod Pourahmad, Andreas Reisenegger, David Reis, Pavlin Savov, Janet Scheel, Yuki Takahashi, Clifford Will, Fun Lim Yee, Yajie Yuan, and Aaron Zimmerman.

For computational advice or assistance, we thank Edward Campbell, Mark Scheel, Chris Mach, and Elizabeth Wood.

Academic support staff who were crucial to our work on this book include Christine Aguilar, JoAnn Boyd, Jennifer Formicelli, and Shirley Hampton.

The editorial and production professionals at Princeton University Press (Peter Dougherty, Karen Fortgang, Ingrid Gnerlich, Eric Henney, and Arthur Werneck) and at Princeton Editorial Associates (Peter Strupp and his freelance associates Paul Anagnostopoulos, Laurel Muller, MaryEllen Oliver, Joe Snowden, and Cyd Westmoreland) have been magnificent, helping us plan and design this book, and transforming our raw prose and primitive figures into a visually appealing volume, with sustained attention to detail, courtesy, and patience as we missed deadline after deadline.

Of course, we the authors take full responsibility for all the errors of judgment, bad choices, and mistakes that remain.

Roger Blandford thanks his many supportive colleagues at Caltech, Stanford University, and the Kavli Institute for Particle Astrophysics and Cosmology. He also acknowledges the Humboldt Foundation, the Miller Institute, the National Science Foundation, and the Simons Foundation for generous support during the completion of this book. And he also thanks the Berkeley Astronomy Department; Caltech; the Institute of Astronomy, Cambridge; and the Max Planck Institute for Astrophysics, Garching, for hospitality.

Kip Thorne is grateful to Caltech—the administration, faculty, students, and staff—for the supportive environment that made possible his work on this book, work that occupied a significant portion of his academic career.